T0237479

This fascinating look at combinatorial games, that is, games not involving chance or hidden information, offers updates on standard games such as Go and Hex, on impartial games such as Chomp and Wythoff's Nim, and on aspects of games with infinitesmal values, plus analyses of the complexity of some games and puzzles and surveys on algorithmic game theory, on playing to lose, and on coping with cycles. The volume is rounded out with an up-to-date bibliography by Aviezri S. Fraenkel and, for readers eager to get their hands dirty, a list of unsolved problems by Richard K. Guy and Richard J. Nowakowski.

Highlights include some of Aaron N. Siegel's groundbreaking work on loopy games, the unveiling by Eric J. Friedman and Adam S. Landsberg of the use of renormalization to give very intriguing results about Chomp, and Teigo Nakamura's "Counting liberties in capturing races of Go."

Like its predecessors, this book should be on the shelf of all serious games enthusiasts.

Mathematical Sciences Research Institute
Publications

56

Games of No Chance 3

Mathematical Sciences Research Institute Publications

Volumes 1–4, 6–8, and 10–27 are published by Springer-Verlag

Games of No Chance 3

Edited by

Michael H. Albert

University of Otago

Richard J. Nowakowski

Dalhousie University

CAMBRIDGE
UNIVERSITY PRESS

Michael H. Albert
Department of Computer Science
University of Otago
New Zealand
malbert@cs.otago.ac.nz

Richard J. Nowakowski
Department of Mathematics and Statistics
Dalhousie University
Halifax, NS, Canada
rjn@mathstat.dal.ca

Silvio Levy (*Series Editor*)
Mathematical Sciences Research Institute
Berkeley, CA 94720
levy@msri.org

The Mathematical Sciences Research Institute wishes to acknowledge support by the National Science Foundation and the *Pacific Journal of Mathematics* for the publication of this series.

CAMBRIDGE
UNIVERSITY PRESS

Shaftesbury Road, Cambridge CB2 8EA, United Kingdom

One Liberty Plaza, 20th Floor, New York, NY 10006, USA

477 Williamstown Road, Port Melbourne, VIC 3207, Australia

314–321, 3rd Floor, Plot 3, Splendor Forum, Jasola District Centre, New Delhi – 110025, India

103 Penang Road, #05–06/07, Visioncrest Commercial, Singapore 238467

Cambridge University Press is part of Cambridge University Press & Assessment, a department of the University of Cambridge.

We share the University's mission to contribute to society through the pursuit of education, learning and research at the highest international levels of excellence.

www.cambridge.org
Information on this title: www.cambridge.org/9780521678544

© Mathematical Sciences Research Institute 2009

This publication is in copyright. Subject to statutory exception and to the provisions of relevant collective licensing agreements, no reproduction of any part may take place without the written permission of Cambridge University Press & Assessment.

First published 2009

A catalogue record for this publication is available from the British Library

Library of Congress Cataloging-in-Publication data
Games of no chance 3 / edited by Michael H. Albert, Richard J. Nowakowski.
    p.   cm. – (Mathematical Sciences Research Institute publications ; 56)
Sequel to: More games of no chance. c2002.
Includes bibliographical references and index.
ISBN 978-0-521-86134-2 (hardback) - ISBN 978-0-521-67854-4 (pbk.)
    1. Game theory –Congresses. 2. Combinatorial analysis –Congresses. I. Albert, Michael H. II. Nowakowski, Richard J. III. Title: Games of no chance three. IV. Series.

QA269.G374 2009
519.3-dc22   2009009347

ISBN   978-0-521-86134-2   Hardback
ISBN   978-0-521-67854-4   Paperback

Cambridge University Press & Assessment has no responsibility for the persistence or accuracy of URLs for external or third-party internet websites referred to in this publication and does not guarantee that any content on such websites is, or will remain, accurate or appropriate.

Games of No Chance 3
MSRI Publications
Volume **56**, 2009

Contents

Games of No Chance 3
MSRI Publications
Volume 56, 2009

Preface

The June 2005 Combinatorial Game Theory Workshop was held at the Banff International Research Station (BIRS) and organized by Elwyn Berlekamp, Martin Mueller, Richard J. Nowakowski, and David Wolfe. It attracted researchers from Asia, Europe, and North America. The highlights were many and the results already have had a great impact in the field.

Aaron N. Siegel had his hand in quite a few of the highlights. Some of his groundbreaking work is presented in "Loopy games," with applications given in "Coping with cycles" and "Backsliding Toads and Frogs." His work with J. P. Grossman, "Reductions of partizan games," showed that the *reduced canonical form* of a game exists. This approximation to the canonical form has already proved very useful; it has been used, for example, in the Mesdal ensemble's analysis of Partizan Splittles as well as in the analysis of other games.

In "Advances in losing," Thane E. Plambeck surveys the state of the art for last-player-to-move-loses games. Even more advances in this area were made during the conference, primarily by Plambeck and Siegel. Indeed, the approach of forming a monoid of game positions in order to discover the structure of winning and losing positions in a misère game has subsequently matured sufficiently that the BIRS Games Workshop in January 2008, had misère games as one of its main themes.

Eric J. Friedman and Adam S. Landsberg unveiled the use of renormalization, a technique from physics, to give very intriguing results about Chomp. This caught the participants off guard and engendered much discussion. It is surprising that a technique known for giving approximations gave such definite results. Like the previous two topics, there is much to be said and explored here.

Another talk that excited many of the participants was Teigo Nakamura's "Counting liberties in Go capturing races." By popular demand, that talk was extended by an hour. *Chilling by 2* is the main idea, but its application is the first ever to appear in analysis of real games, and it appears in a very surprising context.

This volume includes many other interesting papers. To help the reader drill down to a particular topic, the articles are grouped into a number of separate areas of interest.

- *Surveys*, which is self-expanatory;
- *Standards*, which refers to well-known partizan games such as Go, Hex, and others;
- *Complexity* of some games and some puzzles;
- *Impartial games* such as Chomp and Wythoff's Nim;
- *Theory of the small*, that is, aspects of games with infinitesimal values;
- *Columns*, to wit: Aviezri S. Fraenkel's updated bibliography of combinatorial games, and an expanded and reorganized "Unsolved problems in combinatorial games," by Richard K. Guy and Richard J. Nowakowski, for those eager to get their hands dirty.

As editors, we would like to thank the organizers and participants for helping making the workshop a great experience.

As participants and organizers, we would like to recognize and thank the BIRS organization and staff: the former for giving us the chance to hold the workshop in such a wonderful setting and the latter for ensuring that the participants and organizers only had to worry about the scientific aspect of the workshop.

Lastly, a big thanks to Silvio Levy for the final preparation of this document.

Michael H. Albert
Richard J. Nowakowski

Surveys

Playing games with algorithms:
Algorithmic Combinatorial Game Theory

ERIK D. DEMAINE AND ROBERT A. HEARN

ABSTRACT. Combinatorial games lead to several interesting, clean problems in algorithms and complexity theory, many of which remain open. The purpose of this paper is to provide an overview of the area to encourage further research. In particular, we begin with general background in Combinatorial Game Theory, which analyzes ideal play in perfect-information games, and Constraint Logic, which provides a framework for showing hardness. Then we survey results about the complexity of determining ideal play in these games, and the related problems of solving puzzles, in terms of both polynomial-time algorithms and computational intractability results. Our review of background and survey of algorithmic results are by no means complete, but should serve as a useful primer.

1. Introduction

Many classic games are known to be computationally intractable (assuming $P \neq NP$): one-player puzzles are often NP-complete (for instance Minesweeper) or PSPACE-complete (Rush Hour), and two-player games are often PSPACE-complete (Othello) or EXPTIME-complete (Checkers, Chess, and Go). Surprisingly, many seemingly simple puzzles and games are also hard. Other results are positive, proving that some games can be played optimally in polynomial time. In some cases, particularly with one-player puzzles, the computationally tractable games are still interesting for humans to play.

We begin by reviewing some basics of Combinatorial Game Theory in Section 2, which gives tools for designing algorithms, followed by reviewing the

A preliminary version of this paper appears in the *Proceedings of the 26th International Symposium on Mathematical Foundations of Computer Science*, Lecture Notes in Computer Science 2136, Czech Republic, August 2001, pages 18–32. The latest version can be found at http://arXiv.org/abs/cs.CC/0106019.

relatively new theory of Constraint Logic in Section 3, which gives tools for proving hardness. In the bulk of this paper, Sections 4–6 survey many of the algorithmic and hardness results for combinatorial games and puzzles. Section 7 concludes with a small sample of difficult open problems in algorithmic Combinatorial Game Theory.

Combinatorial Game Theory is to be distinguished from other forms of game theory arising in the context of economics. Economic game theory has many applications in computer science as well, for example, in the context of auctions [dVV03] and analyzing behavior on the Internet [Pap01].

2. Combinatorial Game Theory

A *combinatorial game* typically involves two players, often called *Left* and *Right*, alternating play in well-defined *moves*. However, in the interesting case of a *combinatorial puzzle*, there is only one player, and for *cellular automata* such as Conway's Game of Life, there are no players. In all cases, no randomness or hidden information is permitted: all players know all information about gameplay (*perfect information*). The problem is thus purely strategic: how to best play the game against an ideal opponent.

It is useful to distinguish several types of two-player perfect-information games [BCG04, pp. 14–15]. A common assumption is that the game terminates after a finite number of moves (the game is *finite* or *short*), and the result is a unique winner. Of course, there are exceptions: some games (such as Life and Chess) can be *drawn* out forever, and some games (such as tic-tac-toe and Chess) define *ties* in certain cases. However, in the combinatorial-game setting, it is useful to define the *winner* as the last player who is able to move; this is called *normal play*. If, on the other hand, the winner is the first player who cannot move, this is called *misère play*. (We will normally assume normal play.) A game is *loopy* if it is possible to return to previously seen positions (as in Chess, for example). Finally, a game is called *impartial* if the two players (Left and Right) are treated identically, that is, each player has the same moves available from the same game position; otherwise the game is called *partizan*.

A particular two-player perfect-information game without ties or draws can have one of four *outcomes* as the result of ideal play: player Left wins, player Right wins, the first player to move wins (whether it is Left or Right), or the second player to move wins. One goal in analyzing two-player games is to determine the outcome as one of these four categories, and to find a strategy for the winning player to win. Another goal is to compute a deeper structure to games described in the remainder of this section, called the *value* of the game.

A beautiful mathematical theory has been developed for analyzing two-player combinatorial games. A new introductory book on the topic is *Lessons in Play* by Albert, Nowakowski, and Wolfe [ANW07]; the most comprehensive reference is the book *Winning Ways* by Berlekamp, Conway, and Guy [BCG04]; and a more mathematical presentation is the book *On Numbers and Games* by Conway [Con01]. See also [Con77; Fra96] for overviews and [Fra07] for a bibliography. The basic idea behind the theory is simple: a two-player game can be described by a rooted tree, where each node has zero or more *left* branches corresponding to options for player Left to move and zero or more *right* branches corresponding to options for player Right to move; leaves correspond to finished games, with the winner determined by either normal or misère play. The interesting parts of Combinatorial Game Theory are the several methods for manipulating and analyzing such games/trees. We give a brief summary of some of these methods in this section.

2.1. Conway's surreal numbers. A richly structured special class of two-player games are John H. Conway's *surreal numbers*[1] [Con01; Knu74; Gon86; All87], a vast generalization of the real and ordinal number systems. Basically, a surreal number $\{L \mid R\}$ is the "simplest" number larger than all Left options (in L) and smaller than all Right options (in R); for this to constitute a number, all Left and Right options must be numbers, defining a total order, and each Left option must be less than each Right option. See [Con01] for more formal definitions.

For example, the simplest number without any larger-than or smaller-than constraints, denoted $\{\mid\}$, is 0; the simplest number larger than 0 and without smaller-than constraints, denoted $\{0 \mid\}$, is 1; and the simplest number larger than 0 and 1 (or just 1), denoted $\{0, 1 \mid\}$, is 2. This method can be used to generate all natural numbers and indeed all ordinals. On the other hand, the simplest number less than 0, denoted $\{\mid 0\}$, is -1; similarly, all negative integers can be generated. Another example is the simplest number larger than 0 and smaller than 1, denoted $\{0 \mid 1\}$, which is $\frac{1}{2}$; similarly, all dyadic rationals can be generated. After a countably infinite number of such construction steps, all real numbers can be generated; after many more steps, the surreals are all numbers that can be generated in this way.

Surreal numbers form an ordered field, so in particular they support the operations of addition, subtraction, multiplication, division, roots, powers, and even integration in many situations. (For those familiar with ordinals, contrast with surreals which define $\omega - 1$, $1/\omega$, $\sqrt{\omega}$, etc.) As such, surreal numbers are useful in their own right for cleaner forms of analysis; see, e.g., [All87].

What is interesting about the surreals from the perspective of combinatorial game theory is that they are a subclass of all two-player perfect-information

---

[1] The name "surreal numbers" is actually due to Knuth [Knu74]; see [Con01].

Let $x = x^L | x^R$ be a game.

- $x \leq y$ precisely if every $x^L < y$ and every $y^R > x$.
- $x = y$ precisely if $x \leq y$ and $x \geq y$; otherwise $x \neq y$.
- $x < y$ precisely if $x \leq y$ and $x \neq y$, or equivalently, $x \leq y$ and $x \not\geq y$.
- $-x = -x^R | -x^L$.
- $x + y = x^L + y, x + y^L | x^R + y, x + y^R$.
- x is *impartial* precisely if x^L and x^R are identical sets and recursively every position ($\in x^L = x^R$) is impartial.
- A one-pile Nim game is defined by

$$*n = *0, \ldots, *(n-1) | *0, \ldots, *(n-1),$$

together with $*0 = 0$.

Table 1. Formal definitions of some algebra on two-player perfect-information games. In particular, all of these notions apply to surreal numbers.

games, and some of the surreal structure, such as addition and subtraction, carries over to general games. Furthermore, while games are not totally ordered, they can still be compared to some surreal numbers and, amazingly, how a game compares to the surreal number 0 determines exactly the outcome of the game. This connection is detailed in the next few paragraphs.

First we define some algebraic structure of games that carries over from surreal numbers; see Table 1 for formal definitions. Two-player combinatorial games, or trees, can simply be represented as $\{L \mid R\}$ where, in contrast to surreal numbers, no constraints are placed on L and R. The *negation* of a game is the result of reversing the roles of the players Left and Right throughout the game. The *(disjunctive) sum* of two (sub)games is the game in which, at each player's turn, the player has a binary choice of which subgame to play, and makes a move in precisely that subgame. A partial order is defined on games recursively: a game x is *less than or equal to* a game y if every Left option of x is less than y and every Right option of y is more than x. (Numeric) equality is defined by being both less than or equal to and more than or equal to. Strictly inequalities, as used in the definition of less than or equal to, are defined in the obvious manner.

Note that while $\{-1 \mid 1\} = 0 = \{\mid\}$ in terms of numbers, $\{-1 \mid 1\}$ and $\{\mid\}$ denote different games (lasting 1 move and 0 moves, respectively), and in this sense are *equal* in *value* but not *identical* symbolically or game-theoretically. Nonetheless, the games $\{-1 \mid 1\}$ and $\{\mid\}$ have the same outcome: the second player to move wins.

Amazingly, this holds in general: two equal numbers represent games with equal outcome (under ideal play). In particular, all games equal to 0 have the outcome that the second player to move wins. Furthermore, all games equal to a positive number have the outcome that the Left player wins; more generally, all positive games (games larger than 0) have this outcome. Symmetrically, all negative games have the outcome that the Right player wins (this follows automatically by the negation operation). Examples of zero, positive, and negative games are the surreal numbers themselves; an additional example is described below.

There is one outcome not captured by the characterization into zero, positive, and negative games: the first player to move wins. To find such a game we must obviously look beyond the surreal numbers. Furthermore, we must look for games G that are incomparable with zero (none of $G = 0$, $G < 0$, or $G > 0$ hold); such games are called *fuzzy* with 0, denoted $G \parallel 0$.

An example of a game that is not a surreal number is $\{1 \mid 0\}$; there fails to be a number strictly between 1 and 0 because $1 \geq 0$. Nonetheless, $\{1 \mid 0\}$ is a game: Left has a single move leading to game 1, from which Right cannot move, and Right has a single move leading to game 0, from which Left cannot move. Thus, in either case, the first player to move wins. The claim above implies that $\{1 \mid 0\} \parallel 0$. Indeed, $\{1 \mid 0\} \parallel x$ for all surreal numbers x, $0 \leq x \leq 1$. In contrast, $x < \{1 \mid 0\}$ for all $x < 0$ and $\{1 \mid 0\} < x$ for all $1 < x$. In general it holds that a game is fuzzy with some surreal numbers in an interval $[-n, n]$ but comparable with all surreals outside that interval. Another example of a game that is not a number is $\{2 \mid 1\}$, which is positive (> 0), and hence Right wins, but fuzzy with numbers in the range $[1, 2]$.

For brevity we omit many other useful notions in Combinatorial Game Theory, such as additional definitions of summation, superinfinitesimal games $*$ and $\uparrow$, mass, temperature, thermographs, the simplest form of a game, remoteness, and suspense; see [BCG04; Con01].

2.2. Sprague–Grundy theory.

A celebrated early result in Combinatorial Game Theory is the characterization of impartial two-player perfect-information games, discovered independently in the 1930's by Sprague [Spr36] and Grundy [Gru39]. Recall that a game is *impartial* if it does not distinguish between the players Left and Right (see Table 1 for a more formal definition). The Sprague–Grundy theory [Spr36; Gru39; Con01; BCG04] states that every finite impartial game is equivalent to an instance of the game of Nim, characterized by a single natural number n. This theory has since been generalized to all impartial games by generalizing Nim to all ordinals n; see [Con01; Smi66].

Nim [Bou02] is a game played with several *heaps*, each with a certain number of tokens. A Nim game with a single heap of size n is denoted by $*n$ and is called a *nimber*. During each move a player can pick any pile and reduce it to any smaller nonnegative integer size. The game ends when all piles have size 0. Thus, a single pile $*n$ can be reduced to any of the smaller piles $*0$, $*1$, ..., $*(n-1)$. Multiple piles in a game of Nim are independent, and hence any game of Nim is a sum of single-pile games $*n$ for various values of n. In fact, a game of Nim with k piles of sizes $n_1, n_2, \ldots, n_k$ is equivalent to a one-pile Nim game $*n$, where n is the binary XOR of $n_1, n_2, \ldots, n_k$. As a consequence, Nim can be played optimally in polynomial time (polynomial in the encoding size of the pile sizes).

Even more surprising is that *every* impartial two-player perfect-information game has the same value as a single-pile Nim game, $*n$ for some n. The number n is called the *G-value*, *Grundy-value*, or *Sprague–Grundy function* of the game. It is easy to define: suppose that game x has k options $y_1, \ldots, y_k$ for the first move (independent of which player goes first). By induction, we can compute $y_1 = *n_1, \ldots, y_k = *n_k$. The theorem is that x equals $*n$ where n is the smallest natural number not in the set $\{n_1, \ldots, n_k\}$. This number n is called the *minimum excluded value* or *mex* of the set. This description has also assumed that the game is finite, but this is easy to generalize [Con01; Smi66].

The Sprague–Grundy function can increase by at most 1 at each level of the game tree, and hence the resulting nimber is linear in the maximum number of moves that can be made in the game; the encoding size of the nimber is only logarithmic in this count. Unfortunately, computing the Sprague–Grundy function for a general game by the obvious method uses time linear in the number of possible states, which can be exponential in the nimber itself.

Nonetheless, the Sprague–Grundy theory is extremely helpful for analyzing impartial two-player games, and for many games there is an efficient algorithm to determine the nimber. Examples include Nim itself, Kayles, and various generalizations [GS56b]; and Cutcake and Maundy Cake [BCG04, pp. 24–27]. In all of these examples, the Sprague–Grundy function has a succinct characterization (if somewhat difficult to prove); it can also be easily computed using dynamic programming.

The Sprague–Grundy theory seems difficult to generalize to the superficially similar case of misère play, where the goal is to be the first player unable to move. Certain games have been solved in this context over the years, including Nim [Bou02]; see, e.g., [Fer74; GS56a]. Recently a general theory has emerged for tackling misère combinatorial games, based on commutative monoids called "misère quotients" that localize the problem to certain restricted game scenarios. This theory was introduced by Plambeck [Pla05] and further developed by Plambeck and Siegel [PS07]. For good descriptions of the theory, see Plambeck's survey [Plaa], Siegel's lecture notes [Sie06], and a webpage devoted to the topic [Plab].

2.3. Strategy stealing. Another useful technique in Combinatorial Game Theory for proving that a particular player must win is *strategy stealing*. The basic idea is to assume that one player has a winning strategy, and prove that in fact the other player has a winning strategy based on that strategy. This contradiction proves that the second player must in fact have a winning strategy. An example of such an argument is given in Section 4.1. Unfortunately, such a proof by contradiction gives no indication of what the winning strategy actually is, only that it exists. In many situations, such as the one in Section 4.1, the winner is known but no polynomial-time winning strategy is known.

2.4. Puzzles. There is little theory for analyzing combinatorial puzzles (one-player games) along the lines of the two-player theory summarized in this section. We present one such viewpoint here. In most puzzles, solutions subdivide into a sequence of moves. Thus, a puzzle can be viewed as a tree, similar to a two-player game except that edges are not distinguished between Left and Right. With the view that the game ends only when the puzzle is solved, the goal is then to reach a position from which there are no valid moves (normal play). Loopy puzzles are common; to be more explicit, repeated subtrees can be converted into self-references to form a directed graph, and losing terminal positions can be given explicit loops to themselves.

A consequence of the above view is that a puzzle is basically an impartial two-player game except that we are not interested in the outcome from two players alternating in moves. Rather, questions of interest in the context of puzzles are (a) whether a given puzzle is solvable, and (b) finding the solution with the fewest moves. An important open direction of research is to develop a general theory for resolving such questions, similar to the two-player theory.

3. Constraint logic

Combinatorial Game Theory provides a theoretical framework for giving positive algorithmic results for games, but does not naturally accommodate puzzles. In contrast, negative algorithmic results — hardness and completeness within computational complexity classes — are more uniform: puzzles and games have analogous prototypical proof structures. Furthermore, a relatively new theory called Constraint Logic attempts to tie together a wide range of hardness proofs for both puzzles and games.

Proving that a problem is hard within a particular complexity class (like NP, PSPACE, or EXPTIME) almost always involves a reduction to the problem from a known hard problem within the class. For example, the canonical problem to reduce from for NP-hardness is Boolean Satisfiability (SAT) [Coo71]. Reducing SAT to a puzzle of interest proves that that puzzle is NP-hard. Similarly, the canonical problem to reduce from for PSPACE-hardness is Quantified Boolean Formulas (QBF) [SM73].

Constraint Logic [DH08] is a useful tool for showing hardness of games and puzzles in a variety of settings that has emerged in recent years. Indeed, many of the hardness results mentioned in this survey are based on reductions from Constraint Logic. Constraint Logic is a family of games where players reverse edges on a planar directed graph while satisfying vertex in-flow constraints. Each edge has a weight of 1 or 2. Each vertex has degree 3 and requires that the sum of the weights of inward-directed edges is at least 2. Vertices may be restricted to two types: AND vertices have incident edge weights of 1, 1, and 2; and OR vertices have incident edge weights of 2, 2, and 2. A player's goal is to eventually reverse a given edge.

This game family can be interpreted in many game-theoretic settings, ranging from zero-player automata to multiplayer games with hidden information. In particular, there are natural versions of Constraint Logic corresponding to one-player games (puzzles) and two-player games, both of bounded and unbounded length. (Here we refer to whether the length of the game is bounded by a polynomial function of the board size. Typically, bounded games are nonloopy while unbounded games are loopy.) These games have the expected complexities: one-player bounded games are NP-complete; one-player unbounded games and two-player bounded games are PSPACE-complete; and two-player unbounded games are EXPTIME-complete.

What makes Constraint Logic specially suited for game and puzzle reductions is that the problems are already in form similar to many games. In particular, the fact that the games are played on planar graphs means that the reduction does not usually need a crossover gadget, whereas historically crossover gadgets have often been the complex crux of a game hardness proof.

Historically, Constraint Logic arose as a simplification of the "Generalized Rush-Hour Logic" of Flake and Baum [FB02]. The resulting one-player unbounded setting, called *Nondeterministic Constraint Logic* [HD02; HD05], was later generalized to other game categories [Hea06b; DH08].

4. Algorithms for two-player games

Many bounded-length two-player games are PSPACE-complete. This is fairly natural because games are closely related to Boolean expressions with alternating quantifiers (for which deciding satisfiability is PSPACE-complete): there exists a move for Left such that, for all moves for Right, there exists another move for Left, etc. A PSPACE-completeness result has two consequences. First, being in PSPACE means that the game can be played optimally, and typically all positions can be enumerated, using possibly exponential time but only polynomial space. Thus such games lend themselves to a somewhat reasonable exhaustive search for small enough sizes. Second, the games cannot be solved in polynomial time unless P = PSPACE, which is even "less likely" than P equaling NP.

On the other hand, unbounded-length two-players games are often EXPTIME-complete. Such a result is one of the few types of true lower bounds in complexity theory, implying that all algorithms require exponential time in the worst case.

In this section we briefly survey many of these complexity results and related positive results. See also [Epp] for a related survey and [Fra07] for a bibliography.

4.1. Hex. Hex [BCG04, pp. 743–744] is a game designed by Piet Hein and played on a diamond-shaped hexagonal board; see Figure 1. Players take turns filling in empty hexagons with their color. The goal of a player is to connect the opposite sides of their color with hexagons of their color. (In the figure, one player is solid and the other player is dotted.) A game of Hex can never tie, because if all hexagons are colored arbitrarily, there is precisely one connecting path of an appropriate color between opposite sides of the board.

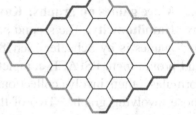

Figure 1. A 5×5 Hex board.

John Nash [BCG04, p. 744] proved that the first player to move can win by using a strategy-stealing argument (see Section 2.3). Suppose that the second player has a winning strategy, and assume by symmetry that Left goes first. Left selects the first hexagon arbitrarily. Now Right is to move first and Left is effectively the second player. Thus, Left can follow the winning strategy for the second player, except that Left has one additional hexagon. But this additional hexagon can only help Left: it only restricts Right's moves, and if Left's strategy suggests filling the additional hexagon, Left can instead move anywhere else. Thus, Left has a winning strategy, contradicting that Right did, and hence the first player has a winning strategy. However, it remains open to give a polynomial characterization of a winning strategy for the first player.

In perhaps the first PSPACE-hardness result for "interesting" games, Even and Tarjan [ET76] proved that a generalization of Hex to graphs is PSPACE-complete, even for maximum-degree-5 graphs. Specifically, in this graph game, two vertices are initially colored Left, and players take turns coloring uncolored vertices in their own color. Left's goal is to connect the two initially Left vertices by a path, and Right's goal is to prevent such a path. Surprisingly, the closely related problem in which players color *edges* instead of vertices can be solved in polynomial time; this game is known as the *Shannon switching game* [BW70]. A special case of this game is *Bridgit* or *Gale*, invented by David Gale [BCG04, p. 744], in which the graph is a square grid and Left's goal is to connect a vertex on the top row with a vertex on the bottom row. However, if the graph in Shannon's switching game has directed edges, the game again becomes PSPACE-complete [ET76].

A few years later, Reisch [Rei81] proved the stronger result that determining the outcome of a position in Hex is PSPACE-complete on a normal diamond-shaped board. The proof is quite different from the general graph reduction of Even and Tarjan [ET76], but the main milestone is to prove that Hex is PSPACE-complete for planar graphs.

4.2. More games on graphs: Kayles, Snort, Geography, Peek, and Interactive Hamiltonicity. The second paper to prove PSPACE-hardness of "interesting" games is by Schaefer [Sch78]. This work proposes over a dozen games and proves them PSPACE-complete. Some of the games involve propositional formulas, others involve collections of sets, but perhaps the most interesting are those involving graphs. Two of these games are generalizations of "Kayles", and another is a graph-traversal game called Edge Geography.

Kayles [BCG04, pp. 81–82] is an impartial game, designed independently by Dudeney and Sam Loyd, in which bowling pins are lined up on a line. Players take turns *bowling* with the property that exactly one or exactly two adjacent pins are knocked down (removed) in each move. Thus, most moves split the game into a sum of two subgames. Under normal play, Kayles can be solved in polynomial time using the Sprague–Grundy theory; see [BCG04, pp. 90–91], [GS56b].

Node Kayles is a generalization of Kayles to graphs in which each bowl "knocks down" (removes) a desired vertex and all its neighboring vertices. (Alternatively, this game can be viewed as two players finding an independent set.) Schaefer [Sch78] proved that deciding the outcome of this game is PSPACE-complete. The same result holds for a partizan version of node Kayles, in which every node is colored either Left or Right and only the corresponding player can choose a particular node as the primary target.

Geography is another graph game, or rather game family, that is special from a techniques point of view: it has been used as the basis of many other PSPACE-hardness reductions for games described in this section. The motivating example of the game is players taking turns naming distinct geographic locations, each starting with the same letter with which the previous name ended. More generally, Geography consists of a directed graph with one node initially containing a token. Players take turns moving the token along a directed edge. In *Edge Geography*, that edge is then erased; in *Vertex Geography*, the vertex moved from is then erased. (Confusingly, in the literature, each of these variants is frequently referred to as simply "Geography" or "Generalized Geography".)

Schaefer [Sch78] established that Edge Geography (a game suggested by R. M. Karp) is PSPACE-complete; Lichtenstein and Sipser [LS80] showed that Vertex Geography (which more closely matches the motivating example above) is also PSPACE-complete. Nowakowski and Poole [NP96] have solved special cases of Vertex Geography when the graph is a product of two cycles.

One may also consider playing either Geography game on an undirected graph. Fraenkel, Scheinerman, and Ullman [FSU93] show that Undirected Vertex Geography can be solved in polynomial time, whereas Undirected Edge Geography is PSPACE-complete, even for planar graphs with maximum degree 3. If the graph is bipartite then Undirected Edge Geography is also solvable in polynomial time.

One consequence of partizan node Kayles being PSPACE-hard is that deciding the outcome in Snort is PSPACE-complete on general graphs [Sch78]. *Snort* [BCG04, pp. 145–147] is a game designed by S. Norton and normally played on planar graphs (or planar maps). In any case, players take turns coloring vertices (or faces) in their own color such that only equal colors are adjacent.

Generalized hex (the vertex Shannon switching game), node Kayles, and Vertex Geography have also been analyzed recently in the context of parameterized complexity. Specifically, the problem of deciding whether the first player can win within k moves, where k is a parameter to the problem, is AW[$*$]-complete [DF97, ch. 14].

Stockmeyer and Chandra [SC79] were the first to prove combinatorial games to be EXPTIME-hard. They established EXPTIME-completeness for a class of logic games and two graph games. Here we describe an example of a logic game in the class, and one of the graph games; the other graph game is described in the next section. One logic game, called Peek, involves a box containing several parallel rectangular plates. Each plate (1) is colored either Left or Right except for one ownerless plate, (2) has circular holes carved in particular (known) positions, and (3) can be slid to one of two positions (fully in the box or partially outside the box). Players take turns either passing or changing the position of one of their plates. The winner is the first player to cause a hole in every plate to be aligned along a common vertical line. A second game involves a graph in which some edges are colored Left and some edges are colored Right, and initially some edges are "in" while the others are "out". Players take turns either passing or changing one edge from "out" to "in" or vice versa. The winner is the first player to cause the graph of "in" edges to have a Hamiltonian cycle.

(Both of these games can be rephrased under normal play by defining there to be no valid moves from positions having aligned holes or Hamiltonian cycles.)

4.3. Games of pursuit: Annihilation, Remove, Capture, Contrajunctive, Blocking, Target, and Cops and Robbers.

The next suite of graph games essentially began study in 1976 when Fraenkel and Yesha [FY76] announced that a certain impartial annihilation game could be played optimally in polynomial time. Details appeared later in [FY82]; see also [Fra74]. The game was proposed by John Conway and is played on an arbitrary directed graph in which some of the vertices contain a token. Players take turns selecting a token and moving it along an edge; if this causes the token to occupy a vertex already containing a token, both tokens are *annihilated* (removed). The winner is determined by normal play if all tokens are annihilated, except that play may be drawn out indefinitely. Fraenkel and Yesha's result [FY82] is that the outcome of the game can be determined and (in the case of a winner) a winning strategy of $O(n^5)$ moves can be computed in $O(n^6)$ time, where n is the number of vertices in the graph.

A generalization of this impartial game, called *Annihilation*, is when two (or more) types of tokens are distinguished, and each type of token can travel along only a certain subset of the edges. As before, if a token is moved to a vertex containing a token (of any type), both tokens are annihilated. Determining the outcome of this game was proved NP-hard [FY79] and later PSPACE-hard [FG87]. For acyclic graphs, the problem is PSPACE-complete [FG87]. The precise complexity for cyclic graphs remains open. Annihilation has also been studied under misère play [Fer84].

A related impartial game, called *Remove*, has the same rules as Annihilation except that when a token is moved to a vertex containing another token, only the moved token is removed. This game was also proved NP-hard using a reduction similar to that for Annihilation [FY79], but otherwise its complexity seems open. The analogous impartial game in which just the *unmoved* token is removed, called *Hit*, is PSPACE-complete for acyclic graphs [FG87], but its precise complexity remains open for cyclic graphs.

A partizan version of Annihilation is *Capture*, in which the two types of tokens are assigned to corresponding players. Left can only move a Left token, and only to a position that does not contain a Left token. If the position contains a Right token, that Right token is *captured* (removed). Unlike Annihilation, Capture allows all tokens to travel along all edges. Determining the outcome of Capture was proved NP-hard [FY79] and later EXPTIME-complete [GR95]. For acyclic graphs the game is PSPACE-complete [GR95].

A different partizan version of Annihilation is *Contrajunctive*, in which players can move both types of tokens, but each player can use only a certain subset of the edges. This game is NP-hard even for acyclic graphs [FY79] but otherwise its complexity seems open.

The *Blocking* variations of Annihilation disallow a token to be moved to a vertex containing another token. Both variations are partizan and played with tokens on directed graph. In *Node Blocking*, each token is assigned to one of the two players, and all tokens can travel along all edges. Determining the outcome of this game was proved NP-hard [FY79], then PSPACE-hard [FG87], and finally EXPTIME-complete [GR95]. Its status for acyclic graphs remains open. In *Edge Blocking*, there is only one type of token, but each player can use only a subset of the edges. Determining the outcome of this game is PSPACE-complete for acyclic graphs [FG87]. Its precise complexity for general graphs remains open.

A generalization of Node Blocking is *Target*, in which some nodes are marked as *targets* for each player, and players can additionally win by moving one of their tokens to a vertex that is one of their targets. When no nodes are marked as targets, the game is the same as Blocking and hence EXPTIME-complete by [GR95]. In fact, general Target was proved EXPTIME-complete earlier by Stockmeyer and Chandra [SC79]. Surprisingly, even the special case in which the graph is acyclic and bipartite and only one player has targets is PSPACE-complete [GR95]. (NP-hardness of this case was established earlier [FY79].)

A variation on Target is *Semi-Partizan Target*, in which both players can move all tokens, yet Left wins if a Left token reaches a Left target, independent of who moved the token there. In addition, if a token is moved to a nontarget vertex containing another token, the two tokens are annihilated. This game is EXPTIME-complete [GR95]. While this game may seem less natural than the others, it was intended as a step towards the resolution of Annihilation.

Many of the results described above from [GR95] are based on analysis of a more complex game called *Pursuit* or *Cops and Robbers*. One player, the *robber*, has a single token; and the other player, the *cops*, have k tokens. Players take turns moving all of their tokens along edges in a directed graph. The cops win if at the end of any move the robber occupies the same vertex as a cop, and the robber wins if play can be forced to draw out forever. In the case of a single cop ($k = 1$), there is a simple polynomial-time algorithm, and in general, many versions of the game are EXPTIME-complete; see [GR95] for a summary. For example, EXPTIME-completeness holds even for undirected graphs, and for directed graphs in which cops and robbers can choose their initial positions. For acyclic graphs, Pursuit is PSPACE-complete [GR95].

4.4. Checkers (Draughts). The standard 8×8 game of Checkers (Draughts), like many classic games, is finite and hence can be played optimally in constant time (in theory). Indeed, Schaeffer et al. [SBB+07] recently computed that optimal play leads to a draw from the initial configuration (other configurations remain unanalyzed). The outcome of playing in a general $n \times n$ board from a natural starting position, such as the one in Figure 2, remains open. On the other hand, deciding the outcome of an arbitrary configuration is PSPACE-hard [FGJ+78]. If a polynomial bound is placed on the number of moves that are allowed in between jumps (which is a reasonable generalization of the drawing rule in standard Checkers

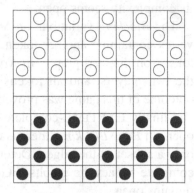

Figure 2. A natural starting configuration for 10×10 Checkers, from [FGJ+78].

[FGJ+78]), then the problem is in PSPACE and hence is PSPACE-complete. Without such a restriction, however, Checkers is EXPTIME-complete [Rob84b].

On the other hand, certain simple questions about Checkers can be answered in polynomial time [FGJ+78; DDE02]. Can one player remove all the other player's pieces in one move (by several jumps)? Can one player king a piece in one move? Because of the notion of parity on $n \times n$ boards, these questions reduce to checking the existence of an Eulerian path or general path, respectively, in a particular directed graph; see [FGJ+78; DDE02]. However, for boards defined by general graphs, at least the first question becomes NP-complete [FGJ+78].

4.5. Go. Presented at the same conference as the Checkers result in the previous section (FOCS'78), Lichtenstein and Sipser [LS80] proved that the classic Asian game of Go is also PSPACE-hard for an arbitrary configuration on an $n \times n$ board. Go has few rules: (1) players take turns either passing or placing stones of their color on positions on the board; (2) if a new black stone (say) causes a collection of white stones to be completely surrounded by black stones, the white stones are removed; and (3) a ko rule preventing repeated configurations. Depending on the country, there are several variations of the ko rule; see [BW94]. Go does not follow normal play: the winner in Go is the player with the highest score at the end of the game. A player's score is counted as either the number of stones of his color on the board plus empty spaces surrounded by his stones (area counting), or as empty spaces surrounded by his stones plus captured stones (territory counting), again varying by country.

The PSPACE-hardness proof given by Lichtenstein and Sipser [LS80] does not involve any situations called kos, where the ko rule must be invoked to avoid infinite play. In contrast, Robson [Rob83] proved that Go is EXPTIME-complete under Japanese rules when kos

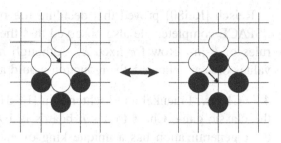

Figure 3. A simple form of ko in Go.

are involved, and indeed used judiciously. The type of ko used in this reduction is shown in Figure 3. When one of the players makes a move shown in the figure, the ko rule prevents (in particular) the other move shown in the figure to be made immediately afterwards.

Robson's proof relies on properties of the Japanese rules for both the upper and lower bounds. For other rulesets, all that is known is that Go is PSPACE-hard and in EXPSPACE. In particular, the "superko" variant of the ko rule (as used in, e.g., the U.S.A. and New Zealand), which prohibits recreation of any former board position, suggests EXPSPACE-hardness, by a result of Robson for no-repeat games [Rob84a]. However, if all dynamical state in the game occurs in kos, as it does in the EXPTIME-hardness construction, then the game is still in EXPTIME, because then it is an instance of Undirected Vertex Geography (Section 4.2), which can be solved in time polynomial in the graph size. (In this case the graph is all the possible game positions, of which there are exponentially many.)

There are also several results for more restricted Go positions. Wolfe [Wol02] shows that even Go endgames are PSPACE-hard. More precisely, a *Go endgame* is when the game has reduced to a sum of Go subgames, each equal to a polynomial-size game tree. This proof is based on several connections between Go and combinatorial game theory detailed in a book by Berlekamp and Wolfe [BW94]. Crâşmaru and Tromp [CT00] show that it is PSPACE-complete to determine whether a ladder (a repeated pattern of capture threats) results in a capture. Finally, Crâşmaru [Crâ99] shows that it is NP-complete to determine the status of certain restricted forms of life-and-death problems in Go.

4.6. Five-in-a-Row (Gobang). *Five-in-a-Row* or *Gobang* [BCG04, pp. 738–740] is another game on a Go board in which players take turns placing a stone of their color. Now the goal of the players is to place at least 5 stones of their color in a row either horizontally, vertically, or diagonally. This game is similar to Go-Moku [BCG04, p. 740], which does not count 6 or more stones in a row, and imposes additional constraints on moves.

Reisch [Rei80] proved that deciding the outcome of a Gobang position is PSPACE-complete. He also observed that the reduction can be adapted to the rules of k-in-a-Row for fixed k. Although he did not specify exactly which values of k are allowed, the reduction would appear to generalize to any $k \geq 5$.

4.7. Chess. Fraenkel and Lichtenstein [FL81] proved that a generalization of the classic game Chess to $n \times n$ boards is EXPTIME-complete. Specifically, their generalization has a unique king of each color, and for each color the numbers of pawns, bishops, rooks, and queens increase as some fractional power of n. (Knights are not needed.) The initial configuration is unspecified; what is EXPTIME-hard is to determine the winner (who can checkmate) from an arbitrary specified configuration.

4.8. Shogi. *Shogi* is a Japanese game along lines similar to Chess, but with rules too complex to state here. Adachi, Kamekawa, and Iwata [AKI87] proved that deciding the outcome of a Shogi position is EXPTIME-complete. Recently, Yokota et al. [YTK+01] proved that a more restricted form of Shogi, *Tsume-Shogi*, in which the first player must continually make oh-te (the equivalent of check in Chess), is also EXPTIME-complete.

4.9. Othello (Reversi). *Othello (Reversi)* is a classic game on an 8×8 board, starting from the initial configuration shown in Figure 4, in which players alternately place pieces of their color in unoccupied squares. Moves are restricted to cause, in at least one row, column, or diagonal, a consecutive sequence of pieces of the opposite color to be enclosed by two pieces of the current player's color. As a result of the move, the enclosed pieces "flip" color into the current player's color. The winner is the player with the most pieces of their color when the board is filled.

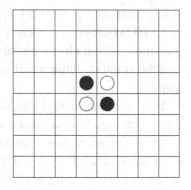

Figure 4. Starting configuration in the game of Othello.

Generalized to an $n \times n$ board with an arbitrary initial configuration, the game is clearly in PSPACE because only $n^2 - 4$ moves can be made. Furthermore, Iwata and Kasai [IK94] proved that the game is PSPACE-complete.

4.10. Hackenbush. *Hackenbush* is one of the standard examples of a combinatorial game in *Winning Ways*; see, e.g., [BCG04, pp. 1–6]. A position is given by a graph with each edge colored either red (Left), blue (Right), or green (neutral), and with certain vertices marked as *rooted*. Players take turns removing an edge of an appropriate color (either neutral or their own color), which also causes all edges not connected to a rooted vertex to be removed. The winner is determined by normal play.

Chapter 7 of *Winning Ways* [BCG04, pp. 189–227] proves that determining the *value* of a red-blue Hackenbush position is NP-hard. The reduction is from minimum Steiner tree in graphs. It applies to a restricted form of hackenbush positions, called *redwood beds*, consisting of a red bipartite graph, with each vertex on one side attached to a red edge, whose other end is attached to a blue edge, whose other end is rooted.

4.11. Domineering (Crosscram) and Cram. *Domineering*, also called *cross-cram* [BCG04, pp. 119–126], is a partizan game involving placement of horizontal and vertical dominoes in a grid; a typical starting position is an $m \times n$ rectangle. Left can play only vertical dominoes and Right can play only horizontal dominoes, and dominoes must remain disjoint. The winner is determined by normal play.

The complexity of Domineering, computing either the outcome or the value of a position, remains open. Lachmann, Moore, and Rapaport [LMR00] have shown that the winner and a winning strategy can be computed in polynomial time for $m \in \{1, 2, 3, 4, 5, 7, 9, 11\}$ and all n. These algorithms do not compute the value of the game, nor the optimal strategy, only a winning strategy.

Cram [Gar86], [BCG04, pp. 502–506] is the impartial version of Domineering in which both players can place horizontal and vertical dominoes. The outcome of Cram is easy to determine for rectangles having an even number of squares [Gar86]: if both sides are even, the second player can win by a symmetry strategy (reflecting the first player's move through both axes); and if precisely one side is even, the first player can win by playing the middle two squares and then applying the symmetry strategy. It seems open to determine the outcome for a rectangle having two odd sides. The complexity of Cram for general boards also remains open.

Linear Cram is Cram in a $1 \times n$ rectangle, where the game quickly splits into a sum of games. This game can be solved easily by applying the Sprague–Grundy theory and dynamic programming; in fact, there is a simpler solution based on proving that its behavior is periodic in n [GS56b]. The variation on Linear Cram in which $1 \times k$ rectangles are placed instead of dominoes can also be solved via dynamic programming, but whether the behavior is periodic remains open even for $k = 3$ [GS56b]. Misère Linear Cram also remains unsolved [Gar86].

4.12. Dots-and-Boxes, Strings-and-Coins, and Nimstring. *Dots-and-Boxes* is a well-known children's game in which players take turns drawing horizontal and vertical edges connecting pairs of dots in an $m \times n$ subset of the lattice. Whenever a player makes a move that encloses a unit square with drawn edges, the player is awarded a point and must then draw another edge in the same move. The winner is the player with the most points when the entire grid has been drawn. Most of this section is based on Chapter 16 of *Winning Ways* [BCG04, pp. 541–

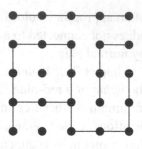

Figure 5. A Dots-and-Boxes endgame.

584]; another good reference is a recent book by Berlekamp [Ber00].

Gameplay in Dots-and-Boxes typically divides into two phases: the *opening* during which no boxes are enclosed, and the *endgame* during which boxes are enclosed in nearly every move; see Figure 5. In the endgame, the "free move" awarded by enclosing a square often leads to several squares enclosed in a single move, following a *chain*; see Figure 6. Most children apply the greedy algorithm of taking the most squares possible, and thus play entire chains of squares. However, this strategy forces the player to open another chain (in the endgame). A simple improved strategy is called *double dealing*, which forfeits the last two squares of the chain, but forces the opponent to open the next chain. The double-dealer is said to *remain in control*; if there are long-enough chains, this player will win (see [BCG04, p. 543] for a formalization of this statement).

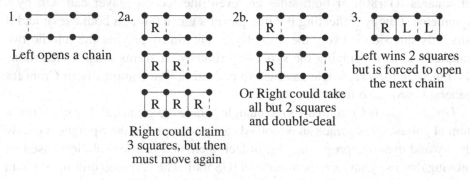

Figure 6. Chains and double-dealing in Dots-and-Boxes.

A generalization arising from the dual of Dots-and-Boxes is *Strings-and-Coins* [BCG04, pp. 550–551]. This game involves a sort of graph whose vertices are *coins* and whose edges are *strings*. The coins may be tied to each other and to the "ground" by strings; the latter connection can be modeled as a loop in the graph. Players alternate cutting strings (removing edges), and if a coin is thereby freed, that player collects the coin and cuts another string in the same move. The player to collect the most coins wins.

Another game closely related to Dots-and-Boxes is *Nimstring* [BCG04, pp. 552–554], which has the same rules as Strings-and-Coins, except that the winner is determined by normal play. Nimstring is in fact a special case of Strings-and-Coins [BCG04, p. 552]: if we add a chain of more than $n+1$ coins to an instance of Nimstring having n coins, then ideal play of the resulting string-and-coins instance will avoid opening the long chain for as long as possible, and thus the player to move last in the Nimstring instance wins string and coins.

Winning Ways [BCG04, pp. 577–578] argues that Strings-and-Coins is NP-hard as follows. Suppose that you have gathered several coins but your opponent gains control. Now you are forced to lose the Nimstring game, but given your initial lead, you still may win the Strings-and-Coins game. Minimizing the number of coins lost while your opponent maintains control is equivalent to finding the maximum number of vertex-disjoint cycles in the graph, basically because the equivalent of a double-deal to maintain control once an (isolated) cycle is opened results in forfeiting four squares instead of two. We observe that by making the difference between the initial lead and the forfeited coins very small (either -1 or 1), the opponent also cannot win by yielding control. Because the cycle-packing problem is NP-hard on general graphs, determining the outcome of such string-and-coins endgames is NP-hard. Eppstein [Epp] observes that this reduction should also apply to endgame instances of Dots-and-Boxes by restricting to maximum-degree-three planar graphs. Embeddability of such graphs in the square grid follows because long chains and cycles (longer than two edges for chains and three edges for cycles) can be replaced by even longer chains or cycles [BCG04, p. 561].

It remains open whether Dots-and-Boxes or Strings-and-Coins are in NP or PSPACE-complete from an arbitrary configuration. Even the case of a $1 \times n$ grid of boxes is not fully understood from a Combinatorial Game Theory perspective [GN02].

4.13. Amazons. *Amazons* is a game invented by Walter Zamkauskas in 1988, containing elements of Chess and Go. Gameplay takes place on a 10×10 board with four *amazons* of each color arranged as in Figure 7 (left). In each turn, Left [Right] moves a black [white] amazon to any unoccupied square accessible by a Chess queen's move, and fires an arrow to any unoccupied square reachable by a Chess queen's move from the amazon's new position. The arrow (drawn as a circle) now occupies its square; amazons and shots can no longer pass over or land on this square. The winner is determined by normal play.

Gameplay in Amazons typically split into a sum of simpler games because arrows partition the board into multiple components. In particular, the *endgame* begins when each component of the game contains amazons of only a single color, at which point the goal of each player is simply to maximize the number of moves in each component. Buro [Bur00] proved that maximizing the num-

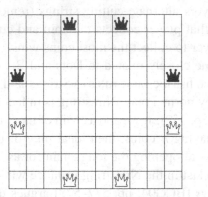
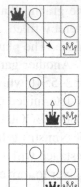

Figure 7. The initial position in Amazons (left) and an example of black trapping a white amazon (right).

ber of moves in a single component is NP-complete (for $n \times n$ boards). In a general endgame, deciding the outcome may not be in NP because it is difficult to prove that the opponent has no better strategy. However, Buro [Bur00] proved that this problem is *NP-equivalent* [GJ79], that is, the problem can be solved by a polynomial number of calls to an algorithm for any NP-complete problem, and vice versa.

Like Conway's Angel Problem (Section 4.16), the complexity of deciding the outcome of a general Amazons position remained open for several years, only to be solved nearly simultaneously by multiple people. Furtak, Kiyomi, Takeaki, and Buro [FKUB05] give two independent proofs of PSPACE-completeness: one a reduction from Hex, and the other a reduction from Vertex Geography. The latter reduction applies even for positions containing only a single black and a single white amazon. Independently, Hearn [Hea05a; Hea06b; Hea08a] gave a Constraint Logic reduction showing PSPACE-completeness.

4.14. Konane. Konane, or Hawaiian Checkers, is a game that has been played in Hawaii since preliterate times. Konane is played on a rectangular board (typically ranging in size from 8×8 to 13×20) which is initially filled with black and white stones in a checkerboard pattern. To begin the game, two adjacent stones in the middle of the board or in a corner are removed. Then, the players alternate making moves. Moves are made as in Peg Solitaire (Section 5.10); indeed, Konane may be thought of as a kind of two-

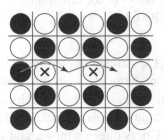

Figure 8. One move in Konane consisting of two jumps.

player peg solitaire. A player moves a stone of his color by jumping it over a horizontally or vertically adjacent stone of he opposite color, into an empty space. (See Figure 8.) Jumped stones are captured and removed from play. A stone may make multiple successive jumps in a single move, so long as they are in a straight line; no turns are allowed within a single move. The first player unable to move wins.

Hearn proved that Konane is PSPACE-complete [Hea06b; Hea08a] by a reduction from Constraint Logic. There have been some positive results for restricted configurations. Ernst [Ern95] derives Combinatorial-Game-Theoretic values for several interesting positions. Chan and Tsai [CT02] analyze the $1 \times n$ game, but even this version of the game is not yet solved.

4.15. Phutball. Conway's game of *Philosopher's Football* or *Phutball* [BCG04, pp. 752–755] involves white and black stones on a rectangular grid such as a Go board. Initially, the unique black stone (the *ball*) is placed in the middle of the board, and there are no white stones. Players take turns either placing a white stone in any unoccupied position, or moving the ball by a sequence of *jumps* over consecutive sequences of white stones each arranged horizontally, vertically, or diagonally. See Figure 9. A jump causes immediate removal of the white stones jumped over, so those stones cannot be used for a future jump in the same move. Left and Right have opposite sides of the grid marked as their *goal lines*. Left's goal is to end a move with the ball on or beyond Right's goal line, and symmetrically for Right.

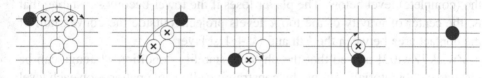

Figure 9. A single move in Phutball consisting of four jumps.

Phutball is inherently loopy and it is not clear that either player has a winning strategy: the game may always be drawn out indefinitely. One counterintuitive aspect of the game is that white stones placed by one player may be "corrupted" for better use by the other player. Recently, however, Demaine, Demaine, and Eppstein [DDE02] found an aspect of Phutball that could be analyzed. Specifically, they proved that determining whether the current player can win in a single move ("mate in 1" in Chess) is NP-complete. This result leaves open the complexity of determining the outcome of a given game position.

4.16. Conway's Angel Problem. A formerly long-standing open problem was Conway's *Angel Problem* [BCG04]. Two players, the Angel and the Devil, alternate play on an infinite square grid. The Angel can move to any valid position within k horizontal distance and k vertical distance from its present position. The Devil can teleport to an arbitrary square other than where the Angel is and "eat" that square, preventing the Angel from landing on (but not leaping over) that square in the future. The Devil's goal is to prevent the Angel from moving.

It was long known that an Angel of power $k = 1$ can be stopped [BCG04], so the Devil wins, but the Angel was not known to be able to escape for any $k > 1$. (In the original open problem statement, $k = 1000$.) Recently, four independent proofs established that a sufficiently strong Angel can move forever, securing the Angel as the winner. Máthé [Mát07] and Kloster [Klo07] showed that $k = 2$ suffices; Bowditch [Bow07] showed that $k = 4$ suffices; and Gács [Gác07] showed that some k suffices. In particular, Kloster's proof gives an explicit algorithmic winning strategy for the $k = 2$ Angel.

4.17. Jenga. *Jenga* is a popular stacked-block game invented by Leslie Scott in the 1970s and now marketed by Hasbro. Two players alternate moving individual blocks in a tower of blocks, and the first player to topple the tower (or cause any additional blocks to fall) loses. Each block is $1 \times 1 \times 3$ and lies horizontally. The initial $3 \times 3 \times n$ tower alternates levels of three blocks each, so that blocks in adjacent levels are orthogonal. (In the commercial game, $n = 18$.) In each move, the player removes any block that is below the topmost complete (3-block) level, then places that block in the topmost level (starting a new level if the existing topmost level is complete), orthogonal to the blocks in the (complete) level below. The player loses if the tower becomes instable, that is, the center of gravity of the top k levels projects outside the convex hull of the contact area between the kth and $(k + 1)$st layer.

Zwick [Zwi02] proved that the physical stability condition of Jenga can be restated combinatorially simply by constraining allowable patterns on each level and the topmost three levels. Specifically, write a 3-bit vector to specify which blocks are present in each level. Then a tower is stable if and only if no level except possibly the top is 100 or 001 and the three topmost levels from bottom to top are none of the following:

$$010, 010, 100; \quad 010, 010, 001; \quad 011, 010, 100; \quad 110, 010, 001.$$

Using this characterization, Zwick proves that the first player wins from the initial configuration if and only if $n = 2$ or $n \geq 4$ and $n \equiv 1$ or 2 (mod 3), and gives a simple characterization of winning moves. It remains open whether such an efficient solution can be obtained in the generalization to odd numbers $k > 3$ of blocks in each level. (The case of even k is a second-player win by a simple mirror strategy.)

5. Algorithms for puzzles

Many puzzles (one-player games) have short solutions and are NP-complete. In contrast, several puzzles based on motion-planning problems are harder — PSPACE-hard. Usually such puzzles occupy a bounded board or region, so they are also PSPACE-complete. A common method to prove that such puzzles are in PSPACE is to give a simple low-space nondeterministic algorithm that guesses the solution, and apply Savitch's theorem [Sav70] that PSPACE = NPSPACE (nondeterministic polynomial space). However, when generalized to the entire plane and unboundedly many pieces, puzzles often become undecidable.

This section briefly surveys some of these results, following the structure of the previous section.

5.1. Instant Insanity. Given n cubes, each face colored one of n colors, is it possible to stack the cubes so that each color appears exactly once on each of the 4 sides of the stack? The case of $n = 4$ is a puzzle called Instant Insanity distributed by Parker Bros. In one of the first papers on hardness of puzzles and games people play, Robertson and Munro [RM78] proved that this *generalized Instant Insanity* problem is NP-complete.

The *cube stacking game* is a two-player game based on this puzzle. Given an ordered list of cubes, the players take turns adding the next cube to the top of the stack with a chosen orientation. The loser is the first player to add a cube that causes one of the four sides of the stack to have a color repeated more than once. Robertson and Munro [RM78] proved that this game is PSPACE-complete, intended as a general illustration that NP-complete puzzles tend to lead to PSPACE-complete games.

5.2. Cryptarithms (Alphametics, Verbal Arithmetic). *Cryptarithms* or *alphametics* or *verbal arithmetic* are classic puzzles involving an equation of symbols, the original being Dudeney's SEND+MORE=MONEY from 1924 [Dud24], in which each symbol (e.g., M) represents a consistent digit (between 0 and 9). The goal is to determine an assignment of digits to symbols that satisfies the equation. Such problems can easily be solved in polynomial time by enumerating all 10! assignments. However, Eppstein [Epp87] proved that it is NP-complete to solve the generalization to base $\Theta(n^3)$ (instead of decimal) and $\Theta(n)$ symbols (instead of 26).

5.3. Crossword puzzles and Scrabble. Perhaps one of the most popular puzzles are *crossword puzzles*, going back to 1913 and today appearing in almost every newspaper, and the subject of the recent documentary *Wordplay* (2006). Here it is easiest to model the problem of designing crossword puzzles, ignoring the nonmathematical notion of clues. Given a list of words (the dictionary), and a rectangular grid with some squares obstacles and others blank, can we place a subset of the words into horizontally or vertically maximal blank strips so that crossing words have matching letters? Lewis and Papadimitriou [GJ79, p. 258] proved that this question is NP-complete, even when the grid has no obstacles so every row and column must form a word.

Alternatively, this problem can be viewed as the ultimate form of crossword puzzle solving, without clues. In this case it would be interesting to know whether the problem remains NP-hard even if every word in the given list must be used exactly once, so that the single clue could be "use these words". A related open problem is *Scrabble*, which we are not aware of having been studied. The most natural theoretical question is perhaps the one-move version: given the pieces in hand (with letters and scores), and given the current board configuration (with played pieces and available double/triple letter/word squares), what move maximizes score? Presumably the decision question is NP-complete. Also open is the complexity of the two-player game, say in the perfect-information variation where both players know the sequence in which remaining pieces will be drawn as well as the pieces in the opponent's hand. Presumably determining a winning move from a given position in this game is PSPACE-complete.

5.4. Pencil-and-paper puzzles: Sudoku and friends. *Sudoku* is a pencil-and-paper puzzle that became popular worldwide starting around 2005 [Del06; Hay06]. American architect Howard Garns first published the puzzle in the May 1979 (and many subsequent) *Dell Pencil Puzzles and Word Games* (without a byline and under the title *Number Place*); then Japanese magazine *Monthly Nikolist* imported the puzzle in 1984, trademarking the name Sudoku ("single numbers"); then the idea spread throughout Japanese publications; finally Wayne Gould published his own computer-generated puzzles in *The Times* in 2004, shortly after which many newspapers and magazines adopted the puzzle. The usual puzzle consists of an 9×9 grid of squares, divided into a 3×3 arrangement of 3×3 tiles. Some grid squares are initially filled with digits between 1 and 9, and some are blank. The goal is to fill the blank squares so that every row, column, and tile has all nine digits without repetition.

Sudoku naturally generalizes to an $n^2 \times n^2$ grid of squares, divided into an $n \times n$ arrangement of $n \times n$ tiles. Yato and Seta [YS03; Yat03] proved that this generalization is NP-complete. In fact, they proved a stronger completeness result, in the class of Another Solution Problems (ASP), where one is given one or more solutions and wishes to find another solution. Thus, in particular, given a Sudoku puzzle and an intended solution, it is NP-complete to determine whether there is another solution, a problem arising in puzzle design. Most Sudoku puzzles give the promise that they have a unique solution. Valiant and Vazirani [VV86] proved that adding such a uniqueness promise keeps a problem

NP-hard under randomized reductions, so there is no polynomial-time solution to uniquely solvable Sudokus unless RP = NP.

ASP-completeness (in particular, NP-completeness) has been established for six other paper-and-pencil puzzles by Japanese publisher Nikoli: Nonograms, Slitherlink, Cross Sum, Fillomino, Light Up, and LITS. In a *Nonogram* or *Paint by Numbers* puzzle [UN96], we are given a sequence of integers on each row and column of a rectangular matrix, and the goal is to fill in a subset of the squares in the matrix so that, in each row and column, the maximal contiguous runs of filled squares have lengths that match the specified sequence. In *Slitherlink* [YS03; Yat03], we are given labels between 0 and 4 on some subset of faces in a rectangular grid, and the goal is to draw a simple cycle on the grid so that each labeled face is surrounded by the specified number of edges. In *Kakuro* or *Cross Sum* [YS03], we are given a polyomino (a rectangular grid where only some squares may be used), and an integer for each maximal contiguous (horizontal or vertical) strip of squares, and the goal is to fill each square with a digit between 1 and 9 such that each strip has the specified sum and has no repeated digit. In *Fillomino* [Yat03], we are given a rectangular grid in which some squares have been filled with positive integers, and the goal is to fill the remaining squares with positive integers so that every maximal connected region of equally numbered squares consists of exactly that number of squares. In *Light Up* (*Akari*) [McP05; McP07], we are given a rectangular grid in which squares are either rooms or walls and some walls have a specified integer between 0 and 4, and the goal is to place lights in a subset of the rooms such that each numbered wall has exactly the specified number of (horizontally or vertically) adjacent lights, every room is horizontally or vertically visible from a light, and no two lights are horizontally or vertically visible from each other. In *LITS* [McP07], we are given a division of a rectangle into polyomino pieces, and the goal is to choose a tetromino (connected subset of four squares) in each polyomino such that the union of tetrominoes is connected yet induces no 2×2 square. As with Sudoku, it is NP-complete to both find solutions and test uniqueness of known solutions in all of these puzzles.

NP-completeness has been established for a few other pencil-and-paper games published by Nikoli: Tentai Show, Masyu, Bag, Nurikabe, Hiroimono, Heyawake, and Hitori. In *Tentai Show* or *Spiral Galaxies* [Fri02d], we are given a rectangular grid with dots at some vertices, edge midpoints, and face centroids, and the goal is to divide the rectangle into exactly one polyomino piece per dot that is two-fold rotationally symmetric around the dot. In *Masyu* or *Pearl Puzzles* [Fri02b], we are given a rectangular grid with some squares containing white or black pearls, and the goal is to find a simple path through the squares that visits every pearl, turns 90° at every black pearl, does not turn immediately before or after black pearls, goes straight through every white pearl, and turns 90° immediately before or after every white pearl. In *Bag* or *Corral Puzzles* [Fri02a], we are given a rectangular grid with some squares labeled with positive integers, and the goal is to find a simple cycle on the grid that encloses all labels and such that the number of squares horizontally and vertically visible from each labeled square equals the label. In *Nurikabe* [McP03; HKK04], we are given a rectangular grid with some squares labeled with positive integers, and the goal is to find a connected subset of unlabeled squares that induces no 2×2 square and whole removal results in exactly one region per labeled square whose size equals that label. McPhail's reduction [McP03] uses labels 1 through 5, while Holzer et al.'s reduction [HKK04] only uses labels 1 and 2 (just 1 would be trivial) and works without the connectivity rule and/or the 2×2 rule. In *Hiroimono* or *Goishi Hiroi* [And07], we are given a collection of stones at vertices of a rectangular grid, and the goal is to find a path that visits all stones, changes directions by $\pm 90°$ and only at stones, and removes stones as they are visited (similar to Phutball in Section 4.15). In *Heyawake* [HR07], we are given a subdivision of a rectangular grid into rectangular rooms, some of which are labeled with a positive integer, and the goal is to paint a subset of unit squares so that the number of painted squares in each labeled room equals the label, painted squares are never (horizontally or vertically) adjacent, unpainted squares are connected (via horizontal and vertical connections), and maximal contiguous (horizontal or vertical) strips of squares intersect at most two rooms. In *Hitori* [Hea08c], we are given a rectangular grid with each square labeled with an integer, and the goal is to paint a subset of unit squares so that every row and every column has no repeated unpainted label (similar to Sudoku), painted squares are never (horizontally or vertically) adjacent, and unpainted squares are connected (via horizontal and vertical connections).

A different kind of pencil-and-paper puzzle is *Morpion Solitaire*, popular in several European countries. The game starts with some configuration of points drawn at the intersections of a square grid (usually in a standard cross pattern). A move consists of placing a new point at a grid intersection, and then drawing a horizontal, vertical, or diagonal line segment connecting five consecutive points that include the new one. Line segments with the same direction cannot share a point (the *disjoint model*); alternatively, line segments with the same direction may overlap only at a common endpoint (the *touching model*). The goal is to maximize the number of moves before no moves are possible. Demaine, De-maine, Langerman, and Langerman [DDLL06] consider this game generalized to moves connecting any number $k + 1$ of points instead of just 5. In addition to bounding the number of moves from the standard cross configuration, they prove complexity results for the general case. They show that, in both game models and for $k \geq 3$, it is NP-hard to find the longest play from a given pattern of n dots, or even to approximate the longest play within $n^{1-\varepsilon}$ for any $\varepsilon > 0$. For $k > 3$, the problem is in fact NP-complete. For $k = 3$, it is open whether the problem is in NP, and for $k = 2$ it could even be in P.

A final NP-completeness result for pencil-and-paper puzzles is the Battleship puzzle. This puzzle is a one-player perfect-information variant on the classic two-player imperfect-information game, *Battleship*. In *Battleships* or *Battleship Solitaire* [Sev], we are given a list of $1 \times k$ ships for various values of k; a rectangular grid with some squares labeled as water, ship interior, ship end, or entire (1×1) ship; and the number of ship (nonwater) squares that should be in each row and each column. The goal is to complete the square labeling to place the given ships in the grid while matching the specified number of ship squares in each row and column.

Several other pencil-and-paper puzzles remain unstudied from a complexity standpoint. For example, Nikoli's English website[2] suggests Hashiwokakero, Kuromasu (Where is Black Cells), Number Link, Ripple Effect, Shikaku, and Yajilin (Arrow Ring); and Nikoli's Japanese website[3] lists more.

5.5. Moving Tokens: Fifteen Puzzle and generalizations.

The *Fifteen Puzzle* or *15 Puzzle* [BCG04, p. 864] is a classic puzzle consisting of fifteen square blocks numbered 1 through 15 in a 4×4 grid; the remaining sixteenth square in the grid is a hole which permits blocks to slide. The

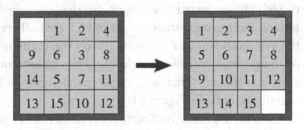

Figure 10. 15 puzzle: Can you get from the left configuration to the right in 16 unit slides?

goal is to order the blocks to be increasing in English reading order. The (six) hardest solvable positions require exactly 80 moves [BMFN99]. Slocum and Sonneveld [SS06] recently uncovered the history of this late 19th-century puzzle, which was well-hidden by popularizer Sam Loyd since his claim of having invented it.

A natural generalization of the Fifteen Puzzle is the $n^2 - 1$ *puzzle* on an $n \times n$ grid. It is easy to determine whether a configuration of the $n^2 - 1$ puzzle can reach another: the two permutations of the block numbers (in reading order) simply need to match in *parity*, that is, whether the number of inversions (out-

---

of-order pairs) is even or odd. See, e.g., [Arc99; Sto79; Wil74]. When the puzzle
is solvable, the required numbers moves is $\Theta(n^3)$ in the worst case [Par95]. On
the other hand, it is NP-complete to find a solution using the fewest possible
slides from a given configuration [RW90]. It is also NP-hard to approximate
the fewest slides within an additive constant, but there is a polynomial-time
constant-factor approximation [RW90].

The parity technique for determining solvability
of the $n^2 - 1$ puzzle has been generalized to a class
of similar puzzles on graphs. Consider an N-vertex
graph in which $N - 1$ vertices have tokens labeled 1
through $N - 1$, one vertex is empty (has no token),
and each operation in the puzzle moves a token to
an adjacent empty vertex. The goal is to reach one
configuration from another. This general puzzle encompasses the $n^2 - 1$ puzzle and several other puzzles involving sliding balls in circular tracks, e.g.,
the Lucky Seven puzzle [BCG04, p. 865] or the puzzle shown in Figure 11. Wilson [Wil74], [BCG04,

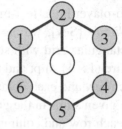

Figure 11. The Tricky
Six Puzzle [Wil74],
[BCG04, p. 868] has six
connected components
of configurations.

p. 866] characterized when these puzzles are solvable, and furthermore characterized their group structure. In most cases, all puzzles are solvable (forming
the symmetric group) unless the graph the graph is bipartite, in which case half
of the puzzles are solvable (forming the alternating group). In addition, there
are three special situations: cycle graphs, graphs having a cut vertex, and the
special example in Figure 11.

Even more generally, Kornhauser, Miller, and Spirakis [KMS84] showed how
to decide solvability of puzzles with any number k of labeled tokens on N
vertices. They also prove that $O(N^3)$ moves always suffice, and $\Omega(N^3)$ moves
are sometimes necessary, in such puzzles. Calinescu, Dumitrescu, and Pach
[CDP06] consider the number of token "shifts" — continuous moves along a
path of empty nodes — required in such puzzles. They prove that finding the
fewest-shift solution is NP-hard in the infinite square grid and APX-hard in
general graphs, even if the tokens are unlabeled (identical). On the positive
side, they present a 3-approximation for unlabeled tokens in general graphs, an
optimal solution for unlabeled tokens in trees, an upper bound of N slides for
unlabeled tokens in general graphs, and an upper bound of $O(N)$ slides for
labeled tokens in the infinite square grid.

Restricting the set of legal moves can make such puzzles harder. Consider a
graph with unlabeled tokens on some vertices, and the constraint that the tokens
must form an independent set on the graph (i.e., no two tokens are adjacent
along an edge). A move is made by sliding a token along an edge to an adjacent

vertex, subject to maintaining the nonadjacency constraint. Then the problem of determining whether a sequence of moves can ever move a given token, called *Sliding Tokens* [HD05], is PSPACE-complete.

Subway Shuffle [Hea05b; Hea06b] is another constrained token-sliding puzzle on a graph. In this puzzle both the tokens and the graph edges are colored; a move is to slide a token along an edge of matching color to an unoccupied adjacent vertex. The goal is to move a specified token (the "subway car you have boarded") to a specified vertex (your "exit station"). A sample puzzle is shown in Figure 12. The complexity of determining whether there is a solution to a given puzzle is open. This open problem is quite fascinating: solving the puzzle empirically seems hard, based on the rapid growth of minimum solution length with graph size

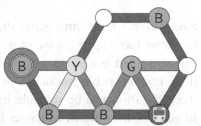

Figure 12. A Subway Shuffle puzzle with one red car (bottom right), four blue cars, one yellow car, and one green car. White nodes are empty. Moving the red car to the circled station requires 43 moves.

[Hea05b]. However, it is easy to determine whether a token may move at all by a sequence of moves, evidently making the proof techniques used for Sliding Tokens and related problems useless for showing hardness. Subway Shuffle can also be seen as a generalized version of 1×1 Rush Hour (Section 5.7).

Another kind of token-sliding puzzle is *Atomix*, a computer game first published in 1990. Game play takes place on a rectangular board; pieces are either *walls* (immovable blocks) or *atoms* of different types. A move is to slide an atom; in this case the atom must slide in its direction of motion until it hits a wall (as in the PushPush family, below (Section 5.8)). The goal is to assemble a particular pattern of atoms (a molecule). Huffner, Edelkamp, Fernau, and Niedermeier [HEFN01] observed that Atomix is as hard as the $(n^2 - 1)$-puzzle, so it is NP-hard to find a minimum-move solution. Holzer and Schwoon [HS04a] later proved the stronger result that it is PSPACE-complete to determine whether there is a solution.

Lunar Lockout is another token-sliding puzzle, similar to Atomix in that the tokens slide until stopped. Lunar Lockout was produced by ThinkFun at one time; essentially the same game is now sold as "Pete's Pike". (Even earlier, the game was called "UFO".) In Lunar Lockout there are no walls or barriers; a token may only slide if there is another token in place that will stop it. The goal is to get a particular token to a particular place. Thus, the rules are fairly simple and natural; however, the complexity is open, though there are partial results. Hock [Hoc01] showed that Lunar Lockout is NP-hard, and that when the target token may not revisit any position on the board, the problem becomes NP-complete. Hartline and Libeskind-Hadas [HLH03] show that a generalization of Lunar Lockout which allows fixed blocks is PSPACE-complete.

5.6. Rubik's Cube and generalizations.

Alternatively, the $n^2 - 1$ puzzle can be viewed as a special case of determining whether a permutation on N items can be written as a product (composition) of given generating permutations, and if so, finding such a product. This family of puzzles also includes *Rubik's Cube* (recently shown to be solvable in 26 moves [KC07]) and its many variations. In general, the number of moves (terms) required to solve such a puzzle can be exponential (unlike the Fifteen Puzzle). Nonetheless, an $O(N^5)$-time algorithm can decide whether a given puzzle of this type is solvable, and if so, find an implicit representation of the solution [Jer86]. On the other hand, finding a solution with the fewest moves (terms) is PSPACE-complete [Jer85]. When each given generator cyclically shifts just a bounded number of items, as in the Fifteen Puzzle but not in a $k \times k \times k$ Rubik's Cube, Driscoll and Furst [DF83] showed that such puzzles can be solved in polynomial time using just $O(N^2)$ moves. Furthermore, $\Theta(N^2)$ is the best possible bound in the worst case, e.g., when the only permitted moves are swapping adjacent elements on a line. See [KMS84; McK84] for other (not explicitly algorithmic) results on the maximum number of moves for various special cases of such puzzles.

5.7. Sliding blocks and Rush Hour.

A classic reference on a wide class of sliding-block puzzles is by Hordern [Hor86]. One general form of these puzzles is that rectangular blocks are placed in a rectangular box, and each block can be moved horizontally and vertically, provided the blocks remain disjoint. The goal is usually either to move a particular block to a particular place, or to rearrange one configuration into another. Figure 13 shows an example which, according to Gardner [Gar64], may be the earliest (1909) and is the most widely sold (after the Fifteen Puzzle, in each case). Gardner [Gar64] first raised the question of whether there is an efficient algorithm to solve such puzzles. Spirakis and Yap [SY83] showed that achieving a specified target configuration is NP-hard, and conjec-

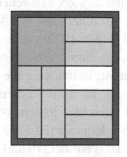

Figure 13. Dad's Puzzle [Gar64]: moving the large square into the lower-left corner requires 59 moves.

tured PSPACE-completeness. Hopcroft, Schwartz, and Sharir [HSS84] proved PSPACE-completeness shortly afterwards, renaming the problem to the "Warehouseman's Problem". In the Warehouseman's Problem, there is no restriction on the sizes of blocks; the blocks in the reduction grow with the size of the containing box. By contrast, in most sliding-block puzzles, the blocks are of small constant sizes. Finally, Hearn and Demaine [HD02; HD05] showed that it is PSPACE-hard to decide whether a given piece can move at all by a sequence of moves, even when all the blocks are 1×2 or 2×1. This result is best possible: the results above about unlabeled tokens in graphs show that 1×1 blocks are easy to rearrange.

A popular sliding-block puzzle is *Rush Hour*, distributed by ThinkFun, Inc. (formerly Binary Arts, Inc.). We are given a configuration of several 1×2, 1×3, 2×1, and 3×1 rectangular blocks arranged in an $m \times n$ grid. (In the commercial version, the board is 6×6, length-two rectangles are realized as *cars*, and length-three rectangles are *trucks*.) Horizontally oriented blocks can slide left and right, and vertically oriented blocks can slide up and down, provided the blocks remain disjoint. (Cars and trucks can drive only forward or reverse.) The goal is to remove a particular block from the puzzle via a one-unit opening in the bounding rectangle. Flake and Baum [FB02] proved that this formulation of Rush Hour is PSPACE-complete. Their approach is also the basis for Nondeterministic Constraint Logic described in Section 3. A version of Rush Hour played on a triangular grid, *Triagonal Slide-Out*, is also PSPACE-complete [Hea06b]. Tromp and Cilibrasi [Tro00; TC04] strengthened Flake and Baum's result by showing that Rush Hour remains PSPACE-complete even when all the blocks have length two (cars). The complexity of the problem remains open when all blocks are 1×1 but labeled whether they move only horizontally or only vertically [HD02; TC04; HD05]. As with Subway Shuffle (Section 5.5), solving the puzzle (by escaping the target block from the grid) empirically seems hard [TC04], whereas it is easy to determine whether a block may move at all by a sequence of moves. Indeed, 1×1 Rush Hour is a restricted form of Subway Shuffle, where there are only two colors, the graph is a grid, and horizontal edges and vertical edges use different colors. Thus, it should be easier to find positive results for 1×1 Rush Hour, and easier to find hardness results for Subway Shuffle. We conjecture that both are PSPACE-complete, but existing proof techniques seem inapplicable.

5.8. Pushing blocks. Similar in spirit to the sliding-block puzzles in Section 5.7 are *pushing-block puzzles*. In sliding-block puzzles, an exterior agent can move arbitrary blocks around, whereas pushing-block puzzles embed a *robot* that can only move adjacent blocks but can also move itself within unoccupied space. The study of this type of puzzle was initiated by Wilfong [Wil91], who proved that deciding whether the robot can reach a desired target is NP-hard when the robot can push and pull L-shaped blocks.

Since Wilfong's work, research has concentrated on the simpler model in which the robot can only push blocks and the blocks are unit squares. Types of puzzles are further distinguished by how many blocks can be pushed at once, whether blocks can additionally be defined to be *unpushable* or *fixed* (tied to the board), how far blocks move when pushed, and the goal (usually for the robot to reach a particular location). Dhagat and O'Rourke [DO92] initiated the exploration of square-block puzzles by proving that PUSH-∗, in which arbitrarily many blocks can be pushed at once, is NP-hard with fixed blocks. Bremner, O'Rourke, and Shermer [BOS94] strengthened this result to PSPACE-completeness. Recently, Hoffmann [Hof00] proved that PUSH-∗ is NP-hard even without fixed blocks, but it remains open whether it is in NP or PSPACE-complete.

Several other results allow only a single block to be pushed at once. In this context, fixed blocks are less crucial because a 2×2 cluster of blocks can never be disturbed. A well-known computer puzzle in this context is *Sokoban*, where the goal is to place each block onto any one of the designated target squares. This puzzle was proved NP-hard by Dor and Zwick [DZ99] and later PSPACE-complete by Culberson [Cul98]. Later this result was strengthened to configurations with no fixed blocks [HD02; HD05]. A simpler puzzle, called PUSH-1, arises when the goal is simply for the robot to reach a particular position, and there are no fixed blocks. Demaine, Demaine, and O'Rourke [DDO00a] prove that this puzzle is NP-hard, but it remains open whether it is in NP or PSPACE-complete. On the other hand, PSPACE-completeness has been established for PUSH-2-F, in which there are fixed blocks and the robot can push two blocks at a time [DHH02].

A variation on the PUSH series of puzzles, called PUSHPUSH, is when a block always slides as far as possible when pushed. Such puzzles arise in a computer game with the same name [DDO00a; DDO00b; OS99]. PUSHPUSH-1 was established to be NP-hard slightly earlier than PUSH-1 [DDO00b; OS99]; the PUSH-1 reduction [DDO00a] also applies to PUSHPUSH-1. PUSHPUSH-k was later shown PSPACE-complete for any fixed $k \geq 1$ [DHH04]. Hoffmann's reduction for PUSH-∗ also proves that PUSHPUSH-∗ is NP-hard without fixed blocks.

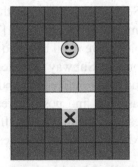

Figure 14. A PUSH-1 or PUSHPUSH-1 puzzle: move the robot to the X by pushing light blocks.

Another variation, called PUSH-X, disallows the robot from revisiting a square (the robot's path cannot cross). This direction was suggested in [DDO00a] because it immediately places the puzzles in NP. Demaine and Hoffmann [DH01] proved that PUSH-1X and PUSHPUSH-1X are NP-complete. Hoffmann's reduction for PUSH-* also establishes NP-completeness of PUSH-*X without fixed blocks.

Friedman [Fri02c] considers another variation, where gravity acts on the blocks (but not the robot): when a block is pushed it falls if unsupported. He shows that PUSH-1-G, where the robot may push only one block, is NP-hard.

River Crossing, another ThinkFun puzzle (originally *Plank Puzzles* by Andrea Gilbert [Gil00]), is similar to pushing-block puzzles in that there is a unique piece that must be used to move the other puzzle pieces. The game board is a grid, with *stumps* at some intersections, and *planks* arranged between some pairs of stumps, along the grid lines. A special piece, the *hiker*, always stands on some plank, and can walk along connected planks. He can also pick up and carry a single plank at a time, and deposit that plank between stumps that are appropriately spaced. The goal is for the hiker to

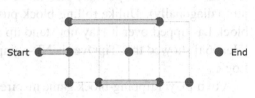

Figure 15. A River Crossing puzzle. Move from start to end.

reach a particular stump. Figure 15 shows a sample puzzle. Hearn [Hea04; Hea06b] proves that River Crossing is PSPACE-complete, by a reduction from Constraint Logic.

5.9. Rolling and tipping blocks. In some puzzles the blocks can change their orientation as well as their position. Rolling-cube puzzles were popularized by Martin Gardner in his *Mathematical Games* columns in *Scientific American* [Gar63; Gar65; Gar75]. In these puzzles, one or more cubes with some labeled sides (often dice) are placed on a grid, and may roll from cell to cell, pivoting on their edges between cells. Some cells may have labels which must match the face-up label of the cube when it visits the cell. The tasks generally involve completing some type of circuit while satisfying some label constraints (e.g., by ensuring that a particular labeled face never points up). Recently Buchin et al. [BBD+07] formalized this type of problem and derived several results. In their version, every labeled cell must be visited, with the label on the top face of the cube matching the cell label. Cells can be labeled, *blocked*, or *free*. Blocked cells cannot be visited; free cells can be visited regardless of cube orientation. Such puzzles turn out to be easy if labeled cells can be visited multiple times. If each labeled cell must be visited exactly one, the problem becomes NP-complete.

Rolling-block puzzles were later generalized by Richard Tucker to puzzles where the blocks no longer need be cubes. In these puzzles, the blocks are $k \times m \times n$ boxes. Typically, some grid cells are blocked, and the goal is to move a block from a start position to an end position by successive rotations into unblocked cells. Buchin and Buchin [BB07] recently showed that these puzzles are PSPACE-complete when multiple rolling blocks are used, by a reduction from Constraint Logic.

A commercial puzzle involving blocks that tip is the ThinkFun puzzle *TipOver* (originally the *Kung Fu Packing Crate Maze* by James Stephens [Ste03]). In this puzzle, all the blocks are $1 \times 1 \times n$ ("crates") and initially vertical. A *tipper* stands on a starting crate, and attempts to reach a target crate. The tipper may tip over a vertical crate it is standing on, if there is empty space in the grid for it to fall into. The tipper may also move between connected crates (but cannot jump diagonally). Unlike rolling-block puzzles, in these tipping puzzles once a block has tipped over it may not stand up again (or indeed move at all). Hearn [Hea06a] showed that TipOver is NP-complete, by a reduction from Constraint Logic.

A two-player tipping-block game inspired by TipOver, called *Cross Purposes*, was invented by Michael Albert, and named by Richard Guy, at the Games at Dalhousie III workshop in 2004. In Cross Purposes, all the blocks are $1 \times 1 \times 2$, and initially vertical. One player, *horizontal*, may only tip blocks over horizontally as viewed from above; the other player, *vertical*, may only tip blocks over vertically as viewed from above. The game follows normal play: the last player to move wins. Hearn [Hea08a] proved that Cross Purposes is PSPACE-complete, by a reduction from Constraint Logic.

5.10. Peg Solitaire (Hi-Q).

The classic *peg solitaire puzzle* is shown in Figure 16. Pegs are arranged in a Greek cross, with the central peg missing. Each move *jumps* a peg over another peg (adjacent horizontally or vertically) to the opposite unoccupied position within the cross, and

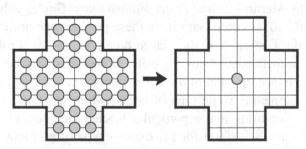

Figure 16. Central peg solitaire (Hi-Q): initial and target configurations.

removes the peg that was jumped over. The goal is to leave just a single peg, ideally located in the center. A variety of similar peg solitaire puzzles are given in [Bea85]. See also Chapter 23 of *Winning Ways* [BCG04, pp. 803–841].

A natural generalization of peg solitaire is to consider pegs arranged in an $n \times n$ board and the goal is to leave a single peg. Uehara and Iwata [UI90] proved that it is NP-complete to decide whether such a puzzle is solvable.

On the other hand, Moore and Eppstein [ME02] proved that the one-dimensional special case (pegs along a line) can be solved in polynomial time. In particular, the binary strings representing initial configurations that can reach a single peg turn out to form a regular language, so they can be parsed using regular expressions. (This fact has been observed in various contexts; see [ME02] for references as well as a proof.) Using this result, Moore and Eppstein build a polynomial-time algorithm to maximize the number of pegs removed from any given puzzle.

Moore and Eppstein [ME02] also study the natural impartial two-player game arising from peg solitaire, *duotaire*: players take turns jumping, and the winner is determined by normal play. (This game is proposed, e.g., in [Bea85].) Surprisingly, the complexity of this seemingly simple game is open. Moore and Eppstein conjecture that the game cannot be described even by a context-free language, and prove this conjecture for the variation in which multiple jumps can be made in a single move. Konane (Section 4.14) is a natural partizan two-player game arising from peg solitaire.

5.11. Card Solitaire. Two solitaire games with playing cards have been analyzed from a complexity standpoint. With all such games, we must generalize the deck beyond 52 cards. The standard approach is to keep the number of suits fixed at four, but increase the number of ranks in each suit to n.

Klondike or *Solitaire* is the classic game, in particular bundled with Microsoft Windows since its early days. In the perfect information of this game, we suppose the player knows all of the normally hidden cards. Longpré and McKenzie [LM07] proved that the perfect-information version is NP-complete, even with just three suits. They also prove that Klondike with one black suit and one red suit is NL-hard; Klondike with any fixed number of black suits and no red suits is in NL; Klondike with one suit is in $AC^0[3]$; among other results.

FreeCell is another common game distributed with Microsoft Windows since XP. We will not attempt to describe the rules here. Helmert [Hel03] proved that FreeCell is NP-complete, for any fixed positive number of free cells.

5.12. Jigsaw, edge-matching, tiling, and packing puzzles. *Jigsaw puzzles* [Wil04] are another one of the most popular kinds of puzzles, dating back to the 1760s. One way to formalize such puzzles is as a collection of square pieces, where each side is either straight or augmented with a tab or a pocket of a particular shape. The goal is to arrange the given pieces so that they form exactly a given rectangular shape. Although this formalization does not explicitly allow for patterns on pieces to give hints about whether pieces match, this information can simply be encoded into the shapes of the tabs and pockets, making them compatible only when the patterns also match. Deciding whether such a puzzle has a solution was recently shown NP-complete [DD07].

A closely related type of puzzles is *edge-matching puzzles* [Hau95], dating back to the 1890s. In the simplest form, the pieces are squares and, instead of tabs or pockets, each edge is colored to indicate compatibility. Squares can be placed side-by-side if the edge colors match, either being exactly equal (*unsigned* edge matching) or being opposite (*signed* edge matching). Again the goal is to arrange the given pieces into a given rectangle. Signed edge-matching puzzles are common in reality where the colors are in fact images of lizards, insects, etc., and one side shows the head while the other shows the tail. Such puzzles are almost identical to jigsaw puzzles, with tabs and pockets representing the sign; jigsaw puzzles are effectively the special case in which the boundary must be uniformly colored. Thus, signed edge-matching puzzles are NP-complete, and in fact, so are unsigned edge-matching puzzles [DD07].

An older result by Berger [Ber66] proves that the infinite generalization of edge-matching puzzles, where the goal is to tile the entire plane given infinitely many copies of each tile type, is undecidable. This result is for unsigned puzzles, but by a simple reduction in [DD07] it holds for signed puzzles as well. Along the same lines, Garey, Johnson, and Papadimitriou [GJ79, p. 257] observe that the finite version with a given target rectangle is NP-complete when given arbitrarily many copies of each tile type. In contrast, the finite result above requires every given tile to be used exactly once, which corresponds more closely to real puzzles.

A related family of tiling and packing puzzles involve polyforms such as *polyominoes*, edge-to-edge joinings of unit squares. In general, we are given a collection of such shapes and a target shape to either tile (form exactly) or pack (form with gaps). In both cases, pieces cannot overlap, so the tiling problem is actually a special case in which the piece areas sum to the target areas. One of the few positive results is for (mathematical) *dominoes*, polyominoes (rectangles) made from two unit squares: the tiling and (grid-aligned) packing problems can be solved in polynomial time for arbitrary polyomino target shapes by perfect and maximum matching, respectively; see also the elegant tiling criterion of Thurston [Thu90]. In contrast, with "real" dominoes, where each square has a color and adjacent dominoes must match in color, tiling (and hence packing) becomes NP-complete [Bie05]. The tiling problem is also NP-complete when the target shape is a polyomino with holes and the pieces are all identical 2×2 squares, or 1×3 rectangles, or 2×2 L shapes [MR01]. The packing problem [LC89] and the tiling problem [DD07] are NP-complete when the given pieces are differently sized squares and the target shape is a square. Finally, the tiling problem is NP-complete when the given pieces are polylogarithmic-area polyominoes and the target shape is a square [DD07]; this result follows by simulating jigsaw puzzles.

5.13. Minesweeper. *Minesweeper* is a well-known imperfect-information computer puzzle popularized by its inclusion in Microsoft Windows. Gameplay takes place on an $n \times n$ board, and the player does not know which squares contain mines. A move consists of uncovering a square; if that square contains a mine, the player loses, and otherwise the player is revealed the number of mines in the 8 adjacent squares. The player also knows the total number of mines.

There are several problems of interest in Minesweeper. For example, given a configuration of partially uncovered squares (each marked with the number of adjacent mines), is there a position that can be safely uncovered? More generally, what is the probability that a given square contains a mine, assuming a uniform distribution of remaining mines? A different generalization of the first question is whether a given configuration is *consistent*, i.e., can be realized by a collection of mines. A consistency checker would allow testing whether a square can be guaranteed to be free of mines, thus answering the first question. An additional problem is to decide whether a given configuration has a unique realization.

Kaye [Kay00b] proves that testing consistency is NP-complete. This result leaves open the complexity of the other questions mentioned above. Fix and McPhail [FM04] strengthen Kaye's result to show NP-completeness of determining consistency when the uncovered numbers are all at most 1. McPhail [McP03] also shows that, given a consistent placement of mines, determining whether there is another consistent placement is NP-complete (ASP-completeness from Section 5.4).

Kaye [Kay00a] also proves that an infinite generalization of Minesweeper is undecidable. Specifically, the question is whether a given finite configuration can be extended to the entire plane. The rules permit a much more powerful level of information revealed by uncovering squares; for example, discovering that one square has a particular label might imply that there are exactly 3 adjacent squares with another particular label. (The notion of a mine is lost.) The reduction is from tiling (Section 5.12).

Hearn [Hea06b; Hea08b] argues that the "natural" decision question for Minesweeper, in keeping with the standard form for other puzzle complexity results, is whether a given (assumed consistent) instance can (definitely) be solved, which is a different question from any of the above. He observes that a simple modification to Kaye's construction shows that this question is coNP-complete, an unusual complexity class for a puzzle. The reduction is from Tautology. (If the instance is not known to be consistent, then the problem may not be in coNP.) Note that this question is not the same as whether a given configuration has a unique realization: there could be multiple realizations, as long as the player is guaranteed that known-safe moves will eventually reveal the entire configuration.

5.14. Mahjong solitaire (Shanghai). *Majong solitaire* or *Shanghai* is a common computer game played with Mahjong tiles, stacked in a pattern that hides some tiles, and shows other tiles, some of which are completely exposed. Each move removes a pair of matching tiles that are completely exposed; there are precisely four tiles in each equivalence class of matching. The goal is to remove all tiles.

Condon, Feigenbaum, Lund, and Shor [CFS97] proved that it is PSPACE-hard to approximate the maximum probability of removing all tiles within a factor of n^ε, assuming that there are arbitrarily many quadruples of matching tiles and that the hidden tiles are uniformly distributed. Eppstein [Epp] proved that it is NP-complete to decide whether all tiles can be removed in the perfect-information version of this puzzle where all tile positions are known.

5.15. Tetris. *Tetris* is a popular computer puzzle game invented in the mid-1980s by Alexey Pazhitnov, and by 1988 it became the best-selling game in the United States and England. The game takes place in a rectangular grid (originally, 20×10) with some squares occupied by blocks. During each move, the computer generates a tetromino piece stochastically and places it at the top of the grid; the player can rotate the piece and slide it left or right as it falls downward. When the piece hits another piece or the floor, its location freezes and the move ends. Also, if there are any completely filled rows, they disappear, bringing any rows above down one level.

To make Tetris a perfect-information puzzle, Breukelaar et al. [BDH$^+$04] suppose that the player knows in advance the entire sequence of pieces to be delivered. Such puzzles appear in *Games Magazine*, for example. They then prove NP-completeness of deciding whether it is possible to stay alive, i.e., always be able to place pieces. Furthermore, they show that maximizing various notions of score, such as the number of lines cleared, is NP-complete to approximate within an $n^{1-\varepsilon}$ factor. The complexity of Tetris remains open with a constant number of rows or columns, or with a stochastically chosen piece sequence as in [Pap85].

5.16. Clickomania (Same Game). *Clickomania* or *Same Game* [BDD$^+$02] is a computer puzzle consisting of a rectangular grid of square blocks each colored one of k colors. Horizontally and vertically adjacent blocks of the same color are considered part of the same *group*. A move selects a group containing at least two blocks and removes those blocks, followed by two "falling" rules; see Figure 17 (top). First, any blocks remaining above created holes fall down in each column. Second, any empty columns are removed by sliding the succeeding columns left.

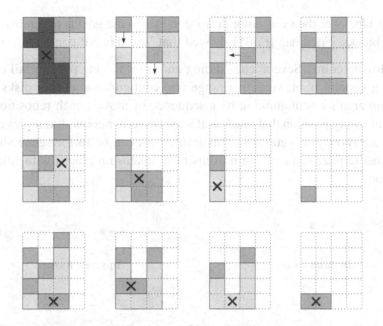

Figure 17. The falling rules for removing a group in Clickomania (top), a failed attempt (middle), and a successful solution (bottom).

The main goal in Clickomania is to remove all the blocks. A simple example for which this is impossible is a checkerboard, where no move can be made. A secondary goal is to maximize the score, typically defined by k^2 points being awarded for removal of a group of k blocks.

Biedl et al. [BDD$^+$02] proved that it is NP-complete to decide whether all blocks can be removed in a Clickomania puzzle. This complexity result holds even for puzzles with two columns and five colors, and for puzzles with five columns and three colors. On the other hand, for puzzles with one column (or, equivalently, one row) and arbitrarily many colors, they show that the maximum number of blocks can be removed in polynomial time. In particular, the puzzles whose blocks can all be removed are given by the context-free grammar $S \rightarrow \Lambda \mid SS \mid cSc \mid cScSc$ where c ranges over all colors.

Various cases of Clickomania remain open, for example, puzzles with two colors, and puzzles with $O(1)$ rows. Richard Nowakowski suggested a two-player version of Clickomania, described in [BDD$^+$02], in which players take turns removing groups and normal play determines the winner; the complexity of this game remains open.

A related puzzle is called *Vexed*, also *Cubic*. In this puzzle there are fixed blocks, as well as the mutually annihilating colored blocks. A move in Vexed is to slide a colored block one unit left or right into an empty space, whereupon gravity will pull the block down until it contacts another block; then any touching blocks of the same color disappear. Again the goal is to remove all the colored blocks. Friedman [Fri01] showed that Vexed is NP-complete.[4]

5.17. Moving coins. Several coin-sliding and coin-moving puzzles fall into the following general framework: rearrange one configuration of unit disks in the plane into another configuration by a sequence of moves, each repositioning a coin in an empty position that touches at least two other coins. Examples of such puzzles are shown in Figure 18. This framework can be further generalized to nongeometric puzzles involving movement of tokens on graphs with adjacency restrictions.

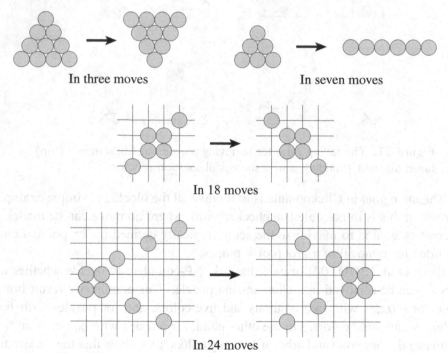

In three moves   In seven moves

In 18 moves

In 24 moves

Figure 18. Coin-moving puzzles in which each move places a coin adjacent to two other coins; in the bottom two puzzles, the coins must also remain on the square lattice. The top two puzzles are classic, whereas the bottom two were designed in [DDV00].

Coin-moving puzzles have been analyzed by Demaine, Demaine, and Verrill [DDV00]. In particular, they study puzzles as in Figure 18 in which the coins' centers remain on either the triangular lattice or the square lattice. Surprisingly, their results for deciding solvability of puzzles are positive.

---

[4]David Eppstein pointed out that all that was shown was NP-hardness; the problem was not obviously in NP (http://www.ics.uci.edu/~eppstein/cgt/hard.html). Friedman and R. Hearn together showed that it is in NP as well (personal communication).

For the triangular lattice, nearly all puzzles are solvable, and there is a polynomial-time algorithm characterizing them. For the square lattice, there are more stringent constraints. For example, the bounding box cannot increase by moves; more generally, the set of positions reachable by moves given an infinite supply of extra coins (the *span*) cannot increase. Demaine, Demaine, and Verrill show that, subject to this constraint, there is a polynomial-time algorithm to solve all puzzles with at least two extra coins past what is required to achieve the span. (In particular, all such puzzles are solvable.)

5.18. Dyson Telescopes. The *Dyson Telescope Game* is an online puzzle produced by the Dyson corporation, whimsically based on their telescoping vacuum cleaners. The goal is to maneuver a ball on a square grid from a starting position to a goal position by extending and retracting telescopes on the grid. When a telescope is extended, it grows to its maximum length in the direction it points (parameters of each telescope), unless it is stopped by another telescope. If the ball is in the way, it is pushed by the end of the telescope. When a telescope is retracted, it shrinks back to unit length, pulling the ball with it if the ball was at the end of the telescope.

Demaine et al. [DDF$^+$08] showed that determining whether a given puzzle has a solution is PSPACE-complete in the general case. On the other hand, the problem is polynomial for certain restricted configurations which are nonetheless interesting for humans to play. Specifically, if no two telescopes face each other and overlap when extended by more than one space, then the problem is polynomial. Many of the game levels in the online version have this property.

5.19. Reflection puzzles. Two puzzles involving reflection of directional light or motion have been studied from a complexity-theoretic standpoint.

In *Reflections* [Kem03], we are given a rectangular grid with one square marked with a laser pointed in one of the four axis-parallel directions, one or more squares marked as light bulbs, some squares marked one-way in an axis-parallel direction, and remaining squares marked either empty or wall. We are also given a number of diagonal mirrors and/or T-splitters which we can place arbitrarily into empty squares. The light then travels from the laser; when it meets a diagonal mirror, it reflects by 90° according to the orientation of the mirror; when it meets a splitter at the base of the T, it splits into both orthogonal directions; when it meets a one-way square, it stops unless the light direction matches the one-way orientation; when it meets a light bulb, it toggles the bulb's state and stops; and when it meets a wall, it stops. The goal is to place the mirrors and splitters so that each light bulb gets hit an odd number of times. This puzzle is NP-complete [Kem03].

In *Reflexion* [HS04b], we are given a rectangular grid in which squares are either walls, mirrors, or diamonds. Also, one square is the starting position for a ball and another square is the target position. We may release the ball in one of the four axis-parallel directions, and we may flip mirrors between their two diagonal orientations while the ball moves. The ball travels like a ray of light, reflecting at mirrors and stopping at walls; at diamonds, it turns around and erases the diamond. The goal is to reach the target position. In this simplest form, Reflexion is SL-complete which actually implies a polynomial-time algorithm [HS04a]. If some of the mirrors can be flipped only before the ball releases, the puzzle becomes NP-complete. If some trigger squares toggle other squares between wall and empty, or if some squares contain horizontally or vertically movable blocks (which also cause the ball to turn around), then the puzzle becomes PSPACE-complete.

5.20. Lemmings. *Lemmings* is a popular computer puzzle game dating back to the early 1990s. Characters called lemmings start at one or more initial locations and behave deterministically according to their mode, initially just walking in a fixed direction, turning around at walls, and falling off cliffs, dying if it falls too far. The player can modify this basic behavior by applying a skill to a lemming; each skill has a limited number of such applications. The goal is for a specified number of lemmings to reach a specified target position. The exact rules, particularly the various skills, are too complicated to detail here. Cormode [Cor04] proved that such puzzles are NP-complete, even with just one lemming. Membership in NP follows from assuming a polynomial upper bound on the time limit in a level (a fairly accurate modeling of the actual game); Cormode conjectures that this assumption does not affect the result.

6. Cellular automata and life

Conway's *Game of Life* is a zero-player cellular automaton played on the square tiling of the plane. Initially, certain cells (squares) are marked *alive* or *dead*. Each move globally evolves the cells: a live cell remains alive if between 2 and 3 of its 8 neighbors were alive, and a dead cell becomes alive if it had precisely 3 live neighbors.

Many questions can be asked about an initial configuration of Life; one key question is whether the population will ever completely die out (no cells are alive). Chapter 25 of *Winning Ways* [BCG04, pp. 927–961] describes a reduction showing that this question is undecidable. In particular, the same question about Life restricted within a polynomially bounded region is PSPACE-complete. More recently, Rendell [Ren05] constructed an explicit Turing machine in Life, which establishes the same results.

There are other open complexity-theoretic questions about Life.[5] How hard is it to tell whether a configuration is a Garden of Eden, that is, a state that cannot result from another? Given a rectangular pattern in Life, how hard is it to extend the pattern outside the rectangle to form a Still Life (which never changes)?

Several other cellular automata, with different survival and birth rules, have been studied; see, e.g., [Wol94].

7. Open problems

Many open problems remain in Combinatorial Game Theory. Guy and Nowakowski [GN02] have compiled a list of such problems.

Many open problems also remain on the algorithmic side, and have been mentioned throughout this paper. Examples of games and puzzles whose complexities remain unstudied, to our knowledge, are Domineering (Section 4.11), Connect Four, Pentominoes, Fanorona, Nine Men's Morris, Chinese checkers, Lines of Action, Chinese Chess, Quoridor, and Arimaa. For many other games and puzzles, such as Dots and Boxes (Section 4.12) and pushing-block puzzles (Section 5.8), some hardness results are known, but the exact complexity remains unresolved. It would also be interesting to consider games of imperfect information that people play, such as Scrabble (Section 5.3, Backgammon, and Bridge. Another interesting direction for future research is to build a more comprehensive theory for analyzing combinatorial puzzles.

Acknowledgments

Comments from several people have helped make this survey more comprehensive, including Martin Demaine, Aviezri Fraenkel, Martin Kutz, and Ryuhei Uehara.

References

[AKI87] Hiroyuki Adachi, Hiroyuki Kamekawa, and Shigeki Iwata. Shogi on $n \times n$ board is complete in exponential time (in Japanese). *Transactions of the IEICE*, J70-D(10):1843–1852, October 1987.

[All87] Norman Alling. *Foundations of Analysis over Surreal Number Fields*. North-Holland Publishing Co., Amsterdam, 1987.

[And07] Daniel Andersson. HIROIMONO is NP-complete. In *Proceedings of the 4th International Conference on FUN with Algorithms*, volume 4475 of *Lecture Notes in Computer Science*, pages 30–39, 2007.

---

[5] These two questions were suggested by David Eppstein.

[ANW07] Michael H. Albert, Richard J. Nowakowski, and David Wolfe. *Lessons in Play: An Introduction to Combinatorial Game Theory*. A K Peters, Wellesley, MA, 2007.

[Arc99] Aaron F. Archer. A modern treatment of the 15 puzzle. *American Mathematical Monthly*, 106(9):793–799, 1999.

[BB07] Kevin Buchin and Maike Buchin. Rolling block mazes are PSPACE-complete. Manuscript, 2007.

[BBD+07] Kevin Buchin, Maike Buchin, Erik D. Demaine, Martin L. Demaine, Dania El-Khechen, Sándor Fekete, Christian Knauer, André Schulz, and Perouz Taslakian. On rolling cube puzzles. In *Proceedings of the 19th Canadian Conference on Computational Geometry*, pages 141–144, Ottawa, Canada, August 2007.

[BCG04] Elwyn R. Berlekamp, John H. Conway, and Richard K. Guy. *Winning Ways for Your Mathematical Plays*. A K Peters, Wellesley, MA, 2nd edition, 2001–2004.

[BDD+02] Therese C. Biedl, Erik D. Demaine, Martin L. Demaine, Rudolf Fleischer, Lars Jacobsen, and J. Ian Munro. The complexity of Clickomania. In R. J. Nowakowski, editor, *More Games of No Chance*, pages 389–404. Cambridge University Press, 2002. Collection of papers from the MSRI Combinatorial Game Theory Research Workshop, Berkeley, California, July 2000.

[BDH+04] Ron Breukelaar, Erik D. Demaine, Susan Hohenberger, Hendrik Jan Hoogeboom, Walter A. Kosters, and David Liben-Nowell. Tetris is hard, even to approximate. *International Journal of Computational Geometry and Applications*, 14(1–2): 41–68, 2004.

[Bea85] John D. Beasley. *The Ins & Outs of Peg Solitaire*. Oxford University Press, 1985.

[Ber66] Robert Berger. The undecidability of the domino problem. *Memoirs of the American Mathematical Society*, 66, 1966.

[Ber00] Elwyn Berlekamp. *The Dots and Boxes Game: Sophisticated Child's Play*. A K Peters, Wellesley, MA, 2000.

[Bie05] Therese Biedl. The complexity of domino tiling. In *Proceedings of the 17th Canadian Conference on Computational Geometry*, pages 187–190, 2005.

[BMFN99] Adrian Brüngger, Ambros Marzetta, Komei Fukuda, and Jurg Nievergelt. The parallel search bench ZRAM and its applications. *Annals of Operations Research*, 90:45–63, 1999.

[BOS94] David Bremner, Joseph O'Rourke, and Thomas Shermer. Motion planning amidst movable square blocks is PSPACE complete. Draft, June 1994.

[Bou02] Charles L. Bouton. Nim, a game with a complete mathematical theory. *Annals of Mathematics*, Series 2, 3:35–39, 1901–02.

[Bow07] Brian H. Bowditch. The angel game in the plane. *Combinatocis, Probability and Computing*, 16(3):345–362, 2007.

[Bur00] Michael Buro. Simple Amazons endgames and their connection to Hamilton circuits in cubic subgrid graphs. In *Proceedings of the 2nd International Conference on Computers and Games*, volume 2063 of *Lecture Notes in Computer Science*, pages 250–261, Hamamatsu, Japan, October 2000.

[BW70] John Bruno and Louis Weinberg. A constructive graph-theoretic solution of the Shannon switching game. *IEEE Transactions on Circuit Theory CT-17*, 1:74–81, February 1970.

[BW94] Elwyn Berlekamp and David Wolfe. *Mathematical Go: Chilling Gets the Last Point*. A. K. Peters, Ltd., 1994.

[CDP06] Gruia Calinescu, Adrian Dumitrescu, and János Pach. Reconfigurations in graphs and grids. In *Proceedings of the 7th Latin American Symposium on Theoretical Informatics*, pages 262–273, 2006.

[CFS97] Anne Condon, Joan Feigenbaum, and Carsten Lund Peter Shor. Random debaters and the hardness of approximating stochastic functions. *SIAM Journal on Computing*, 26(2):369–400, 1997.

[Con77] J. H. Conway. All games bright and beautiful. *American Math. Monthly*, 84: 417–434, 1977.

[Con01] John H. Conway. *On Numbers and Games*. A K Peters, Wellesley, MA, 2nd edition, 2001.

[Coo71] Stephen A. Cook. The complexity of theorem-proving procedures. In *Proceedings of the 3rd IEEE Symposium on the Foundations of Computer Science*, pages 151–158, 1971.

[Cor04] Graham Cormode. The hardness of the Lemmings game, or Oh no, more NP-completeness proofs. In *Proceedings of the 3rd International Conference on FUN with Algorithms*, pages 65–76, Isola d'Elba, Italy, May 2004.

[Crâ99] Marcel Crâsmaru. On the complexity of Tsume-Go. In *Proceedings of the 1st International Conference on Computers and Games*, volume 1558 of *Lecture Notes in Computer Science*, pages 222–231, Springer, London, 1999.

[CT00] Marcel Crâşmaru and John Tromp. Ladders are PSPACE-complete. In *Proceedings of the 2nd International Conference on Computers and Games*, volume 2063 of *Lecture Notes in Computer Science*, pages 241–249, 2000.

[CT02] Alice Chan and Alice Tsai. $1 \times n$ Konane: a summary of results. In R. J. Nowakowski, editor, *More Games of No Chance*, pages 331–339, 2002.

[Cul98] Joseph Culberson. Sokoban is PSPACE-complete. In *Proceedings of the International Conference on Fun with Algorithms*, pages 65–76, Elba, Italy, June 1998.

[DD07] Erik D. Demaine and Martin L. Demaine. Jigsaw puzzles, edge matching, and polyomino packing: Connections and complexity. *Graphs and Combinatorics*, 23 (Supplement):195–208, 2007. Special issue on Computational Geometry and Graph Theory: The Akiyama–Chvatal Festschrift.

[DDE02] Erik D. Demaine, Martin L. Demaine, and David Eppstein. Phutball endgames are NP-hard. In R. J. Nowakowski, editor, *More Games of No Chance*, pages 351–360. Cambridge University Press, 2002. Collection of papers from the MSRI

Combinatorial Game Theory Research Workshop, Berkeley, California, July 2000. http://www.arXiv.org/abs/cs.CC/0008025.

[DDF+08] Erik D. Demaine, Martin L. Demaine, Rudolf Fleischer, Robert A. Hearn, and Timo von Oertzen. The complexity of Dyson telescopes. In *Games of No Chance III*. 2008.

[DDLL06] Erik D. Demaine, Martin L. Demaine, Arthur Langerman, and Stefan Langerman. Morpion solitaire. *Theory of Computing Systems*, 39(3):439–453, June 2006.

[DDO00a] Erik D. Demaine, Martin L. Demaine, and Joseph O'Rourke. PushPush and Push-1 are NP-hard in 2D. In *Proceedings of the 12th Annual Canadian Conference on Computational Geometry*, pages 211–219, Fredericton, Canada, August 2000. http://www.cs.unb.ca/conf/cccg/eProceedings/26.ps.gz.

[DDO00b] Erik D. Demaine, Martin L. Demaine, and Joseph O'Rourke. PushPush is NP-hard in 2D. Technical Report 065, Department of Computer Science, Smith College, Northampton, MA, January 2000. http://arXiv.org/abs/cs.CG/0001019.

[DDV00] Erik D. Demaine, Martin L. Demaine, and Helena Verrill. Coin-moving puzzles. In *MSRI Combinatorial Game Theory Research Workshop*, Berkeley, California, July 2000.

[Del06] Jean-Paul Delahaye. The science behind Sudoku. *Scientific American*, pages 80–87, June 2006.

[DF83] James R. Driscoll and Merrick L. Furst. On the diameter of permutation groups. In *Proceedings of the 15th Annual ACM Symposium on Theory of Computing*, pages 152–160, 1983.

[DF97] R. G. Downey and M. R. Fellows. *Parameterized Complexity*. Springer, 1997.

[DH01] Erik D. Demaine and Michael Hoffmann. Pushing blocks is NP-complete for noncrossing solution paths. In *Proceedings of the 13th Canadian Conference on Computational Geometry*, pages 65–68, Waterloo, Canada, August 2001. http://compgeo.math.uwaterloo.ca/~cccg01/proceedings/long/eddemaine-24711.ps.

[DH08] Erik D. Demaine and Robert A. Hearn. Constraint logic: A uniform framework for modeling computation as games. In *Proceedings of the 23rd Annual IEEE Conference on Computational Complexity*, College Park, Maryland, June 2008. To appear.

[DHH02] Erik D. Demaine, Robert A. Hearn, and Michael Hoffmann. Push-2-F is PSPACE-complete. In *Proceedings of the 14th Canadian Conference on Computational Geometry*, pages 31–35, Lethbridge, Canada, August 2002.

[DHH04] Erik D. Demaine, Michael Hoffmann, and Markus Holzer. PushPush-k is PSPACE-complete. In *Proceedings of the 3rd International Conference on FUN with Algorithms*, pages 159–170, Isola d'Elba, Italy, May 2004.

[DO92] Arundhati Dhagat and Joseph O'Rourke. Motion planning amidst movable square blocks. In *Proceedings of the 4th Canadian Conference on Computational Geometry*, pages 188–191, 1992.

[Dud24] Henry E. Dudeney. *Strand Magazine*, 68:97 and 214, July 1924.

[dVV03] Sven de Vries and Rakesh Vohra. Combinatorial auctions: A survey. *IN-FORMS Journal on Computing*, 15(3):284–309, Summer 2003.

[DZ99] Dorit Dor and Uri Zwick. SOKOBAN and other motion planning problems. *Computational Geometry: Theory and Applications*, 13(4):215–228, 1999.

[Epp] David Eppstein. Computational complexity of games and puzzles. See http://www.ics.uci.edu/~eppstein/cgt/hard.html.

[Epp87] David Eppstein. On the NP-completeness of cryptarithms. *SIGACT News*, 18(3):38–40, 1987.

[Ern95] Michael D. Ernst. Playing Konane mathematically: A combinatorial game-theoretic analysis. *UMAP Journal*, 16(2):95–121, Spring 1995.

[ET76] S. Even and R. E. Tarjan. A combinatorial problem which is complete in polynomial space. *Journal of the Association for Computing Machinery*, 23(4):710–719, 1976.

[FB02] Gary William Flake and Eric B. Baum. *Rush Hour* is PSPACE-complete, or "Why you should generously tip parking lot attendants". *Theoretical Computer Science*, 270(1–2):895–911, January 2002.

[Fer74] T. S. Ferguson. On sums of graph games with last player losing. *International Journal of Game Theory*, 3(3):159–167, 1974.

[Fer84] Thomas S. Ferguson. Misère annihilation games. *Journal of Combinatorial Theory*, Series A, 37:205–230, 1984.

[FG87] Aviezri S. Fraenkel and Elisheva Goldschmidt. PSPACE-hardness of some combinatorial games. *Journal of Combinatorial Theory*, Series A, 46:21–38, 1987.

[FGJ+78] A. S. Fraenkel, M. R. Garey, D. S. Johnson, T. Schaefer, and Y. Yesha. The complexity of checkers on an $N \times N$ board - preliminary report. In *Proceedings of the 19th Annual Symposium on Foundations of Computer Science*, pages 55–64, Ann Arbor, Michigan, October 1978.

[FKUB05] Timothy Furtak, Masashi Kiyomi, Takcaki Uno, and Michael Buro. Generalized Amazons is PSPACE-complete. In *Proceedings of the 19th International Joint Conference on Artificial Intelligence*, pages 132–137, 2005.

[FL81] Aviezri S. Fraenkel and David Lichtenstein. Computing a perfect strategy for $n \times n$ chess requires time exponential in n. *Journal of Combinatorial Theory*, Series A, 31:199–214, 1981.

[FM04] James D. Fix and Brandon McPhail. Offline 1-Minesweeper is NP-complete. Manuscript, 2004. http://people.reed.edu/~jimfix/papers/1MINESWEEPER.pdf.

[Fra74] Aviezri S. Fraenkel. Combinatorial games with an annihilation rule. In *The Influence of Computing on Mathematical and Research and Education*, volume 20 of *Proceedings of the Symposia in Applied Mathematics*, pages 87–91, 1974.

[Fra96] Aviezri S. Fraenkel. Scenic trails ascending from sea-level Nim to alpine Chess. In R. J. Nowakowski, editor, *Games of No Chance*, pages 13–42. Cambridge University Press, 1996.

[Fra07] Aviezri S. Fraenkel. Combinatorial games: Selected bibliography with a succinct gourmet introduction. *Electronic Journal of Combinatorics*, 1994–2007. Dynamic Survey DS2, http://www.combinatorics.org/Surveys/. One version also appears in *Games of No Chance*, pages 493–537, 1996.

[Fri01] Erich Friedman. Cubic is NP-complete. In *Proceedings of the 34th Annual Florida MAA Section Meeting*, 2001.

[Fri02a] Erich Friedman. Corral puzzles are NP-complete. Unpublished manuscript, August 2002. http://www.stetson.edu/~efriedma/papers/corral/corral.html.

[Fri02b] Erich Friedman. Pearl puzzles are NP-complete. Unpublished manuscript, August 2002. http://www.stetson.edu/~efriedma/papers/pearl/pearl.html.

[Fri02c] Erich Friedman. Pushing blocks in gravity is NP-hard. Unpublished manuscript, March 2002. http://www.stetson.edu/~efriedma/papers/gravity/gravity.html.

[Fri02d] Erich Friedman. Spiral galaxies puzzles are NP-complete. Unpublished manuscript, March 2002. http://www.stetson.edu/~efriedma/papers/spiral/spiral.html.

[FSU93] Aviezri S. Fraenkel, Edward R. Scheinerman, and Daniel Ullman. Undirected edge geography. *Theoretical Computer Science*, 112(2):371–381, 1993.

[FY76] A. S. Fraenkel and Y. Yesha. Theory of annihilation games. *Bulletin of the American Mathematical Society*, 82(5):775–777, September 1976.

[FY79] A. S. Fraenkel and Y. Yesha. Complexity of problems in games, graphs and algebraic equations. *Discrete Applied Mathematics*, 1:15–30, 1979.

[FY82] A. S. Fraenkel and Y. Yesha. Theory of annihilation games, I. *Journal of Combinatorial Theory*, Series B, 33:60–86, 1982.

[Gác07] Peter Gács. The angel wins. Manuscript, 2007. http://arXiv.org/abs/0706.2817.

[Gar63] Martin Gardner. Mathematical games column. *Scientific American*, 209(6): 144, December 1963. Solution in January 1964 column.

[Gar64] Martin Gardner. The hypnotic fascination of sliding-block puzzles. *Scientific American*, 210:122–130, 1964. Appears as Chapter 7 of *Martin Gardner's Sixth Book of Mathematical Diversions*, University of Chicago Press, 1984.

[Gar65] Martin Gardner. Mathematical games column. *Scientific American*, 213(5): 120–123, November 1965. Problem 9. Solution in December 1965 column.

[Gar75] Martin Gardner. Mathematical games column. *Scientific American*, 232(3): 112–116, March 1975. Solution in April 1975 column.

[Gar86] Martin Gardner. Cram, bynum and quadraphage. In *Knotted Doughnuts and Other Mathematical Entertainments*, chapter 16. W. H. Freeman, 1986.

[Gil00] Andrea Gilbert. Plank puzzles, 2000. See http://www.clickmazes.com/planks/ixplanks.htm.

[GJ79] Michael R. Garey and David S. Johnson. *Computers and Intractability: A Guide to the Theory of NP-Completeness*. W. H. Freeman & Co., 1979.

[GN02] Richard K. Guy and Richard J. Nowakowski. Unsolved problems in combinatorial games. In R. J. Nowakowski, editor, *More Games of No Chance*, pages 457–473. Cambridge University Press, 2002.

[Gon86] Harry Gonshor. *An Introduction to the Theory of Surreal Numbers*. Cambridge University Press, 1986.

[GR95] Arthur S. Goldstein and Edward M. Reingold. The complexity of pursuit on a graph. *Theoretical Computer Science*, 143:93–112, 1995.

[Gru39] P. M. Grundy. Mathematics and games. *Eureka*, 2:6–8, Oct 1939. Reprinted in *Eureka: The Archimedeans' Journal*, 27:9–11, Oct 1964.

[GS56a] P. M. Grundy and C. A. B. Smith. Disjunctive games with the last player losing. *Proceedings of the Cambridge Philosophical Society*, 52:527–533, 1956.

[GS56b] Richard K. Guy and Cedric A. B. Smith. The *G*-values of various games. *Proceedings of the Cambridge Philosophical Society*, 52:514–526, 1956.

[Hau95] Jacques Haubrich. *Compendium of Card Matching Puzzles*. Self-published, May 1995. Three volumes.

[Hay06] Brian Hayes. Unwed numbers. *American Scientist*, 94(1):12, Jan–Feb 2006.

[HD02] Robert A. Hearn and Erik D. Demaine. The nondeterministic constraint logic model of computation: Reductions and applications. In *Proceedings of the 29th International Colloquium on Automata, Languages and Programming*, volume 2380 of *Lecture Notes in Computer Science*, pages 401–413, Malaga, Spain, July 2002.

[HD05] Robert A. Hearn and Erik D. Demaine. PSPACE-completeness of sliding-block puzzles and other problems through the nondeterministic constraint logic model of computation. *Theoretical Computer Science*, 343(1–2):72–96, October 2005. Special issue "Game Theory Meets Theoretical Computer Science".

[Hea04] Robert A. Hearn. The complexity of sliding block puzzles and plank puzzles. In *Tribute to a Mathemagician*, pages 173–183. A K Peters, 2004.

[Hea05a] Robert A. Hearn. Amazons is PSPACE-complete. Manuscript, February 2005. http://www.arXiv.org/abs/cs.CC/0008025.

[Hea05b] Robert A. Hearn. The subway shuffle puzzle, 2005. www.subwayshuffle.com.

[Hea06a] Robert Hearn. TipOver is NP-complete. *Mathematical Intelligencer*, 28(3): 10–14, 2006.

[Hea06b] Robert A. Hearn. *Games, Puzzles, and Computation*. PhD thesis, Department of Electrical Engineering and Computer Science, Massachusetts Institute of Technology, Cambridge, Massachusetts, May 2006. http://www.swiss.ai.mit.edu/~bob/hearn-thesis-final.pdf.

[Hea08a] Robert A. Hearn. Amazons, Konane, and Cross Purposes are PSPACE-complete. In R. J. Nowakowski, editor, *Games of No Chance 3*, 2008. To appear.

[Hea08b] Robert A. Hearn. The complexity of Minesweeper revisited. Manuscript in preparation, 2008.

[Hea08c] Robert A. Hearn. Hitori is NP-complete. Manuscript in preparation, 2008.

[HEFN01] Falk Hüffner, Stefan Edelkamp, Henning Fernau, and Rolf Niedermeier. Finding optimal solutions to Atomix. In *Proceedings of the Joint German/Austrian Conference on AI: Advances in Artificial Intelligence*, volume 2174 of *Lecture Notes in Computer Science*, pages 229–243, Vienna, Austria, 2001.

[Hel03] Malte Helmert. Complexity results for standard benchmark domains in planning. *Artificial Intelligence*, 143(2):219–262, 2003.

[HKK04] Markus Holzer, Andreas Klein, and Martin Kutrib. On the NP-completeness of the NURIKABE pencil puzzle and variants thereof. In *Proceedings of the 3rd International Conference on FUN with Algorithms*, pages 77–89, Isola d'Elba, Italy, May 2004.

[HLH03] Jeffrey R. Hartline and Ran Libeskind-Hadas. The computational complexity of motion planning. *SIAM Review*, 45:543–557, 2003.

[Hoc01] Martin Hock. Exploring the complexity of the ufo puzzle. Undergraduate thesis, Carnegie Mellon University, 2001. http://www.cs.cmu.edu/afs/cs/user/mjs/ftp/thesis-02/hock.ps.

[Hof00] Michael Hoffmann. Push-* is NP-hard. In *Proceedings of the 12th Canadian Conference on Computational Geometry*, pages 205–210, Fredericton, Canada, August 2000. http://www.cs.unb.ca/conf/cccg/eProceedings/13.ps.gz.

[Hor86] Edward Hordern. *Sliding Piece Puzzles*. Oxford University Press, 1986.

[HR07] Markus Holzer and Oliver Ruepp. The troubles of interior design — a complexity analysis of the game Heyawake. In *Proceedings of the 4th International Conference on FUN with Algorithms*, volume 4475 of *Lecture Notes in Computer Science*, pages 198–212, 2007.

[HS04a] Markus Holzer and Stefan Schwoon. Assembling molecules in ATOMIX is hard. *Theoretical Computer Science*, 313(3):447–462, February 2004.

[HS04b] Markus Holzer and Stefan Schwoon. Reflections on REFLEXION — computational complexity considerations on a puzzle game. In *Proceedings of the 3rd International Conference on FUN with Algorithms*, pages 90–105, Isola d'Elba, Italy, May 2004.

[HSS84] J. E. Hopcroft, J. T. Schwartz, and M. Sharir. On the complexity of motion planning for multiple independent objects: PSPACE-hardness of the 'Warehouseman's Problem'. *International Journal of Robotics Research*, 3(4):76–88, 1984.

[IK94] Shigeki Iwata and Takumi Kasai. The Othello game on an $n \times n$ board is PSPACE-complete. *Theoretical Computer Science*, 123:329–340, 1994.

[Jer85] Mark R. Jerrum. The complexity of finding minimum-length generator sequences. *Theoretical Computer Science*, 36(2–3):265–289, 1985.

[Jer86] Mark Jerrum. A compact representation for permutation groups. *Journal of Algorithms*, 7(1):60–78, 1986.

[Kay00a] Richard Kaye. Infinite versions of minesweeper are Turing-complete. Manuscript, August 2000. http://www.mat.bham.ac.uk/R.W.Kaye/minesw/infmsw.pdf.

[Kay00b] Richard Kaye. Minesweeper is NP-complete. *Mathematical Intelligencer*, 22(2):9–15, 2000.

[KC07] Daniel Kurkle and Gene Cooperman. Twenty-six moves suffice for Rubik's cube. In *Proceedings of International Symposium on Symbolic and Algebraic Computation*, pages 235–242, 2007.

[Kem03] David Kempe. On the complexity of the reflections game. Unpublished manuscript, January 2003. http://www-rcf.usc.edu/~dkempe/publications/reflections.pdf.

[Klo07] Oddvar Kloster. A solution to the Angel Problem. *Theoretical Computer Science*, 389(1–2):152–161, December 2007.

[KMS84] Daniel Kornhauser, Gary Miller, and Paul Spirakis. Coordinating pebble motion on graphs, the diameter of permutation groups, and applications. In *Proceedings of the 25th Annual Symposium on Foundations of Computer Science*, pages 241–250, 1984.

[Knu74] Donald Knuth. *Surreal Numbers*. Addison-Wesley, Reading, Mass., 1974.

[LC89] K. Li and K. H. Cheng. Complexity of resource allocation and job scheduling problems on partitionable mesh connected systems. In *Proceedings of the 1st Annual IEEE Symposium on Parallel and Distributed Processing*, pages 358–365, May 1989.

[LM07] Luc Longpré and Pierre McKenzie. The complexity of solitaire. In *Proceedings of the 32nd International Symposium on Mathematical Foundations of Computer Science*, volume 4708 of *Lecture Notes in Computer Science*, pages 182–193, 2007.

[LMR00] Michael Lachmann, Cristopher Moore, and Ivan Rapaport. Who wins domineering on rectangular boards? In *MSRI Combinatorial Game Theory Research Workshop*, Berkeley, California, July 2000.

[LS80] David Lichtenstein and Michael Sipser. GO is polynomial-space hard. *Journal of the Association for Computing Machinery*, 27(2):393–401, April 1980.

[Mát07] András Máthé. The angel of power 2 wins. *Combinatocis, Probability and Computing*, 16(3):363–374, 2007.

[McK84] Pierre McKenzie. Permutations of bounded degree generate groups of polynomial diameter. *Information Processing Letters*, 19(5):253–254, November 1984.

[McP03] Brandon McPhail. The complexity of puzzles. Undergraduate thesis, Reed College, Portland, Oregon, 2003.

[McP05] Brandon McPhail. Light Up is NP-complete. Unpublished manuscript, 2005. http://www.cs.umass.edu/~mcphailb/papers/2005lightup.pdf.

[McP07] Brandon McPhail. Metapuzzles: Reducing SAT to your favorite puzzle. Lecture at UMass Amherst, December 2007. See http://www.cs.umass.edu/~mcphailb/papers/2007metapuzzles.pdf.

[ME02] Cristopher Moore and David Eppstein. One-dimensional peg solitaire, and duotaire. In R. J. Nowakowski, editor, *More Games of No Chance*, pages 341–350. Cambridge University Press, 2002.

[MR01] Cristopher Moore and John Michael Robson. Hard tiling problems with simple tiles. *Discrete and Computational Geometry*, 26(4):573–590, 2001.

[NP96] Richard J. Nowakowski and David G. Poole. Geography played on products of directed cycles. In R. J. Nowakowski, editor, *Games of No Chance*, pages 329–337. Cambridge University Press, 1996.

[OS99] Joseph O'Rourke and the Smith Problem Solving Group. PushPush is NP-hard in 3D. Technical Report 064, Department of Computer Science, Smith College, Northampton, MA, November 1999. http://arXiv.org/abs/cs/9911013.

[Pap85] Christos H. Papadimitriou. Games against nature. *Journal of Computer and System Sciences*, 31(2):288–301, 1985.

[Pap01] Christos H. Papadimitriou. Algorithms, games, and the Internet. In *Proceedings of the 33rd Annual ACM Symposium on Theory of Computing*, Crete, Greece, July 2001.

[Par95] Ian Parberry. A real-time algorithm for the $(n^2 - 1)$-puzzle. *Information Processing Letters*, 56(1):23–28, 1995.

[Plaa] Thane E. Plambeck. Advances in losing. In *Games of No Chance 3*. To appear. http://arxiv.org/abs/math/0603027.

[Plab] Thane E. Plambeck. miseregames.org.

[Pla05] Thane E. Plambeck. Taming the wild in impartial combinatorial games. *INTEGERS: Electronic Journal of Combinatorial Number Theory*, 5(G05), 2005.

[PS07] Thane E. Plambeck and Aaron N. Siegel. Misère quotients for impartial games. arXiv:math/0609825v5, August 2007. http://arxiv.org/abs/math/0609825.

[Rei80] Stefan Reisch. Gobang ist PSPACE-vollständig. *Acta Informatica*, 13:59–66, 1980.

[Rei81] Stefan Reisch. Hex ist PSPACE-vollständig. *Acta Informatica*, 15:167–191, 1981.

[Ren05] Paul Rendell. A Turing machine implemented in Conway's Game of Life. http://rendell-attic.org/gol/tm.htm, January 2005.

[RM78] Edward Robertson and Ian Munro. NP-completeness, puzzles and games. *Utilitas Mathematica*, 13:99–116, 1978.

[Rob83] J. M. Robson. The complexity of Go. In *Proceedings of the IFIP 9th World Computer Congress on Information Processing*, pages 413–417, 1983.

[Rob84a] J. M. Robson. Combinatorial games with exponential space complete decision problems. In *Proceedings of the 11th Symposium on Mathematical Foundations of Computer Science*, volume 176 of *Lecture Notes in Computer Science*, pages 498–506, 1984.

[Rob84b] J. M. Robson. N by N Checkers is EXPTIME complete. *SIAM Journal on Computing*, 13(2):252–267, May 1984.

[RW90] Daniel Ratner and Manfred Warmuth. The $(n^2 - 1)$-puzzle and related relocation problems. *Journal of Symbolic Computation*, 10:111–137, 1990.

[Sav70] Walter J. Savitch. Relationships between nondeterministic and deterministic tape complexities. *Journal of Computer and System Sciences*, 4(2):177–192, 1970.

[SBB+07] Jonathan Schaeffer, Neil Burch, Yngvi Björnsson, Akihiro Kishimoto, Martin Müller, Robert Lake, Paul Lu, and Steve Sutphen. Checkers is solved. *Science*, 317(5844):1518–1522, September 2007.

[SC79] Larry J. Stockmeyer and Ashok K. Chandra. Provably difficult combinatorial games. *SIAM Journal on Computing*, 8(2):151–174, 1979.

[Sch78] Thomas J. Schaefer. On the complexity of some two-person perfect-information games. *Journal of Computer and System Sciences*, 16:185–225, 1978.

[Sev] Merlijn Sevenster. Battleships as a decision problem. *ICGA Journal*, 27(3):142–147.

[Sie06] Aaron N. Siegel. Misère games and misère quotients. arXiv:math/0612616v2, December 2006. http://arxiv.org/abs/math/0612616.

[SM73] L. J. Stockmeyer and A. R. Meyer. Word problems requiring exponential time(preliminary report). In *Proceedings of the 5th Annual ACM Symposium on Theory of Computing*, pages 1–9, Austin, Texas, 1973.

[Smi66] Cedric A. B. Smith. Graphs and composite games. *Journal of Combinatorial Theory*, 1:51–81, 1966.

[Spr36] R. Sprague. Über mathematische Kampfspiele. *Tôhoku Mathematical Journal*, 41:438–444, 1935–36.

[SS06] Jerry Slocum and Dic Sonneveld. *The 15 Puzzle*. Slocum Puzzle Foundation, 2006.

[Ste03] James W. Stephens. The Kung Fu Packing Crate Maze, 2003. See http://www.puzzlebeast.com/crate/.

[Sto79] W. E. Story. Note on the '15' puzzle. *American Mathematical Monthly*, 2:399–404, 1879.

[SY83] Paul Spirakis and Chee Yap. On the combinatorial complexity of motion coordination. Report 76, Computer Science Department, New York University, 1983.

[TC04] John Tromp and Rudy Cilibrasi. Limits of Rush Hour Logic complexity. Manuscript, June 2004. http://www.cwi.nl/~tromp/rh.ps.

[Thu90] William P. Thurston. Conway's tiling groups. *American Math. Monthly*, 97(8): 757–773, October 1990.

[Tro00] John Tromp. On size 2 Rush Hour logic. Manuscript, December 2000. http://turing.wins.uva.nl/~peter/teaching/tromprh.ps.

[UI90] Ryuhei Uehara and Shigeki Iwata. Generalized Hi-Q is NP-complete. *Transactions of the IEICE*, E73:270–273, February 1990.

[UN96] Nobuhisa Ueda and Tadaaki Nagao. NP-completeness results for NONO-GRAM via parsimonious reductions. Technical Report TR96-0008, Department of Computer Science, Tokyo Institute of Technology, Tokyo, Japan, May 1996.

[VV86] L. G. Valiant and V. V. Vazirani. NP is as easy as detecting unique solutions. *Theoretical Computer Science*, 47:85–93, 1986.

[Wil74] Richard M. Wilson. Graph puzzles, homotopy, and the alternating group. *Journal of Combinatorial Theory*, Series B, 16:86–96, 1974.

[Wil91] Gordon Wilfong. Motion planning in the presence of movable obstacles. *Annals of Mathematics and Artificial Intelligence*, 3(1):131–150, 1991.

[Wil04] Anne D. Williams. *The Jigsaw Puzzle: Piecing Together a History*. Berkley Books, New York, 2004.

[Wol94] Stephen Wolfram. *Cellular Automata and Complexity: Collected Papers*. Perseus Press, 1994.

[Wol02] David Wolfe. Go endgames are PSPACE-hard. In R. J. Nowakowski, editor, *More Games of No Chance*, pages 125–136. Cambridge University Press, 2002.

[Yat03] Takayuki Yato. Complexity and completeness of finding another solution and its application to puzzles. Master's thesis, University of Tokyo, Tokyo, Japan, January 2003.

[YS03] Takayuki Yato and Takahiro Seta. Complexity and completeness of finding another solution and its application to puzzles. *IEICE Transactions on Fundamentals of Electronics, Communications, and Computer Sciences*, E86-A(5):1052–1060, 2003. Also IPSJ SIG Notes 2002-AL-87-2, 2002.

[YTK+01] Masaya Yokota, Tatsuie Tsukiji, Tomohiro Kitagawa, Gembu Morohashi, and Shigeki Iwata. Exptime-completeness of generalized Tsume-Shogi (in Japanese). *Transactions of the IEICE*, J84-D-I(3):239–246, 2001.

[Zwi02] Uri Zwick. Jenga. In *Proceedings of the 13th Annual ACM-SIAM Symposium on Discrete Algorithms*, pages 243–246, San Francisco, California, 2002.

ERIK D. DEMAINE
MIT COMPUTER SCIENCE AND ARTIFICIAL INTELLIGENCE LABORATORY,
32 VASSAR ST., CAMBRIDGE, MA 02139, UNITED STATES
edemaine@mit.edu

ROBERT A. HEARN
NEUKOM INSTITUTE FOR COMPUTATIONAL SCIENCE, DARTMOUTH COLLEGE,
SUDIKOFF HALL, HB 6255, HANOVER, NH 03755, UNITED STATES
robert.a.hearn@dartmouth.edu

Games of No Chance 3
MSRI Publications
Volume 56, 2009

Advances in losing

THANE E. PLAMBECK

ABSTRACT. We survey recent developments in the theory of impartial combinatorial games in misere play, focusing on how Sprague–Grundy theory of
normal-play impartial games generalizes to misere play via the *indistinguishability quotient construction* [P2]. This paper is based on a lecture given on 21
June 2005 at the Combinatorial Game Theory Workshop at the Banff International Research Station. It has been extended to include a survey of results on
misere games, a list of open problems involving them, and a summary of *MisereSolver* [AS2005], the excellent Java-language program for misere indistinguishability quotient construction recently developed by Aaron Siegel. Many
wild misere games that have long appeared intractable may now lie within the
grasp of assiduous losers and their faithful computer assistants, particularly
those researchers and computers equipped with *MisereSolver*.

1. Introduction

We've spent a lot of time teaching you how to win games
by being the last to move. But suppose you are baby-sitting
little Jimmy and want, at least occasionally, to make sure you
lose? This means that instead of playing the normal play rule
in which whoever can't move is the *loser*, you've switched to
misere play rule when he's the *winner*. Will this make much
difference? Not *always*. . .

That's the first paragraph from the thirteenth chapter ("Survival in the Lost
World") of Berlekamp, Conway, and Guy's encyclopedic work on combinatorial
game theory, *Winning Ways for your Mathematical Plays* [WW].

And why "not *always?*" The misere analysis of an impartial combinatorial
game often proves to be far more difficult than it is in normal play. To take a
typical example, the normal play analysis of **Dawson's Chess** [D] was published

as early as 1956 by Guy and Smith [GS], but even today, a complete misere analysis hasn't been found (see Section 10.1). Guy tells the story [Guy91]:

> [Dawson's chess] is played on a $3 \times n$ board with white pawns on the first rank and black pawns on the third. It was posed as a *losing* game (last-player-losing, now called **misere**) so that capturing was obligatory. Fortunately, (because we *still* don't know how to play misere Dawson's Chess) I assumed, as a number of writers of that time and since have done, that the misere analysis required only a trivial adjustment of the normal (last-player-winning) analysis. This arises because Bouton, in his original analysis of Nim [B1902], had observed that only such a trivial adjustment was necessary to cover both normal and misere play. . .
>
> But even for *impartial* games, in which the same options are available to both players, regardless of whose turn it is to move, Grundy & Smith [GrS1956] showed that the general situation in misere play soon gets very complicated, and Conway [ONAG], (p. 140) confirmed that the situation can only be simplified to the microscopically small extent noticed by Grundy & Smith.
>
> At first sight Dawson's Chess doesn't look like an impartial game, but if you know how pawns move at Chess, it's easy to verify that it's equivalent to the game played with rows of skittles in which, when it's your turn, you knock down any skittle, together with its immediate neighbors, if any.

So misere play can be difficult. But is it a hopeless situation? It has often seemed so. Returning to chapter 13 in [WW], one encounters the *genus theory of impartial misere disjunctive sums*, extended significantly from its original presentation in chapter 7 ("How to Lose When You Must") of Conway's *On Numbers and Games* [ONAG]. But excluding the *tame games* that play like Nim in misere play, there's a remarkable paucity of example games that the genus theory completely resolves. For example, the section "Misere Kayles" from the 1982 first edition of [WW] promises

> Although several tame games arise in Kayles (see Chapter 4), wild game's abounding and we'll need all our [genus-theoretic] resources to tackle it. . .

However, it turns out Kayles isn't "tackled" at all — after an extensive table of genus values to heap size 20, one finds the slightly embarrassing question

> Is there a larger single-row P-position?

It was left to the amateur William L. Sibert [SC] to settle misere Kayles using completely different methods. One finds a description of his solution at end of the updated Chapter 13 in the second edition of [WW], and also in [SC]. In 2003, [WW] summarized the situation as follows (p. 451):

> Sibert's remarkable *tour de force* raises once again the question: are misere analyses really so difficult? A referee of a draft of the Sibert–Conway paper wrote "the actual solution will have no bearing on other problems," while another wrote "the ideas are likely to be applicable to some other games. . . "

1.1. Misere play — the natural impartial game convention? When nonmathematicians play impartial games, they tend to choose the misere play convention[1]. This was already recognized by Bouton in his classic paper "Nim, A Game with a Complete Mathematical Theory," [B1902]:

> The game may be modified by agreeing that the player who takes the last counter from the table *loses*. This modification of the three pile [Nim] game seems to be more widely known than that first described, but its theory is not quite so simple. . .

But why do people prefer the misere play convention? The answer may lie in Fraenkel's observation that impartial games lack *boardfeel*, and simple *Schadenfreude*[2]:

> For many MathGames, such as Nim, a player without prior knowledge of the strategy has no inkling whether any given position is "strong" or "weak" for a player. Even two positions before ultimate defeat, the player sustaining it may be in the dark about the outcome, which will stump him. The player has no boardfeel. . . [Fraenkel, p. 3].

If both players are "in the dark," perhaps it's only natural that the last player compelled to make a move in such a pointless game should be deemed the *loser*. Only when a mathematician gets involved are things ever-so-subtly shifted toward the normal play convention, instead — but this is only because there is a simple and beautiful theory of normal-play impartial games, called Sprague–Grundy theory. Secretly computing nim-values, mathematicians win normal-play impartial games time and time again. Papers on normal play impartial games outnumber misere play ones by a factor of perhaps fifty, or even more[3].

---

[1] "Indeed, if anything, misere Nim is more commonly played than normal Nim. . ." [ONAG], p. 136.

[2] The joy we take in another's misfortune.

[3] Based on an informal count of papers in the [Fraenkel] CGT bibliography.

In the last twelve months it has become clear how to generalize such Sprague–Grundy nim-value computations to misere play via *indistinguishability quotient construction* [P2]. As a result, many misere game problems that have long appeared intractable, or have been passed over in silence as too difficult, have now been solved. Still others, such as a Dawson's Chess, appear to remain out of reach and await new ideas. The remainder of this paper surveys this largely unexplored territory.

2. Two wild games

We begin with two impartial games: **Pascal's Beans** — introduced here for the first time — and **Guiles** (the octal game **0.15**). Each has a relatively simple normal-play solution, but is *wild*[4] in misere play. Wild games are characterized by having misere play that differs in an essential way[5] from the play of misere Nim. They often prove notoriously difficult to analyze completely. Nevertheless, we'll give complete misere analyses for both Pascal's Beans and Guiles by using the key idea of the *misere indistinguishability quotient*, which was first introduced in [P2], and which we take up in earnest in Section 5.

3. Pascal's Beans

Pascal's Beans is a two-player impartial combinatorial game. It's played with heaps of beans placed on Pascal's triangle, which is depicted in Figure 1. A legal move in the game is to slide a single bean either up a single row and to the left one position, or alternatively up a single row and to the right one position in the triangle. For example, in Figure 1, a bean resting on the cell marked 20 could be moved to either cell labelled 10.

The actual numbers in Pascal's triangle are not relevant in the play of the game, except for the 1's that mark the border positions of the board. In play of Pascal's Beans, a bean is considered out of play when it first reaches a border position of the triangle. The game ends when all beans have reached the border.

3.1. Normal play. In *normal play* of Pascal's Beans, the last player to make a legal move is declared the *winner* of the game. Figure 2 shows the pattern of *nim values* that arises in the analysis of the game. Using the figure, it's possible to quickly determine the best-play outcome of an arbitrary starting position in Pascal's Beans using *Sprague–Grundy theory* and the *nim addition* operation $\oplus$. Provided one knows the $\mathbb{Z}_2 \times \mathbb{Z}_2$ addition table in Figure 3, all is well — the

---

[4] See Chapter 13 ("Survival in the Lost World") in [WW] and Section 7 in this paper for more information on wild misere games.

[5] To be made precise in Section 7.

```
                              1
                          1       1
                      1       2       1
                  1       3       3       1
              1       4       6       4       1
          1       5      10      10       5       1
      1       6      15      20      15       6       1
  1       7      21      35      35      21       7       1
      ⋮               ⋮               ⋮
```

Figure 1. The Pascal's Beans board.

P-positions (second-player winning positions) are precisely those that have nim value zero (that is, $*0$), and every other position is an *N-position* (or next-player win), of nim value $*1$, $*2$, or $*3$.

```
                              *0
                          *0      *0
                      *0      *1      *0
                  *0      *2      *2      *0
              *0      *1      *0      *1      *0
          *0      *2      *2      *2      *2      *0
      *0      *1      *0      *0      *0      *1      *0
  *0      *2      *2      *1      *1      *2      *2      *0
*0    *1    *0    *0    *0    *0    *0    *1    *0
          ⋮               ⋮               ⋮
```

Figure 2. The pattern of single-bean nim-values in normal play of Pascal's Beans. Each interior value is the *minimal excludant* (or *mex*) of the two nim values immediately above it. The boldface entries form the first three rows of an infinite subtriangle whose rows alternate between $*0$ and $*1$.

$\oplus$	$*0$	$*1$	$*2$	$*3$
$*0$	$*0$	$*1$	$*2$	$*3$
$*1$	$*1$	$*0$	$*3$	$*2$
$*2$	$*2$	$*3$	$*0$	$*1$
$*3$	$*3$	$*2$	$*1$	$*0$

Figure 3. Addition for normal play of Pascal's Beans.

3.2. Misere play. In *misere play* of Pascal's Beans, the last player to make a move is declared the *loser* of the game. Is it possible to give an analysis of misere Pascal's Beans that resembles the normal play analysis? The answer is yes — but the positions of the triangle can no longer be identified with nim heaps $*k$, and the rule for the misere addition is no longer given by nim addition. Instead, both the values to be identified with particular positions of the triangle and the desired misere addition are given by a particular twelve-element commutative monoid $\mathcal{M}$, the *misere indistinguishability quotient*[6] of Pascal's Beans. The monoid $\mathcal{M}$ has an identity 1 and is presentable using three generators and relations:

$$\mathcal{M} = \langle\, a, b, c \mid a^2 = 1,\ c^2 = 1,\ b^3 = b^2 c \,\rangle.$$

Assiduous readers might enjoy verifying that the identity $b^4 = b^2$ follows from these relations, and that a general word of the form $a^i b^j c^k$ $(i, j, k \geq 0)$ will always reduce to one of the twelve *canonical words*

$$\mathcal{M} = \{1,\ a,\ b,\ ab,\ b^2,\ ab^2,\ c,\ ac,\ bc,\ b^2c,\ abc,\ ab^2c\}.$$

Amongst the twelve canonical words, three represent P-position types

$$\mathcal{P} = \{a,\ b^2, ac\},$$

and the remaining nine represent N-position types:

$$\mathcal{N} = \{1,\ b,\ ab,\ ab^2,\ c,\ bc,\ b^2c,\ abc,\ ab^2c\}.$$

Figure 4 shows the identification of triangle positions with elements of $\mathcal{M}$.

$$\begin{array}{ccccccccccccccccc}
&&&&&&&&1&&&&&&&& \\
&&&&&&&1&&1&&&&&&& \\
&&&&&&1&&\boxed{a}&&1&&&&&& \\
&&&&&1&&b&&b&&1&&&&& \\
&&&&1&&a&&\boxed{b^2}&&a&&1&&&& \\
&&&1&&b&&c&&c&&b&&1&&& \\
&&1&&a&&b^2&&\mathbf{b^2}&&b^2&&a&&1&& \\
&1&&b&&c&&\mathbf{ab^2}&&\mathbf{ab^2}&&c&&b&&1& \\
1&&a&&b^2&&\mathbf{b^2}&&\mathbf{b^2}&&\mathbf{b^2}&&b^2&&a&&1 \\
&&&&\vdots&&&&\vdots&&&&\vdots&&&&
\end{array}$$

Figure 4. Identifications for single-bean positions in misere play of Pascal's Beans. The values are elements of the misere indistinguishability quotient $\mathcal{M}$ of Pascal's Beans. The boldface entries form the first three rows of an infinite subtriangle whose rows alternate between the values b^2 and ab^2.

[6] See Section 5.

Although we've used multiplicative notation to represent the addition operation in the monoid $\mathcal{M}$, we use it to analyze general misere-play Pascal's Beans positions just as we used the nim values of Figure 2 and nim addition in normal play. For example, suppose a Pascal's Beans position involves just two beans — one placed along the central axis of the triangle at each of the two boxed positions in Figure 4. Combining the corresponding entries a and b^2 as monoid elements, we obtain the element ab^2, which we've already asserted is an N-position. What is the winning misere-play move? From the lower bean, at the position marked b^2, the only available moves are both to a cell marked b. This move is of the form

$$ab^2 \to ab,$$

that is, the result is another misere N-position type (here ab). So this option is not a winning misere move. But the cell marked a has an available move is to the border. The resulting winning move is of the form

$$ab^2 \to b^2,$$

that is, the result is b^2, a P-position type.

4. Guiles

Guiles can be played with heaps of beans. The possible moves are to remove a heap of 1 or 2 beans completely, or to take two beans from a sufficiently large heap and partition what is left into two smaller, nonempty heaps. This is the octal game **0.15**.

4.1. Normal play. The nim values of the octal game **Guiles** fall into a period 10 pattern. See Figure 5.

	1	2	3	4	5	6	7	8	9	10
0+	1	1	0	1	1	2	2	1	2	2
10+	1	1	0	1	1	2	2	1	2	2
20+	1	1	0	1	1	2	2	1	2	2
30+	1	1	0	1	1	$\cdots$				

Figure 5. Nim values for normal play **0.15**.

4.2. Misere play. Using his recently-developed Java-language computer program *MisereSolver*, Aaron Siegel [PS] found that the misere indistinguishability quotient $\mathfrak{Q}$ of misere Guiles is a (commutative) monoid of order 42. It has the presentation

$$\mathfrak{Q} = \langle a, b, c, d, e, f, g, h, i \mid$$
$$a^2 = 1, \ b^4 = b^2, \ bc = ab^3, \ c^2 = b^2, \ b^2d = d,$$
$$cd = ad, \ d^3 = ad^2, \ b^2e = b^3, \ de = bd, \ be^2 = ace,$$
$$ce^2 = abe, \ e^4 = e^2, \ bf = b^3, \ df = d, \ ef = ace,$$
$$cf^2 = cf, \ f^3 = f^2, \ b^2g = b^3, \ cg = ab^3, \ dg = bd,$$
$$eg = be, \ fg = b^3, \ g^2 = bg, \ bh = bg, \ ch = ab^3,$$
$$dh = bd, \ eh = bg, \ fh = b^3, \ gh = bg, \ h^2 = b^2,$$
$$bi = bg, \ ci = ab^3, \ di = bd, \ ei = be, \ fi = b^3,$$
$$gi = bg, \ hi = b^2, \ i^2 = b^2 \rangle.$$

In Figure 6 we show the single-heap misere equivalences for Guiles. It is a remarkable fact that this sequence is also periodic of length ten — it's just that the (aperiodic) *preperiod* is longer (length 66), and a person needs to know the monoid $\mathfrak{Q}$! The P-positions of Guiles are precisely those positions equivalent to one of the words

$$P = \{ a, \ b^2, \ bd, \ d^2, \ ae, \ ae^2, \ ae^3, \ af, \ af^2, \ ag, \ ah, \ ai \}.$$

	1	2	3	4	5	6	7	8	9	10
0+	a	a	1	a	a	b	b	a	b	b
10+	a	a	1	c	c	b	b	d	b	e
20+	c	c	f	c	c	b	g	d	h	i
30+	ab^2	abg	f	abg	abe	b^3	h	d	h	h
40+	ab^2	abe	f^2	abg	abg	b^3	h	d	h	h
50+	ab^2	abg	f^2	abg	abg	b^3	b^3	d	b^3	b^3
60+	ab^2	abg	f^2	abg	abg	b^3	b^3	d	b^3	b^3
70+	ab^2	ab^2	f^2	ab^2	ab^2	b^3	b^3	d	b^3	b^3
80+	ab^2	ab^2	f^2	ab^2	ab^2	b^3	b^3	d	b^3	b^3
90+	ab^2	ab^2	f^2	ab^2	ab^2	b^3	b^3	d	b^3	b^3
100+										

Figure 6. Misere equivalences for Guiles.

Knowledge of the monoid presentation $\mathfrak{Q}$, its partition into N- and P-position types, and the single-heap equivalences in Figure 6 suffices to quickly determine the outcome of an arbitrary misère Guiles position. For example, suppose a position contains four heaps of sizes 4, 58, 68, and 78. Looking up monoid values in Figure 6, we obtain the product

$$a \cdot d \cdot d \cdot d = ad^3$$
$$= a \cdot ad^2 \quad (\text{relation } d^3 = ad^2)$$
$$= d^2 \quad (\text{relation } a^2 = 1)$$

We conclude that $4 + 58 + 68 + 78$ is a misère Guiles P-position.

5. The indistinguishability quotient construction

What do these two solutions have in common? They were both obtained via a computer program called *MisereSolver*, by Aaron Siegel. Underpinning *MisereSolver* is the notion of the *indistinguishability quotient construction*. Here, we'll sketch the main ideas of the indistinguishability quotient construction only. They are developed in detail in [P2].

Suppose $\mathcal{A}$ is a set of (normal, or alternatively, misère) impartial game positions that is closed under the operations of game addition and taking options (that is, making moves). Unless we say otherwise, we'll always be taking $\mathcal{A}$ to be the set of all positions that arise in the play of a specific game Γ, which we fix in advance. For example, one might take

$$\Gamma = \text{Normal-play Nim,}$$
$$\mathcal{A} = \text{All positions that arise in normal-play Nim,}$$

or

$$\Gamma = \text{Misère-play Guiles,}$$
$$\mathcal{A} = \text{All positions that arise in misère-play Guiles.}$$

Two games $G, H \in \mathcal{A}$ are then said to be *indistinguishable*, and we write the relation $G \rho H$, if for every game $X \in \mathcal{A}$, the sums $G + X$ and $H + X$ have the same outcome (that is, are both N-positions, or are both P-positions). Note in particular that if G and H are indistinguishable, then they have the same outcome (choose X to be the *endgame* — that is, the terminal position, with no options).

The indistinguishability relation ρ is easily seen to be an equivalence relation on $\mathcal{A}$, but in fact more is true — it's a *congruence* on $\mathcal{A}$ [P2]. This follows because indistinguishability is *compatible* with addition; that is, for every set of three games $G, H, X \in \mathcal{A}$:

$$G \rho H \implies (G + X) \rho (H + X). \tag{5-1}$$

Now let's make the definition

$$\rho G = \{\, H \in \mathcal{A} \mid G \, \rho \, H \,\}.$$

We'll call ρG the *congruence class of $\mathcal{A}$ modulo ρ containing G*. Because ρ is a congruence, there is a well-defined addition operation

$$\rho G + \rho H = \rho(G + H)$$

on the set $\mathcal{A}/\rho$ of all congruence classes ρG of $\mathcal{A}$ modulo ρ

$$\mathcal{Q} = \mathcal{Q}(\Gamma) = \mathcal{A}/\rho = \{\, \rho G \mid G \in \mathcal{A}. \,\} \tag{5-2}$$

The monoid $\mathcal{Q}$ is called the *indistinguishability quotient* of Γ. It captures the essential information of "how to add" in the play of game Γ, and is the central figure of our drama.

The natural mapping

$$\Phi : G \mapsto \rho G$$

from $\mathcal{A}$ to $\mathcal{A}/\rho$ is called a *pretending function* (see [P2]). Figures 4 and 6 illustrate the (as it happens, provably periodic [P2]) pretending functions of Pascal's Beans and Guiles, respectively. We shall gradually come to see that the recovery of $\mathcal{Q}$ and Φ from Γ is the essence of impartial combinatorial game analysis in both normal and misere play.

When Γ is chosen as a normal-play impartial game, the elements of $\mathcal{Q}$ work out to be in 1-1 correspondence with the *nim-heap values* (or *G-values*) that occur in the play of the game Γ. For if G and H are normal-play impartial games with $G = *g$ and $H = *h$, one easily shows that G and H are indistinguishable if and only if $g = h$. Additionally, in normal play, every position G satisfies the equation

$$G + G = 0.$$

As a result, the addition in a normal-play indistinguishability quotient is an abelian group in which every element is its own additive inverse. The addition operation in the quotient $\mathcal{Q}$ is *nim addition*. Every normal play indistinguishability quotient is therefore isomorphic to a (possibly infinite) direct product

$$\mathbb{Z}_2 \times \mathbb{Z}_2 \times \cdots,$$

and a position is a P-position precisely if it belongs the congruence class of the identity (that is, $*0$) of this group. In this sense "nothing new" is learned about normal play impartial games via the indistinguishability quotient construction — instead, we've simply recast Sprague–Grundy theory in new language. The fun begins when the construction is applied in *misere play*, instead.

6. Misere indistinguishability quotients

In misere play, the indistinguishability quotient $\mathfrak{Q}$ turns out to be a commutative monoid whose structure intimately depends upon the particular game Γ that is chosen for analysis. We need to cover some background material first.

6.1. Preliminaries. Consider the following three concepts in impartial games:

(i) The notion of the *endgame* (or *terminal position*), that is, a game that has no options at all.
(ii) The notion of a *P-position*, that is, a game that is a second-player win in best play of the game.
(iii) The notion of the *sum of two identical games*, that is, $G + G$.

In normal play, these three notions are *indistinguishable* — wherever a person sees (1) in a sum S, he could freely substitute (2) or (3) (or vice-versa, or any combination of such substitutions) without changing the outcome of S.

The three notions do not coincide in misere play. Let's see what happens instead.

The misere endgame. In misere play, the endgame is an N-position, not a P-position: even though there is no move available from the endgame, a player still wants it to be his *turn to move* when facing the endgame in misere play, because that means his opponent just *lost*, on his previous move.

Misere outcome calculation. After the special case of the endgame is taken care of, the recursive rule for outcome calculation in misere play is exactly as it is in normal play: a non-endgame position G is a P-position if and only if all its options are N-positions. Misere games cannot be identified with nim heaps, in general, however — instead, a typical misere game looks like a complicated, usually unsimplifiable tree of options [ONAG], [GrS1956].

Misere P-positions. Since the endgame is not a misere P-position, the simplest misere P-position is the *nim-heap of size one*, that is, the game played using one bean on a table, where the game is to take that bean. To avoid confusion both with what happens in normal play, and with the algebra of the misere indistinguishability quotient to be introduced in the sequel, let's introduce some special symbols for the three simplest misere games:

$\mathfrak{o}$ = The misere *endgame*, that is, a position with no moves at all.

$\mathbb{1}$ = The misere *nim heap of size one*, that is, a position with one move (to $\mathfrak{o}$).

$\mathbb{2}$ = The misere *nim heap of size two*, that is, the game $\{\mathfrak{o}, \mathbb{1}\}$.

Two games that we've intentionally left off this list are $\{\mathbb{1}\}$ and $\mathbb{1} + \mathbb{1}$. Assiduous readers should verify they are both indistinguishable from $\mathfrak{o}$.

Misere sums involving P-positions. Suppose that G is an arbitrary misere P-position. Consider the misere sum

$$S = 1 + G. \tag{6-1}$$

Who wins S? It's an N-position — a winning first-player move is to simply take the nim heap of size one, leaving the opponent to move first in the P-position G. In terms of outcomes, equation (6-1) looks like

$$N = P + P. \tag{6-2}$$

Equation (6-2) does not remind us of normal play very much — instead, we always have $P + P = P$ in normal play. On the other hand, it's not true that sum of two misere P-positions is *always* a misere N-position — in fact, when two typical misere P-positions G and H are added together with *neither* equal to 1, it *usually* happens that their sum is a P-position, also. But that's not *always* the case — it's also possible that two misere impartial P-positions, neither of which is 1, can nevertheless result in an N-position when added together. Without knowing the details of the misere P-position involved, little more can be said in general about the outcome when it's added to another game.

Misere sums of the form $G + G$. In normal play, a sum $G + G$ of two identical games is always indistinguishable from the endgame. In misere play, it's true that both $o + o$ and $1 + 1$ are indistinguishable from o, but beyond those two sums, positions of the form $G + G$ are rarely indistinguishable from o. It frequently happens that a position G in the play of a game Γ has no $H \in \mathscr{A}$ such that $G + H$ is indistinguishable from o. This lack of natural inverse elements makes the structure of a typical misere indistinguishability quotient a *commutative monoid* rather than an *abelian group*.

The game $2 + 2$. The sum

$$2 + 2$$

is an important one in the theory of impartial misere games. It's a P-position in misere play: for if you move first by taking 1 bean from one summand, I'll take two from the other, forcing you to take the last bean. Similarly, if you choose to take 2 beans, I'll take 1 from the other. So whereas in normal play one has the equation

$$(*2 + *2) \ \rho \ *0,$$

it's certainly **not** the case in misere play that

$$(2 + 2) \ \rho \ o,$$

since the two sides of that proposed indistinguishability relation don't even have the same outcome. But perhaps

$$2 + 2 \overset{?}{\rho} 1 \qquad (6\text{-}3)$$

is valid? The indistinguishability relation (6-3) looks plausible at first glance — at least the positions on both sides are P-positions. To decide whether it's possible to distinguish between $2+2$ and 1, we might try adding various fixed games X to both, and see if we ever get differing outcomes:

Misere game X	Misere outcome of $2+2+X$	Misere outcome of $1+X$
0	P	P
1	N	N
2	N	N
$1+2$	N	N
$2+2$	P	N

The two positions look like they *might* be indistinguishable, until we reach the final row of the table. It reveals that $(2+2)$ distinguishes between $(2+2)$ and 1. So equation (6-3) fails. Since a set of misere game positions $\mathscr{A}$ that includes 2 and is closed under addition and taking options must contain all of the games 1, 2, and $2+2$, we've shown that a game that isn't She-Loves-Me-She-Loves-Me-Not *always* has at least *two* distinguishable P-position types. In normal play, there's just one P-position type up to indistinguishability — the game $*0$.

6.2. Indistinguishability versus canonical forms.

In normal play, Sprague–Grundy theory describes how to determine the outcome of a sum $G + H$ of two games G and H by computing *canonical* (or *simplest*) forms for each summand — these turn out to be *nim-heap equivalents* $*k$. In both normal and misere play, canonical forms are obtained by pruning reversible moves from game trees (see [GrS1956], [ONAG] and [WW]).

In [ONAG], Conway succinctly gives the rules for misere game tree simplification to canonical form:

> When H occurs in some sum we should naturally like to replace it by [a] simpler game G. Of course, we will normally be given only H, and have to find the simpler game G for ourselves. How do we do this? Here are two observations which make this fairly easy:
>
> (i) G must be obtained by deleting certain options of H.

(ii) G itself must be an option of any of the deleted options of H, and so G must be itself be a *second option* of H, if we can delete any option at all.

On the other hand, if we obey (1) and (2), the deletion is permissible, except that we can only delete *all* the options of H (making $G = 0$ [the endgame]) if one of the them is a second-player win.

Unlike in normal play, the canonical form of a misere game is not a nim heap in general. In fact, many misere game trees hardly simplify at all under the misere simplification rules. Figure 7, which duplicates information in [ONAG] (its Figure 32), shows the 22 misere game trees born by day 4.

$$
\begin{array}{lll}
0 = \{\} & 2_{++} = \{2_+\} & 2_+3o = \{2_+, 3, o\} \\
1 = \{o\} & 2_+o = \{2_+, o\} & 2_+31 = \{2_+, 3, 1\} \\
2 = \{o, 1\} & 2_+1 = \{2_+, 1\} & 2_+32 = \{2_+, 3, 2\} \\
3 = \{o, 1, 2\} & 2_+2 = \{2_+, 2\} & 2_+32o = \{2_+, 3, 2, o\} \\
4 = \{o, 1, 2, 3\} & 2_+2o = \{2_+, 2, o\} & 2_+321 = \{2_+, 3, 2, 1\} \\
2_+ = \{2\} & 2_+21 = \{2_+, 2, 1\} & 2_+321o = \{2_+, 3, 2, 1, o\} \\
3_+ = \{3\} & 2_+21o = \{2_+, 2, 1, o\} & \\
2 + 2 = \{3, 2\} & 2_+3 = \{2_+, 3\} &
\end{array}
$$

Figure 7. Canonical forms for misere games born by day 4.

Whereas only one normal-play nim-heap is born at each birthday n, over 4 million nonisomorphic misere canonical forms are born by day five. The number continues to grow very rapidly, roughly like a tower of exponentials of height n ([ONAG]). This very large number of mutually distinguishable trees has often made misere analysis look like a hopeless activity.

Indistinguishability identifies games with different misere canonical forms. The key to the success of the indistinguishability quotient construction is that it is a *construction localized to the play of a particular game* Γ. It therefore has the possibility of identifying misere games with different canonical forms. While it's true that for misere games G, H with different canonical forms that there must be a game X such that $G + X$ and $H + X$ have different outcomes, such an X *might possibly never occur* in play of the fixed game Γ that we've chosen to analyze. Indistinguishability quotients are often *finite*, even for games Γ that involve an infinity of different canonical forms amongst their position sums.

7. What is a wild misere game?

Roughly speaking, a misere impartial game Γ is said to be *tame* when a complete analysis of it can be given by identifying each of its positions with some position that arises in the misere play of Nim. Tameness is therefore an attribute of a *set* of positions, rather than a *particular* position. Games Γ that are not tame are said to be *wild*. Unlike tame games, wild games cannot be completely analyzed by viewing them as disguised versions of misere Nim.

7.1. Tame games. Conway's *genus theory* was first described in chapter 12 of [ONAG]. It describes a method for calculating whether all the positions of particular misere game Γ are tame, and how to give a complete analysis of Γ, if so. For completeness, we've summarized the genus theory in the Appendix (page 81).

For misere games Γ that genus theory identifies as tame, a complete analysis can be given without reference to the indistinguishability quotient construction. Various efforts to extend genus theory to wider classes of games have been made. Example settings where progress has been made are the main subject of papers by of Ferguson [F2], [F3] and Allemang [A1], [A2], [A3].

Indistinguishability quotients for tame games. In this section, we reformulate the genus theory of tame games in terms of the indistinguishability quotient language.

Suppose S is some finite set of misere combinatorial games. We'll use the notation $\mathrm{cl}(S)$ (the *closure* of S) to stand for the smallest set of games that includes every element of S and is closed under addition and taking options. Putting $\mathcal{A} = \mathrm{cl}(S)$ and defining the indistinguishability quotient

$$\mathcal{Q} = A/\rho,$$

the natural question arises, what is the monoid $\mathcal{Q}$? Figure 8 shows answers for $S = \{1\}$ and $S = \{2\}$.

S	Presentation for monoid $\mathcal{Q}$	Order	Symbol	Name
$\{1\}$	$\langle\, a \mid a^2 = 1 \,\rangle$	2	$\mathcal{T}_1$	First tame quotient
$\{2\}$	$\langle\, a,\, b \mid a^2 = 1,\ b^3 = b \,\rangle$	6	$\mathcal{T}_2$	Second tame quotient

Figure 8. The first and second tame quotients.

$\mathcal{T}_1$ is called the *first tame quotient*. It represents the misere play of *She-Loves-Me, She-Loves-Me-Not*. In $\mathcal{T}_1$, misere P-positions are represented by the monoid (in fact, group) element a, and N-positions by 1.

$\mathcal{T}_2$, the *second tame quotient*, has the presentation

$$\langle\, a, b \mid a^2 = 1,\ b^3 = b \,\rangle.$$

It is a six-element monoid with two P-position types $\{a, b^2\}$. The prototypical game Γ with misere indistinguishability quotient $\mathcal{T}_2$ is the game of Nim, played with heaps of 1 and 2 only. See Figures 9 and 10.

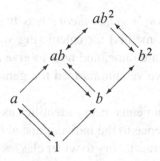

Figure 9. *The misere impartial game theorist's coat of arms,* or the Cayley graph of $\mathcal{T}_2$. Arrows have been drawn to show the action of the generators a (the doubled rungs of the ladder) and b (the southwest-to-northeast-oriented arrows) on $\mathcal{T}_2$. See also Figure 10.

The general tame quotient. For $n \geq 2$, the n-th *tame quotient* is the monoid $\mathcal{T}_n$ with $2^n + 2$ elements and the presentation

$$\mathcal{T}_n = \langle\, a, \underbrace{b, c, d, e, f, g, \ldots}_{n-1 \text{ generators}} \mid a^2 = 1, b^3 = b,\ c^3 = c,\ d^3 = d,\ e^3 = e,\ \ldots,$$
$$b^2 = c^2 = d^2 = e^2 = \cdots \,\rangle.$$

$\mathcal{T}_n$ is a disjoint union of its two maximal subgroups $\mathcal{T}_n = U \cup V$. The set

$$U = \{1, a\}$$

is isomorphic to $\mathbb{Z}_2$. The remaining 2^n elements of $\mathcal{T}_n$ form the set

$$V = \{\ a^{a_i} b^{b_i} c^{c_i} d^{d_i} e^{e_i} \cdots \ \mid \ \begin{aligned}&a_i = 0 \text{ or } 1\\&b_i = 1 \text{ or } 2\\&\text{Each of } c_i, d_i, e_i, \ldots = 0 \text{ or } 1\ \}.\end{aligned}$$

and have an addition isomorphic to

$$\underbrace{\mathbb{Z}_2 \times \cdots \times \mathbb{Z}_2}_{n \text{ copies}}.$$

Position type	Misere indistinguishability quotient element	Outcome	Genus
Even #1's only	1	N	0^{120}
Odd #1's only	a	P	1^{031}
Odd #2's and Even #1's	b	N	2^{20}
Odd #2's and Odd #1's	ab	N	3^{31}
Even #2's (≥ 2) and Even #1's	b^2	P	0^{02}
Even #2's (≥ 2) and Odd #1's	ab^2	N	1^{13}

Figure 10. When misere Nim is played with heaps of size 1 and 2 only, the resulting misere indistinguishability quotient is the tame six-element monoid $\mathcal{T}_2$. For more on genus symbols and tameness, see Section 7. See also Figure 9.

The elements a and b^2 are the only P-position types in $\mathcal{T}_n$.

8. More wild quotients

8.1. The commutative monoid $\mathcal{R}_8$. The smallest *wild* misere indistinguishability quotient $\mathcal{R}_8$ has eight elements, and is unique up to isomorphism [S1] amongst misere quotients with eight elements. Its monoid presentation is

$$\mathcal{R}_8 = \langle\, a,\, b,\, c \mid a^2 = 1,\, b^3 = b,\, bc = ab,\, c^2 = b^2 \,\rangle.$$

The P-positions are $\{a, b^2\}$.

0.75. An example game with misere quotient $\mathcal{R}_8$ is the octal game **0.75**. The first complete analysis of **0.75** was given by Allemang using his *generalized genus theory* [A1]. Alternative formulations of the **0.75** solution are also discussed at length in the appendix of [P] and in [A2]. See Figure 11, left.

	1	2
0+	1	a
2+	b	a
4+	b	c
6+	b	c
8+	b	ab^2
10+	b	ab^2
12+	b	ab^2
14+	...	

	1	2	3	4	5	6	7	8
0+	a	1	a	b	1	a	1	ab
8+	a	c	a	b	1	ac	1	ab
16+	a	c	a	b	1	ac	1	ab

Figure 11. The pretending function for misere play of **0.75** (left) and **0.34**.

8.2. Flanigan's games. Jim Flanigan found solutions to the wild octal games **0.34** and **0.71**; a description of them can be found in the "Extras" of chapter 13 in [WW]. It's interesting to write down the corresponding misere quotients.

0.34. The misere indistinguishability quotient of **0.34** has order 12. There are three P-position types. The pretending function has period 8 (see Figure 11, right).

$$\mathcal{Q}_{0.34} = \langle\, a, b, c \mid a^2 = 1,\ b^4 = b^2,\ b^2 c = b^3,\ c^2 = 1 \,\rangle, \qquad P = \{a, b^2, ac\}$$

0.71. The game **0.71** has a misere quotient of order 36 with the presentation

$$\mathcal{Q}_{0.71} = \langle\, a, b, c, d \mid a^2 = 1,\ b^4 = b^2,\ b^2 c = c,\ c^4 = ac^3,\ c^3 d = c^3,\ d^2 = 1 \,\rangle.$$

The P-positions are $\{a, b^2, bc, c^2, ac^3, ad, b^3 d, cd, bc^2 d\}$. The pretending function appears in Figure 12.

	1	2	3	4	5	6
0+	a	b	a	1	c	1
6+	a	d	a	1	c	1
12+	a	d	a	1	c	1
18+	...					

Figure 12. The pretending function for misere play of **0.71**.

8.3. Other quotients. Hundreds more such solutions have been found amongst the octal games. The forthcoming paper [PS] includes a census of such results.

9. Computing presentations and MisereSolver

How are such solutions computed? Aaron Siegel's recently developed Java program *MisereSolver* [AS2005] will do it for you! Some details on the algorithms used in *MisereSolver* are included in [PS]. Here, we simply give a flavor of the some ideas underpinning it and how the software is used.

9.1. Misere periodicity. At the center of Sprague–Grundy theory is the equation $G + G = 0$, which always holds for an arbitrary normal play combinatorial game G. One consequence of $G + G = 0$ is the equation

$$G + G + G = G,$$

in which all we've done is add G to both sides. In general, in normal play,

$$(k + 2) \cdot G = k \cdot G.$$

holds for every $k \geq 0$.

In *misere play*, the relation

$$(G + G) \, \rho \, \text{\o}$$

happens to be true for $G = \text{\o}$ and $G = \mathbb{1}$, but beyond that, it is only seldom true for occasional rule sets Γ and positions G. On the other hand,

$$(G + G + G) \, \rho \, G$$

is *very often* true in misere play, and it is *always* true, for all G, if Γ is a tame game. And in *wild* games Γ for which the latter equation fails, often a weaker equation such as

$$(G + G + G + G) \, \rho \, (G + G),$$

is still valid, regardless of G.

These considerations suggest that a useful place to look for misere quotients is inside commutative monoids having some (unknown) number of generators x each satisfying a relation of the form

$$x^{k+2} = x^k$$

for each generator x and some value of $k \geq 0$.

9.2. Partial quotients for heap games. A *heap game* is an impartial game Γ whose rules can be expressed in terms of play on separated, noninteracting heaps of beans. In constructing misere quotients for heap games, it's useful to introduce the *n-th partial quotient*, which is just the indistinguishability quotient of Γ when all heaps are required to have n or fewer beans.

9.3. MisereSolver output of partial quotients. Here is an (abbreviated) log of *MisereSolver* output of partial quotients for **0.123**, an octal game that is studied in great detail in [P2]. In this output, monomial exponents have been juxtaposed with the generator names (so that b^2c, for example, appears as b2c). The program stops when it discovers the entire quotient — the partial quotients stabilize in a monoid of order 20, whose single-heap pretending function Φ is periodic of length 5.

```
C:\work>java -jar misere.jar 0.123
=== Normal Play Analysis of 0.123 ===
Max    : G(3) = 2
Period: 5 (5)
=== Misere Play Analysis of 0.123 ===
-- Presentation for 0.123 changed at heap 1 --
Size 2: TAME
P = {a}
Phi = 1 a 1
-- Presentation for 0.123 changed at heap 3 --
Size 6: TAME
P = {a,b2}
Phi = 1 a 1 b b a b2 1
-- Presentation for 0.123 changed at heap 8 --
Size 12: {a,b,c | a2=1,b4=b2,b2c=b3,c2=1}
P = {a,b2,ac}
Phi = 1 a 1 b b a b2 1 c
-- Presentation for 0.123 changed at heap 9 --
Size 20: {a,b,c,d | a2=1,b4=b2,b2c=b3,c2=1,b2d=d,cd=bd,d3=ad2}
P = {a,b2,ac,bd,d2}
Phi = 1 a 1 b b a d2 1 c d a d2 1 c d a d2 1 c d a d2 1
=== Misere Play Analysis Complete for 0.123 ===
Size 20: {a,b,c,d | a2=1,b4=b2,b2c=b3,c2=1,b2d=d,cd=bd,d3=ad2}
P = {a,b2,ac,bd,d2}
Phi = 1 a 1 b b a d2 1 c d a d2 1 c d a d2 1 c d a d2 1
Standard Form : 0.123
Normal Period : 5
Normal Ppd    : 5
Normal Max G  : G(3) = 2
Misere Period : 5
Misere Ppd    : 5
Quotient Order: 20
Heaps Computed: 22
Last Tame Heap: 7
```

9.4. Partial quotients and pretending functions. Let's look more closely at the *MisereSolver* partial quotient output in order to illustrate some of the subtlety of misere quotient presentation calculation.

In Figure 13, we've shown three pretending functions for **0.123**. The first is just the normal play pretending function (that is, the nim-sequence) of the game, to heap six. The second table shows the corresponding misere pretending function for the partial quotient to heap size 6, and the final table shows the initial portion of the pretending function for the entire game (taken over arbitrarily large heaps).

With these three tables in mind, consider the following question:

When is $4 + 4$ indistinguishable from 6 in **0.123**?

Normal **0.123**

n	1	2	3	4	5	6	7	8	9	10
$G(n)$	$*1$	$*0$	$*2$	$*2$	$*1$	$*0$	$\cdots$	$\cdots$	$\cdots$	$\cdots$

Misere **0.123** to heap 6: $\langle a, b \mid a^2 = 1,\ b^3 = b \rangle$, order 6

n	1	2	3	4	5	6
$\Phi(n)$	a	1	b	b	a	b^2

Complete misere **0.123** quotient, order 20

$$\langle a, b, c, d \mid a^2 = 1,\ b^4 = b^2,\ b^2 c = b^3,\ c^2 = 1,\ b^2 d = d,\ cd = bd,\ d^3 = ad^2 \rangle$$

n	1	2	3	4	5	6	7	8	9	10
$\Phi(n)$	a	1	b	b	a	d^2	1	c	d	$\cdots$

Figure 13. Iterative calculation of misere partial quotients differs in a fundamental way from normal play nim-sequence calculation because sums at larger heap sizes (for example, $8 + 9$) may distinguish between positions that previously were indistinguishable at earlier partial quotients (e.g., $4 + 4$ and 6, to heap size six).

Let's answer the question. In normal play (the top table), $4 + 4$ is indistinguishable from 6 because

$$G(4 + 4) = G(4) + G(4) = *2 + *2 = *0 = G(6).$$

And in the middle table, $4 + 4$ is also indistinguishable from 6, since both sums evaluate to b^2. But in the final table,

$$\Phi(4 + 4) = \Phi(4) + \Phi(4) = b \cdot b = b^2 \neq d^2 = \Phi(6),$$

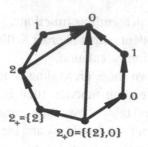

Figure 14. Misere coin-sliding on a directed heptagon with two additional edges. An arbitrary number of coins are placed at the vertices, and two players take turns sliding a single coin along a single directed edge. Play ends when the final coin reaches the topmost (sink) node (labelled o). Whoever makes the last move loses the game. The associated indistinguishability quotient is a commutative monoid of order 14 with presentation $\langle\, a,\ b,\ c \mid a^2 = 1,\ b^3 = b,\ b^2c = c,\ c^3 = ac^2\,\rangle$ and P-positions $\{\, a,\ b^2,\ bc,\ c^2 \}$. See Section 9.5 and Figure 15.

that is, $4 + 4$ can be distinguished from 6 in play of **0.123** *when no restriction is placed on the heap sizes.* In fact, one verifies that the sum $8 + 9$, a position of type cd, distinguishes between $4 + 4$ and 6 in **0.123**.

The fact that the values of partial misere pretending functions may change in this way, as larger heap sizes are encountered, makes it highly desirable to carry out the calculations via computer programs that know how to account for it.

9.5. Quotients from canonical forms. In addition to computing quotients directly from the Guy–Smith code of octal games [GS], *MisereSolver* also can take as input the a canonical form of a misere game G. It then computes the indistinguishability quotient of its closure $\mathrm{cl}(G)$. This permits more general games than simply heap games to be analyzed.

A coin-sliding game. For example, suppose we take $G = \{2_+, o\}$, a game listed in Figure 7. In the output script below, *MisereSolver* calculates that the indistinguishability quotient of $\mathrm{cl}(G)$ is a monoid of order 14 with four P-position types:

```
-- Presentation for 2+0 changed at heap 1 --
Size 2: TAME
P = {a}
Phi = 1 a
-- Presentation for 2+0 changed at heap 2 --
Size 6: TAME
P = {a,b2}
Phi = 1 a b b2
-- Presentation for 2+0 changed at heap 4 --
Size 14: {a,b,c | a2=1,b3=b,b2c=c,c3=ac2}
```

```
P = {a,b2,bc,c2}
Phi = 1 a b c2 c
```

Figure 9.5 shows a coin-sliding game that can be played perfectly using this information. Figure 15 shows how the canonical forms at each vertex correspond to elements of the misere quotient.

Canonical form	0	1	2	2_+	$\{2_+, 0\}$
Quotient element	1	a	b	c^2	c

Figure 15. Assignment of single-coin positions in the heptagon game to misere quotients elements.

10. Outlook

At the time of this writing (December 2005), the indistinguishability quotient construction is only one year old. Several aspects of the theory are ripe for further development, and the misere versions of many impartial games with complete normal play solutions remain to be investigated. We have space only to describe a few of the many interesting topics for further investigation.

10.1. Infinite quotients. Misere quotients are not always finite. Today, it frequently happens that *MisereSolver* will "hang" at a particular heap size as it discovers more and more distinguishable position types. Is it possible to improve upon this behavior and discover algorithms that can handle infinite misere quotients?

Dawson's chess. One important game that seems to have an infinite misere quotient is Dawson's Chess. In the equivalent form **0.07**, (called Dawson's Kayles), Aaron Siegel [PS] found that the order of its misere partial quotients $\mathcal{Q}$ grows as indicated in Figure 16:

Heap size	24	26	29	30	31	32	33	34
$|Q|$	24	144	176	360	520	552	638	$\infty(?)$

Figure 16. Is **0.07** infinite at heap 34?

Since Redei's Theorem (see [P2] for discussion and additional references) asserts that a finitely generated commutative monoid is always finitely presentable, the object being sought in Figure 16 (the misere quotient presentation to heap size 34) certainly exists, although it most likely has a complicated structure of P- and N-positions. New ideas are needed here.

Infinite, but not at bounded heap sizes. Other games seemingly exhibit infinite behavior, but appear to have finite order (rather than simply finitely presentable) partial quotients at all heap sizes. One example is **.54**, which shows considerable structure in the partial misere quotients output by *MisereSolver*. Progress on this game would resolve difficulties with an incorrect solution of this game that appears in the otherwise excellent paper [A3]. Siegel calls this behavior *algebraic periodicity*.

10.2. Classification problem. The *misere quotient classification problem* asks for an enumeration of the possible nonisomorphic misere quotients at each order $2k$, and a better understanding of the category of commutative monoids that arise as misere quotients[7]. Preliminary computations by Aaron Siegel suggest that the number of nonisomorphic misere quotients grows as follows:

Order	2	4	6	8	10	12
# quotients	1	0	1	1	1?	6?

Figure 17. Conjectured number of nonisomorphic misere quotients at small orders.

Evidently misere quotients are far from general commutative semigroups — by comparison, the number of nonisomorphic commutative semigroups at orders 4, 6, and 8 are already 58, 2143, and 221805, respectively [Gril, p. 2].

10.3. Relation between normal and misere play quotients. If a misere quotient is finite, does each of its elements x necessarily satisfy a relation of the form $x^{k+2} = x^k$, for some $k \geq 0$? The question is closely related to the structure of maximal subgroups inside misere finite quotients. Is every maximal subgroup of the form $(\mathbb{Z}_2)^m$, for some m?

At the June 2005 Banff conference on combinatorial games, the author conjectured that an octal game, if misere periodic, had a periodic normal play nim sequence with the two periods (normal and misere) equal. Then Aaron Siegel pointed out that **0.241**, with normal period two, has misere period 10. Must the normal period length *divide* the misere one, if both are periodic?

10.4. Quaternary bounties. Again at the Banff conference, the author distributed the list of wild misere quaternary games in Figure 18.

The author offered a bounty of $25 dollars/game to the first person to exhibit the misere indistinguishability quotient and pretending function of the games in the list. Aaron Siegel swept up 17 of the bounties [PS], but **.3102, .3122, .3123**, and **.3312** are still open.

---

[7] It can be shown that a finite misere quotient has even order [PS].

$(.0122, 1^{20}, 12)$ $(.0123, 1^{20}, 12)$ $(.1023, 2^{1420}, 11)$ $(.1032, 2^{1420}, 12)$

$(.1033, 1^{20}, 11)$ $(.1231, 2^{1420}, 8)$ $(.1232, 2^{1420}, 9)$ $(.1233, 2^{1420}, 9)$

$(.1321, 2^{1420}, 9)$ $(.1323, 2^{1420}, 10)$ $(.1331, 1^{20}, 8)$ $(.2012, 1^{20}, 5)$

$(.2112, 1^{20}, 5)$ $(.3101, 1^{20}, 4)$ $(.3102, 0^{20}, 5)$ $(.3103, 1^{20}, 4)$

$(.3112, 2^{1420}, 7)$ $(.3122, 2^{1420}, 4)$ $(.3123, 1^{31}, 6)$ $(.3131, 2^{1420}, 6)$

$(.3312, 2^{1420}, 5)$

Figure 18. The twenty-one wild four-digit quaternary games (with first wild genus value and corresponding heap size).

10.5. Misere sprouts endgames. Misere Sprouts (see [WW], 2nd edition, Vol III) is perhaps the only misere combinatorial game that is played competitively in an organized forum, the *World Game of Sprouts Association*. It would be interesting to assemble a database of misere sprout endgames and compute the indistinguishability quotient of their misere addition.

10.6. The misere mex mystery. In normal play game computations for heap games, the *mex rule* allows the computation of the heap $n + 1$ nim-heap equivalent from the equivalents at heaps of size n and smaller. The *misere mex mystery* asks for the analogue of the normal play mex rule, in misere play. It is evidently closely related to the partial quotient computations performed by *MisereSolver*.

10.7. Commutative algebra. A beginning at application of theoretical results on commutative monoids to misere quotients was begun in [P2]. What more can be said?

Appendix: Genus theory

We summarize Conway's *genus theory*, first described in [ONAG, chapter 12] and used extensively in *Winning Ways*. It describes a method for calculating whether all the positions of particular game Γ are tame, and how to give a complete analysis of Γ, if so. The genus theory assigns to each position G a particular symbol

$$\mathrm{genus}(G) = \mathrm{G}^*(G) = g^{g_0 g_1 g_2 \cdots}. \tag{A-1}$$

where the g and the g_i's are always nonnegative integers. We'll define this genus value precisely and illustrate how to calculate genus values for some example games G, below.

To look at this in more detail, we need some preliminary definitions before giving definition of genus values.

A.8. Grundy numbers. Let $*k$ represent the nim heap of size k. The *Grundy number* (or *nim value*) of an impartial game position G is the unique number k such that $G + *k$ is a second-player win. Because Grundy numbers may be defined relative to normal or misere play, we distinguish between the *normal play Grundy number* $G^+(G)$ and its counterpart $G^-(G)$, the *misere Grundy number*.

In normal play, Grundy numbers can be calculated using the rules $G^+(0) = 0$, and otherwise, $G^+(G)$ is the least number (from $0,1,2,\ldots$) that is *not* the Grundy number of an option of G (the so-called *minimal excludant*, or *mex*).

When normal play is in effect, every game with Grundy number $G^+(G) = k$ can be thought of as the nim heap $*k$. No information about best play of the game is lost by assuming that G is in fact precisely the nim heap of size k. Moreover, in normal play, the Grundy number of a sum is just the nim-sum of the Grundy numbers of the summands.

The misere Grundy number is also simple to define [p. 140, bottom][ONAG]:

> $G^-(0) = 1$. Otherwise, $G^-(G)$ is the least number (from $0,1,2,\ldots$) which is not the G^--value of any option of G. Notice that this is just like the ordinary "mex" rule for computing G^+, except that we have $G^-(0) = 1$, and $G^+(0) = 0$.

Misere P-positions are precisely those whose first genus exponent is 0.

A.9. Indistinguishability vs misere Grundy numbers.

When misere play is in effect, Grundy numbers can still be defined — as we've already said — but many *distinguishable* games are assigned the *same* Grundy number, and the outcome of a sum is *not* determined by Grundy numbers of the summands. These unfortunate facts lead directly to the apparent great complexity of many misere analyses.

Here is the definition of the genus, directly from [ONAG], now at the bottom of page 141:

> In the analysis of many games, we need even more information than is provided by either of these values [G^+ and G^-], and so we shall define a more complicated symbol that we call the G^*-value, [or *genus*], $G^*(G)$. This is the symbol
>
> $$g^{g_0 g_1 g_2 \cdots}$$
>
> where
>
> $$
> \begin{aligned}
> g &= G^+(G) \\
> g_0 &= G^-(G) \\
> g_1 &= G^-(G+2) \\
> g_2 &= G^-(G+2+2) \\
> \cdots &= \cdots
> \end{aligned}
> $$

where in general g_n is the G⁻-value of the sum of G with n other games all equal to [the nim-heap of size] 2.

At first sight, the genus symbol looks to be an potentially infinitely long symbol in its "exponent." In practice, it can be shown that the g_i's always fall into an eventual period two pattern. By convention, a genus symbol is written down with a finite exponent with the understanding that its final two values repeat indefinitely.

The only genus values that arise in misere Nim are the *tame genera*

$$\underbrace{0^{120}, 1^{031}}$$

Genera of normal play $*0$ (resp, $*1$) Nim positions involving nim heaps of size 1 only

and

$$\underbrace{0^{02}, 1^{13}, 2^{20}, 3^{31}, 4^{46}, \ldots, n^{n(n \oplus 2)}, \ldots}$$

Genera of $*n$ normal-play Nim positions involving at least one nim heap of size ≥ 2.

Figure 19. Correspondence between normal play nim positions and tame genera.

The value of genus theory lies in the following result [ONAG, Theorem 73]:

Theorem: If all the positions of some game Γ have tame genera, the genus of a sum $G + H$ can be computed by replacing the summands by Nim-positions of the same genus values, and taking the genus value of the resulting sum.

In order to apply the theorem to analyze a tame game Γ, a person needs to know several things:

(i) How to compute genus symbols for positions G of a game Γ;
(ii) That every position of the game Γ does have a tame genus;
(iii) The correspondence between the tame genera and Nim positions.

We've already given the correspondence between normal-play Nim positions and their misere genus values, in Figure (19). We'll defer the most complicated part — how to compute genera, and verify that they're all tame — to the next section.

The addition rule for tame genera is not complicated. The first two symbols have the $\mathbb{Z}_2$ addition

$$0^{120} + 0^{120} = 0^{120}$$
$$0^{120} + 1^{031} = 1^{031}$$
$$1^{031} + 1^{031} = 0^{120}$$

Two positions with genus symbols of the form $n^{n(n\oplus 2)}$ add just like Nim heaps of $*n$. For example,

$$2^{20} + 3^{31} = 1^{13}.$$

The symbol 0^{120} adds like an identity, for example:

$$4^{46} + 0^{120} = 4^{46}.$$

When 1^{031} is added to a $n^{n(n\oplus 2)}$, it acts like 1^{13}:

$$4^{46} + 1^{031} = 5^{57}.$$

It has to emphasized that these rules work *only if all positions in play of Γ are known to have tame genus values*. If, on the other hand, even a single position in a game Γ does *not* have a tame genus, the game is wild and *nothing can be said in general about the addition of tame genera*.

A.10. Genus calculation in octal game 0.123. Let's press on with genus theory, illustrating it in an example game, and keeping in mind the end of Chapter 13 in [WW]:

> The misere theory of impartial games is the last and most complicated theory in this book. Congratulations if you've followed us so far...

Genus computations, and the nature of the conclusions that can be drawn from them, are what makes Chapter 13 in *Winning Ways* complicated. In this section we illustrate genus computations by using them to initiate the analysis of a particular wild octal game (**0.123**). Because the game **0.123** is wild, genus theory will *not* lead to a complete analysis of it. A complete analysis can nevertheless be obtained via the indistinguishability quotient construction; for details, see [P2].

The octal game **0.123** can be played with counters arranged in heaps. Two players take turns removing one, two or three counters from a heap, subject to the following additional conditions:

(i) Three counters may be removed from any heap;
(ii) Two counters may be removed from a heap, but only if it has more than two counters; and

+	1	2	3	4	5
0+	1	0	2	2	1
5+	0	0	2	1	1
10+	0	0	2	1	1
15+	$\cdots$				

Figure 20. Normal play nim values of **0.123**.

(iii) One counter may be removed only if it is the only counter in that heap.

Normal play of 0.123. The nim sequence of $\mathbf{0.123}^8$ is periodic of length 5, beginning at heap 5. See Figure 20.

Misere play genus computations for 0.123. We exhibit single-heap genus values of **0.123** in Figure 21. It's possible to prove that this sequence is also periodic of length 5. However, a periodic genus sequence is not the same thing as a complete misere analysis. Let's see what happens instead.

+	1	2	3	4	5
0+	1^{031}	0^{120}	2^{20}	2^{20}	1^{031}
5+	0^{02}	0^{120}	2^{1420}	1^{20}	1^{031}
10+	0^{02}	0^{120}	2^{1420}	1^{20}	1^{031}
15+	$\cdots$				

Figure 21. G*-values of **0.123**.

There are some tame genus symbols in Figure 21. They are

$$0 = 0^{1202020\cdots} = 0^{120}$$
$$1 = 1^{0313131\cdots} = 1^{031}$$
$$2 = 2^{2020202\cdots} = 2^{20}$$

Despite the presence of these tame genera, the game is still wild — the first wild genus value, 2^{1420}, occurs at heap 8. Conway's Theorem 73 on tame games therefore does *not* apply, since it requires *all* positions to have tame genera in order for the game to be treated as misere Nim. We can say nothing about how genera add — even the tame genera — without examining the game more closely.

Here's what we can (and cannot) do with Figure 21.

---

[8] See *Winning Ways*, Chapter 4, p. 97, "Other Take-Away Games;" also Table 7(b), p. 104.

+	h_1	h_2	h_3	h_4	h_5	h_6	h_7	h_8	h_9
h_1	0^{120}	1^{031}	3^{31}	3^{31}	0^{120}	1^{13}	1^{031}	3^{0531}	0^{31}
h_2		0^{120}	2^{20}	2^{20}	1^{031}	0^{02}	0^{120}	2^{1420}	1^{20}
h_3			0^{02}	0^{02}	3^{31}	2^{20}	2^{20}	0^{420}	3^{02}
h_4				0^{02}	3^{31}	2^{20}	2^{20}	0^{420}	3^{02}
h_5					0^{120}	1^{13}	1^{031}	3^{0531}	0^{31}
h_6						0^{02}	0^{02}	2^{20}	1^{13}
h_7							0^{120}	2^{1420}	1^{20}
h_8								0^{120}	3^{02}
h_9									0^{02}

Figure 22. Some genus values of games $h_i + h_j$ in **0.123**.

Single heaps. We *can* determine the outcome class of *single-heap* **0.123** positions. The first superscript in a heap's genus symbol is 0 if and only if that heap size is a P-position. The single heap P-positions of **0.123** therefore occur at heap sizes

$$1, 5, 6, 10, 11, 15, 16, 20, 21, \ldots$$

For example, the genus of the heap of size 7 has its first superscript $= 1$. It is therefore an N-position. The winning move is $7 \to 5$.

Multiple heaps. Using Figure 21, we *cannot* immediately determine the outcome class of **0.123** positions involving *multiple heaps*. However, the figure does provide a basis for investigating multiheap positions. For example, Figure 22 is a table that shows the genera of two-heap positions up to heap size nine.

A.11. Genus calculation algorithm. Here's how the genus of a particular sum $G = h_8 + h_5$ was computed from the earlier single-heap values in Figure 21. First, we rewrote genus(G) in terms of its options:

$$\text{genus}(G) = \text{genus}(h_8 + h_5) = \text{genus}(\{h_6 + h_5, h_5 + h_5, h_8 + h_3, h_8 + h_2\})$$

The genus of a nonempty game $G = \{A, B, \ldots\}$ can be calculated from the genus of its options $A, B, \ldots$ using the *mex-with-carrying algorithm* ($\diamond$ symbols represent positions with no carry):

$$
\begin{aligned}
\text{carry}(\gamma) &= \diamond^{\diamond 05313} \\
\text{carry}(\gamma \oplus 1) &= \diamond^{\diamond 14202} \\
\text{genus}(h_6 + h_5) &= 1^{131313\ldots} \\
\text{genus}(h_5 + h_5) &= 0^{120202\ldots} \\
\text{genus}(h_8 + h_3) &= 0^{420202\ldots} \\
\text{genus}(h_8 + h_2) &= \underline{2^{142020\ldots}}
\end{aligned}
$$

$$\text{genus}(G) = 3^{053131\cdots}$$

The result $\text{genus}(G) = 3^{053131\cdots} = 3^{0531}$ was computed columnwise, working from left to right. First, the "base" and "first superscript" results

$$G^+(G) = \text{mex}(\{1, 0, 0, 2\}) = 3$$

and

$$G^-(G) = \text{mex}(\{1, 1, 4, 1\}) = 0$$

were computed from the corresponding four positions in each option of G, with no carries present. The "carry out" is then $\gamma = 0$. The second superscript result

$$G^-(G + *2) = \text{mex}(\{3, 2, 2, 4, \mathbf{0}, \mathbf{1}\}) = 5$$

involved a similar computation, but with two *carry values*

$$\{\gamma, \gamma \oplus 1\} = \{0, 1\}.$$

thrown into the mex calculation (they're shown in bold). See the more complete description of this algorithm in the section titled *"But What if They're Wild?" asks the Bad Child* ([WW], page 410). It's also illustrated in [ONAG, p. 143].

References

[A1] D. T. Allemang, "Machine Computation with Finite Games," MSc Thesis, Trinity College (Cambridge), 1984. See http://www.plambeck.org/oldhtml/mathematics/games/misere/allemang/index.htm.

[A2] D. T. Allemang, "Solving misere games quickly without search," unpublished research (2002).

[A3] D. T. Allemang, "Generalized genus sequences for misere octal games," International Journal of Game Theory **30** (2002) 4, 539–556.

[AS2005] Aaron Siegel, *MisereSolver*. (A standalone Java language program for indistinguishability quotient calculation in misere impartial games). Private communication, August 2005.

[B1902] Charles L. Bouton, Nim, a game with a complete mathematical theory, *Ann. Math., Princeton (2)*, **3** (1901–02) 35–39.

[D] T. R. Dawson (1935) "Caissa's Wild Roses," in *Five Classics of Fairy Chess*, Dover Publications Inc, New York (1973).

[F1] Thomas S Ferguson, "A Note on Dawson's Chess," unpublished research note. See http://www.math.ucla.edu/~tom/papers/unpublished/DawsonChess.pdf.

[F2] Thomas S. Ferguson, "Misere Annihilation Games," *Journal of Combinatorial Theory*, Series A **37**, 205–230 (1984).

[F3] Thomas S. Ferguson, "On Sums of Graph Games with Last Player Losing," *Int. Journal of Game Theory* Vol. 3, Issue 3, pp. 159–167.

[Fraenkel] Aviezri S. Fraenkel, *Combinational Games: Selected Bibliography with a Succinct Gourmet Introduction*, Electronic Journal of Combinatorics, #DS2

[Gril] P. A. Grillet, *Commutative Semigroups*. Kluwer Academic Publishers, 2001. ISBN 0–7923–7067–8.

[GrS1956] P. M. Grundy and Cedric A. B. Smith, Disjunctive games with the last player losing, *Proc. Cambridge Philos. Soc.*, **52** (1956) 527–533.

[Guy89] Richard K Guy, *Fair Game: How to Play Impartial Combinatorial Games*, COMAP, Inc., Arlington, MA 02174.

[Guy91] R. K. Guy (1991), Mathematics from fun & fun from mathematics: an informal autobiographical history of combinatorial games, in: *Paul Halmos: Celebrating 50 Years of Mathematics* (J. H. Ewing and F. W. Gehring, editors). Springer, New York, pp. 287–295.

[Guy] R. K. Guy, "Unsolved Problems in Combinatorial Games," in *Games of No Chance*, R. J. Nowakowski (editor). Cambridge University Press, 1994.

[GN] Richard K. Guy and Richard J. Nowakowski, "Unsolved Problems in Combinatorial Games," in *More Games of No Chance*, MSRI Publications, **42** 2002.

[GS] R. K. Guy and C. A. B. Smith (1955) "The G-values of various games," *Proc Camb. Phil. Soc.* **52**, 512–526.

[ONAG] J. H. Conway (1976) *On Numbers and Games*, Academic Press, New York.

[P] Thane E. Plambeck, *Misere Games*. (Web pages devoted to problems, computer software, and theoretical results in impartial combinatorial games in misere play). See http://www.plambeck.org/oldhtml/mathematics/games/misere.

[P1] Thane E. Plambeck, "Daisies, Kayles, and the Sibert–Conway decomposition in misere octal games", *Theoretical Computer Science* (Math Games) **96**, pp. 361–388.

[P2] Thane E. Plambeck, "Taming the Wild in Impartial Combinatorial Games", IN-TEGERS: Electronic J. of Combinatorial Number Theory 5 (2005) #G5, 36 pages. Also available at http://arxiv.org/abs/math.CO/0501315.

[PS] Thane E. Plambeck and Aaron Siegel, "The Φ-values of various games", in preparation.

[S1] Aaron Siegel, personal communication, November 2005.

[SC] W. L. Sibert and J. H. Conway, "Mathematical Kayles," *International Journal of Game Theory* (1992) 237–246.

[Si] William L. Sibert, *The Game of Misere Kayles: The "Safe Number" vs "Unsafe Number" Theory*, unpublished manscript, October 1989.

[WW] E. R. Berlekamp, J. H. Conway and R. K. Guy [1982], *Winning Ways for your Mathematical Plays*, vol. I and II, Academic Press, London. 2nd edition: vol. 1 (2001), vol. 2 (2003), vol. 3 (2003), vol. 4 (2004), A K Peters, Natick, MA; German translation: *Gewinnen, Strategien für Mathematische Spiele*, Vieweg, Braunschweig (four volumes), 1985.

[Y] Yohei Yamasaki, "On misere Nim-type games," *J. Math. Soc. Japan* **32** No. 3, 1980, pp. 461–475.

THANE E. PLAMBECK
tplambeck@gmail.com

Games of No Chance 3
MSRI Publications
Volume **56**, 2009

Coping with cycles

AARON N. SIEGEL

ABSTRACT. *Loopy games* are combinatorial games in which repetition is permitted. The possibility of nonterminating play inevitably raises difficulties, and several theories have addressed these by imposing a variety of assumptions on the games under consideration. In this article we survey some significant results on partizan loopy games, focusing on the theory developed in the 1970s by Conway, Bach and Norton.

1. Introduction

A substantial portion of combinatorial games research focuses on games without repetition — those that are guaranteed to terminate after some finite number of moves. Such games are highly tractable, both theoretically and computationally, and the full force of the classical partizan theory can be brought to bear upon them. The great success of this theory has produced a vast body of splendid results, but it has also resulted in an unjust neglect of games *with* repetition.

In the late 1970s, John Conway and his students, Clive Bach and Simon Norton, introduced a disjunctive theory of partizan games with repetition — called *loopy games* because their game graphs may contain cycles. They showed that in many interesting cases, such games admit canonical forms. The past few years have witnessed some significant applications of this theory, to games as diverse as Fox and Geese, Hare and Hounds, Entrepreneurial Chess, and one-dimensional Phutball. In light of these advances, it is time for a reappraisal of the theory with an eye to the future.

A short history. The first disjunctive theory of loopy games is due to Cedric A. B. Smith and Aviezri Fraenkel. They showed (independently) that the usual Sprague–Grundy theory generalizes well to loopy games. In particular, many impartial loopy games are equivalent to nimbers, and the remainder are characterized by their nimber-valued options. Over a period of several decades,

Fraenkel and his students explored this theory in depth. They constructed numerous examples and studied both their solutions and their computational complexity.

The partizan theory was introduced by Robert Li, who studied *Zugzwang games*, those in which it is a disadvantage to move. Li showed that Zugzwang games are completely characterized by a certain pair of ordinary numbers. Soon thereafter, Conway, Bach and Norton extended Li's theory to a much broader class of games. They showed that many loopy games γ — including most positions encountered in actual play — decompose into a pair of much simpler games, called the *sides* of γ. Their theory was published in the first edition of *Winning Ways*, together with a handful of examples, most notably the children's game *Fox and Geese*.

Intermittent progress was made over the next twenty years, but it was not until 2003 that loopy games saw a full-fledged revival. John Tromp and Jonathan Welton had recently detected an error in the *Winning Ways* analysis of Fox and Geese, and Berlekamp set out to repair it. His corrected analysis appears in the second edition of *Winning Ways*. Berlekamp's effort led to the development of new algorithms, which in turn paved the way for a re-examination of several other loopy games mentioned in *Winning Ways*.

In this survey, the *Winning Ways* theory is introduced first, so that earlier developments — notably those of Smith, Fraenkel and Li — can be presented in the modern context. Section 2 is an expository overview of some interesting properties of loopy games, with a focus on Fox and Geese. Much of that material is formalized in Section 3, and in Section 4 we tackle the theory of sides as it appears in *Winning Ways*. Each of these sections also addresses some related topics. Section 5 discusses several specific partizan games that have been successfully analyzed with this theory. In Section 6, we discuss the generalized Sprague–Grundy theory and its relationship to partizan games. Section 7 gives an overview of the Smith–Flanigan results on conjunctive and selective sums. Finally, in Section 8 we survey the development of algorithms for loopy games.

Two topics are notably absent from this survey. The first is the immense body of work on loopy impartial games, assembled over several decades by Aviezri Fraenkel and his students. Their work includes an extensive theoretical and algorithmic analysis of the generalized Sprague–Grundy theory; many beautiful examples; and connections to other fields, including combinatorial number theory and error-correcting codes. The present article is focused mainly on the partizan theory, and so does not do justice to their achievement; a forthcoming book by Fraenkel surveys this material in far more detail and accuracy than we could hope to achieve here.

The combinatorial theory of Go is another major omission. This might seem surprising, since Go is without question the most significant loopy game that has been subjected to a combinatorial analysis. However, there is good reason for its omission. Although Go is fundamentally disjunctive in nature, its unique *koban* rule implies an interrelationship between all components on the board. This gives rise to a rich and fascinating temperature theory that has been explored by many researchers, including Berlekamp, Fraser, Kao, Kim, Müller, Nakamura, Snatzke, Spight, and Takizawa, to list just a few. However, this temperature theory appears to be incompatible with the canonical theory that is the focus of our discussion. Because Go is so prominent, its body of results is vast; yet because it is so singular, these appear disconnected from other theories of loopy games. Thus while Go desperately deserves its own survey, this article is not the appropriate place for it.

This apparent dichotomy also raises the first — and arguably the most important — open problem of this survey.

OPEN PROBLEM. *Formulate a temperature theory that applies to all loopy games.*

Notation and preliminaries. Following *Winning Ways*, we denote loopy games by loopy letters $\gamma, \delta, \alpha, \beta, \ldots$. If γ is loopy, we define the associated *game graph* $\mathcal{G}$ as follows. $\mathcal{G}$ has one vertex, V_α, for each subposition α of γ (including γ itself), and there is an edge directed from V_α to V_β just if there is a legal move from α to β. When γ is partizan, we color the edge bLue, Red, or grEen, depending on whether *Left*, *Right*, or *Either* player may move from α to β.

An abbreviated notation is often useful. In many loopy games, repetition is limited to simple pass moves. In such cases we can borrow the usual brace-and-slash notation used to describe loopfree games, enhanced with the additional symbol **pass**. For example, if we write $\gamma = \{0 \mid \textbf{pass}\}$, we mean that Left has a move from γ to 0, and that Right has a move from γ back to γ. Likewise, if $\delta = \{0 \mid \textbf{pass} \parallel -1\}$, this means that Right has a move from δ to -1, and that Left has a move from δ to $\{0 \mid \textbf{pass}\} = \gamma$. For comparison, the game graph of δ is shown in Figure 1.

The main complication introduced by loopy games is the possibility of non-terminating play. The simplest way to resolve this issue is to declare all infinite

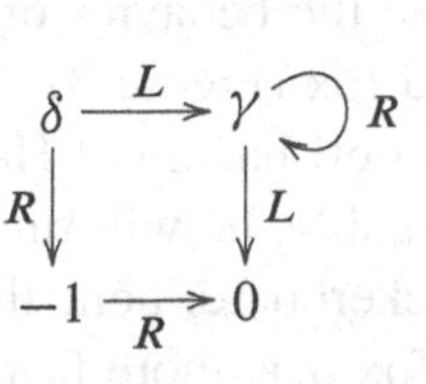

Figure 1. The game graph of $\delta = \{0 \mid \textbf{pass} \parallel -1\}$.

plays *drawn*, and this will be our assumption throughout Sections 2 and 3. We
will often say that a player *survives* the play of a game if he achieves at least a
draw.

2. Loops large and small

Fox and Geese is an old children's game played on an ordinary checkerboard.
Four geese are arranged against a single fox as in Figure 2. The geese (controlled
by Left) move as ordinary checkers, one space diagonally in the forward direction, while the fox (controlled by Right) moves as a checker king — one space
in any diagonal direction. Neither animal may move onto an occupied square,
and there are no jumps or captures. The geese try to trap the fox, while the fox
tries to escape.

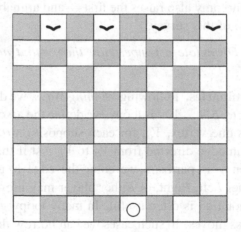

Figure 2. The usual starting position in Fox and Geese.

Fox and Geese has a curious feature: the game must end if played in isolation,
because the geese will eventually run out of moves, whether or not they trap the
fox. However, from a combinatorial perspective the game is certainly loopy.
The *fox* may return to a previous location, and this results in local repetition if
Left's intervening moves occur in a different component.

Before turning to a more formal treatment of loopy games and canonical
forms, let us briefly investigate the behavior of Fox and Geese. Consider first
the happy affair of an escaped fox (Figure 3). The geese have exhausted their
supply of moves, and though Left has a tall Hackenbush stalk at his disposal,
his situation is hopeless. Inevitably, he will run out of moves, and the fox will
still be dancing about the checkerboard, none the worse for wear.

It is clear that an escaped fox α is more favorable to Right than any (finite)
Hackenbush stalk we might assemble. In an informal sense, we have established

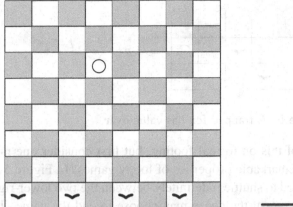

Figure 3. Hackenbush is hopeless facing an escaped fox.

that

$$\alpha < -n,$$

for every integer n.

It is equally clear that the fox's precise location on the checkerboard is irrelevant; all that matters is that she has an indefinite supply of moves at her disposal. The many distinct positions that arise as she moves about the board are all equivalent, and α can be written as a single pass move for Right: $\alpha = \{ \mid \textbf{pass} \}$, with game graph shown in Figure 4.

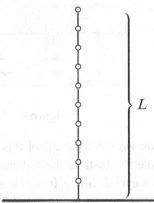

Figure 4. The games $\alpha = \textbf{off}$, $\beta = \textbf{on}$, and $\delta = \textbf{dud}$.

The game α is normally known as **off**, and its inverse — from which *Left* can pass — is naturally enough called **on**.[1] One might expect that **on** + **off** = 0, but this is not the case: in their sum either player may pass, so that **on** + **off** is a draw, while 0 is a second-player win.

In fact it is easy to see that **on** + **off** + γ is drawn, no matter what game γ we include in the sum: both players have an inexhaustible supply of moves; so neither has anything to fear. Therefore **on** + **off** is a deathless universal draw, which we abbreviate by **dud**, and we have the identity

$$\textbf{dud} + \gamma = \textbf{dud}$$

for all γ.

---

[1] The name is set-theoretic in origin: **ON** is standard notation for the class of all ordinal numbers, and the game **on** behaves much like an ill-founded relation, an entity that exceeds all the ordinals.

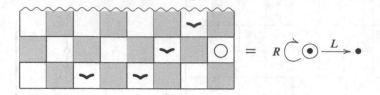

Figure 5. A trapped fox has value **over**.

Soon we shall put all of this on formal footing, but first consider one more example to illustrate the remarkable properties of loopy games. In Figure 5 the Fox is trapped. She is forced to shuttle indefinitely between the two lower-right-hand spaces, and at any moment the geese may choose to end the game. It is clear this game is positive, for Left may win at any time. Its abbreviated graph, known as **over**, is also pictured in Figure 5.

Just how large is **over**? The reader might wish to confirm that, for any n,

$$n \cdot \uparrow < \textbf{over} < \frac{1}{2^n},$$

by showing that Left can win the appropriate differences. **over** is larger than every loopfree infinitesimal, but smaller than every positive number.

3. Stoppers

When γ is loopy, there are typically three possible outcomes: Left wins (if he gets the last move); Right wins (if she gets the last move); or a Draw (if play never terminates). This divides loopy games into nine outcome classes, since the outcome might depend on who moves first:

		Left moves first		
		Left wins	Draw	Right wins
Right	Left wins	$\mathscr{L}$	$\hat{\mathscr{P}}$	$\mathscr{P}$
moves	Draw	$\hat{\mathscr{N}}$	$\mathscr{D}$	$\check{\mathscr{P}}$
first	Right wins	$\mathscr{N}$	$\check{\mathscr{N}}$	$\mathscr{R}$

We denote by $o(\gamma)$ the outcome class of γ. The outcome classes are naturally partially-ordered as shown in Figure 6.

As always in combinatorial game theory, we define equality by indistinguishability in sums:

$$\gamma = \delta \quad \text{if} \quad o(\gamma + \alpha) = o(\delta + \alpha) \text{ for all loopy games } \alpha.$$

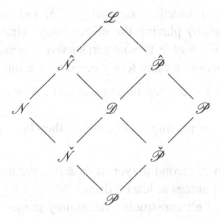

Figure 6. The partial order of loopy outcome classes.

As remarked in Section 2, it is *not* always true that $\gamma - \gamma = 0$. Second player can always assure a draw by playing the mirror image strategy, but in general this does not guarantee a win. For this reason, loopy games do not form a group, and we are forced to consider instead the monoid of loopy games, equipped with the natural partial order:

$$\gamma \geq \delta \quad \text{if} \quad o(\gamma + \alpha) \geq o(\delta + \alpha) \text{ for all loopy games } \alpha.$$

The theory of loopy games is motivated by two fundamental questions.

- Does every loopy game admit a unique simplest form, analogous to the canonical form of a loopfree game?
- Can one specify an *effective* equivalent definition of $\gamma \geq \delta$?

It turns out that both of these questions are easiest to answer for an important special class of loopy games called *stoppers*. They can also be resolved quite nicely for a larger class, the *stopper-sided games*, that encompasses most positions arising in studies of actual (playable) games.

A loopy game γ is a *stopper* if there is no infinite *alternating* sequence of play proceeding from any subposition of γ. The games **on**, **off**, and **over**, which we met in Section 2, are all stoppers, but **dud** is not. Further, *every* position that arises in Fox and Geese is a stopper, since the geese are constrained to make finitely many moves throughout the game.

If γ is a stopper, then γ is guaranteed to terminate when played in isolation. This property is central to the following characterization.

THEOREM 1 (CONWAY). *Let γ, δ be stoppers. Then*

$$\gamma \geq \delta \text{ iff Left, playing second, can survive } \gamma - \delta.$$

PROOF. For the forward direction, suppose $\gamma \geq \delta$, and let $\alpha = -\delta$. Certainly Left can survive $\delta + \alpha$, by playing the mirror image strategy; then it follows directly from the definition of $\geq$ that he can survive $\gamma + \alpha$.

For the reverse direction, fix any loopy game α. We must show that:

(i) If Left can survive $\delta + \alpha$ playing first (second), then he can survive $\gamma + \alpha$ playing first (second).

(ii) If Left can win $\delta + \alpha$ playing first (second), then he can win $\gamma + \alpha$ playing first (second).

First suppose that Left is second player in case (i). We describe a strategy for playing $\gamma + \alpha$ that guarantees at least a draw.

Before play begins, Left constructs two dummy games: one copy of $\delta + \alpha$, and one copy of $\gamma - \delta$. Whenever Right makes a move in $\gamma + \alpha$, Left copies the move to the appropriate dummy game: if Right moves in the γ component, Left copies the move to $\gamma - \delta$; while if Right moves in the α component, Left copies the move to $\delta + \alpha$.

Now Left responds with his survival move in the dummy game. If this move is in the δ or $-\delta$ component, Left immediately makes the mirror image move in the *other* dummy game, and responds accordingly. Successive responses in the δ and $-\delta$ components produce an alternating sequence of moves proceeding from a subposition of δ. Since δ is a stopper, this cannot go on forever, and eventually Left's response must occur in the γ or α component. At that point Left copies it back to $\gamma + \alpha$ and awaits Right's next move.

If Left keeps to this strategy, he will never run out of moves in $\gamma + \alpha$. This proves case (i). In case (ii), Left uses the same technique, but follows his *winning* strategy in $\delta + \alpha$. This guarantees that eventually, $\delta + \alpha$ will reach a terminal position. At that point the α component of $\gamma + \alpha$ is terminal; therefore, since γ is a stopper, it must eventually terminate as well. So the game will necessarily end, and since Left has survived, Right cannot have made the last move.

If Left is first player, the argument is exactly the same. He makes his initial move in the $\delta + \alpha$ dummy component, according to his first-player survival (or winning) strategy for $\delta + \alpha$, and continues accordingly.   □

Stoppers also admit a clean canonical theory: if γ is a stopper, then we can eliminate dominated options and bypass reversible ones, just as for loopfree games. The proofs are straightforward applications of Theorem 1.

A stopper is in *simplest form* if it has no dominated or reversible options.

THEOREM 2 (CONWAY). *If γ and δ are stoppers in simplest form with $\gamma = \delta$, then for every γ^L there is a δ^L with $\gamma^L = \delta^L$, and vice versa; and likewise for Right options.*

PROOF. See [5, Section 10].   □

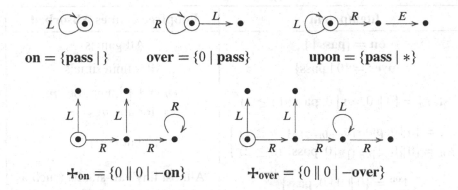

Figure 7. Simple stoppers.

Several simple stoppers. Figure 7 shows some of the simplest stoppers in canonical form. The reader might wish to verify some of their remarkable properties, which clarify their behavior in the partial-order of games:

- $\mathbf{on} \geq \gamma$ for all γ.
- $n \cdot \uparrow < \mathbf{over} < 2^{-n}$ for all n.
- $\uparrow^{\to n} < \mathbf{upon} < \uparrow^{\to n} + \uparrow^{n}$ for all n.
- $+_{\mathbf{over}} \lesssim \gamma$ for every all-small game $\gamma > 0$, but $+_{\mathbf{over}} > +_x$ for every number $x > 0$.
- $+_{\mathbf{on}}$ is the smallest positive game: if $\gamma > 0$, then $+_{\mathbf{on}} \leq \gamma$.

With the exception of $+_{\mathbf{over}}$, all of these values arise frequently in playable games. Also common is $\mathbf{upon} + *$, which has the canonical form $\{0, \mathbf{pass} \mid 0\}$.

In all of these examples, the only loops are simple pass moves (1-cycles). Stoppers with longer cycles exist, but are much less common in nature. A typical example is the game τ shown in Figure 8, which has a 4-cycle in canonical form.

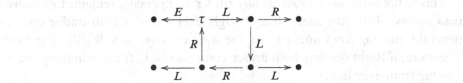

Figure 8. A stopper with a canonical 4-cycle.

Stoppers in canonical form can never have 2-cycles or 3-cycles; see [30] (this volume) for a proof, together with examples of stoppers with canonical n-cycles for all $n \geq 4$.

Idempotent	Loopfree Games Absorbed
$\textbf{on} = \{\textbf{pass} \mid \}$	All games
$\textbf{over} = \{0 \mid \textbf{pass}\}$	All infinitesimals
$\textbf{star}_n = \{0 \parallel 0, *n \mid 0, \textbf{pass}\} \; (n \geq 2)$	$*n$ and $\uparrow^2$, but not $*m$ for any $m \neq n$
$\begin{aligned}\mathscr{I}_n &= \{0 \parallel 0, \textbf{pass} \mid 0, \downarrow_{[n-2]}*\} \; (n \geq 2) \\ \mathscr{J}_n &= \{0 \parallel 0, \downarrow_{[n-1]}* \mid 0, \textbf{pass}\} \; (n \geq 2)\end{aligned}$	$\uparrow^n$ but not $\uparrow^{n-1}$
$\uparrow^{\textbf{on}} = \{0 \parallel 0 \mid 0, \textbf{pass}\}$	"Almost tiny" all-smalls (such as $\{0\|0\|\Downarrow\}$), but not $\uparrow^n$ for any n
$+_{\textbf{over}} = \{0 \parallel 0 \mid \textbf{under}\}$	All tinies, but no all-smalls
$+_{x\textbf{under}} = \{0 \parallel 0 \mid -x\textbf{over}\} \; (x > 0)$	$+_{x\downarrow n}$ for all n, but not $+_{x-2^{-n}}$
$\mathscr{T}_x = \{0 \parallel 0 \mid -x, \textbf{pass}\} \; (x > 0)$	$+_y$ for all $y > x$, but not $+_x$
$+_{x\textbf{over}} = \{0 \parallel 0 \mid -x\textbf{under}\} \; (x > 0)$	$+_{x+2^{-n}}$ for all n, but not $+_{x\uparrow n}$
$+_{\textbf{on}} = \{0 \parallel 0 \mid \textbf{off}\}$	None (except 0)

Figure 9. A variety of idempotents.

Idempotents. It is easy to see that $\textbf{on} + \textbf{on} = \textbf{on}$: certainly $\textbf{on} + \textbf{on} \geq \textbf{on}$, but we also know that $\textbf{on} \geq \gamma$ for all γ. Slightly less obvious is the fact that $\textbf{over} + \textbf{over} = \textbf{over}$, and here Theorem 1 is useful. To show that $\textbf{over} + \textbf{over} \leq \textbf{over}$, we need simply exhibit a second-player survival strategy for Left in

$$\textbf{over} + \textbf{under} + \textbf{under},$$

where $\textbf{under} = -\textbf{over} = \{\textbf{pass} \mid 0\}$.

This is not difficult: so long as any **under** components remain, Left makes pass moves. This guarantees that, if Right ever destroys both **under** components (by moving from **under** to 0), the **over** component will still be present. Therefore, if Right destroys both **under** components, Left can win the game by moving from **over** to 0.

This example illustrates a striking feature of the monoid of loopy games: the presence of explicit idempotents. Figure 9, reproduced from [27], lists many more. Each idempotent ι is listed together with some of the loopfree games that it absorbs (where ι *absorbs* γ if $\iota + \gamma = \iota$). It's also worth noting that each idempotent ι in Figure 9 has a "negative variant" $-\iota$ and a "neutral variant" $\iota - \iota$, both of which are also idempotents (though of course, $\iota - \iota$ is not a stopper).

Berlekamp [2] describes several other idempotents that do not appear to have explicit representations as loopy games. These include $\stackrel{\star}{\star}$ and $\mathscr{E}_t$, which play central roles in the atomic weight and orthodox theories, respectively. It would be interesting to describe a formal system that encompasses these in addition to the idempotents of Figure 9.

Pseudonumbers. The *psuedonumbers* form an interesting subclass of infinite stoppers.

DEFINITION 3. A stopper x is said to be a *pseudonumber* if, for every follower y of x (including x itself), we have $y^L < y^R$ for all y^L, y^R.

So a surreal number is just a well-founded pseudonumber. It is not hard to show that x is a pseudonumber if and only if, for every follower y of x, each $y^L \leq y$ and $y \leq$ each y^R. Then as a consequence of Li's Theorem (Theorem 9 in Section 4, below), the only *finite* pseudonumbers are **on**, **off**, and the dyadic rationals and their sums with **over** and **under**. However, there are many infinite pseudonumbers. A typical example is the game

$$\widehat{\mathbb{Z}} = \{0, 1, 2, \ldots \mid \textbf{pass}\} = \omega : \textbf{off}.$$

It is not hard to check that $\widehat{\mathbb{Z}} \geq n$ for any integer n. Furthermore, it is the *least* pseudonumber with this property: if $y \geq n$ for all n, then $y \geq \widehat{\mathbb{Z}}$. Therefore $\widehat{\mathbb{Z}}$ is a least upper bound for the integers. This generalizes:

THEOREM 4. *The pseudonumbers are totally ordered by $\geq$. Furthermore, every set $X = \{x, y, z, \ldots\}$ of pseudonumbers has a least upper bound, given by*

$$\widehat{X} = \{x, y, z, \ldots \mid \textbf{pass}\} = \{x, y, z, \ldots \mid\} : \textbf{off}.$$

PROOF. See [27, Section 1.8]. □

Contrast this with surreal numbers, which certainly do *not* admit tight bounds. However, while they acquire some analytic structure, pseudonumbers lose the rich algebraic structure of the surreal numbers: they are not even closed under addition, since (say) **on** + **off** is not a stopper.

Pseudonumbers might seem fanciful, but astonishingly, Berlekamp and Pearson recently discovered positions in Entrepreneurial Chess with offside $\widehat{\mathbb{Z}}$ (see Section 5 for a description). Like all good numbers, $\widehat{\mathbb{Z}}$ also makes an appearance in Blue-Red Hackenbush (Figure 10).

4. Sides

As we have seen, stoppers generalize the canonical theory of loopfree games in a straightforward way. Most loopy games, however, are not stoppers.

A typical example is the game *Hare and Hounds*, which has experienced occasional bouts of popularity dating back to the late nineteenth century. The

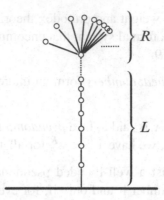

Figure 10. A Blue-Red Hackenbush position of value $\widehat{\mathbb{Z}} = \omega$: **off**.

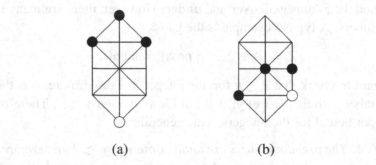

(a)   (b)

Figure 11. Hare and Hounds: (a) the starting position; (b) an endgame position.

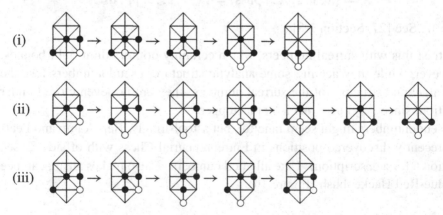

Figure 12. The position of Figure 11(b) is a second-player win.

game can played on an $n \times 3$ board for any odd $n \geq 5$; the starting position on the 5×3 board is shown in Figure 11(a).

The play resembles Fox and Geese. Left controls three hounds (black circles) and Right the lone hare (white circle). Each player, on his turn, may move any one of his units to an adjacent unoccupied intersection. The only restriction is that the hounds may never retreat — they can only advance or move sideways. There are no jumps or captures. The hounds win if they trap the hare (that is, if it is Right's turn and she has no moves available); the hare wins if this never happens.

Since the hounds are allowed to move sideways, Hare and Hounds is not always a stopper. It has another notable feature: if play never terminates, then the game is declared a win for Right. This differs from the other games we have studied, in which infinite plays are drawn. However, we will see that it actually makes the game simpler, since it causes many positions to reduce to stoppers that otherwise would not.

For example, consider the endgame position γ of Figure 11(b). If Right makes either of her available moves, then the hounds can certainly trap her; see Figure 12(i) and (ii). Conversely, if Left moves first, then the hare can evade capture indefinitely by following the pattern shown in Figure 12(iii). (The reader might wish to verify that if the *hounds* ever deviate from this pattern, then the hare can escape outright.) Therefore γ is a second-player win, and we conclude that $\gamma = 0$.

In the late 1970s, Conway, Bach and Norton made a breakthrough in the study of loopy games [5]. They observed, first of all, that games such as Hare and Hounds — where infinite plays are wins for one of the players — can often be brought into the theory of stoppers in a coherent way. Furthermore, their presence actually simplifies the analysis of games where infinite plays are drawn.

To understand this relationship, let γ be an arbitrary loopy game with infinite plays drawn, and suppose we wish to know whether Left can win γ. Then we might as well assume that infinite plays are wins for Right. Likewise, if we wish to know whether Left can survive γ, then we might as well assume that infinite plays are wins for Left. Therefore, we can determine the outcome class of γ by considering each of these two variants in turn. As it turns out, the variants often reduce to stoppers, even when γ itself does not; and in such cases, this reduction yields a substantial simplification.

Therefore, we now drop the assumption that all infinite plays are drawn. We assume that each game γ comes equipped with one of three winning conditions: Left wins infinite plays; Right wins infinite plays; or infinite plays drawn. We say that γ is *free* if infinite plays are draws and *fixed* otherwise.

When γ is free, we denote by γ^+ and γ^- the matching fixed games with infinite plays redefined as wins for Left and Right, respectively. When γ is fixed, we simply put $\gamma^+ = \gamma^- = \gamma$.

If infinite play occurs in a sum

$$\alpha + \beta + \cdots + \gamma,$$

we assume that Left (Right) wins the sum if he wins on *every* component in which play is infinite. If there are any draws, or if several components with infinite play are split between the players, the outcome of the sum is a draw.

When we consider the definition of $\geq$, we suppose now that α ranges over all fixed games in addition to free ones:

$$\gamma \geq \delta \quad \text{if} \quad o(\gamma + \alpha) \geq o(\delta + \alpha) \text{ for all } \textit{fixed or free} \text{ loopy games } \alpha.$$

The main result is the following, called the *Swivel Chair Theorem* in Winning Ways. It is a direct generalization of Theorem 1.

THEOREM 5 (SWIVEL CHAIR THEOREM). *The following are equivalent, for any loopy games* γ, δ:

(i) $\gamma \geq \delta$;
(ii) *Left, playing second, can survive both* $\gamma^+ - \delta^+$ *and* $\gamma^- - \delta^-$.

PROOF. See [3, Chapter 11] or [5, Section 2]; it's very similar to the proof of Theorem 1. $\qquad\square$

Note the key implication of Theorem 5: how γ compares with other games depends only on γ^+ and γ^-. Thus when γ^+ and γ^- are equivalent to stoppers s^+ and t^-, the behavior of γ is completely characterized by s and t. In such cases we call s and t the *sides* of γ (the *onside* and *offside* respectively), and we say that γ is *stopper-sided*. It is customary to write

$$\gamma = s \,\&\, t,$$

and with s and t in simplest form, this should be regarded as a genuine canonical representation for γ.

For example, consider the game **dud** = {**pass** | **pass**}. We know that $\mathbf{on}^+ \geq \mathbf{dud}^+$ (since $\mathbf{on}^+$ is the largest game of all). But also, Left can survive the game

$$\mathbf{dud}^+ - \mathbf{on}^+$$

by passing indefinitely in the **dud** component, where he wins infinite plays. We conclude that $\mathbf{dud}^+ = \mathbf{on}^+$, and by a symmetric argument $\mathbf{dud}^- = \mathbf{off}^-$. This gives the identity

$$\mathbf{dud} = \mathbf{on} \,\&\, \mathbf{off}.$$

If $\gamma = s \,\&\, t$, then the outcome class of γ is determined by those of s and t. Since s and t are stoppers, their outcomes fall into the usual classes: positive, negative, fuzzy or zero. This yields a total of sixteen possibilities for γ. However, since $\gamma^+ \geq \gamma^-$, we know that $s^+ \geq t^-$; and since s and t are stoppers, this implies that $s \geq t$. This restriction rules out seven possibilities, leaving the remaining nine in one-to-one correspondence with the nine outcome classes discussed in Section 3. This correspondence is summarized in Figure 13.

	s			
	> 0	$\not\gtrless 0$	$= 0$	< 0
< 0	$\mathscr{D}$	$\check{\mathscr{N}}$	$\check{\mathscr{P}}$	$\mathscr{R}$
$\not\gtrless 0$	$\hat{\mathscr{N}}$	$\mathscr{N}$	$-$	$-$
$t \quad = 0$	$\hat{\mathscr{P}}$	$-$	$\mathscr{P}$	$-$
> 0	$\mathscr{L}$	$-$	$-$	$-$

Figure 13. The outcome class of $\gamma = s$ & t is determined by those of s and t.

The sides of γ therefore carry a great amount of information. Given their applicability, it is natural to ask how they might be computed in general. *Winning Ways* introduced a method called *sidling* that yields a sequence of increasingly good approximations to the sides of γ. Sometimes this sequence converges to the true onside and offside; but more often than not, it fails to converge. Nonetheless, sidling has been applied to obtain some interesting results, notably by David Moews in his 1993 thesis [21] and a subsequent article on Go [22].[2]

More recent discoveries include effective methods for computing sides (when they exist); see [30] in this volume for discussion.

Carousels. Stopper-sided decompositions are both useful and extremely common. However, there do exist loopy games that are not stopper-sided. In the 1970s, Clive Bach produced the first example of such a game, known as *Bach's Carousel*, by specifying its game graph explicitly. Much more recently, similar "carousels" have been discovered on 11×3 boards in Hare and Hounds. See Figure 14 for an example and [27] for further discussion.

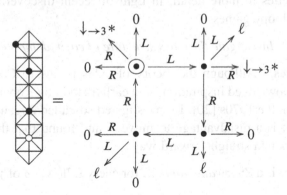

Figure 14. A carousel in Hare and Hounds. Here $\ell = \{0, \downarrow_{\to 2}*|0, \downarrow_{\to 2}*\}$.

---

[2]In order to bring Go into the canonical theory, Moews considered Go positions together with explicitly kobanned moves. Means and temperatures as defined by Berlekamp cannot be recovered from the resulting canonical forms. Nonetheless, Moews' analysis yields interesting information about Go positions that is not captured by thermography.

What about sums of stoppers? *Winning Ways* gives an example, due to Bach, of certain *infinite* stoppers whose sum is not stopper-sided. But the following question remains open:

QUESTION. *Is the sum of* finite *stoppers necessarily stopper-sided?*

Finally, the following question was posed in *Winning Ways* and remains open.

QUESTION (BERLEKAMP–CONWAY–GUY). *Is there an alternative notion of simplest form that works for all finite loopy games?*

Degrees, classes, and varieties. When γ is loopy, it is often the case that $\gamma - \gamma \neq 0$. Provided $\gamma - \gamma$ is stopper-sided, we define the *degree of loopiness* (or *degree*) γ° by

$$\gamma^\circ = \mathrm{Onside}(\gamma - \gamma).$$

If γ is equivalent to a loopfree game, then $\gamma^\circ = 0$; otherwise $\gamma^\circ > 0$. For example, it is not hard to check that $\mathbf{on}^\circ = \mathbf{on}$, $\mathbf{over}^\circ = \mathbf{over}$, and $\mathbf{upon}^\circ = \uparrow^{\mathbf{on}} = \{0 \mid -\mathbf{upon}*\}$.

For a fixed idempotent ι, the games of degree ι tend to group naturally into *classes* and *varieties* that interact in predictable ways. These were investigated in *Winning Ways* for the idempotent

$$\spadesuit = \{0^2 \parallel \{\mathbf{on} \mid 0^4\}\}.$$

However, since the publication of the first edition of *Winning Ways*, there has been little effort to move the theory forward. For this reason, we omit a full discussion and instead refer the reader to *Winning Ways*. It is perhaps time to study classes and varieties in more detail, in light of recent discoveries concerning other aspects of loopy games.

OPEN PROBLEM. *Investigate the class structure of each idempotent in Figure 9.*

Zugzwang games. Although the theory of sides is due to Conway and his students, its acknowledged inspiration is an earlier study by Robert Li, a student of Berlekamp's in the 1970s [20]. Li investigated so-called *Zugzwang games*— those in which it is a disadvantage to move— and found that they generalize ordinary numbers in a straightforward way.

DEFINITION 6. γ is a *Zugzwang game* if, for every follower δ of γ, each $\delta^L < \delta$ and $\delta < $ each δ^R.

Li's Theorem completely classifies all loopy Zugzwang games:

THEOREM 7 (LI). *Let γ be a loopy game. Then the following are equivalent:*

(a) *γ is equal to some Zugzwang game;*

(b) *There exist dyadic rationals x and y, $x \geq y$, such that*

$$\gamma = x \mathbin{\&} y.$$

PROOF. See [20, Section 4]. □

Li also studied a mild generalization of Zugzwang games, which he called *weak Zugzwang games*.

DEFINITION 8. γ is a *weak Zugzwang game* if, for every follower δ of γ, each $\delta^L \leq \delta$ and $\delta \leq$ each δ^R.

Note that for *loopfree* games G, the weak and strong Zugzwang notions coincide, since necessarily $G \neq G^L, G^R$. For loopy games, however, there are several further weak Zugzwang games.

THEOREM 9 (LI). *Let γ be a loopy game. Then the following are equivalent*:

(a) γ *is equal to some weak Zugzwang game*;
(b) $\gamma = x \mathbin{\&} y$, *where $x \geq y$ and each of x, y is one of the following*:

(i) **on**;
(ii) **off**;
(iii) *A dyadic rational*;
(iv) $z +$ **over** *for some dyadic rational z; or*
(v) $z +$ **under** *for some dyadic rational z.*

PROOF. See [20, Section 6]. □

Li's results are intrinsically interesting, and also quite remarkable, given that he had none of the modern machinery of loopy games at his disposal.

5. Some specific partizan games

Several partizan games have been successfully analyzed using the disjunctive theory. We briefly survey the most important examples.

Fox and Geese. This game has been largely solved by Berlekamp, who showed that the *critical position* of Figure 15 has the exact value $1 + 2^{-(n-8)}$, where $n \geq 8$. *CGSuite* has confirmed that the 8×8 starting position (Figure 2) has value $2 +$ **over**. Many other interesting values arise; these are summarized in *Winning Ways* and in slightly more detail in [27].

Berlekamp's analysis leaves little to be discovered about Fox and Geese proper. Nonetheless, we can ask interesting questions about certain variants of the game. Murray [23] describes a variant from Ceylon, *Koti keliya*, which is played with six geese ("dogs" or "cattle") on the 12×12 board, with the fox

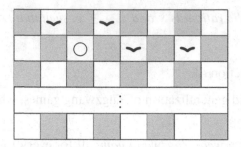

Figure 15. This critical position on an $n \times 8$ board has value $1 + 2^{-(n-8)}$ $(n \geq 8)$.

("leopard") permitted two moves per turn. It is unclear whether these moves must be in the same direction. Although a full solution to the 12×12 board appears to be out of reach computationally, it is interesting to observe how the fox's increased mobility affects play on smaller boards. As one might expect, it is far easier for the fox to escape, and positions whose values are large *negative* numbers become quite common. In fact, the following conjecture seems justified:

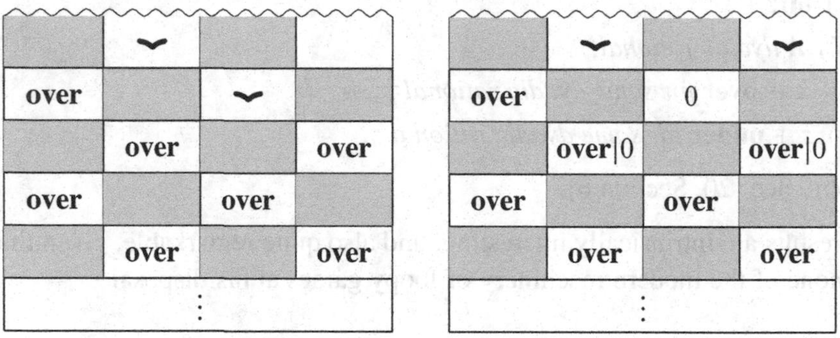

Figure 16. Conjectured values of $n \times 4$ Fox and Geese $(n \geq 5)$.

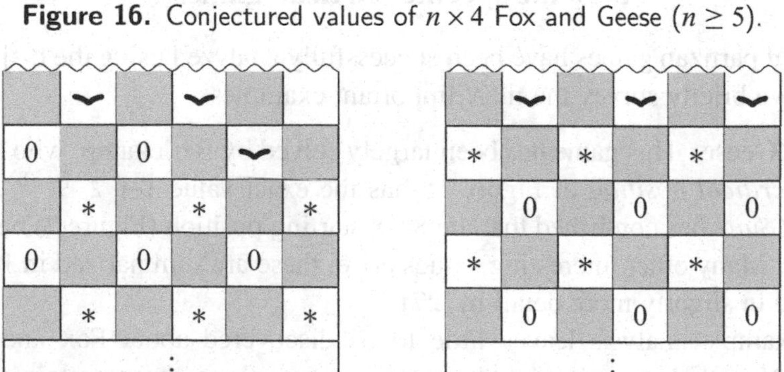

Figure 17. Conjectured values of $n \times 6$ Fox and Geese $(n \geq 8)$.

CONJECTURE. *The critical position of Figure 15, played with Ceylonese rules, has value* $-2n + 11$ *for all* $n \geq 6$.

Finally, there is overwhelming experimental evidence for the following two conjectures.

CONJECTURE. *The diagrams of Figure 16 are valid on the* $n \times 4$ *Fox and Geese board, for all* $n \geq 5$. *Furthermore, the range of values that appear on the* $n \times 4$ *board can be classified completely.*

(In the diagrams of this and the next figure, the geese are fixed, and conjectured values are shown for each possible placement of the fox.)

CONJECTURE. *The diagrams of Figure 17 are valid on the* $n \times 6$ *Fox and Geese board, for all* $n \geq 8$.

Backsliding Toads and Frogs. *Backsliding Toads and Frogs* was introduced in *Winning Ways*. The game is played on a $1 \times n$ strip populated by several toads (controlled by Left, facing right) and frogs (controlled by Right, facing left). See Figure 18 for a typical starting position. There are two types of moves. Either player may *slide* one of his animals one space in either direction. Alternatively, he may choose to *jump* in the facing direction (toads to the right, frogs to the left). Players must jump over exactly one enemy (never a friendly animal) and must land on an unoccupied space. Jumps do not result in capture.

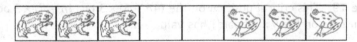

Figure 18. A typical starting position in Backsliding Toads and Frogs.

Readers familiar with ordinary Toads and Frogs will recognize the only difference between the two games: in the ordinary version, the animals are constrained to slide in the facing direction; in the loopy variant, they may slide backwards as well. This single difference has a monumental impact on the values that arise. The most obvious effect is that almost all positions in the Backsliding variant are loopy; for example, the position of Figure 18 has the remarkable value

$$\{\textbf{on} \parallel 0 \mid -\tfrac{1}{2}\} \ \& \ \{\tfrac{1}{2} \mid 0 \parallel \textbf{off}\}.$$

Positions in the Backsliding variant tend to have substantially *simpler* canonical forms than those in the loopfree version. For example, Erickson [8] noted that in ordinary Toads and Frogs, the "natural starting positions" of the form $T^m \square^k F^n$ are often quite complicated. In the Backsliding version, the *only* values (among all possibilities for k, m, n) are 0, $*$, **on**, **off**, **dud**, **on** & $\{\textbf{on} \mid \textbf{off}\}$, $\{\textbf{on} \mid \textbf{off}\}$ & **off**, and the single anomalous value given above.

Nonetheless, Backsliding Toads and Frogs exhibits positions of value n and 2^{-n}, as well as positions of *temperature* n and 2^{-n}, for all $n \geq 0$. See [27, Chapter 3] or [29] for a complete discussion.

Hare and Hounds. Hare and Hounds exhibits asymptotic behavior much like Fox and Geese: the position shown in Figure 19, on a $(4n+5) \times 3$ or $(4n+7) \times 3$ board, has the exact value $-n$.

The mathematical analysis of Hare and Hounds began in the 1960s, when Berlekamp demonstrated a winning strategy for the hare on large boards. He was close to proving that Figure 19 has value $-n$, but the canonical theory had not yet been invented.

Hare and Hounds exhibits many interesting values, including $*2$ (rare among partizan games); $\uparrow^2$, $\uparrow^3$, and $\uparrow^4$ (but not, it seems, $\uparrow^5$); and a bewildering variety of stoppers. See [27, Chapter 4] or [28] (this volume) for examples of these, as well as a proof that Figure 19 has value $-n$.

Figure 19. This critical position on the $(4n + 5) \times 3$ or $(4n + 7) \times 3$ board (shown here on the 9×3 board) has value $-n$.

Chess. Noam Elkies has observed several loopy values in Chess (in addition to many loopfree ones). See [6] for his constructions of **over** and **tis** $= 1 \ \& \ 0$. More recently, Elkies has produced positions of values **upon** and $\twoheaddownarrow_{\mathbf{on}}$ [7]; see Figure 20. (The kings have been omitted from these diagrams in order to focus on the essential features of each position, but they can easily be restored without affecting the positions' values, using techniques outlined by Elkies [6].)

Entrepreneurial Chess. Entrepreneurial Chess is played on a quarter-infinite chessboard, with just the two kings and a White rook (Figure 21). In addition to his ordinary king moves, Left (Black) has the additional option of "cashing out." When he cashes out, the entire position is replaced by the integer n, where n is the sum of the *rank* and *column* values indicated in the diagram. Thus Left stands to gain by advancing his king as far to the upper-right as possible; and Right, with his rook, will eventually be able to stop him.

Entrepreneurial Chess was invented by Berlekamp, and has been studied extensively by Berlekamp and Pearson [4]. They have discovered many interesting

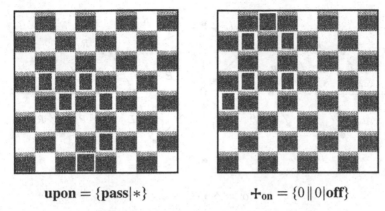

upon $= \{\mathbf{pass}|*\}$ $\mathbf{+_{on}} = \{0\,\|\,0\,|\,\mathbf{off}\}$

Figure 20. Loopy values in chess.

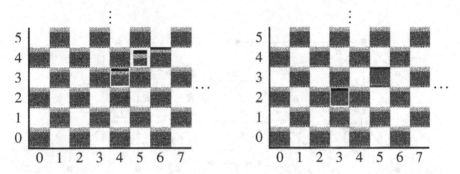

Figure 21. Entrepreneurial Chess.

values. For example, the position shown in Figure 21, left, has value $7 + \mathbf{over}$: Left can cash out for 7 points at any time, and in the meantime Right is constrained to shuttle his king between the squares adjacent to his rook. Berlekamp and Pearson's results also include a detailed temperature analysis of a wide class of positions.

A particularly interesting position γ arises in the pathological case when Left has captured Right's rook, as in Figure 21, right. The onside of γ is **on**, since Left need never cash out. Now consider the offside. Left must cash out *eventually*, since infinite plays are wins for Right, but he can defer this action for as long as necessary. Thus we have the remarkable identity

$$\gamma = \mathbf{on}\ \&\ \widehat{\mathbb{Z}},$$

where $\widehat{\mathbb{Z}} = \{0, 1, 2, \ldots \mid \mathbf{pass}\}$ is the pseudonumber defined in Section 3. This identity can be verified formally using the theory presented in Section 4.

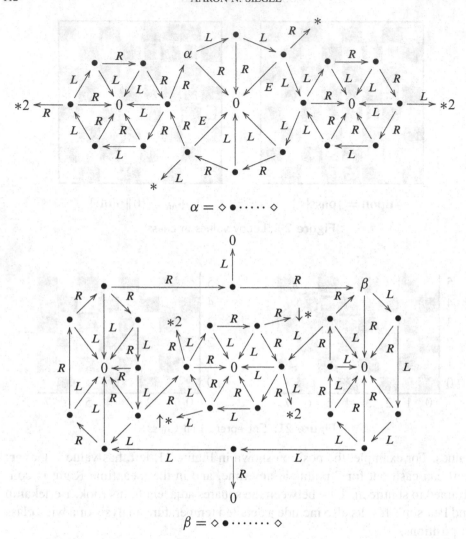

Figure 22. Some cycles that arise naturally in One-Dimensional Phutball.

One-Dimensional Phutball. Some extraordinary loopy positions in 1D Phutball were discovered jointly by Richard Nowakowski, Paul Ottaway, and myself. A few of these are shown in Figure 22. The game of Phutball, and the notation used here to describe the positions, are explained in [19]. It is interesting that although Phutball obviously allows for alternating cycles, all positions yet studied are equivalent to stoppers.

QUESTION. *Is every position in One-Dimensional Phutball equivalent to a stopper?*

These Phutball positions contain the most complicated loops yet detected. Moreover, the corresponding position on the 1×12 board ($\diamond \bullet \cdots\cdots \diamond$) is a stopper whose canonical game graph has 168 vertices and a 23-cycle. However, all of these examples are "tame" in the sense that every cycle alternates just once between Left and Right edges. It is possible to construct "wild" stoppers with more complicated cycles (see [30] in this volume), but nonetheless we have the following open problem.

OPEN PROBLEM. *Find a position in an actual combinatorial game (Phutball or otherwise) whose canonical form is a stopper containing a wild cycle.*

6. Impartial loopy games

Not surprisingly, *impartial* loopy games were studied long before partizan ones. In 1966, ten years before the publication of *On Numbers and Games*, Cedric A. B. Smith generalized the Sprague–Grundy theory to games with cycles.

For γ to be impartial, of course, infinite plays must be considered draws. We therefore have three outcome classes: the usual $\mathcal{N}$- and $\mathcal{P}$-positions, and also $\mathcal{D}$-positions (called $\mathcal{O}$-positions in *Winning Ways*).

Now consider an arbitrary impartial game γ. If all the options of γ are known to be nimbers $*a, *b, *c, \ldots$, then certainly $\gamma = *n$, where $n = \mathrm{mex}(a, b, c, \ldots)$: the usual Sprague–Grundy argument applies. But some games γ are equivalent to nimbers even though some of their options are not.

For instance, consider the example of Figure 23. It is not hard to see that $\gamma = *2$: in $\gamma + *2$, second player wins by mirroring moves to 0 or $*$; while if first player moves to $\delta + *2$, second player reverses to $*2 + *2 = 0$. However, the subposition δ is not equivalent to any nimber, since first player can always *draw* $\delta + *n$ by moving to the infinite loop.

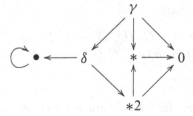

Figure 23. δ is not a nimber, but $\gamma = *2$.

Roughly speaking, $\gamma = *2$ because 2 is the mex of its nimber-valued options, and all *other* options reverse out, in the usual sense, to positions of value $*2$. Care is needed, however, to avoid circular definitions: the analysis of Figure 23

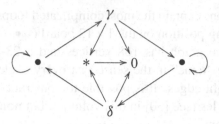

Figure 24. It's tempting to declare $\gamma = \delta = *2$ (cf. Figure 23); but $\gamma + *2$
and $\delta + *2$ are both draws.

works because the reversing move is "already known" to be $*2$. Indeed, Figure 24 shows that we cannot indiscriminately draw conclusions about the value of γ without a definite starting point.

These concerns led Smith to formulate the key notion of a rank function. The idea, motivated by Figures 23 and 24, is that we can safely assign Grundy values to subpositions of γ provided they are ranked in order of precedence. Formally,

DEFINITION 10. Let γ be a game, let $\mathscr{A}$ be the set of all followers of γ, and fix a partial function $G : \mathscr{A} \to \mathbb{N}$. Then G is a *Grundy function* if there exists a map $R : \mathscr{A} \to \mathbb{N}$ (a *rank function* for G) such that:

(i) If $G(\alpha) = n$ and $k < n$, then there is some option β of α with $R(\beta) < R(\alpha)$ and $G(\beta) = k$.

(ii) If $G(\alpha) = n$ and β is any option of α with $R(\beta) < R(\alpha)$, then $G(\beta) \neq n$.

(iii) Suppose $G(\alpha) = n$ and β is any option of α. If $G(\beta)$ is undefined, or if $R(\beta) \geq R(\alpha)$, then there exists an option δ of β with $R(\delta) < R(\alpha)$ and $G(\delta) = n$.

Conditions (i) and (ii) imply that $G(\alpha)$ obeys the mex rule, taken over all options of α with strictly lower rank. Condition (iii) implies that any remaining options reverse out to positions of lower rank than α. The main result is that there is a unique *maximal* Grundy function associated to γ (where G is maximal if its domain cannot be expanded).

THEOREM 11 (SMITH). *Let $G, H : \mathscr{A} \to \mathbb{N}$ be two Grundy functions for γ. If G and H are maximal, then $G = H$.*

PROOF. See [31, Section 9]. □

So we can safely refer to *the* Grundy function G of γ. It is a remarkable fact that G completely characterizes the behavior of γ.

LEMMA 12 (SMITH). *Let γ be a game with Grundy function G. If $G(\gamma) = n$, then $\gamma = *n$; if $G(\gamma)$ is undefined, then γ is not equal to any nimber.*

PROOF. See [31, Section 9]. □

When $G(\gamma)$ is undefined, we write

$$\gamma = \infty_{abc\cdots}$$

to mean that the nimber-valued followers of γ are exactly $*a$, $*b$, $*c$, ... We can now describe the outcome class of any sum of impartial games.

THEOREM 13 (SMITH).

(a) $\infty_{abc\cdots} + *n$ is an $\mathcal{N}$-position if n is one of $a, b, c, \ldots$; otherwise it's a $\mathcal{D}$-position.

(b) $\infty_{abc\cdots} + \infty_{def\cdots} + \cdots$ is always a $\mathcal{D}$-position.

PROOF. See [31, Section 9]. □

The parallel between Smith's theory and the classical Sprague–Grundy theory breaks down in one important respect. If $\gamma = *n$, then we can be quite certain that $\gamma + X$ and $*n + X$ have the same outcomes, *even when X is partizan*. However, there exist games α and β, both "equal to" ∞_0, whose outcomes are distinguished by a certain partizan game (see Figure 25). There is no contradiction: α and β indeed behave identically, provided they occur in sums *comprised entirely of impartial games*. One could say that the Sprague–Grundy theory embeds nicely in the partizan theory, while the Smith generalization does not.

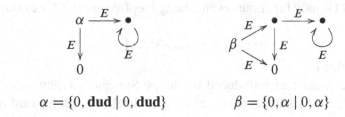

$$\alpha = \{0, \mathbf{dud} \mid 0, \mathbf{dud}\} \qquad\qquad \beta = \{0, \alpha \mid 0, \alpha\}$$

Figure 25. Right can draw $\alpha + 1$ moving first, while $\beta + 1$ is a win for Left, no matter who moves first. Therefore $\alpha \neq \beta$; yet no *impartial* game distinguishes them.

This is an interesting fact, one that does not seem to appear elsewhere in the literature; and it raises an equally intriguing question:

OPEN PROBLEM. *Classify all impartial loopy games, relative to all partizan ones.*

Additional subtraction games. *Additional subtraction games* are just like ordinary subtraction games, except that their subtraction sets may contain *negative* numbers (so that players are permitted to *add* to a nonempty heap in certain fixed quantities). Such games are more interesting than one might expect. Several examples are mentioned in *Winning Ways*, and several related classes of games were studied by Fraenkel and Perl [12] and Fraenkel and Tassa [14] in the 1970s. The additional subtraction games cry out for further investigation.

OPEN PROBLEM. *Extend the analysis of additional subtraction games.*

The annihilation game. The *annihilation game* is an impartial game played on an arbitrary directed graph. At the start of the game, tokens are placed on the vertices of the graph, at most one per vertex. A move consists of sliding a token to an adjacent vertex, and whenever two tokens occupy the same vertex, they are both immediately removed from the game (the annihilation rule).

If the game is played on a loopfree graph, then the annihilation rule has no effect, since identical loopfree games ordinarily sum to zero. On loopy graphs, however, the effect is significant.

The annihilation game was proposed by Conway in the 1970s. Shortly thereafter, it was solved by Aviezri Fraenkel and his student Yaacov Yesha [16]. They specified a polynomial-time algorithm for determining the generalized Sprague–Grundy values of arbitrary positions. Interested readers should consult Fraenkel and Yesha's 1982 paper on the subject [17].

Infinite impartial games. The Smith–Fraenkel results completely resolve the disjunctive theory of finite impartial games. It is therefore natural to seek generalizations of the theory to infinite games. In the infinite case, one must allow ordinal-valued Grundy functions, even among loopfree games: for example, the game

$$*\omega = \{0, *, *2, *3, \dots\}$$

has Grundy value ω.

In the same paper that introduced the loopy Sprague–Grundy theory [31], Smith noted that his results generalize in a completely straightforward manner to infinite games with ordinal-valued Grundy functions. The definitions and theorems are essentially the same, with the functions G and R permitted to take on arbitrary ordinal values.

A more substantive result is due to Fraenkel and Rahat [13]. They identified a class of infinite loopy games whose Grundy values are nonetheless guaranteed to be finite. Their result can be summarized as follows:

DEFINITION 14. Let G be a graph. A *path* of G (of length n) is a sequence of *distinct* vertices

$$V_0, V_1, V_2, \dots, V_n$$

such that there is an edge directed from each V_i to V_{i+1}. We say that the path *starts at* V_0.

DEFINITION 15. Let G be a graph. A vertex V is said to be *path-bounded* if there is an integer N such that every path starting at V has length $\leq N$. G is said to be *locally path-bounded* if every vertex of $\mathcal{G}$ is path-bounded. (There need not exist a single bound that extends uniformly over all vertices.)

Note that all loopfree graphs are locally path-bounded.

THEOREM 16 (FRAENKEL–RAHAT). *Let γ be a (possibly infinite) impartial game. If the graph of γ is locally path-bounded, then the Grundy function for γ is finite wherever it is defined.*

PROOF. See [13, Section 3]. $\qquad\qquad\qquad\qquad\qquad\qquad\qquad\qquad\qquad\qquad\square$

7. Conjunctive and selective sums

Although disjunctive sums have received the most attention, several authors have investigated the behavior of loopy games under other types of compound. The two most prominent are *conjunctive* and *selective* sums:

- In the *conjunctive sum* $\alpha \wedge \beta \wedge \cdots \wedge \gamma$, a player must move in every component. If any component is terminal, then there are no legal moves.
- In the *selective sum* $\alpha \vee \beta \vee \cdots \vee \gamma$, a player may move in any number of components (but at least one).

This line of research, like so many others, was pioneered by Cedric Smith [31], who focused on the impartial case. Smith's results are best described in terms of the *Steinhaus remoteness* of a position. If γ is a loopy game, we define the remoteness $R(\delta)$, for each follower δ of γ, as follows:

DEFINITION 17. Let γ be an impartial game, let $\mathscr{A}$ be the set of all followers of γ, and fix a function $R : \mathscr{A} \to \mathbb{N} \cup \{\infty\}$. Then R is a *remoteness function* provided that, for each $\delta \in \mathscr{A}$:

- If δ is terminal, then $R(\delta) = 0$.
- If $R(\alpha)$ is *even* for at least one option α of δ, then

$$R(\delta) = 1 + \min\{R(\alpha) : \alpha \text{ is an option of } \delta \text{ with } R(\alpha) \text{ even}\}.$$

- If $R(\alpha)$ is *odd* for every option α of δ, then

$$R(\delta) = 1 + \max\{R(\alpha) : \alpha \text{ is an option of } \delta\}.$$

It is not hard to check that every game admits a unique remoteness function R. The remoteness function tells us quite a bit about γ: it's a $\mathscr{P}$-position if $R(\gamma)$ is even, an $\mathscr{N}$-position if $R(\gamma)$ is odd, and a $\mathscr{D}$-position if $R(\gamma) = \infty$.

Furthermore, if the winning player strives to achive victory as quickly as possible, and the losing player tries to postpone defeat for as long as possible, then the magnitude of $R(\gamma)$ determines exactly how long the game will last.

Smith's main results are summarized by the following theorem.

THEOREM 18 (SMITH). *Let α, β, ..., γ be impartial loopy games. Then*:

(a) $R(\alpha \wedge \beta \wedge \cdots \wedge \gamma) = \min\{R(\alpha), R(\beta), \ldots, R(\gamma)\}$.

(b) *If $R(\alpha), R(\beta), \ldots, R(\gamma)$ are all even, then*

$$R(\alpha \vee \beta \vee \cdots \vee \gamma) = R(\alpha) + R(\beta) + \cdots + R(\gamma).$$

If $R(\alpha), R(\beta), \ldots, R(\gamma)$ are all finite, and k of them are odd ($k \geq 1$), then

$$R(\alpha \vee \beta \vee \cdots \vee \gamma) = R(\alpha) + R(\beta) + \cdots + R(\gamma) - k + 1.$$

Finally, if any of $R(\alpha), R(\beta), \ldots, R(\gamma)$ is infinite, then

$$R(\alpha \vee \beta \vee \cdots \vee \gamma) = \infty.$$

PROOF. See [31, Sections 6 and 7].   □

Theorem 18 enables us to find the outcome of any conjunctive or selective sum, provided we know the remoteness of each component. The remoteness function can therefore be regarded as an analogue of the Grundy function.

Partizan games. Smith's results were substantially extended by Alan Flanigan, who studied partizan loopy games under conjunctive and selective sums, as well as two additional types of compound, the *continued conjunctive* and *shortened selective* sums. We summarize Flanigan's results for conjunctive sums here. The remaining cases are beyond the scope of this paper; interested readers should consult Flanigan's 1979 thesis [9] and two subsequent papers [10; 11].

First note that we can define *partizan remoteness functions R^L and R^R* for γ. They are defined just as in the impartial case; but we only consider moves for the player in question, minimaxing over the *opponent's* remoteness function applied to each option.

DEFINITION 19. Let γ be a partizan game, let $\mathscr{A}$ be the set of all followers of γ, and fix functions $R^L, R^R : \mathscr{A} \to \mathbb{N} \cup \{\infty\}$. Then R^L, R^R are *partizan remoteness functions* provided that the following conditions (and their equivalents with Left and Right interchanged) are satisfied for each $\delta \in \mathscr{A}$:

- If δ has no Left options, then $R^L(\delta) = 0$.
- If $R^R(\delta^L)$ is *even* for at least one δ^L, then

$$R^L(\delta) = 1 + \min\{R^R(\delta^L) : R^R(\delta^L) \text{ is even}\}.$$

- If $R^R(\delta^L)$ is *odd* for every δ^L, then

$$R^L(\delta) = 1 + \max\{R^R(\delta^L)\}.$$

Smith's result for conjunctive sums is virtually unchanged in the partizan context.

THEOREM 20 (SMITH–FLANIGAN). *Let $\alpha, \beta, \ldots, \gamma$ be (partizan) loopy games. Then for $X = L, R$, we have*

$$R^X(\alpha \wedge \beta \wedge \cdots \wedge \gamma) = \min\{R^X(\alpha), R^X(\beta), \ldots, R^X(\gamma)\}.$$

PROOF. See [9, Chapter II.2]. □

Since the outcome class of γ is determined by the parities of $R^L(\gamma)$ and $R^R(\gamma)$, this is all we need to know.

Flanigan also noted that the analysis of conjunctive sums (but not selective sums) extends to infinite games: one can suitably define ordinal-valued remoteness functions, taking suprema instead of maxima when R is odd; then Theorem 20 generalizes verbatim.

8. Algorithms and computation

Computation is an essential part of combinatorial game theory. This is particularly true in the study of loopy games, since they are especially difficult to analyze by hand.

The basic algorithm for determining the outcome class of an impartial loopy game was introduced by Fraenkel and Perl [12] in 1975. The strategy is to iterate over all vertices V of the game graph of γ, assigning labels as summarized in Algorithm 1.

THEOREM 21 (FRAENKEL–PERL). *Algorithm 1 correctly labels the subpositions of γ according to their outcome classes, and concludes in time $O(n^2)$ in the number of vertices.*

PROOF. See [12, Section 3]. □

In fact, Fraenkel observes that we can improve slightly upon Algorithm 1: traverse the vertices of γ just once; and whenever a label is assigned to V, reexamine all unlabeled predecessors of V. With this modification, the algorithm runs in time $O(n)$ in the number of edges. Since game graphs tend to have relatively low edge density, this will usually be an improvement.

For each vertex V of the game graph of γ:

- If all options of V have been labeled $\mathcal{N}$, then label V by $\mathcal{P}$. (This includes the case where V is terminal.)

- If any option of V has labeled $\mathcal{P}$, then label V by $\mathcal{N}$.

The algorithm continues until no more vertices can be labeled, whereupon all remaining vertices are labeled by $\mathcal{D}$.

Algorithm 1. Computing the outcome class of an impartial game γ.

Let $\mathcal{G}$ be the game graph of γ.

(i) Put $k = 0$.

(ii) For each vertex V of $\mathcal{G}$:

• If all options of V have been labeled $\mathcal{N}$, then label V by $\mathcal{P}$.
• If any option of V has been labeled $\mathcal{P}$, then label V by $\mathcal{N}$.

(iii) For each unlabeled vertex V, all of whose options are now labeled: if each option of V has an option labeled $\mathcal{P}$, then label V by $\mathcal{P}$ as well.

(iv) Label all remaining (unlabeled) vertices by $\mathcal{D}$.

(v) For each vertex V labeled $\mathcal{P}$, define $G(V) = k$ and remove V from $\mathcal{G}$.

(vi) If all remaining vertices of $\mathcal{G}$ are labeled $\mathcal{D}$, then stop: we are done.

(vii) Clear all $\mathcal{N}$ labels (but retain all $\mathcal{D}$ labels).

(viii) Put $k = k + 1$ and return to Step 2.

Algorithm 2. Computing the generalized Sprague–Grundy value of γ.

Fraenkel and Perl have also given an algorithm for computing the generalized Sprague–Grundy values of impartial loopy games (Algorithm 2); see Fraenkel and Yesha [18] for further discussion.

THEOREM 22 (FRAENKEL–PERL). *Algorithm 2 correctly defines the maximal Grundy function for γ, and concludes in time $O(n^3)$ in the number of vertices.*

PROOF. See [12, Section 4].   □

Algorithm 1 is virtually unchanged in the partizan case. Given a game γ with graph $\mathcal{G}$, one first constructs the corresponding *state graph* S. The vertices of S consist of pairs (V, X), where V is a vertex of $\mathcal{G}$ and X is either L or R. There is an edge directed from (U, L) to (V, R) just if there is a Left edge directed from U to V, and so on. Algorithm 1 can then be applied directly to S. This was noticed independently by Shaki [26], Fraenkel and Tassa [15], and Michael Albert [1].

Comparison. Algorithm 1 suffices to compare stoppers. Recall from Section 3 that if γ and δ are stoppers, then $\gamma \geq \delta$ if and only if Left, playing second, can survive $\gamma - \delta$. So to test whether $\gamma \geq \delta$, we simply compute the state graph of $\gamma - \delta$ and apply Algorithm 1. If V is the start vertex (corresponding to $\gamma - \delta$ itself), then $\gamma \geq \delta$ if and only if (V, R) is not marked $\mathcal{N}$.

One can extend these ideas in order to compare arbitrary games, but the algorithms are somewhat more involved. See [30] in this volume for a discussion.

Simplification and strong equivalence. Fraenkel and Tassa [15] studied various simplification techniques in detail. They identified certain situations in which one can safely simplify an arbitrary (free) loopy game γ. These techniques yield a good algorithm for determining whether γ is equivalent to a loopfree game. We summarize their results.

DEFINITION 23. Let γ be a free loopy game.

(a) A Left option γ^L is *strongly dominated* if Left, playing second, can *win* the game $\gamma^{L'} - \gamma^L$ for some other Left option $\gamma^{L'}$.
(b) A Left option γ^L is *strongly reversible* if Left, playing second, can *win* the game $\gamma - \gamma^{LR}$ for some Right option γ^{LR}.
(c) If δ is any free loopy game, then γ and δ are *strongly equivalent* if either player can *win* $\gamma - \delta$ playing second. In this case we write $\gamma \overset{*}{=} \delta$.

Strongly dominated and strongly reversible Right options are defined analogously.

Note that $\gamma \overset{*}{=} \gamma$ if and only if $\gamma - \gamma = 0$, i.e., if and only if γ is equivalent to a loopfree game.

THEOREM 24 (FRAENKEL–TASSA). *Let γ be a free loopy game and let δ be any follower of γ. Let γ' be obtained from γ by either:*

(a) *Replacing δ with a strongly equivalent game δ'; or*
(b) *Eliminating a strongly dominated option of δ; or*
(c) *Bypassing a strongly reversible option of δ.*

Then $\gamma = \gamma'$.

THEOREM 25 (FRAENKEL–TASSA). *Let γ be a free loopy game and assume that:*

(i) *γ is equivalent to a loopfree game (i.e., $\gamma - \gamma = 0$); and*
(ii) *No follower of γ has any strongly dominated or strongly reversible options.*

Then γ is itself loopfree.

THEOREM 26 (FRAENKEL–TASSA). *Let γ be a free loopy game. If, for each subposition of γ, we repeatedly eliminate strongly dominated options and bypass strongly reversible ones, then the process is guaranteed to terminate. We will eventually arrive at a form for γ that contains no strongly dominated or strongly reversible options.*

Thus if γ is equivalent to a loopfree game, then Theorems 24 through 26 yield an algorithm for computing its canonical form: eliminate strongly dominated options and bypass strongly reversible ones until none remain.

Theorem 24 fails if the strong notions of domination and reversibility are replaced by their naive weakenings. This is a major obstacle to developing a general canonical theory of loopy games. These issues are discussed at length, and partially resolved, in [30] in this volume.

References

[1] M. Albert. Personal communication, 2004.

[2] E. R. Berlekamp. Idempotents among partisan games. In Nowakowski [25], pages 3–23.

[3] E. R. Berlekamp, J. H. Conway, and R. K. Guy. *Winning Ways for Your Mathematical Plays*. A. K. Peters, Ltd., Natick, MA, second edition, 2001.

[4] E. R. Berlekamp and M. Pearson. Entrepreneurial chess. In this volume.

[5] J. H. Conway. Loopy games. In B. Bollobás, editor, *Advances in Graph Theory*, number 3 in Ann. Discrete Math., pages 55–74, 1978.

[6] N. D. Elkies. On numbers and endgames: Combinatorial game theory in chess endgames. In Nowakowski [24], pages 135–150.

[7] N. D. Elkies. Personal communication, 2005.

[8] J. Erickson. New Toads and Frogs results. In Nowakowski [24], pages 299–310.

[9] J. A. Flanigan. *An Analysis of Some Take-Away and Loopy Partizan Graph Games*. PhD thesis, University of California, Los Angeles, 1979.

[10] J. A. Flanigan. Selective sums of loopy partizan graph games. *Internat. J. Game Theory*, 10:1–10, 1981.

[11] J. A. Flanigan. Slow joins of loopy games. *J. Combin. Theory, Ser. A*, 34(1):46–59, 1983.

[12] A. S. Fraenkel and Y. Perl. Constructions in combinatorial games with cycles. In A. Hajnal, R. Rado, and V. T. Sós, editors, *Infinite and Finite Sets, Vol. 2*, number 10 in Colloq. Math. Soc. János Bolyai, pages 667–699. North-Holland, 1975.

[13] A. S. Fraenkel and O. Rahat. Infinite cyclic impartial games. *Theoretical Comp. Sci.*, 252:13–23, 2001.

[14] A. S. Fraenkel and U. Tassa. Strategy for a class of games with dynamic ties. *Comput. Math. Appl.*, 1:237–254, 1975.

[15] A. S. Fraenkel and U. Tassa. Strategies for compounds of partizan games. *Math. Proc. Cambridge Philos. Soc.*, 92:193–204, 1982.

[16] A. S. Fraenkel and Y. Yesha. Theory of annihilation games. *Bull. Amer. Math. Soc.*, 82:775–777, 1976.

[17] A. S. Fraenkel and Y. Yesha. Theory of annihilation games, I. *J. Combin. Theory, Ser. B*, 33:60–82, 1982.

[18] A. S. Fraenkel and Y. Yesha. The generalized Sprague–Grundy function and its invariance under certain mappings. *J. Combin. Theory, Ser. A*, 43:165–177, 1986.

[19] J. P. Grossman and R. J. Nowakowski. One-dimensional Phutball. In Nowakowski [25], pages 361–367.

[20] S.-Y. R. Li. Sums of zuchswang games. *J. Combin. Theory, Ser. A*, 21:52–67, 1976.

[21] D. J. Moews. *On Some Combinatorial Games Connected with Go*. PhD thesis, University of California at Berkeley, 1993.

[22] D. J. Moews. Loopy games and Go. In Nowakowski [24], pages 259–272.

[23] H. J. R. Murray. *A History of Board-Games Other Than Chess*. Oxford University Press, 1952.

[24] R. J. Nowakowski, editor. *Games of No Chance*. Number 29 in MSRI Publications. Cambridge University Press, Cambridge, 1996.

[25] R. J. Nowakowski, editor. *More Games of No Chance*. Number 42 in MSRI Publications. Cambridge University Press, Cambridge, 2002.

[26] A. Shaki. Algebraic solutions of partizan games with cycles. *Math. Proc. Cambridge Philos. Soc.*, 85(2):227–246, 1979.

[27] A. N. Siegel. *Loopy Games and Computation*. PhD thesis, University of California at Berkeley, 2005.

[28] A. N. Siegel. Loopy and loopfree canonical values in Hare and Hounds. To appear.

[29] A. N. Siegel. Backsliding Toads and Frogs. In this volume.

[30] A. N. Siegel. New results in loopy games. In this volume.

[31] C. A. B. Smith. Graphs and composite games. *J. Combin. Theory, Ser. A*, 1:51–81, 1966.

AARON N. SIEGEL
aaron.n.siegel@gmail.com

Games of No Chance 3
MSRI Publications
Volume 56, 2009

On day n

DAVID WOLFE

ABSTRACT. We survey the work done to date about games born by day n.

1. Introduction

Both number theory and combinatorial game theory are interesting in large part because of the wonderful interplay between algebraic and combinatorial structure. Here we survey some general results that investigate either the additive structure or the partial order of the games born by day n.

The games born by day n, $\mathcal{G}_n$, have game trees of height at most n. More formally, $\mathcal{G}_n$ is defined inductively:

$$\mathcal{G}_0 \stackrel{\text{def}}{=} \{0\},$$
$$\mathcal{G}_n \stackrel{\text{def}}{=} \{\{G^L \mid G^R\} : G^L, G^R \subseteq \mathcal{G}_{n-1}\}.$$

$\mathcal{G}_1$ consists of games whose left and right options are subsets of $\mathcal{G}_0$, i.e., either $\{\}$ or $\{0\}$. This yields four games born by day 1, those being

$$0 = \{ \mid \}, \quad 1 = \{0 \mid \}, \quad -1 = \{ \mid 0\}, \quad * = \{0 \mid 0\}.$$

We can draw the partial order of these four games to get

On day 2, left and right options are subsets of the day 1 lattice. Since there are 16 subsets of $\mathcal{G}_1$, this yields at most $16 \cdot 16 = 256$ games born by day 2.

Right

	-1	$0, *$	0	$*$	1	$\varnothing$
1	± 1	$1\|0, *$	$1\|0$	$1\|*$	$1*$	2
0, *	$0,*\|-1$	$*2$	$\uparrow*$	$\uparrow$	$\frac{1}{2}$	1
0	$0\|-1$	$\downarrow*$	$*$			
*	$*\|-1$	$\downarrow$				
−1	$-1*$	$-\frac{1}{2}$		0		
∅	-2	-1				

(Left labels the rows at left.)

Figure 1. The 22 games born by day 2 organized by Left and Right options.

Note, however, that we can restrict our attention to only those subsets without dominated options, i.e., the antichains in $\mathcal{G}_1$. There are six such antichains

$$\{1\}, \{0, *\}, \{0\}, \{*\}, \{-1\}, \{\}$$

roughly sorted so that those Left most wishes to be her option list are listed first. This leaves us with at most 36 games born by day 2. Of these 36, many are equal, leaving the 22 distinct games shown in Figure 1.

2. Games as a group

Under game addition, although the games born by day $n > 0$ do not form a group, it is natural to investigate the group generated by the games born by day n, which we will denote J_n. On day 0, we have just the singleton $J_0 = \{0\}$. $\mathcal{G}_1 = \{0, 1, -1, *\}$, and sums of these games consist of integers n and $n*$. Since $* + * = 0$, we have that J_1 is isomorphic to $\mathbb{Z} \times \mathbb{Z}_2$.

Moews [1991] investigates J_2 and J_3. He shows that J_2 has the basis

$$1/2, *2, A, \uparrow, \alpha, \pm\tfrac{1}{2}, \pm 1,$$

where

$$A = \{1|0\} - \{1|*\}, \quad \alpha = \{1|0\} - \{1|0, *\}.$$

A has order 4 since $A + A = *$, while $\alpha > 0$ has atomic weight 0 and is therefore linearly independent with $\uparrow$. So, we have that J_2 is isomorphic to $\mathbb{Z}^3 \times \mathbb{Z}_4 \times \mathbb{Z}_2^3$.

Let I_n be the group of infinitesimal games within J_n. Then I_2 is $\mathbb{Z}^2 \times \mathbb{Z}_4 \times \mathbb{Z}_2$ and J_2/I_2 is $\mathbb{Z} \times \mathbb{Z}_2^2$.

Moews employs a combination of computation and mathematical ingenuity to describe J_3/I_3, but leaves open I_3 (and therefore J_3.) His key result that

$$J_3/I_3 = \mathbb{Z}^7 \times \mathbb{Z}_4 \times \mathbb{Z}_2^8$$

has a reasonably technical proof.

3. Games as a partial order

Games born by day n form a distributive lattice, but the collection of all short games, $\mathcal{G} = \bigcup_{n \geq 0} \mathcal{G}[n]$, is not a lattice [Calistrate et al. 2002]. The key to identifying the lattice structure is to explicitly construct the *join* (or least upper bound) and *meet* (or greatest lower bound) of two elements. Since the partial order is self-dual (i.e., each game has a negative and $G \geq H$ exactly when $-H \geq -G$), we will only state theorems in terms of the join operation, and leave it to the reader to construct the symmetric assertions concerning the meet operation.

For the day n lattice, define the join in terms of the operation

$$\lceil G \rceil \overset{\text{def}}{=} \{H \in \mathcal{G}_{n-1} : H \not\leq G\}$$

The notation $\lceil G \rceil$, and $G_1 \vee G_2$ below, take the current day n for granted. Then the join of two games is given by

$$G_1 \vee G_2 \overset{\text{def}}{=} \{G_1^L, G_2^L \mid \lceil G_1 \rceil \cap \lceil G_2 \rceil\}$$

Note that $G_1 \vee G_2$ is in $\mathcal{G}_n$ since its left and right options are all in $\mathcal{G}_{n-1}$.

It is now a reasonable graduate level exercise to prove that the join operation above exactly reflects the partial order of games born by day n, and that join distributes over a symmetrically defined meet.

The *Hasse diagram* of the lattices for days 1 and 2 is shown on the left side of Figure 2. One property of distributive lattices is that they are *graded* or *ranked*, where the partial order can be drawn with edges only going between adjacent levels. The lattices for days $n \leq 3$ all share the property that the middle level is the widest (i.e., has the most games). It is still open, but should be computationally feasible, to organize and describe the exact structure of the day 3 lattice of 1474 games.

In a lattice, the *join irreducible* elements are those elements that cannot be formed by the join of other elements. Looking at the Hasse diagram of the lattice, a join irreducible element has exactly one element immediately below it in the lattice. (The single element at the bottom is not considered a join irreducible for it is the join of the empty set.) The right side of Figure 2 shows the partial order of the day 2 join irreducibles.

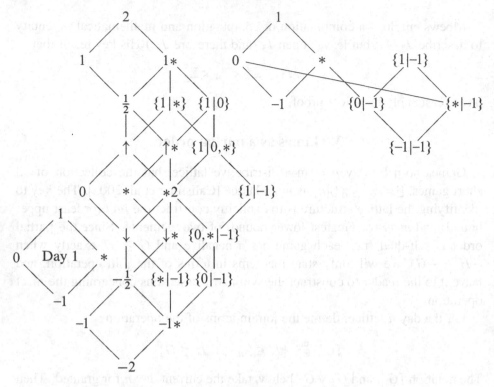

Figure 2. Left: day 1 and day 2 lattices. Right: join irreducibles from day 2.

Birkhoff [1940] showed (amazingly) that there is a natural one-to-one correspondence of finite partial orders with finite distributive lattices, where that correspondence is via the partial order on join irreducibles. As shown in [Fraser et al. 2005], the join irreducibles from the $\mathcal{G}_{n+1}$ lattice are exactly those games of the form g or $\{g \mid -n\}$ where $g \in \mathcal{G}_n$.

As an immediate corollary of this fact (and Birkhoff's construction of the distributive lattice from its join irreducibles), all maximal chains on day n are of length exactly one plus double the number of games born by day $n-1$.

Aaron Siegel [2005] showed that the distributive lattice for $\mathcal{G}_n$ has exactly two automorphism, i.e., one order-preserving symmetry. In particular, he defines a companion $g^\bullet$ of each element $g \in \mathcal{G}_n$ by

$$g^\bullet = \begin{cases} * & \text{if } G = 0, \\ \{0, (G^L)^\bullet \mid (G^R)^\bullet\} & \text{if } G > 0, \\ \{(G^L)^\bullet \mid (G^R)^\bullet\} & \text{if } G \text{ is incomparable with } 0, \\ \{(G^L)^\bullet \mid 0, (G^R)^\bullet\} & \text{if } G < 0, \end{cases}$$

This is the only nontrivial automorphism which preserves the partial order on $\mathcal{G}_n$. Further, this automorphism also preserves birthday (for games other than

0 and ∗) and atomic weight of all-small games. He defines the *longitude* of a game G by the difference in ranks between G and $G \vee G^{\bullet}$; this is some measure of how far G is from the "spine" of self-companions.

4. The all-small lattice

An all-small game is one in which Left has an option if and only if Right has one as well. On day 1, 0 and ∗ are all-small, while 1 and −1 are not. Day 2 has 7 all-small games:

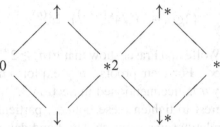

In his thesis, Aaron Siegel [2005] proved that, subject to a minor caveat, the all-small games born on day n also form a distributive lattice. The caveat is that one must adjoin a single element to the top (and, symmetrically, bottom) of the lattice which is the join of the two maximal elements $(n-1) \cdot \uparrow$ and $(n-1) \cdot \uparrow\ast$. This lattice also has the unique nontrivial automorphism given by $g^{\bullet}$ above. There are 67 all-smalls born by day 3, and a figure of the lattice appears in Siegel's thesis. He also computes the 534,483 all-smalls born on day 4 and has found that while the middle level of this lattice remains the largest, its thickest level, as measured by maximum longitude, is not the middle level.

5. Counting games

The fact that there are 1474 games born by day 3 has been known for some time. Dean Hickerson found them by hand sometime around 1974, though he may not have been the first. The best known upper and lower bounds on the number of games born by day n for larger values of n are given in [Wolfe and Fraser 2004], and depend upon observations made (in personal communications) by Dean Hickerson and Dan Hoey.

Consider the lattice of games born by $\mathcal{G}_n$. Call a pair $(\mathcal{T}, \mathcal{B})$ of antichains in this lattice *admissible* if $\mathcal{T} > \mathcal{B}$ (i.e., each game in $\mathcal{T}$ exceeds each game in $\mathcal{B}$.) The new games born by day $n+1$ are in one-to-one correspondence with admissible pairs from day n. This fact can be used to bound the number of games $g(n)$ born by day n recursively by,

$$g(n+1) \le g(n) + 2^{1+g(n)}.$$

The bound can be tightened somewhat to

$$g(n+1) \leq g(n) + 2^{g(n)} + 2,$$

or even further to

$$g(n+1) \leq g(n) + \left(g(n-1)^2 + \tfrac{5}{2}g(n-1) + 2\right) \cdot 2^{g(n)-2g(n-1)}$$

For $n \geq 2$, the right-hand side is upper bounded by

$$\left(2g(n-1)^2/4^{g(n-1)}\right) \cdot 2^{g(n)}.$$

For lower bounds, Wolfe and Fraser show that $g(n) \geq 2^{g(n-1)^\alpha}$ where $\alpha > .51$ and $\alpha \to 1$ as $n \to \infty$. For their proofs, they exploit knowledge of the join irreducibles of the day n lattice mentioned in Section 3.

It would be of interest to tighten these bounds, particularly if doing so entailed describing the relationships between day n and day $n+1$ in more detail. Is the middle level of each lattice the widest? Are the level sizes monotonic nondecreasing down to the middle level? (There are four levels with 5 games in the day 3 all-smalls.) Determine bounds on the number of all-smalls born by day n.

6. Further work

There are several other directions for further work besides those mentioned in the body of the survey.

While all of the above results were stated for short games (i.e., games born by day n for $n < \omega$), proofs by induction imply similar results for $\mathcal{G}_\alpha$ where α is a transfinite ordinal [Siegel 2006]. However, Aaron Siegel's results concerning the all-small lattice do not generalize so easily, for it is not clear what ordinal multiples of $\uparrow$ should be.

Berlekamp (personal communication) has suggested other possible definitions for games born by day n, $\mathcal{G}_n$, depending on how one defines $\mathcal{G}_0$. The usual definition is 0-based, as $\mathcal{G}_0 = \{0\}$. Other natural definitions are integer-based (where $\mathcal{G}_0$ are integers) or number-based. While these two alternatives do not yield distributive lattices, perhaps there is still combinatorial structure worth investigating.

Acknowledgment

Aaron Siegel provided insightful feedback on multiple drafts.

References

[Birkhoff 1940] G. Birkhoff, *Lattice theory*, American Mathematical Society, New York, 1940. Third edition, 1967.

[Calistrate et al. 2002] D. Calistrate, M. Paulhus, and D. Wolfe, "On the lattice structure of finite games", pp. 25–30 in *More games of no chance* (Berkeley, 2000), edited by R. Nowakowski, Math. Sci. Res. Inst. Publ. **42**, Cambridge Univ. Press, Cambridge, 2002.

[Fraser et al. 2005] W. Fraser, S. Hirshberg, and D. Wolfe, "The structure of the distributive lattice of games born by day n", *Integers* **5**:2 (2005), A6.

[Moews 1991] D. Moews, "Sums of games born on days 2 and 3", *Theoret. Comput. Sci.* **91**:1 (1991), 119–128.

[Siegel 2005] A. Siegel, *Loopy games and computation*, Ph.D. thesis, University of California, Berkeley, 2005.

[Siegel 2006] A. N. Siegel, Notes on $\mathcal{L}_\alpha$, 2006. Unpublished.

[Wolfe and Fraser 2004] D. Wolfe and W. Fraser, "Counting the number of games", *Theoret. Comput. Sci.* **313**:3 (2004), 527–532. Algorithmic combinatorial game theory.

DAVID WOLFE
wolfe@gustavus.edu

Standards

Games of No Chance 3
MSRI Publications
Volume 56, 2009

Goal threats, temperature
and Monte-Carlo Go

TRISTAN CAZENAVE

ABSTRACT. Keeping the initiative, i.e., playing sente moves, is important in
the game of Go. This paper presents a search algorithm for verifying that
reaching a goal is sente on another goal. It also presents how goals are evaluated. The evaluations of the goals are based on statistics performed on almost
random games. Related goals, such as goals and associated threatened goals,
are linked together to form simple subgames. An approximation of the temperature is computed for each move that plays in a simple subgame. The move
with the highest temperature is chosen. Experimental results show that using
the method improves a Go program.

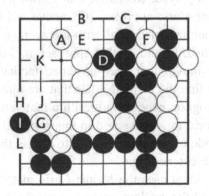

Figure 1. Examples of connections and connection threats.

1. Introduction

In Figure 1, if White plays at H, G and H are connected. Playing at H also threatens to connect H and L. However the connection between G and H is not sente on the connection between H and L: if White plays at H, Black answers at L, and White has to play at J to keep G and H connected. We say the connection between G and H is not sente on the connection between H and L.

On the other hand, for Black, the connection between I and J is sente on the connection between J and K. If Black plays at J, I and J are connected. Playing at J also threatens to connect J and K. The connection (I,J) is sente on the connection (J,K) because whatever White plays after Black J, either it does not threaten the connection between I and J, or it threatens it, but Black has an answer that both connects I and J, and keeps the threat of connecting J and K.

Moreover, all these connections are related. If White connects G and H, it prevents Black from connecting I and J. If Black connects I and J, it prevents White from connecting G and H. We aggregate them in a single structure in order to evaluate the White move at H and the Black move at J. If $Val_{(G,H)}$ is the evaluation of the connection (G,H), $Val_{(I,J)}$ the evaluation of the connection (I,J), and $Val_{(I,J),(J,K)}$ the evaluation of connecting both (I,J) and (J,K), we approximate the temperature of the White move at H with the temperature of the subgame $\{Val_{(G,H)} \| Val_{(I,J)} | Val_{(I,J),(J,K)}\}$.

A common approach to Go programming is to compute the status of tactical goals. Examples of tactical goals are connecting two strings, capturing a string or making a group live. The status of a tactical goals is assessed using heuristic search. Once unsettled goals are found, they are evaluated and the one with the highest evaluation is played. Recently, I have shown how to evaluate unsettled goals using a Monte-Carlo approach [Cazenave and Helmstetter 2005a]. It consists in evaluating an unsettled goal with the average of the random games where it has been reached. I build on this approach in this paper.

Besides finding the moves that play unsettled tactical goals, an important aspect of Go is to also find the tactical goals that are threatened by each move. In order to do this a program needs an algorithm that verifies a goal is sente on another goal. This algorithm is presented in this paper as a search algorithm. It uses search at each node of the main search to assess the statuses of the goal and of the associated threatened goal.

The evaluation of a threat uses the Monte-Carlo method, it consists in computing the average of all the random games where the goal and the threatened goal have been reached. Once the goals and the associated threats have been evaluated, they are aggregated in a single structure that is used to approximate the temperature of moves.

The rest of the paper is organized as follows. The second section discusses related work. The third section details the tactical goals used and how the program computes their status. The fourth section gives a search algorithm that verifies if reaching a goal is sente on another goal. The fifth section explains how goals are evaluated with a Monte-Carlo algorithm. The sixth section details the evaluation of moves given the evaluation of goals. The seventh section gives experimental results.

2. Related work

The use of search for assessing dependencies between goals in the game of Go [Cazenave and Helmstetter 2005b] is related to the search algorithm we present. The evaluation of goals with Monte-Carlo Go [Cazenave and Helmstetter 2005a] is related to the evaluation of goals and threatened goals in this paper. Related goals are aggregated in a structure. The structure and the evaluation of goals are used to build a combinatorial game. Thermography [Berlekamp et al. 1982] can be used to play in a sum of combinatorial games. In Go endgames, it has already been used to find better than professional play [Spight 2002], relying on a computer assisted human analysis. A simple and efficient strategy based on thermography is Hotstrat, it consists in playing in the hottest game. Hotstrat competes well with other strategies on random games [Cazenave 2002]. Another approach used to play in a sum of hot games is to use locally informed global search [Müller and Li 2005; Müller et al. 2004]. In this paper, we use Hotstrat to evaluate the subgames built with goals evaluations.

3. Tactical goals

This section deals with tactical goals of the game of Go and the related search algorithms that compute the status of the goals. Examples of tactical goals are connecting two strings, or capturing a string. Traditional Go programs spend most of their time searching tactical goals. The search finds moves that reach the goals, and these moves are then used by Go programs to choose the best move according to the evaluation of the associated goals.

3.1. Possible goals. Goals that appear frequently in a Go game are connection, separation [Cazenave 2005], capture and life. Goals are associated to evaluation functions that take values in the interval [Lost,Won]. Usually Won is a large integer and returning Won means the goal is reached, Lost is the opposite of Won and returning Lost means the goal cannot be reached. We make the distinction between positive and negative goals. A positive goal is well defined and when the associated evaluation function returns Won, it is certain that the goal is reached. Connection, separation, capture and life are positive goals. For example, the evaluation function for connections returns Won when the two stones to connect are part of the same string. Negative goals are the opposite of positive goals. The opposite of connection is disconnection, the opposite of separating is unseparating, the opposite of capturing is escaping and the opposite of living is killing. Negative goals are often ill-defined: when the evaluation returns Won for a negative goal it is not sure that it is reached. For example, the evaluation function returns Won for disconnections when the two strings to disconnect have a distance greater than four. However, there are cases when strings have a distance greater than four and can still be connected. Symmetrically, when the evaluation function returns Lost for a positive goal, it is not sure it cannot be reached.

Positive and negative goals can also be mixed. For example for the connection goal, our algorithm also verifies that the string that contains the two intersections to connect cannot be simply captured.

The empty connection goal consists in finding if an empty intersection can be connected to a string.

3.2. Finding relevant goals. For each possible goal, the program assesses if it has chances to be reached. For connections, it selects pairs of strings that are at a distance less than four. For empty connections, for all strings, it selects liberties of order one to four, as well as liberties of adjacent strings that have less than three liberties. Once the goals are selected, the program uses search to assess their status.

3.3. Searching goals. The algorithm we use to search goals is the Generalized Threats Search algorithm [Cazenave 2003]. It is fast and ensures that when a search with a positive goal returns Won the goal can be reached. For all possible goals, the program first searches if the color of the positive goal can reach the goal by playing first. If it is the case the program performs a second search to detect if the color of the positive goal can still reach the goal even if the opposite color starts playing. If both searches return Won or if the first search returns Lost, the goal is not settled and it is not necessary to play in relation to it. If the first search returns Won and the second search returns Lost then the goal is unsettled.

3.4. The traces. Two traces are associated to each search. The positive trace is a set of intersections that can possibly invalidate the result of a search that returns Won. The negative trace is a set of intersections that can possibly change the result of a search that returns Lost.

Figure 2 gives an example of a positive trace: the trace of the search where White tries to disconnect the two black stones but Black succeeds connecting them.

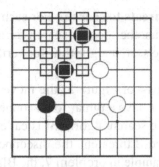

Figure 2. Positive trace of a won connection.

3.5. Finding all moves related to unsettled goals. In order to evaluate the
threats associated to the moves that reach the goal, a program has to find all
the moves that reach an unsettled goal, and all the moves that prevent it being
achieved. The program currently uses heuristic functions to generate a set of
moves that can possibly reach the goal. It could also use the negative trace of
the search that returns Lost when the opposite color of the goal plays first. It
uses the positive trace of the search that returns Won when the color of the goal
plays first, in order to generate all the moves that can possibly prevent the goal
being achieved. For each of these moves, it plays it and then performs a search
to verify if it achieves the goal for moves of the goal color, or prevents it being
achieved for moves of the opposite color.

3.6. Finding threatening moves for goals. It is interesting to look for threats
in two cases. The first case is when the positive goal search returns Lost when
the goal color plays first. In this case, all the moves of the color of the goal on
the intersections of the negative trace are tried. For each move, it is played and
a search for the goal with the color of the goal starting first is tried. If the search
returns Won, then the move is a threat to reach the goal. The second case is when
the goal search returns Won when the opposite color of the goal plays first. In
this case all the moves of the opposite color of the goal, on the intersections of
the positive trace are tried. For each move, it is played and a search for the goal
with the opposite color playing first is performed. If the search does not return
Won then the move is a threat to prevent from reaching the goal.

4. A search algorithm for verifying threats

This section describes how the potential threats are found and details the
search algorithm used to verify that reaching a goal threatens another goal.

4.1. Finding moves and threats to search. For each move that reaches a goal and that is also a threat on another goal, a search for a threat is performed. The move is played and then the search algorithm for verifying threats is called at a min node.

For example in problem 14 of Figure 5, the move at E achieves the goal of connecting A and B, and it is also a threat to disconnect C and D. So the White move at E is played and the search algorithm is called at a min node to verify that the connection of A and B threatens the disconnection of C and D. This is not always the case. For example in problem 7, the Black move at B connects A and B, and the move at B threatens to connect B and C. However the connection of A and B is not sente on the connection of B and C, because when Black plays B, White answers at C both threatening to disconnect A and B, and removing the threat of connecting B to C.

4.2. The search algorithm. A simple idea that comes to mind when trying to model threats is to play two moves in a row and use search on a double goal [Cazenave and Helmstetter 2005b] to verify whether both goals are achieved even if the opponent moves first. However this approach does not work as can be seen in Figure 3. In this figure, if Black plays at B and C in a row, White cannot prevent him from connecting A, B and C. However playing at B, connecting A and B, is not sente on the connection of B and C. If Black plays at B, White can prevent the threat in sente as in the right hand diagram of figure 3.

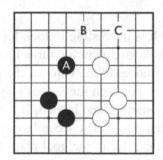 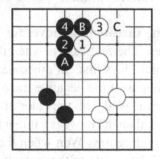

An empty connection problem   White prevents the threat in sente

Figure 3. Black B does not threaten Black C.

As the search algorithm on double goals does not work, we have designed a new search algorithm that verifies that when a first goal can be achieved, it also threatens a second goal. The first goal is called the *goal*, the threatened goal is called the *threat goal*.

The search algorithm is a selective alpha-beta, where verifications of the status of goals are searched at each node.

The pseudocode for max nodes is given below:

```
1.   MaxNode (int alpha, int beta) {
2.      search goal with max playing first;
3.      if (goal cannot be reached)
4.         return Lost;
5.      search goal with min playing first;
6.      if (goal can be reached with min playing first)
7.         return Won;
8.      if (threatGoal is Lost)
9.         return Lost;
10.     for max moves m that reach the goal {
11.        try move m;
12.        tmp = MinNode (alpha, beta);
13.        undo move;
14.        if (tmp > alpha) alpha = tmp;
15.        if (alpha >= beta) return alpha;
16.     }
17.     return alpha;
18. }
```

At max nodes, the first important thing is to verify that the goal can be achieved (lines 1–4). If not, the search stops as it is necessary to reach the goal for the algorithm to send back Won. If the goal can be achieved when Min plays first, then Min has just played a move that does not threaten the goal, and has therefore lost the initiative. So Max has reached the goal and has kept the initiative (lines 5–7). In this case the algorithm returns Won because the goal has been achieved keeping the initiative and while continuing to threaten the threat goal as it has been verified in the upper min node (compare the MinNode pseudocode). If the goal is not achieved and the threat goal is lost, then Min has succeeded in preventing Max from threatening the threat goal, so the algorithm returns Lost (lines 8–9). The only Max moves to try are the moves that achieve the goal, so the algorithm tries them and calls the MinNode function after each of them (lines 10–16).

The code for the MinNode function is as follows:

```
1.   MinNode (int alpha, int beta) {
2.      search goal with max playing first;
3.      if (goal cannot be reached)
4.         return Lost;
5.      search goal with min playing first->traceMin;
6.      if (goal cannot be reached with min playing first)
```

```
7.      return Lost;
8.      search threatGoal with max playing first->traceMax;
9.      if (threatGoal cannot be reached)
10       return Lost;
11.     if (intersection of traceMin and traceMax is empty)
12.        return Won;
13.     for min moves m in traceMin and traceMax {
14.        try move m;
15.        tmp = MaxNode (alpha, beta);
16.        undo move;
17.        if (tmp < beta) beta = tmp;
18.        if (alpha >= beta) return beta;
19.     }
20.     return beta;
21. }
```

At min nodes, the program starts verifying Max can achieve the goal if it plays first (lines 2–4), then it verifies that he can achieve the goal even if Min plays first (lines 5–7). The positive trace of this search is memorized in traceMin as it will be useful later to select the Min moves to try. Then the program verifies if the threat goal can still be achieved by Max in order to verify that Max keeps threatening it (lines 8–10). The positive trace of the threat goal is memorized in traceMax. The only moves that Min tries are on the intersection of TraceMin and TraceMax (lines 13–19).

For negative goals, the search is performed for the opposite positive goal, and the algorithm takes the opposite of the result of the search. For example, if the threat goal is to disconnect, at min nodes, the algorithm searches the connection goal with the disconnecting color playing first. If the search does not return Won, it considers that the disconnection can be achieved.

5. Evaluation of goals

This section deals with the approximate evaluation of how many points reaching a goal can gain. We use Monte-Carlo simulations to evaluate the importance of goals.

5.1. Standard Monte-Carlo Go. Standard Monte-Carlo Go consists in playing a large number of random games. The moves of the random games are chosen randomly among the legal moves that do not fill the player's eyes. A player passes in a random game when his only legal moves are on his own eyes. The game ends when both players pass. At the end of each random game, the score of the game is computed using Chinese rules (in our case, it consists in counting one point for each stone and each eye of the player's color, and subtracting the player's count from its opponent count). The program computes, for each intersection, the mean results of the random games where it has been played first by one player, and the mean for the other player. The value of a move is the difference between the two means. The program plays the move of highest value.

5.2. Evaluation of unsettled goals.

The only simple goals that need to be evaluated are the unsettled ones. For each of these, the program computes the mean score of the games where it has been achieved during the game. It also computes the mean score of the games where it has not been achieved. The difference between the two means gives an evaluation of the importance of the goal.

5.3. Evaluation of threatened goals.

For each move of each unsettled goal, the program searches for all the possible associated threat goals. For each combination of a goal and a threat goal, the program computes the mean value of the random games where they have both been achieved.

For a combination of two positive goals, things are simple. The program tests at each move of each random games if both goals have been achieved. If it is the case the game counts. If it is not the case at the end of the game, the game does not count.

For goals that are a combination of a positive goal and of a negative goal we have a special treatment. For example for connection moves that threaten a disconnection, there are four intersections to take into account. The intersections to connect are s_1 and s_2, the intersections to disconnect are s_3 and s_4. The following tests are performed at each move of each random game :

- if s_1 or s_2 are empty or of the opposite color of the connection, the random game does not count.
- if s_1 and s_2 are in the same string, and if s_3 or s_4 are empty or of the color of the connection, the random game counts.
- if s_3 and s_4 are in the same string, the random game does not count.

At the end of a random game, if the two strings s_1 and s_2 have been connected, and s_3 and s_4 have not, the random game counts.

6. Evaluation of moves

This section explains the different values computed for each move, details how the values are computed for the empty connection goal and for the connection goal, and eventually gives the evaluation of the moves based on these values.

6.1. Values computed for each move. There are basically four values that may be computed for a move. The value of achieving the associated goal (the FriendValue), the value for the opposite color of preventing the goal from being achieved (the EnemyValue), the value of reaching both the goal and the best associated threat goal (the FriendThreatValue), and the value for the opposite color of preventing the goal from being achieved and achieving another enemy threat goal (the EnemyThreatValue).

6.2. Values for empty connection moves. For each empty connection move, the program finds the set of empty connections of the opposite color that are invalidated by the move. The EnemyValue is set to the value of the highest invalidated enemy empty connection. The EnemyThreatValue is set to the highest threat associated to the selected enemy empty connection. The FriendValue is set to the evaluation of reaching the empty connection, and the FriendThreatValue is set to the value of the highest threat goal associated to the empty connection. The only empty connection threats computed for a friend empty connection goal, are the threats of empty connection to an intersection that is not already empty connected to a friend group. Similarly, enemy empty connection threats do not empty connect to intersections already empty connected to enemy groups.

6.3. Values for connection moves. For connection moves, the FriendValue is the evaluation of achieving the connection, the EnemyValue is the value of not achieving the connection, the FriendThreatValue is the best evaluation among all the threat goals associated to the connection move, and the EnemyThreatValue is the best threat associated to the disconnection.

6.4. Evaluation of moves given their associated values. Once the four values are computed for each move of each unsettled goal, an approximation of the temperature of the moves can be computed. The computation of an approximate temperature given these values is based on the computation of the temperature of the game {{FriendThreatValue | FriendValue} || {EnemyValue | EnemyThreatValue}}.

The thermograph is exact for connection values since all the enemy's options have the same EnemyValue, and all the friend's options have the same FriendValue. However, for empty connections there are different options for the enemy that lead to different EnemyValue, instead of reflecting this in the thermograph, we only take the best the subgame with the best EnemyValue. So we only compute an approximation of the temperature for empty connections.

When there are no friend threats, FriendThreatValue is set to FriendValue, and when there are no enemy threats, EnemyThreatValue is set to EnemyValue. The code for computing the approximate temperature given the four values is as follows (ABS is the absolute value):

```
1. temperature (EnemyThreatValue, EnemyValue,
                    FriendValue, FriendThreatValue) {
2.    tempEnemy = ABS(EnemyThreatValue - EnemyValue)/2;
3.    tempFriend = ABS(FriendThreatValue - FriendValue)/2;
4.    width = ABS(FriendValue-EnemyValue);
5.    if (tempFriend - tempEnemy > width)
6.       return tempEnemy + width;
7.    else if (tempEnemy - tempFriend > width)
8.       return tempFriend + width;
9.    else
10.      return ABS(FriendThreatValue/4 + FriendValue/4 -
                    EnemyValue/4 - EnemyThreatValue/4);
11. }
```

When the program does not use threats, it returns ABS(FriendValue - Enemy-Value)/2.

7. Experimental results

We have measured the speed and the correctness of the search algorithm that verifies if goals are sente on other goals. We have also measured the benefits a program gets when taking into account the values of threats, and the average time it takes to compute threats.

For the experiments we have only used the connection goal and the empty connection goal. The machine used is a 3.0 GHz Pentium 4. The search algorithm that detects threats has been programmed using templates for the goals. It means that the same search code is used for any combination of goals. The search uses iterative deepening, transposition tables, two killer moves and the history heuristic.

7.1. The test suite for threats. The problems used to test the algorithms are given in Figure 5. In the first eleven problems, an empty connection between A and B threatens (or not) an empty connection between B and C. Problem number twelve is an example of a problem where an empty connection between A and B threatens a connection between A and C. Problem thirteen shows an empty connection between A and B that threatens to disconnect C and D. In problem fourteen, the black move at E connects A and B, and threatens to disconnect C and D.

Table 1 gives the number of moves played in the search algorithm and the time it takes for each of the problems of the test suite. All the problems are correctly solved by the search algorithm.

Problem	moves	time (ms)
1	2,095	10
2	34,651	50
3	59,704	90
4	6,063	30
5	68,209	90
6	2,522	30
7	24	10
8	4,982	20
9	91,774	120
10	2,083	10
11	17	20
12	263	10
13	10,529	30
14	1,341	20
total	284,405	540

Table 1. Nodes and time for the threat problems.

7.2. Integration in a Go program. The use of connection threats has been tested in a Monte-Carlo based program which evaluates goals. The program is restricted to connection and empty connection goals. The threats used are only connection threats: the empty connections that threaten empty connections, the empty connections that threaten connections, the empty connections that threaten disconnections and the connections that threaten disconnections.

The experiment consists in playing one hundred 9×9 games against Gnugo 3.6, fifty as Black and fifty as White. At each move of each game, the program plays ten thousand random games before choosing its move. Table 2 gives the mean score, the variance, the number of won games and the average time per move of the program with and without connection threats against Gnugo 3.6.

Using connection threats enables to gain approximately nine points per 9×9 game, and sixteen more games on a total of one hundred games. This is an

Algorithm	mean	std deviation	won games	time (s)
Without threats	−22.33	24.80	11/100	5.7
With connection threats	−13.02	25.06	27/100	11.9

Table 2. Score and time for the Go program with and without connection threats.

encouraging result given that the program only uses connections and connection threats. An interesting point is that the mean of the games where the Monte-Carlo based program was black, is -6.9 including komi, and the mean with white is -19.1 also including komi. The program is much better with black than with white. Looking at the games, a possible explanation is that the Monte-Carlo program does not use any life and death search, it loses eight games by 75.5 as white, and only one game by 86.5 as black. Many of the games it loses completely are due to lack of life and death search and evaluation.

7.3. Over-evaluation of threats. The values computed for goals with the Monte-Carlo algorithm already take into account, to a certain extent, the threats associated to the goals. In some cases, this can be misleading for the program as it overevaluates the value of some threats. An example is given in Figure 1. White B threatens White C, and invalidates Black E. The connection (A,B) evaluates to -2.2, (D,E) to 12.4, (A,B) and (B,C) to -14.2. So the temperature of White B is evaluated to 10.3. The connection (G,H) evaluates to -5.8, (I,J) to 8.1, (G,H) and (II,L) to -9.4, (I,J) and (J,K) to 15.5. So the temperature of White H evaluates to 9.7. The program prefers White B to White H, which is bad.

One problem here is overevaluation of the threat for White of connecting (A,B) and (B,C). In the random games where both connections are reached, one half of the games also have C and F connected, which makes a big difference in the final score. However, if Black plays well, C and F never get connected, and the value of the threat is much lower. A possible solution to the problem of the overevaluation of threats could be to make the program play better during the Monte-Carlo games [Bouzy 2005].

8. Conclusion and future work

An algorithm to verify if reaching a goal threatens another goal has been described, as well as its incorporation in a Go program. It has also been shown how unsettled goals and the associated threat goals, found by this algorithm, can be evaluated in a Monte-Carlo Go framework. An approximation of the temperature has been used to evaluate moves given the related goal evaluations. Results on a test suite for threats have been detailed. The use of connection threats in a Go program improves its results by approximately nine points for 9×9 games against Gnugo 3.6.

There are many points left for future work. First, it would be interesting to test the algorithm with other goals than connection ones. Second, the program currently often overestimates the values of threats. An improvement of the Monte-Carlo algorithm in order to make it play less randomly could address this

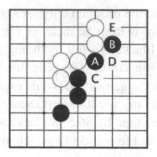

Figure 4. Moves can avoid threats.

point. Third, the interactions between goal threats and combination of goals are also interesting to explore. Fourth, the program currently evaluates the value of playing threats but does not take into account the value of playing moves that invalidate threats: for example, in Figure 4, the Black moves at C and D both connect A and B, Black D invalidates the threat of White E, but Black C does not invalidate the threat of White E. Eventually, work remains to be done to take into account the different options of the enemy when building the thermograph for empty connections.

References

[Berlekamp et al. 1982] E. Berlekamp, J. H. Conway, and R. K. Guy, *Winning Ways*, Academic Press, 1982.

[Bouzy 2005] B. Bouzy, "Associating domain-dependent knowledge and Monte Carlo approaches within a go program", *Information Sciences* **175**:4 (2005), 247–257.

[Cazenave 2002] T. Cazenave, "Comparative evaluation of strategies based on the value of direct threats", in *Board Games in Academia V* (Barcelona, Spain), 2002.

[Cazenave 2003] T. Cazenave, "A generalized threats search algorithm", pp. 75–87 in *Computers and Games 2002* (Edmonton, Canada), Lecture Notes in Computer Science **2883**, Springer, New York, 2003.

[Cazenave 2005] T. Cazenave, "The separation game", in *JCIS 2005* (Salt Lake City, UT), 2005.

[Cazenave and Helmstetter 2005a] T. Cazenave and B. Helmstetter, "Combining tactical search and Monte-Carlo in the game of Go", in *CIG'05* (Colchester, UK), 2005.

[Cazenave and Helmstetter 2005b] T. Cazenave and B. Helmstetter, "Search for transitive connections", *Information Sciences* **175**:4 (2005), 284–295.

[Müller and Li 2005] M. Müller and Z. Li, "Locally informed global search for sums of combinatorial games", in *Computers and Games 2004* (Ramat-Gan, Israel), Lecture Notes in Computer Science **3864**, Springer, 2005.

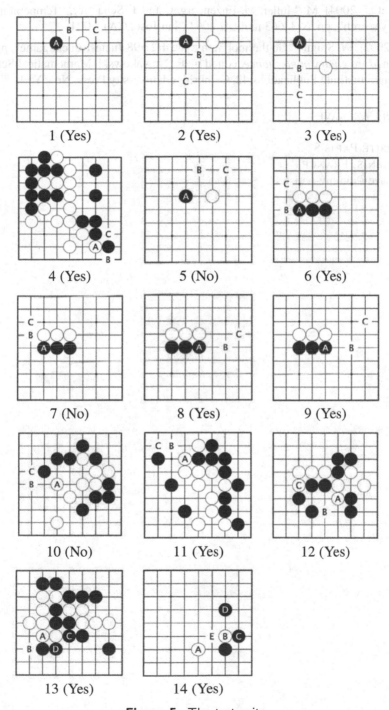

Figure 5. The test suite.

[Müller et al. 2004] M. Müller, M. Enzenberger, and J. Schaeffer, "Temperature discovery search", pp. 658–663 in *AAAI 2004* (San Jose, CA), 2004.

[Spight 2002] W. Spight, "Go thermography: the 4/21/98 Jiang-Rui endgame", pp. 89–105 in *More games of no chance*, edited by R. Nowakowski, Mathematical Sciences Research Institute Publications **42**, Cambridge University Press, New York, 2002.

TRISTAN CAZENAVE
LIASD
UNIVERSITÉ PARIS 8
SAINT DENIS, FRANCE
   cazenave@ai.univ-paris8.fr

Games of No Chance 3
MSRI Publications
Volume **56**, 2009

A puzzling Hex primer

RYAN B. HAYWARD

ABSTRACT. We explain some analytic methods that can be useful in solving Hex puzzles.

1. Introduction

Solving Hex puzzles can be both fun and challenging. In this paper — a puzzling companion to *Hex and Combinatorics* [5] and *Dead Cell Analysis in Hex and the Shannon Game* [2], both written in tribute to Claude Berge — we illustrate some theoretical concepts that can be useful in this regard.

We begin with a quick review of the rules, history, and classic results of Hex. For an in depth treatment of these topics, see [5].

The parallelogram-shaped board consists of an $m \times n$ array of hexagonal cells. The two players, say Black and White, are each assigned a set of coloured stones, say black and white respectively, and two opposing sides of the board, as indicated in our figures by the four stones placed off the board. In alternating turns, each player places a stone on an unoccupied cell. The first player to connect his or her two sides wins.

In the fall of 1942 Piet Hein introduced the game, then called Polygon, to the Copenhagen University student science club *Parenthesis*. Soon after, he penned an article on the game for the newspaper *Politiken* [6; 8; 9]. In 1948 John Nash independently reinvented the game in Princeton [4; 10], and in 1952 he wrote a classified document on it for the Rand Corporation [11]. In 1957 Martin Gardner introduced Hex to a wide audience via his *Mathematical Games* column [3], later reprinted with an addendum as a book chapter [4].

For Hex played on an $m \times n$ board, the game cannot end in a draw (Hein [6], Nash [11]); for $m = n$, there exists a winning strategy for the first player

The support of NSERC is gratefully acknowledged.

(Hein, Nash [11]; see also [3]); for $m < n$, there exists a winning strategy for the player whose sides are closer together, even if the other player moves first (Gardner/Shannon [4]); for arbitrary Hex positions, determining the winner is PSPACE-complete (Reisch [12]).

To start our discussion, consider Puzzle 1:

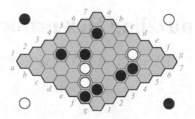

Puzzle 1. An easy warm-up. White to play and win.

2. Virtual connections

One useful Hex concept is that of a *virtual connection*, namely a subgame in which one player can establish a connection even if the opponent moves first. In Puzzle 1, as shown in the left diagram of in Figure 1 below, the cell set {d7, e7} forms a 'bridge' virtual connection between the white stone at e6 and the white border on the upper right side. If Black ever plays at one of these two bridge cells, White can then make the connection by playing at the other. Similarly, the white border on the lower left side is virtually connected to the two white stones at {d3, e2} via the cell set {c1, c2, c3, d1, d2, e1}: if Black plays at any of c1, c2, c3, d1, d2 White can then play at e1, whereas if Black plays at e1 White can then play at c2 and subsequently make use of the resulting bridge cell sets {c1, d1} and {c3, d2}.

As suggested by Figure 1, left, the gap between the two white groups is an obvious place to look for a winning move; the right diagram shows such a move at e4. After this move, the new stone is virtually connected by the upper eight

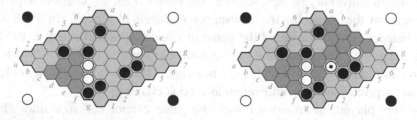

Figure 1. Two white virtual connections (left) and, after a winning move, a side-to-side white virtual connection (right).

marked cells to the upper white side, and by a bridge to {d3, e2}, and so then by the lower six marked cells to the lower white side, yielding a virtual connection joining the two white sides. Thus e4 is a winning move for Puzzle 1.

3. Mustplay regions

Are there any other winning moves for Puzzle 1?

Hex is a game in which it is easy to blunder. Even from obviously won positions, there are usually many moves that lead to quick losses. Since there are no draws in Hex, one way to answer the above question is to first check whether any losing moves can be identified. A *weak connection* is a subgame in which one player can force a connection if allowed to play first. Does the opponent have any side-to-side, and so win-threatening, weak connections?

A virtual connection for a player is *winning* if it connects the player's two sides; a *win-set* is the set of cells of a winning virtual connection. Analogously, a weak connection for a player is *win-threatening* if it connects the player's two sides; a *weak win-set* is the cell set of a win-threatening weak connection. The first three parts of Figure 2 show three black weak win-sets for Puzzle 1.

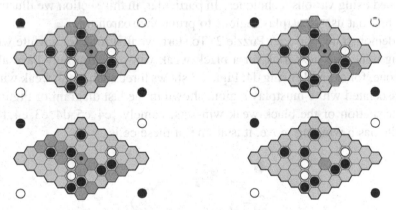

Figure 2. For Puzzle 1, three black weak win-sets and the resulting white mustplay region. This region has only one cell, so White has only one possible winning move.

Notice in the figure that, in order to prevent Black from winning, White's next move must intersect each of Black's weak win-sets, since any weak connection that is not intersected by White's move can be turned into a virtual connection on Black's subsequent move. More generally, at any point in a Hex game, *a move is winning if it intersects all of an opponent's weak win-sets.*[1]

---

[1] The converse of this statement holds as long as the opponent has at least one weak win-set; then a move is winning if and only if it intersects all of an opponent's weak win-sets. However, if the player about to

A *gamestate* specifies a *boardstate*, or board configuration, and whose turn it is to move. With respect to a player, a gamestate, and a collection of opponent weak win-sets, we call the combined intersection of these weak win-sets the *mustplay region*, since a player 'must play' there or lose the game.

As shown in Figure 2, the white mustplay region associated with the three weak connections is {e4}. We have already seen that e4 is a winning move for Puzzle 1; our mustplay analysis tells us that every other move loses. So, to answer the question from the start of this section, there are no other winning moves for Puzzle 1.

4. A Hex solver based on mustplay analysis

There is a straightforward way to solve any Hex puzzle: completely explore the search tree resulting from all possible continuations of the puzzle. This approach is usually impractical, as the number of different gamestates in the search tree is exponential in the number of unoccupied cells. Since solving Hex puzzles is PSPACE-complete, there is unlikely to be any 'fast', namely polynomial time, Hex-solving algorithm. Nonetheless, the search tree can often be pruned using various techniques. In particular, in this section we illustrate an algorithm that uses mustplay regions to prune the search tree.

To demonstrate, consider Puzzle 2. To start, we first look for a white win-set. Finding none, we next look for a black weak win-set. You may have already found one, for example using d4; Figure 3 shows three such black weak win-sets. The associated white mustplay region, shown in the last diagram of Figure 3, is the intersection of the black weak win-sets, namely {c4, c5, d4, e3, e4, f2, f4}. If White has a winning move, it is at one of these cells.

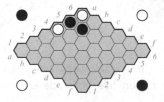

Puzzle 2. A more challenging problem. White to play and win.

Figure 4 shows what happens as, in no particular order, we next consider the moves of this mustplay region. In the first diagram we make the white move at c5; by continuing to recursively apply our algorithm, we eventually discover that Black wins the resulting gamestate with the black win-set as shown. At this

---

move is so far ahead in the game that the opponent has no weak win-set, then the intersection of all of the opponent's weak win-sets is the empty set; thus the converse does not hold in such cases.

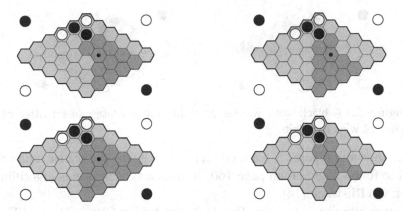

Figure 3. For Puzzle 2, three black weak win-sets and the resulting white mustplay region.

point we undo the white move, so the black win-set becomes a black weak win-set. We next use this black weak win-set to update the white mustplay region; it becomes reduced to {c4, d4, e3, e4, f2, f4}. In similar fashion, we eventually discover that the next three white moves considered, namely d4, e3, e4, also lose for White; the resulting black weak win-sets are shown in Figure 4. Notice that the last of these weak win-sets does not contain f4, so by this point the white mustplay region has been reduced to {c4, f2}.

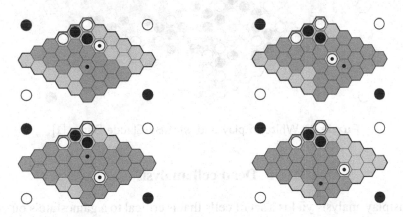

Figure 4. Black weak win-sets after moves c5, d4, e3, e4 respectively.

Figure 5 shows what happens as we consider these last two possible moves. The white move at f2 loses, but the white move at c4 wins. Thus c4 is the unique winning move for Puzzle 2.

We have omitted all the details from the recursive calls of this algorithm. We leave as exercises for the reader to verify that the five weak win-sets and the one

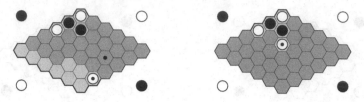

Figure 5. A black weak win-set after f2, and a white win-set after c4. Thus c4 wins for White.

win-set shown in Figures 3–5 are correct.[2] As a guide, the reader might find it useful to follow Figure 6 on page 160; it gives a version of this algorithm due to Jack van Rijswijck [14].

Another exercise is to solve Puzzle 3, created by Claude Berge. There is more than one solution; running down the upper-left region is straightforward, while breaking through to the upper-right side is more difficult. Try to find a win-set with no unnecessary cells. One such win-set appears in the last section (page 159).

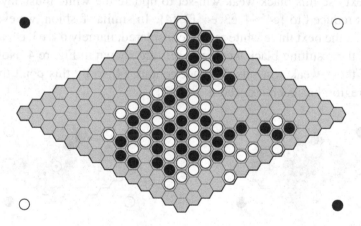

Puzzle 3. White to play and win. By Claude Berge [1].

5. Dead cell analysis

Mustplay analysis yields a set of cells that is critical to a gamestate's outcome. A different form of analysis is based on recognizing individual cells that are irrelevant. We illustrate this 'dead cell analysis' by working through Puzzle 4, created by Piet Hein.

---

[2]The most challenging of these exercises is the last one, namely to show that c4 wins for White. The strongest next moves for Black include c3, c5, c6, d3, and e2; respective winning replies for White include d3, e4, e5, e4, and *d3*. For other exercises on small boards, see the opening theory link on Jack van Rijswijck's Queenbee webpage [13].

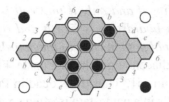

Puzzle 4. White to play and win. By Piet Hein [7].

A *completion* of a boardstate is any boardstate obtained by filling all vacant cells of the given boardstate with any combination of black and/or white stones. A cell of a boardstate is *dead* if, for every possible completion, changing the colour of the stone on the given cell does not alter the winner of the completion. A cell is *live* if it is not dead.

For example, the boardstate of Puzzle 4 has 25 vacant cells and so has 2^{25} completions. We leave it to the reader to consider a sample of these completions and verify that in each case, changing the colour of the stone at cell d1 does not change the winner of the completion. Thus, in this boardstate d1 is dead.

A gamestate is *undecided* if neither player has yet won. A useful feature of dead cells is that *placing or removing a stone of either colour at a dead cell does not alter the gamestate's winner.* Therefore *every undecided gamestate with a winning move has a winning move to a live cell.*

Thus, dead cells can be safely pruned from the search tree of a gamestate.

Happily for Hex puzzlers, dead cells can be recognized without having to consider all of a boardstate's completions. The left diagram in Figure 6 is the *white adjacency graph* for the Puzzle 4 boardstate. The nodes of the graph correspond to the vacant board cells; additionally, two terminal nodes represent the white borders. In the graph, a pair of nodes is joined by an edge if the corresponding cells touch or are joined by connecting white stones.

A path is *induced* if it has no 'shortcuts', namely if the only edges among vertices of the path are between pairs of vertices that are consecutive in the path. The following characterization is an easy consequence of the definition of dead.

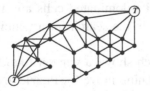
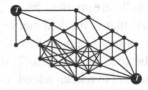

Figure 6. White and black adjacency graphs for Puzzle 4.

A cell with a stone is live if and only if that cell is live after removing that stone. A vacant cell of a boardstate is live if and only if the cell is in some induced terminal-to-terminal path in each of the boardstate's adjacency graphs.

Notice that the white adjacency graph for Puzzle 4 has no induced terminal-to-terminal path that contains d1. Thus d1 is dead in Puzzle 4, as are a1 and c1.

The number of dead cells in a gamestate is often small. However, considering cells that can be 'killed' allows further possible moves to be ignored. In Puzzle 4 it would be pointless for White to play at the *white-vulnerable* cell e2, since a Black response at d3 would kill a white stone at f2.

This line of reasoning can be continued. Black has a 'second-player kill' strategy for {f2, f3}: if White ever plays at one of these cells, Black can reply at the other, leaving one cell black and the other dead. We say this set is *black-captured*, since assuming that these cells are already occupied by black stones does not change the theoretical outcome of the game. As an exercise, the reader should verify that {f1, e2, f2, f3} is black-captured. It suffices to find, for the subgame played on these cells, a second-player strategy for Black that leaves every stone black or dead.

The notion of *dominated* is analogous to the notion of captured. In Puzzle 4 {a6, b5, b6} is *white-dominated*, since White has a first-player strategy for the subgame on these cells that leaves every stone white or dead. The first move in this strategy is to b5, so for this strategy b5 is *white-dominating* and the remaining cells are *white-dominated*. When White is searching for a winning move, it is sufficient to consider among the cells of a white-dominated set only the dominating cell since after moving there the remaining cells become white-captured.

To summarize these ideas, let us complete our analysis of Puzzle 4. It is White's turn to move. The cells in {a1, c1, d1}, {f1, e2, f2, f3}, and {a2, b2} are respectively dead, black-captured, and white-captured. After white- and black-captured stones have been added to the board, the cells in {d3, f4, f5} are white-vulnerable, as they would be killed by respective responses, and subsequent black-capturing, at d4, e4, e5. The sets {b4, a4, b3}, {b5, a6, b6}, {e5, d6, e6}, {f5, e6, f6} are white-dominated by b4, b5, e5, f5 respectively.

This analysis is illustrated in the first diagram of Figure 7, where dead cells are indicated with grey circles, captured stones are marked with dots, white-vulnerable cells are marked by 'v', and white-dominated cells are marked by 'x'. Any cell that is marked can be ignored in the search for a winning move, so there are only six cells left to consider.

As can be seen from Figure 7, right, which shows a win-set found after the captured stones have been added, a4 is a winning move for Puzzle 4. We leave it to the reader to check whether there are any other winning moves.

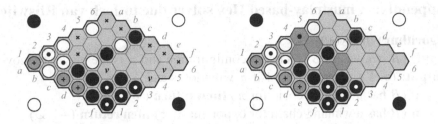

Figure 7. Dead, captured, white-dominated, and white-vulnerable cells of Puzzle 4 (left), and, after dead and captured stones are added, a black weak win-set (right).

6. A win-set for Puzzle 3

Berge designed Puzzle 3 (page 156) to be a study rather than a puzzle, so there is more than one winning move. A solution that involves play in the upper right region of the board appears in [5].

Another solution is to start at c11, and use the threat of connecting the top white group of three stones with the white line ending at e5 to force play towards the lower white border. A win-set for this solution, verified by a computer program written by Van Riswijck, is shown in Figure 8. This win-set is minimal, in that it contains no unnecessary cells; if any cell of the win-set is removed and black stones are then placed at all vacant cells and the one removed cell of the win-set, then White can no longer win. As a final exercise, we leave it to the reader to find a winning strategy that uses only the cells of this win-set. An answer appears in Van Rijswijck's doctoral thesis [15].

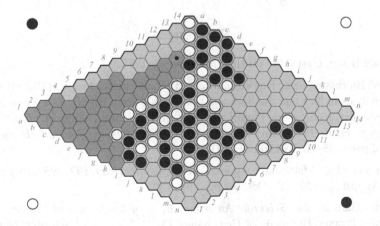

Figure 8. A white win-set for Puzzle 3.

Appendix: A mustplay-based Hex solver due to Jack van Rijswijck

Algorithm WINVALUE
Input: (B, π), where B is a board configuration and π is the player to move
Output: (v, X), where v is 1/-1 if π wins/loses and X is a win-set
    if (B has a winning chain for π) **then return** $(+1, \varnothing)$
    if (B has a winning chain for opponent of π) **then return** $(-1, \varnothing)$
    $W \leftarrow \varnothing$ [W is the cell set of a winning virtual connection]
    $M \leftarrow$ unoccupied cells of B [M is the must-play]
    while ($M \neq \varnothing$)
        $m \leftarrow$ any cell in M
        $B' \leftarrow$ board configuration after adding to B at cell m a stone of π's
        $\pi' \leftarrow$ opponent of π
        $(v, S) \leftarrow$ WINVALUE(B', π')
        if ($v = -1$) **then return** $(+1, S \cup \{m\})$
        $W \leftarrow W \cup S; \quad M \leftarrow M \cap S$
    endwhile
    return $(-1, W)$

Acknowledgements

I thank Jack van Rijswijck for his solution to the Berge puzzle, Cameron Browne and Jack for supplying Hex-drawing software, Mike Johanson, Morgan Kan, and Broderick Arneson for their programming support, Bjarne Toft and Thomas Maarup for providing Hein references and translations, Philip Henderson for critical feedback, and Richard Nowakowski for organizing the BIRS conference and encouraging me to finish this chapter.

References

[1] Claude Berge. L'art subtil du Hex. Manuscript, 1977.

[2] Yngvi Björnsson, Ryan Hayward, Michael Johanson, and Jack van Rijswijck. Dead cell analysis in Hex and the Shannon Game. In Adrian Bondy, Jean Fonlupt, Jean-Luc Fouquet, Jean-Claude Fournier, and Jorge L. Ramirez Alfonsin, editors, *Graph Theory in Paris: Proceedings of a Conference in Memory of Claude Berge (GT04 Paris)*, pages 45–60. Birkhäuser, 2007.

[3] Martin Gardner. Mathematical games. *Scientific American*, 197, 1957. July pp. 145–150; August pp. 120–127; October pp. 130–138.

[4] Martin Gardner. *The Scientific American Book of Mathematical Puzzles and Diversions*, chapter The game of Hex, pages 73–83. Simon and Schuster, New York, 1959.

[5] Ryan Hayward and Jack van Rijswijck. Hex and combinatorics (formerly Notes on Hex). *Discrete Mathematics*, 306:2515–2528, 2006.

[6] Piet Hein. Vil de lacre Polygon? Article in *Politiken* newspaper, 26 December 1942.

[7] Piet Hein. Polygon. Article in *Politiken* newspaper, 3 February 1943.

[8] Thomas Maarup. Hex – everything you always wanted to know about Hex but were afraid to ask. Master's thesis, Department of Mathematics and Computer Science, University of Southern Denmark, Odense, Denmark, 2005.

[9] Thomas Maarup. Hex webpage, 2005. See http://maarup.net/thomas/hex/.

[10] Sylvia Nasar. *A Beautiful Mind*. Touchstone, New York, 1998.

[11] John Nash. Some games and machines for playing them. Technical Report D-1164, Rand Corp., 1952.

[12] Stefan Reisch. Hex ist PSPACE-vollständig. *Acta Informatica*, 15:167–191, 1981.

[13] Jack van Rijswijck. Queenbee's home page, 2000. See http://www.cs.ualberta.ca/~queenbee.

[14] Jack van Rijswijck. Search and evaluation in Hex. Technical report, University of Alberta, 2002. See http://www.javhar.net/javharpublications.

[15] Jack van Rijswijck. *Set colouring games*. PhD thesis, University of Alberta, Edmonton, Canada, 2006.

RYAN B. HAYWARD
DEPT. OF COMPUTING SCIENCE
UNIVERSITY OF ALBERTA
EDMONTON, AB
CANADA
hayward@cs.ualberta.ca

Games of No Chance 3
MSRI Publications
Volume **56**, 2009

Tigers and Goats is a draw

LIM YEW JIN AND JURG NIEVERGELT

ABSTRACT. Bagha Chal, or "Moving Tiger", is an ancient Nepali board game
also known as *Tigers and Goats*. We briefly describe the game, some of its
characteristics, and the results obtained from an earlier computer analysis. As
in some other games such as Merrill's, play starts with a placement phase
where 20 pieces are dropped on the board, followed by a sliding phase during
which pieces move and may be captured. The endgame sliding phase had
been analyzed exhaustively using retrograde analysis, yielding a database consisting of 88,260,972 positions, which are inequivalent under symmetry. The
placement phase involves a search of 39 plies whose game tree complexity is
estimated to be of the order 10^{41}. This search has now been completed with the
help of various optimization techniques. The two main ones are: confronting a
heuristic player with an optimal opponent, thus cutting the search depth in half;
and constructing a database of positions halfway down the search tree whose
game-theoretic value is determined exhaustively. The result of this search is
that *Tigers and Goats* is a draw if played optimally.

1. Introduction

Bagha Chal, or "Moving Tiger", is an ancient Nepali board game, which
has recently attracted attention among game fans under the name *Tigers and
Goats*. This game between two opponents, whom we call "Tiger" and "Goat",
is similar in concept to a number of other asymmetric games played around
the world — asymmetric in the sense that the opponents fight with weapons of
different characteristics, a feature whose entertainment value has been known
since the days of Roman gladiator combat.

On the small, crowded board of 5 x 5 grid points shown in Figure 1, four
tigers face up to 20 goats. A goat that strays away from the safety of the herd
and ventures next to a tiger gets eaten, and the goats lose if too many of them get
swallowed up. A tiger that gets trapped by a herd of goats is immobilized, and
the tigers lose if none of them can move. Various games share the characteristic

163

that a multitude of weak pieces tries to corner a few stronger pieces, such as "Fox and Geese" in various versions, as described in "Winning ways" [BCG 2001] and other sources.

The rules of *Tigers and Goats* are simple. The game starts with the four tigers placed on the four corner spots (grid points), followed by alternating moves with Goat to play first. In a *placement phase*, which lasts 39 plies, Goat drops his 20 goats, one on each move, on any empty spot. Tiger moves one of his tigers according to either of the following two rules:

- A tiger can slide from his current spot to any empty spot that is adjacent and connected by a line.
- A tiger may jump in a straight line over any single adjacent goat, thereby killing the goat (removing it from the board), provided the landing spot beyond the goat is empty.

If Tiger has no legal move, he loses the game; if a certain number of goats have been killed (typically five), Goat loses.

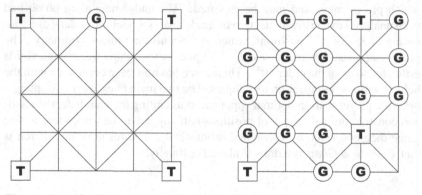

Figure 1. Left: The position after the only (modulo symmetry) first Goat move that avoids an early capture of a goat. At Right: Tiger to move can capture a goat, but thereafter Goat suffocates the tigers with a forcing sequence of 5 plies (challenge: find it).

These rules are illustrated in Figure 2, which also show that Goat loses a goat within 10 plies unless his first move is on the center spot of a border.

The 39-ply placement phase is followed by the *sliding phase* that can last forever. Whereas the legal Tiger moves remain the same, the Goat rule changes: on his turn to play, Goat must slide any of his surviving goats to an adjacent empty spot connected by a line. If there are 17 or fewer goats on the board, 4 tigers cannot block all of them and such a move always exists. In some exceptional cases (which arise only if Goat cooperates with Tiger) with 18 or more goats, the 4 tigers can surround and block off a corner and prevent any goat moves. Since Goat has no legal moves, he loses the game.

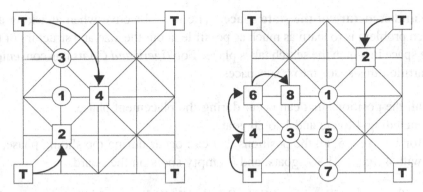

Figure 2. Goat has 5 distinct first moves (ignoring symmetric variants). All but the first move shown in Figure 1 lead to the capture of a goat within at most 10 plies, as these two forcing sequences show. At right, Tiger's last move 8 sets up a double attack against the two goats labeled 1 and 3.

Although various web pages that describe *Tigers and Goats* offer advice on how to play the game, we have found no expert know-how about strategy and tactics. Plausible rules of thumb about play include the following. First, it is obvious that the goats have to hug the border during the placement phase — any goat that strays into the center will either get eaten or cause the demise of some other goat. Goat's strategy sounds simple: first populate the borders, and when at full strength, try to advance in unbroken formation, in the hope of suffocating the tigers. Unfortunately, this recipe is simpler to state than to execute. In contrast, we have found no active Tiger strategy. It appears that the tigers cannot do much better than to wait, "doing nothing" (just moving back and forth), until near the end of the placement phase. Their goal is to stay far apart from each other, for two reasons: to probe the full length of the goats' front line for gaps, and to make it hard for the goats to immobilize all four tigers at the same time. Tiger's big chance comes during the sliding phase, when the compulsion to move causes some goat to step forward and offers Tiger a forcing sequence that leads to capture. Thus, it seems that Tiger's play is all tactics, illustrating chess Grandmaster Tartakover's famous pronouncement: "Tactics is what you do when there is something to do. Strategy is what you do when there is nothing to do".

2. Results of a previous investigation

Our earlier investigation with the goal of solving *Tigers and Goats* had given us partial results and a good understanding of the nature of this game, but we fell short of achieving an exhaustive analysis in the sense of determining the outcome: win, loss or draw, under optimal play. Here we summarize the main insights reported in [Lim 2004].

Size and structure of the state space. The first objective when attacking any search problem is to learn as much as possible about the size and structure of the state space in which the search takes place. For *Tigers and Goats* it is convenient to partition this space into 6 subspaces:

S_0: all the positions that can occur during the placement phase,
     including 4 tigers and 1 to 20 goats.
S_k: for $k = 1 \ldots 5$, all the positions that can occur during the sliding phase,
     with 4 tigers, $21 - k$ goats, and k empty spots on the board.

Notice that any position in any S_1 to S_5 visually looks exactly like some position in S_0, yet the two are different positions: in S_0, the legal Goat moves are to drop a goat onto an empty spot, whereas in S_1 to S_5, the legal moves are to slide one of the goats already on the board. For each subspace S_1 to S_5, a position is determined by the placement of pieces on the board, which we call the board image, and by the player whose turn it is to move. Thus, the number of positions in S_1 to S_5 is twice the number of distinct board images.

   For S_0, however, counting positions is more difficult, since the same board image can arise from several different positions, depending on how many goats have been captured. As an example consider an arbitrary board image in S_5, hence with 16 goats and 5 empty spots. This same board image could have arisen, as an element of S_0, from ten different positions, in which 0, 1, 2, 3, or 4 goats have been captured, and in each case, it is either Tiger's or Goat's turn to move. Although for board images in S_1 through S_4 the multiplier is less than 10, these small subspaces do not diminish the average multiplier by much. Thus, we estimate that the number of positions in S_0 is close to 10 times the number of board images in S_0, which amounts to about 33 billion.

   Since the game board has all the symmetries of a square that can be rotated and flipped, many board positions have symmetric "siblings" that behave identically for all game purposes. Thus, all the spaces S_0 to S_5 can be reduced in size by roughly a factor of 8, so as to contain only positions that are pairwise inequivalent. Using Polya's counting theory [Polya 1937] we computed the exact size of the symmetry-reduced state spaces S_1 to S_5, and of the board images of S_0, as shown in Table 1.

   S_0 is very much larger than all of S_1 to S_5 together, and has a more complex structure. Due to captures during the placement phase, play in S_0 can proceed back and forth between more or fewer goats on the board, whereas play in the sliding phase proceeds monotonically from S_k to S_{k+1}. These two facts suggest that the subspaces are analyzed differently: S_1 to S_5 are analyzed exhaustively using retrograde analysis, whereas S_0 is probed selectively using forward search [Gasser 1996].

	# of board images	# of positions
S_0	3,316,529,500	~33,000,000,000
S_1	33,481	66,962
S_2	333,175	666,350
S_3	2,105,695	4,211,390
S_4	9,469,965	18,939,930
S_5	32,188,170	64,376,340

Table 1. Number of distinct board images and positions for corresponding subspaces

Database and statistics for the sliding phase. Using retrograde analysis [Wu 2002] we determined the game-theoretic value of each of the 88,260,972 positions in the spaces S_1 to S_5, i.e., during the sliding phase. A Tiger win is defined as the capture of 5 goats, a Tiger loss as the inability to move, and a draw (by repetition) is defined as a position where no opponent can force a win, and each can avoid a loss. Table 2 shows the distribution of won, drawn and lost positions.

		Number of goats captured				
		4	3	2	1	0
Goat to move	Wins	913,153 (2.8%)	1,315,111 (13.9%)	882,523 (41.9%)	252,381 (75.8%)	30,609 (91.4%)
	Draws	8,045,787 (25.0%)	6,226,358 (65.7%)	1,199,231 (57.0%)	80,706 (24.2%)	2,812 (8.4%)
	Losses	23,229,230 (72.2%)	1,928,496 (20.4%)	23,941 (1.1%)	88 (0.03%)	60 (0.2%)
	Total	32,188,170	9,469,965	2,105,695	333,175	33,481
Tiger to move	Wins	30,469,634 (94.7%)	6,260,219 (66.1%)	465,721 (22.1%)	6,452 (1.9%)	146 (0.4%)
	Draws	1,569,409 (4.9%)	2,918,104 (30.8%)	1,353,969 (64.3%)	197,537 (59.3%)	9,468 (28.3%)
	Losses	149,127 (0.5%)	291,642 (3.1%)	286,005 (13.6%)	129,186 (38.8%)	23,867 (71.3%)

Table 2. Endgame database statistics. Percentages are relative to totals for a given player to move.

Goats captured	1	2	3	4	5
Complexity	1.28×10^{24}	4.23×10^{36}	8.92×10^{38}	3.09×10^{40}	4.88×10^{41}

Table 3. Estimated tree complexity for various winning criteria.

Game tree complexity. The search space S_0, with approximately 33 billion positions, is too large for a static data structure that stores each position exactly once. Hence it is generated on the fly, with portions of it stored in hash tables. As a consequence, the same position may be generated and analyzed repeatedly. A worst case measure of the work thus generated is called game tree complexity. The size of the full search tree can be estimated by a Monte Carlo technique as described by [Knuth 1975]. For each of a number of random paths from the root to a leaf, we evaluate the quantity $F = 1 + f_1 + f_1 \times f_2 + f_1 \times f_2 \times f_3 + \cdots$, where f_j is the fan out, or the number of children, of the node at level j encountered along this path. The average of these values F, taken over the random paths sampled, is the expected number of nodes in the full search tree. Table 3 lists the estimated game tree complexity (after the removal of symmetric positions) of five different "games", where the game ends by capturing 1 to 5 goats during the placement phase. These estimates are based on 100,000 path samples.

Cutting search trees in half. A 39-ply search with a branching factor that often exceeds a dozen legal moves is a big challenge. Therefore, the key to successful forward searches through the state space S_0 of the placement phase is to replace a 39-ply search with a number of carefully designed searches that are effectively only 20 plies deep. This is achieved by 1) formulating hypotheses of the type "player X can achieve result Y", 2) programming a competent and efficient heuristic player X that generates only one or a few candidate moves in each position, and 3) confronting the selective player X with his exhaustive opponent who tries all his legal moves. If this search that alternates selective and exhaustive move generation succeeds, the hypothesis Y is proven. If not, one may try to develop a stronger heuristic player X, or weaken the hypothesis, e.g. from "X wins" to "X can get at least a draw". Using such searches designed to verify a specific hypothesis we were able to prove several results including the following:

(i) Tiger can force the capture of a single goat within 30 plies, but no sooner.
(ii) Tiger can force the capture of two goats within 40 plies, i.e., by the end of the placement phase, but not earlier.
(iii) After the most plausible first two moves (the first by Goat, the second by Tiger) Goat has a drawing strategy.

Heuristic attackers and defenders. In order to make these searches feasible we had to develop strong heuristic players. Given our lack of access to human expertise, we developed player programs that learn from experience by being pitted against each other — a topic described in [Lim 2005]. For example, the proof that Tiger can kill a certain number of goats requires a strong Tiger that tries to overcome an exhaustive Goat. Conversely, the proof that Goat has a drawing strategy after the most plausible opening requires a strong heuristic Goat that defies an exhaustive Tiger.

Goat has at least a draw. After Goat's most reasonable first move, Tiger has 6 symmetrically distinct replies at ply 2. Using the same techniques and software as described above, further computer runs that stretched over a couple of months proved that Goat has a successful defense against all of them. Having shown that Goat can ensure at least a draw, the next question is "does Goat have a winning strategy?".

Insights into the nature of the game. We were unable to discover easily formulated advice to players beyond plausible rules-of-thumb such as "goats cautiously hug the border, tigers patiently wait to spring a surprise attack". On the other hand, our database explains the seemingly arbitrary number "five" in the usual winning criterion "Tiger wins when 5 goats have been killed". This magic number "5" must have been observed as the best way to balance the chances. We know that Tiger can kill some goats, so Tiger's challenge must be more ambitious than "kill any one goat". On the other hand, we see from Table 2 that there is a significant jump in number of lost positions for Goat from three goats captured to four goats captured. It is therefore fairly safe to conjecture that once half a dozen goats are gone, they are all gone — Goat lacks the critical mass to put up resistance. But as long as there are at least 16 goats on the board (at most 4 goats have been captured), the herd is still large enough to have a chance at trapping the tigers.

Table 2 also shows that unless Tiger succeeds in capturing at least two goats during the placement phase, he has practically no chance of winning. If he enters the sliding phase facing 19 goats, less than 2% of all positions are won for Tiger, regardless of whether it is his turn to move or not. The fact that Tiger can indeed force the capture of 2 goats within 40 plies, that is, by the end of the placement phase (see page 168), is another example of how well-balanced the opponents' chances are.

3. Proving Tiger's draw

The previous investigation, with the result that Goat has at least a draw, had brought us tantalizingly close to determining the game-theoretic value of *Tigers and Goats*. Computing the endgame database had been relatively straightforward, but the 39-ply forward search had not yielded to the judicious application of established techniques. Experience had shown that by approximating a 39-ply search by various 19-ply and 20-ply searches (see "Cutting search trees in half", page 168), we were able to answer a variety of questions. It appeared plausible that by formulating sufficiently many well-chosen hypotheses this approach would eventually yield a complete analysis of the game. We conjectured that Tiger also has a drawing strategy, and set out to try to prove this using the same techniques that had yielded Goat's drawing strategy.

The asymmetric role of the two opponents, however, made itself felt at this point: the searches pitting a heuristic Tiger player against an exhaustive Goat progressed noticeably more slowly than those involving a heuristic Goat versus an exhaustive Tiger. In retrospect we interpret this different behavior as due to the phenomenon "Tiger's play is all tactics". Positional considerations — keep the goats huddled together — make it easy to generate one or a few "probably safe" Goat's moves, even without any look-ahead at the immediate consequences. For Tiger, on the other hand, neither we nor apparently the neural network that trained the player succeeded in recognizing "good moves" without a local search. An attempt to make Tiger a stronger hunter (by considering the top 3 moves suggested by the neural network followed by a few plies of full-width search) is inconsistent with the approach of "cutting the tree in half" and made the search unacceptably slow.

Thus, a new approach had to be devised. The experience that 20-ply forward searches proved feasible suggests a more direct approach: compute a database of positions of known value halfway down the search tree. Specifically, we define *halfway position* as one arising after 19 plies, i.e., after the placement of 10 goats, with Tiger to move next. The value of any such position can be computed with a search that ends in the endgame database after at most 20 plies. If sufficiently many such "halfway positions" are known and stored, searches from the root of the tree (the starting position of the game) will run into them and terminate the search after at most 19 plies.

The problem with this approach is that the number of halfway positions is large, even after symmetric variants have been eliminated. Because of captures not all 10 goats placed may still be on the board, hence a halfway position has anywhere between 6 and 10 goats, and correspondingly, 15 to 11 empty spots. Using the terminology of Section 2, the set of halfway positions is (perhaps a subset of) the union of S_{11}, S_{12}, S_{13}, S_{14} and S_{15}, where S_k is the set of all symmetrically inequivalent positions containing 4 tigers, $21 - k$ goats, and k empty spaces. S_{11}, with about equally as many goats as empty spots, is particularly large. On the assumption that in any subspace S_k the number of symmetrically inequivalent positions is close to 1/8 of the total, S_{11} contains about 550 million inequivalent positions. The union of S_{11} through S_{15} contains about 1.6×10^9 positions. This number is about 25 times larger than the largest endgame database we had computed before, namely S_5.

The approach to overcome the problem of constructing a large halfway database exploits two ideas. First, the database of halfway positions of known value need not necessarily include all halfway positions. In order to prove that Tiger has a drawing strategy, the database need only include a sufficient number of positions known to be drawn or a win for Tiger so that any forward search is trapped by the filter of these positions. Second, the database of halfway positions is built on the fly: whenever a halfway position is encountered whose value is unknown, this position is entered into the database and a full-width search continues until its value has been computed.

Although there was no a priori certainty that this approach would terminate within a reasonable time, trial and error and repeated program optimization over a period of five months led to success. Table 4 contains the statistics of the halfway database actually constructed. For each of S_{15} through S_{11}, it shows the number of positions whose value was actually computed, broken down into the two categories relevant from Tiger's point of view, win-or-draw vs. loss.

# Captured	# Win or Draw	# Loss	Total	Estimated state space size
4	17,902,335	0	17,902,335	85,804,950
3	33,152,214	0	33,152,214	183,867,750
2	64,336,692	17,944	64,354,636	321,768,563
1	84,832,697	329,183	85,161,880	464,776,813
0	15,857,243	91,676	15,948,919	557,732,175
Total	216,081,181	438,803	216,519,984	1,613,950,251

Table 4. Halfway database statistics: the number of positions computed and their value from Tiger's point of view: win-or-draw vs. loss

Although the construction of the halfway database is intertwined with the forward searches — a position is added and evaluated only as needed — logically it is clearest to separate the two. We discuss details of the forward searches in the next section.

4. Implementation, optimization, verification

Our investigation of *Tigers and Goats* has been active, on and off, for the past three years. The resources used have varied form a Pentium 4 personal computer to a cluster of Linux PC workstations. Hundreds of computer runs were used to explore the state space, test and confirm hypotheses, and verify results. The longest continuous run lasted for five months as a background process on an Apple PowerMac G5 used mainly for web surfing.

The algorithmic search techniques used are standard, but three main challenges must be overcome in order to succeed with an extensive search problem such as *Tigers and Goats*. First, efficiency must be pushed to the limit by adapting general techniques to the specific problem at hand, such as the decision described above on how to combine different search techniques. Second, programs must be optimized for each of the computer systems used. Third, the results obtained must be verified to insure they are indeed correct. We address these three issues as follows.

Domain-specific optimizations. The two databases constructed, of endgame positions and halfway positions, limit all forward searches to at most 20 plies. Still, performing a large number of 20-ply searches in a tree with an average branching factor of 10 remains a challenge that calls for optimization wherever possible.

The most profitable source of optimizations is the high degree of symmetry of the game board. Whereas the construction of the two databases of endgame and halfway positions is designed to avoid symmetric variants, this same desirable goal proved not to be feasible during forward searches — it would have meant constructing a database consisting of all positions.

Instead, the goal is to avoid generating some, though not necessarily all, symmetrically equivalent positions when this can be done quickly, namely during move generation. Although the details are cumbersome to state, in particular for Tiger moves, the general idea is straightforward. Any position that arises during the search is analyzed to determine all active symmetries. Thereafter, among all the moves that generate symmetric outcomes, only that one is retained that generates the resulting position of lowest index. This analysis guarantees that all immediate successors to any given position are inequivalent. Because of transpositions, of course, symmetric variants will appear among successor positions further down in the tree. Table 5 shows the effect of this symmetry-avoiding move generation for the starting position. Although there is a considerable reduction in the number of positions generated, the relative savings diminish with an expanding horizon.

Ply	Naïve move generator	Symmetry-avoiding move generator	Number of distinct positions
1	21	5	5
2	252	36	33
3	5,052	695	354
4	68,204	9,245	2,709
5	1,304,788	173,356	18,906
6	18,592,000	2,441,126	93,812

Table 5. Number of positions created by different move generators.

System-specific optimization. Our previous result for *Tigers and Goats* used a cluster of eight Linux PC workstations with a simple synchronous distributed game-tree search algorithm. However, there are fundamental problems with synchronous algorithms, discussed in [Brockington 1997], that limit their efficiency. Furthermore, the cluster was becoming more popular and was constantly overloaded. We therefore decided against implementing a more sophisticated asynchronous game-tree search and instead relied on a sequential program running on a single dedicated processor.

We focused our attention on improving the sequential program to run on an Apple PowerMac G5 1.8 GHz machine running Mac OS-X. Firstly, the neural network code was optimized using the Single Instruction Multiple Data (SIMD) unit in the PowerPC architecture called AltiVec. AltiVec consists of highly parallel operations which allow simultaneous execution of up to 16 operations in a single clock cycle. This provided a modest improvement of about 15% to the efficiency of neural network evaluations of the board, but sped up the overall efficiency of the search much more as the neural network is used repeatedly within the search to evaluate and reorder the moves.

Next, we moved many of the computations off-line. For example, the moves for Tiger at each point on the board in every combination of surrounding pieces were precomputed into a table so that the program simply retrieved the table and appended it to the move list during search. Operations like the indexing of the board and symmetry transformation were also precomputed so that the program only needed to retrieve data from memory to get the result. Finally, we recompiled the software with G5-specific optimizations.

Verification. Two independent re-searches confirm different components of the result. They used separately coded programs written in C, and took 2 months to complete.

The first verification search used the database of halfway positions to confirm the result at the root, namely, "Tiger has a drawing strategy". Notice that this verification used only the positions marked as win-or-draw in the database.

The second verification search confirmed the halfway positions marked as win-or-draw by searching to the endgame database generated by the retrograde analysis described in [Lim 2004]. All other positions can be ignored, as they have no effect on the first search.

Another program was written in C to 'reprove' the results. This program had the benefit of a posteriori knowledge that the game is a draw, and this fact allowed us to concentrate on using aggressive forward pruning techniques to verify the result. The program used the same domain-specific optimizations such as symmetry reduction and the halfway databases.

The halfway database was optimized for size by storing the boolean evaluation of each position using a single bit. Depending on the type of search, this boolean evaluation could mean "Goat can at least draw" or "Tiger can at least draw". Due to this space optimization the halfway positions and endgame databases could be stored in memory, thereby avoiding disk accesses and speeding up the search by orders of magnitude.

As Tiger is able to force the capture of two goats only by the end of the placement phase, at ply 40, the search for "Goat can at least draw" used an aggressive forward pruning strategy of pruning positions which had two or more goats already captured. The halfway database was set at ply 23, when 12 goats have already been placed and it is Tiger's turn to move. The search confirmed that "Goat can at least draw" in approximately 7 hours while visiting 7,735,443,119 nodes.

The program was also able to confirm that "Tiger can at least draw". Due to the large game-tree complexity of this search, two intermediate databases were placed at ply 21 and ply 31. These databases contribute towards efficiency in two ways: first, they terminate some searches early, and second, they generate narrower search trees. The latter phenomenon is due to the fact that these databases are free of symmetrically equivalent positions. In exchange for a large memory footprint of approximately 2 GB, search performance was dramatically improved. The searched confirmed that "Tiger can at least draw" in approximately 48 hours while visiting 40,521,418,103 nodes.

5. Conclusion

The theory of computation has developed powerful techniques for estimating the asymptotic complexity of problem classes. By contrast, there is little or no theory to help in estimating the concrete complexity of computationally hard problem instances, such as determining the game-theoretic value of *Tigers and Goats*. Although the general techniques for attacking such problems have been well-known for decades, there are only rules of thumb to guide us in adapting them to the specific problem at hand in an attempt to optimize their efficiency [Nievergelt 2000].

The principal rule of thumb we have followed in our approach to solving *Tigers and Goats* is to precompute the solutions of as many subproblems as can be handled efficiently with the storage available, both in main memory (hashtables) and disks (position data bases). If the net of these known subproblems is dense enough, it serves to truncate the depth of many forward searches, an effect that plays a decisive role since the computation time tends to grow exponentially with search depth. Beyond such rules of thumb, at the present state of knowledge about exhaustive search there is not much more we can do than persistent experimentation. Developing a technology that gives us quantitative estimates of the complexity of computationally hard problems remains a challenge.

Acknowledgment

Elwyn Berlekamp pointed out *Tigers and Goats* and got us interested in trying to solve this game — an exhaustive search problem whose solution stretched out over three years. We are grateful to Elwyn, Tony Tan, Thomas Lincke and H. J. van den Herik for helpful comments that improved this paper. Some of the present text is taken from our earlier paper [Lim 2004].

Note about references

We are not aware of any widely available publications on *Tigers and Goats*. Searching the web for *Tigers and Goats*, or Bagha Chal in various spellings, readily leads to a collection of web sites that describe the game and/or let you play against a computer program.

References

[BCG 2001] E. Berlekamp, J. H. Conway, R. K. Guy, *Winning Ways For Your Mathematical Plays*, A K Peters, 4 volumes, 2nd edition, 2001.

[Brockington 1997] M. G. Brockington, *Asynchronous parallel game-tree search*, Ph.D. Thesis, Department of Computing Science, University of Alberta, 1997.

[Gasser 1996] R. Gasser, *Solving Nine Men's Morris*, pp. 101–113 in Games of No Chance, edited by Richard Nowakowski, MSRI Publications 29, Cambridge University Press, New York, 1996.

[Knuth 1975] D. E. Knuth, *Estimating the efficiency of backtrack programs*, Math. Comp. 29, 1975, 121–136.

[Lim 2004] Y. J. Lim and J. Nievergelt, *Computing* Tigers and Goats, ICGA Journal 27:3, 131–141, Sep 2004.

[Lim 2005] Y. J. Lim, *Using biased two-population co-evolution to evolve heuristic game players for* Tigers and Goats, Unpublished manuscript.

[Nievergelt 2000] J. Nievergelt, *Exhaustive search, combinatorial optimization and enumeration: Exploring the potential of raw computing power*, pp. 18–35 in Sofsem 2000: Theory and Practice of Informatics, edited by V. Hlavac, K.G. Jeffery and J. Wiedermann, Lecture Notes in Computer Science 1963, Springer, Berlin, 2000.

[Polya 1937] G. Polya, *Kombinatorische Anzahlbestimmungen für Gruppen, Graphen und chemische Verbindungen*, Acta Mathematica 68, 1937, 145–253.

[Wu 2002] R. Wu and D. F. Beal, *A Memory efficient retrograde algorithm and its application to Chinese Chess endgames*, pp. 213–227 in More Games of No Chance, edited by Richard Nowakowski, MSRI Publications 42, Cambridge University Press, New York, 2002.

LIM YEW JIN
SCHOOL OF COMPUTING
NATIONAL UNIVERSITY OF SINGAPORE
SINGAPORE
limyewjin@gmail.com

JURG NIEVERGELT
INFORMATIK ETH
8092 ZURICH
SWITZERLAND
jn@inf.ethz.ch

Games of No Chance 3
MSRI Publications
Volume **56**, 2009

Counting liberties in Go capturing races

TEIGO NAKAMURA

ABSTRACT. Applications of combinatorial game theory to Go have, so far, been focused on endgames and eyespace values, but CGT can be applied to any situation that involves counting. In this paper, we will show how CGT can be used to count liberties in Go *semeai* (capturing races).

Our method of analyzing capturing races applies when there are either no shared liberties or only simple shared liberties. It uses combinatorial game values of external liberties to give an evaluation formula for the outcome of the capturing races.

1. Introduction

Combinatorial game theory (CGT) [1][2] has been applied to many kinds of existing games and has produced a lot of excellent results. In the case of Go, applications have focused on endgames [3][4][6][7][10] and eyespace values [5] so far, but CGT can be applied to any situation that involves counting. In this paper, we will show another application of CGT to Go, that is, to count liberties in capturing races. A capturing race, or *semeai*, is a particular kind of life and death problem in which two adjacent opposing groups fight to capture each other's group. In addition to the skills involved in openings and endgames, skills in winning capturing races are an important factor in a player's strength at Go. In order to win a complicated capturing race, various techniques, such as counting liberties, taking away the opponent's liberties and extending self-liberties, are required in addition to wide and deep reading. Expert human players usually count liberties for each part of blocks involved in semeai and sum them up. A position of capturing races can also be decomposed into independent subpositions, as in the cases of endgames and eyespaces.

We propose a method to analyze capturing races having no shared liberty or only simple shared liberties. The method uses combinatorial game values of

external liberties. We also prove an evaluation formula to find the outcome of
the capturing races.

2. Capturing races

Terminology. Figure 1 shows an example of capturing races in Go. Both Black
and White blocks of circled stones have only one eye and want to capture the
opponent's block to make the second eye. In order to describe capturing races,
we use the following terminology.

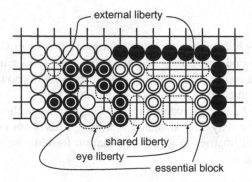

Figure 1. Example of a capturing race.

ESSENTIAL BLOCK: A *block* is a maximal connected set of stones of the same
   color and the adjacent empty points of a block are called *liberties*. If a block
   loses all its liberties, it is captured and removed from the board. An *essential
   block* is a block of Black or White stones which must be saved from capture.
   Capturing an essential block immediately decides a semeai.

SAFE BLOCK: A block which is alive or assumed to be safe. Nonessential
   blocks surrounding an essential block are usually regarded as safe blocks.

NEUTRAL BLOCK: A block other than an essential block and a safe block.
   Saving or capturing such blocks does not decide a semeai.

LIBERTY REGION: A region which is surrounded by at least one essential block
   and some other essential blocks and safe blocks. Liberty regions contain only
   empty points and neutral blocks. A liberty region is called an *external region*
   if its boundary does not consist of essential blocks of different color. A liberty
   region is called an *eye region* if its boundary consists entirely of essential
   blocks of the same color.

EXTERNAL LIBERTY: A liberty of an essential block in an external liberty region.

SHARED LIBERTY: A common liberty of a Black essential block and a White essential block.

PLAIN LIBERTY: A liberty of an essential block that is also adjacent to an opponent's safe block. A plain liberty can be filled without any additional approach moves by the opponent.

EYE LIBERTY: A liberty in an eye liberty region. An amount of liberty count in an eye liberty region is greater than or equal to the number of eye liberties [8]. Eye liberties behave like external liberties in capturing races.

ATTACKER AND DEFENDER: In an external liberty region, the player who owns the essential block is called the *defender* and the opponent is called the *attacker*.

Related research. Müller [8] gives a detailed discussion of capturing races. He classifies semeais into nine classes in terms of types of external and shared liberties and eye status. Figure 2 shows some examples of his classification.

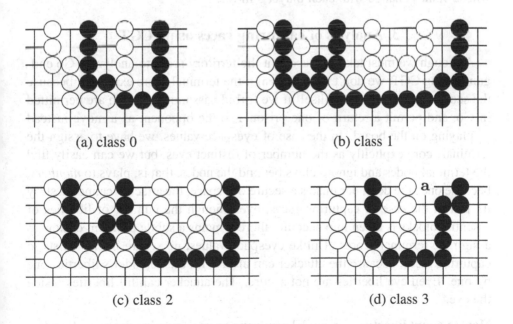

(a) class 0   (b) class 1

(c) class 2   (d) class 3

Figure 2. Müller's semeai classes.

He gives an evaluation formula for class 0 and class 1 semeais, which have exactly one essential block of each color and only plain external and shared liberties, and have no eye or only one eye in *nakade* shape.

Formula 1: Müller's semeai formula ──────────────

Δ: Advantage of external liberties for the attacker
  (difference in the number of liberties for each player's essential blocks)

S: Number of shared liberties

$$F = \begin{cases} S & \text{if } S = 0 \text{ or defender has an eye,} \\ S - 1 & \text{if } S > 0 \text{ and defender has no eye.} \end{cases}$$

> The attacker can win the semeai if $\Delta \geq F$.

This is a really simple formula but it is enough to decide complicated semeais such as *one eye versus no eye semeais* and *big eye versus small eye semeais*. Nevertheless, it is only applicable to the restricted classes of semeais whose liberties are all plain liberties, that is, all the liberty counts are *numbers*. Generally, liberty counts are not numbers in the higher class of semeais, but are *games* whose values change with each player's move.

3. Analysis of capturing races using CGT

Although we must take into account the territory score to analyze a Go endgame using CGT, we don't need to assign the terminal score explicitly because the score comes out of CGT itself if we forbid pass moves, which are permitted in Go, and permit a return of one prisoner to the opponent as a move instead of playing on the board. In the case of eyespace values, we have to assign the terminal score explicitly as the number of distinct eyes, but we can easily find the terminal nodes and ignore plays beyond the nodes, that is, plays to *numbers*, because once the defender makes a secure eye space, the attacker cannot destroy it. But in the case of capturing races, even though the number of liberties of essential blocks is taken into account, there remains a subtle problem of how to assign the terminal scores. Unlike eyespace values, no secure liberty exists in capturing races, because the attacker can always fill the defender's liberties one by one. Even eye liberties are not secure. The attacker can fill liberties inside the eye.

How to count liberties. To model capturing races, we define the game **SemGo**, which has the same rules as Go except for scoring. In **SemGo**, the terminal score is basically the number of liberties of essential blocks, but it is exactly the number of the opponent's moves which are required to take away all the liberties. By convention, Black is Left and White is Right; Black scores are positive and White scores are negative.

Figure 3 shows some examples of CGT values of **SemGos**. In Figure 3(a), White's essential block has three liberties, but Black cannot directly attack White's external liberty, because if he simply fills the liberty, Black's attacking block gets to be in *atari*, and White can capture Black's five stones and White's essential block is alive. So, Black needs to spend one move to protect at the point of his false eye prior to attacking. Generally, liberty scores in external liberty regions are greater than or equal to the number of liberties of essential blocks. In Figure 3(b), if White plays first, the score is zero, but if Black plays first, the number of liberties becomes three. In Figure 3(c), Black can connect his two stones and the score becomes four, if he plays first.

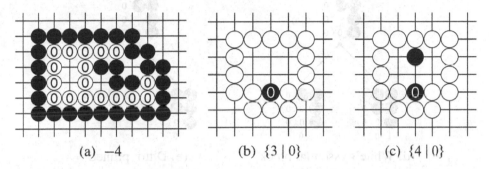

(a) −4   (b) {3 | 0}   (c) {4 | 0}

Figure 3. CGT values of **SemGos**.

Although it seems easy to obtain a CGT expression from a position of **SemGo**, we have to resolve the subtle problem of how we can find the terminal nodes and assign scores to them. As long as liberties of essential blocks exist, both players still have legal moves in **SemGo**, but not all moves are useful for both players. The attacker always has good moves, because he can fill the defender's liberties one by one. The defender, on the other hand, may not have any useful moves. If the defender cannot extend his own liberties by at least one, he should not play any more. It is always a bad move to take away his own liberties. In order to obtain CGT expressions, we have to prune the useless moves explicitly in contrast to endgames and eyespace values.

Figure 4 on the next page shows the key idea of pruning in **SemGo**. The position in Figure 4(a) has exactly one *dame* and the ordinary CGT value is {0 | 0} = *. In part (b) of the figure, Black's essential block has one liberty and Black's option of reducing his own liberty should be pruned. We assign the value one to the root node of the resulting game tree in part (c). Part (d) and (e) show the case of White's essential block.

The process for assigning the terminal value is summarized in the sidebar on the next page.

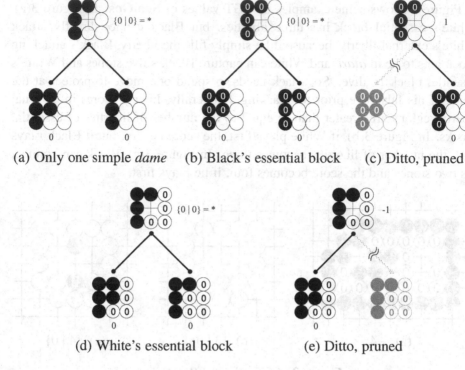

(a) Only one simple *dame*   (b) Black's essential block   (c) Ditto, pruned

(d) White's essential block   (e) Ditto, pruned

Figure 4. How should we assign the terminal score?

Process for Assigning Terminal Score

(i) Play out all legal moves for both players until the essential block has no liberty.

- At present, it doesn't matter whether the move is good or bad.
- All the leaf nodes are zero.
- In practice, however, we can count the number of liberties when they become plain liberties.

(ii) Execute the following operations from bottom to top.

- If the temperature of a node is less than or equal to one, prune the defender's branch.
- If the resulting node becomes either of the following forms, replace it:

$$\{ \ | \ n\} \Longrightarrow n+1,$$
$$\{-n \ | \ \} \Longrightarrow -n-1.$$

- This replacement operation is exactly contrary to conventional CGT.

Figures 5 and 6 illustrate the process for assigning the terminal score. A number between parentheses denotes the temperature of the node. All three positions in Figure 5 have the same score of 1, even if Black poses a threat to gain the number of liberties at the bottom position. In Figure 6 on the next page, the threat on White is immediately reversed by Black's fill, and White's option is pruned.

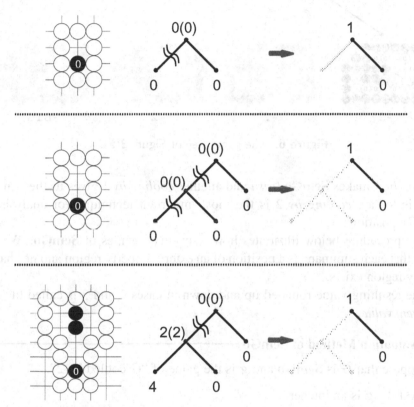

Figure 5. Process for assigning the terminal score: some positions having only one liberty.

Evaluation method. The winning condition of **SemGo** is different from that of endgames and eyespaces, because we cannot define the winner as the player who plays the last move. In **SemGo** the player who fills all the liberties of all of the opponent's essential blocks in all summands is the winner. Suicidal moves are forbidden except for the last winning move. In the case of endgames and eyespaces, the fact that the smallest incentive for a move is infinitesimal, that is 0-*ish* and cooling by one, or *chilling*, plays a very important role in analyzing positions. But in the case of **SemGo**, the smallest incentive is 1-*ish*, because the attacker always has a move to fill the opponent's liberties one by one. Therefore,

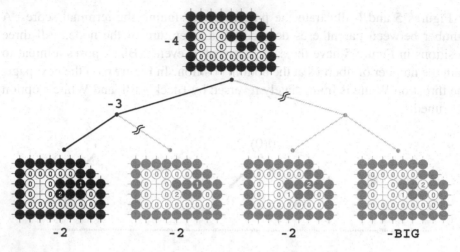

Figure 6. The game tree of Figure 3(a).

cooling by 1 makes **SemGo** 0-*ish* and another *cooling by* 1 gives us the winner. That is to say, *cooling by* 2 is the most important technique for analysis of **SemGo** positions.

The procedure below illustrates how to evaluate games of **SemGo**. We assume that each summand is a position of an external liberty region and no shared liberty region exists.

The resulting value rounded up and down in cases 2 and 3 is called the *adjustment value*.

Evaluation Method of SemGo

Suppose that G is **SemGo** and g is the game of "G cooled by 2".

CASE 1: g is an integer:

If $g > 0$, Black wins. If $g < 0$, White wins. If $g = 0$, the first player wins.

CASE 2: $n < g < n + 1$ (for an integer n):

If Black plays first, he can round up g to $n + 1$ in keeping his turn.
If White plays first, he can round down g to n in keeping his turn.
Check the resulting value using the conditions of case 1.

CASE 3: $g <> n$ (g is not comparable to an integer n):

If Black plays first, he can round up g to $n + 1$ in keeping his turn.
If White plays first, he can round down g to $n - 1$ in keeping his turn.
Check the resulting value using the conditions of case 1.

---

4. Examples

4.1. Partial board problems. Figure 7 shows some simple (!) problems of capturing races and Figure 9 shows the analyses of these problems using our method described in the previous section. The recommended winning move is ▉ and the other winning move is ▣ in Figure 9.

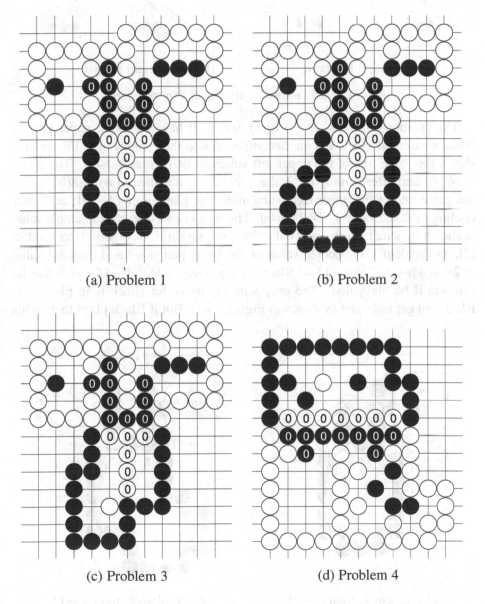

(a) Problem 1

(b) Problem 2

(c) Problem 3

(d) Problem 4

Figure 7. Semeai problems (Black plays first).

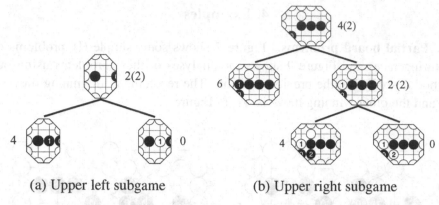

(a) Upper left subgame   (b) Upper right subgame

Figure 8. Game trees of the upper common parts.

The positions in parts (a)–(c) of Figure 7 all have the same upper part of Black's essential block and are decomposed into three subgames. Figure 8(a) shows the game tree of the upper left subgame of Figure 7(a)–(c). The game is {4 | 0} and after cooling by 2, the value 2∗ is obtained. Figure 8(b) shows the game tree of the upper right subgame. The game is {6 | {4 | 0}} and after cooling by 2, the value 4↑ is obtained. The integer part of each subgame's value is shown by small Black and White dots, or *markings*, in Figure 9, as used in [3]. In Problem 1, the cooled value of the lower part is −7 and the total value is 2∗ + 4↑ + (−7) = −1↑∗. Black's adjustment value of −1↑∗ is 0 and he can win if he plays first. The only winning move for Black is to play to ∗. Black can get *tedomari* as shown in Figure 10(a). But if Black plays to ↑ in his

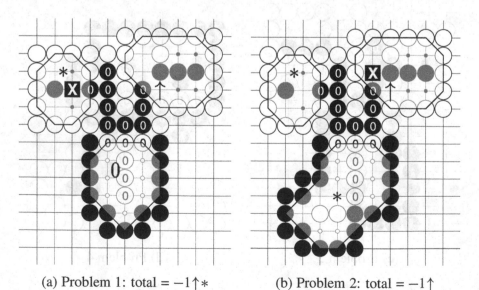

(a) Problem 1: total = −1↑∗   (b) Problem 2: total = −1↑

Figure 9. Analysis using CGT (continued on next page).

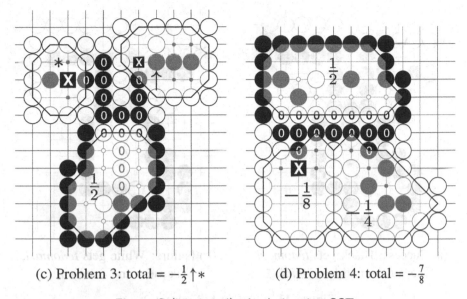

(c) Problem 3: total $= -\frac{1}{2}\uparrow*$ (d) Problem 4: total $= -\frac{7}{8}$

Figure 9 (continued). Analysis using CGT.

first move, White gets *tedomari* and White wins as shown in Figure 10(b). In Problem 2, the lower subgame is $\{-5 \mid -9\}$ and the cooled value is $-7*$. The total value is $2* + 4\uparrow - 7* = -1\uparrow$. Black can round up the value to 0 and win. Unlike Problem 1, Black should play to $\uparrow$, which is the only winning move in this problem. Figure 11(a) shows a winning sequence and Figure 11(b) shows a failure. In Problem 3, the lower subgame is $\{-5 \mid -8\}$ and the cooled value

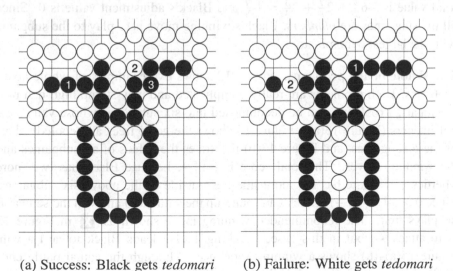

(a) Success: Black gets *tedomari*   (b) Failure: White gets *tedomari*

Figure 10. Solution of Problem 1.

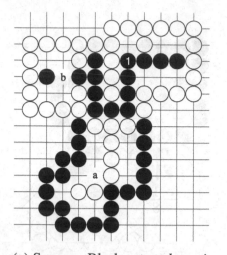
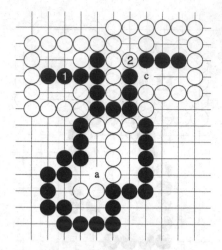

(a) Success: Black gets *tedomari*,
because *a* and *b* are *miai*.

(b) Failure: White gets *tedomari*,
because *a* and *c* are *miai*.

Figure 11. Solution of Problem 2.

is $-6\frac{1}{2}$. The total value is $2* + 4\uparrow - 6\frac{1}{2} = -\frac{1}{2}\uparrow*$. Since $0 > -\frac{1}{2}\uparrow* > -1$, Black's adjustment value is 0 and he can win the semeai. In this problem, there are two winning moves for Black. Both plays to $\uparrow$ and $*$ lead Black toward a win. In Problem 4, the upper subgame is $\{-5 \mid -8\}$ and the cooled value is $-6\frac{1}{2}$. The subgame of lower left is $\{\{\{8 \mid 5\} \mid 3\} \mid 1\}$ and the cooled value is $2\frac{7}{8}$. The subgame of lower right is $\{\{6 \mid 3\} \mid 1\}$ and the cooled value is $2\frac{3}{4}$. So the total value is $-6\frac{1}{2} + 2\frac{7}{8} + 2\frac{3}{4} = -\frac{7}{8}$ and Black's adjustment value is 0. Since all the subgames are *numbers*, Black's winning move is to play to the subgame with the largest denominator.

4.2. Whole board problems. Figure 12 shows two examples of whole board problems. Both problems are really complicated and may stump high-dan professionals. Problem 5 can be decomposed into six subgames whose values are not integers (Figure 13(a)). Figure 14 shows the game trees and cooled values of these subgames, and Figure 13(b) illustrates the values of all subgames and the winning moves. The total score is $-1\uparrow\rightarrow$, because Black has two more liberties and White has ten more liberties outside of the above six subgames. Since $0 > -1\uparrow\rightarrow > -1$, Black can round up the value to 0 and win the semeai if he plays first. The recommended winning move shown as $\boxed{X}$ in Figure 13 is to attack $\rightarrow$, but in this case, attacking $\downarrow$ also leads Black toward a win. Figure 15(a)–(h) shows a winning sequence. Although the actual battle ends at Black's move 11 (Figure 15(b)), a total of 49 moves is required by the end of the capturing race where the essential block is finally captured. Figure 15(i)

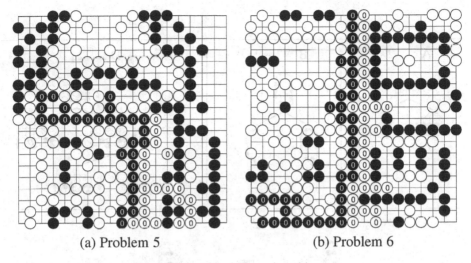

(a) Problem 5   (b) Problem 6

Figure 12. Whole board problems.

is an example of failure. If Black plays to ⇑*, Black cannot win even if Black plays both — and ↓ afterward.

The position of Problem 6 is decomposed into thirteen subgames as shown in Figure 16(a). The game value and its atomic weight for each subgame is shown in Figure 16(b). The total value is −1-*ish* and is fuzzy with −1. Black's adjustment value is 0 and Black can win. Black's only winning move is to play to the subgame C. Figure 16(c) illustrates a winning sequence. The battle continues up to Black's move 27. At this point, the number of liberties of Black's

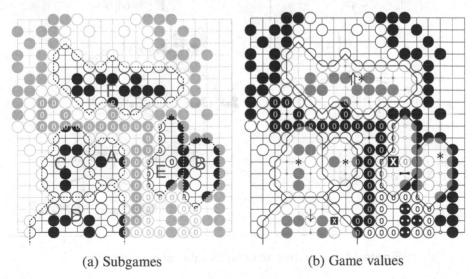

(a) Subgames   (b) Game values

Figure 13. Solution of Problem 5.

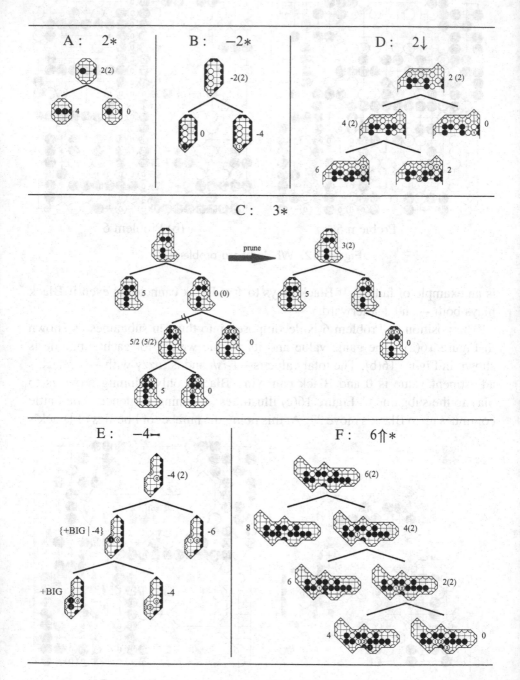

Figure 14. Game tree and cooled value of each subgame.

essential block is 22 and the number of liberties of White's essential block is 21, so Black wins the semeai regardless of who moves first from here. Figure 16(d) shows an interesting game tree whose cooled value has ↑* as an infinitesimal.

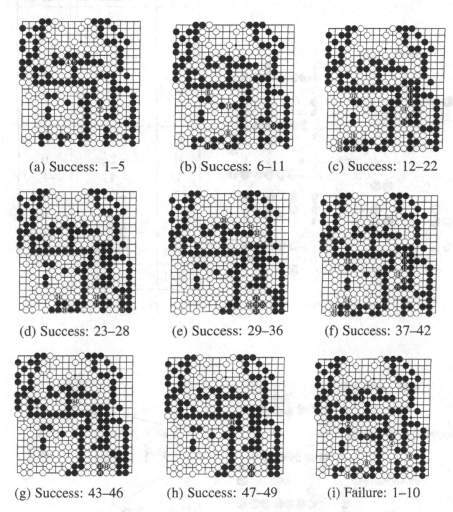

(a) Success: 1–5   (b) Success: 6–11   (c) Success: 12–22

(d) Success: 23–28   (e) Success: 29–36   (f) Success: 37–42

(g) Success: 43–46   (h) Success: 47–49   (i) Failure: 1–10

Figure 15. Example sequence to win.

5. Corridors

Corridors are simple positions that are precisely analyzable and hence give us good examples. In **SemGo**, corridors of width three exactly correspond to the corridors of width one in endgames, and corridors of width one or two have integer values. Figure 17(a) shows some examples of corridors of width three. The game tree of the corridor is shown in Figure 18 and all the corridors in

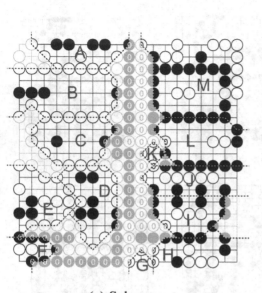

	value value	atomic weight
A	$1\frac{3}{4}$	0
B	$2\{-_3\mid 0^3\}$	-3
C	$3\{-_1\mid 0\}$	-1
D	$3{\uparrow}*$	1
E	$1\frac{3}{4}$	0
F	3	0
G	-1	0
H	$-2\frac{1}{2}$	0
I	$-4{\downarrow}$	-1
J	$-2*$	0
K	-1	0
L	$-3\{0\mid +_2\}$	1
M	$-2\{0^2\mid +_3\}$	2
Total	-1 ish	-1

(a) Subgames   (b) Game values

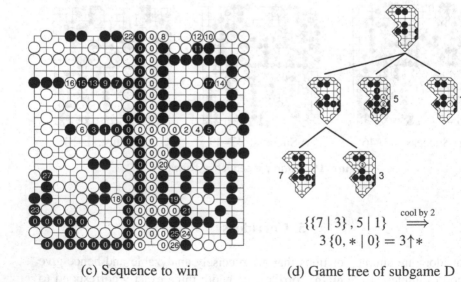

(c) Sequence to win   (d) Game tree of subgame D

$$\{\{7\mid 3\},5\mid 1\}\ \overset{\text{cool by 2}}{\Longrightarrow}$$
$$3\{0,*\mid 0\}=3{\uparrow}*$$

Figure 16. Analysis of Problem 6.

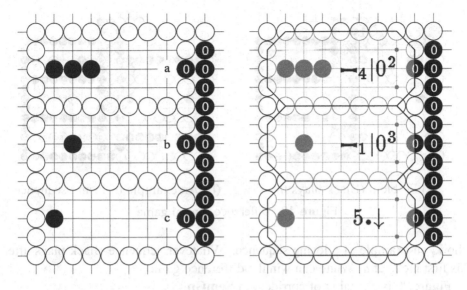

(a) Three corridors of width three   (b) Cooled value of each corridor

Figure 17. Which move is the largest?

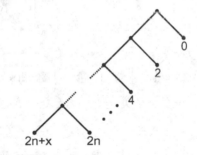

Figure 18. Game tree of the corridor.

Figure 17(a) are described using n and x: $n = 3$, $x = 8$ for the top corridor, $n = 4$, $x = 5$ for the middle, and $n = 5$, $x = 4$ for the bottom.

Now we can answer the question "Which move is best: a, b, or c?" Using the cooled value of each corridor shown in Figure 17(b), we conclude $b > a > c$. In order to verify $b > a$, we use the difference game shown in Figure 19. In Figure 19(a), the right side is a mirror image of the left side except that Black's essential block has one more liberty in the top center and White has just played at the point corresponding to b. It is obvious that Black can win the game if he plays at b in his first move and follows a symmetric strategy supposing that an essential block cannot make two eyes without capturing the opponent's essential block. But if Black plays at a, White gets *tedomari* and Black loses. Figure 19(b)

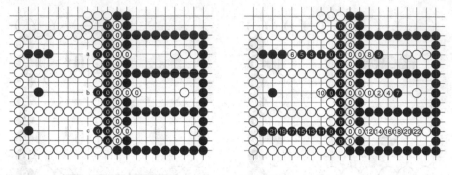

(a) Problem (Black plays first)   (b) An example of losing sequence

Figure 19. Difference semeai game.

shows an example of a losing sequence. Whatever sequence Black plays after his first move at *a*, White can win the difference game.

Figure 20 is a catalog of corridors in **SemGo**.

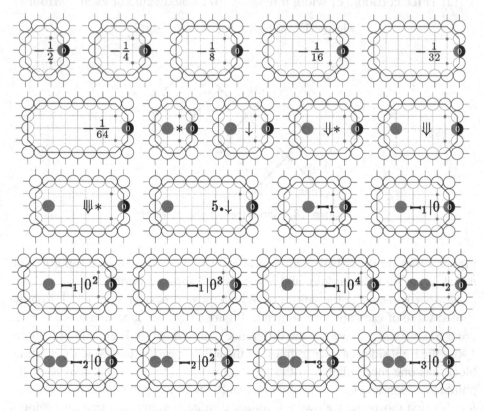

Figure 20. Catalog of corridors.

6. Summary and future work

In this paper, we described a new genre of applications of CGT to the game of Go. We proposed a method of counting liberties in capturing races using CGT and analyzed some complicated semeai problems which seem to be hard to solve even for the most experienced players. Although we have assumed that **SemGo** positions have no shared liberty regions so far, we can easily extend the applicability of our method to **SemGo** positions with some simple shared liberties.

Formula 2 : Extended semeai formula

G : Entire semeai game with the exclusion of shared liberty regions

Δ': Adjustment value of $\mathrm{Cool}(G, 2)$ for the attacker

S : Number of shared liberties

$$F = \begin{cases} S & \text{if } S = 0 \text{ or defender has an eye,} \\ S - 1 & \text{if } S > 0 \text{ and defender has no eye.} \end{cases}$$

The attacker can win the semeai if $\Delta' \geq F$.

Formula 2 is a straightforward improvement of Müller's semeai formula to be applied to the higher class of semeais. In Formula 2, we use the adjustment value of the sum of external liberty regions instead of a simple number of liberties.

Analyzing capturing races poses various problems. In shared liberty regions, a player can be an attacker and a defender at the same time. Properties of complicated shape of shared liberty regions are still unknown. Kos in capturing races have different characteristics from kos in endgames and capturing races with kos are also difficult to analyze. In order to truly analyze Go positions, we have to integrate all the analyses of eyespaces, capturing races and endgames into a unified description.

Acknowledgements

The basic idea of this research came into my mind in the summer of 2001, when I visited Berkeley to study Combinatorial Game Theory and its application to the game of Go with Professor Berlekamp and his research group. I am very grateful to Elwyn Berlekamp, William Fraser and William Spight for their kind advice and useful suggestions. I also want to thank Rafael Caetano dos Santos for many discussions about the game of Go and valuable comments on this paper.

References

[1] John H. Conway: "On Numbers and Games", Academic Press (1976).

[2] Elwyn Berlekamp, John H. Conway and Richard K. Guy: "Winning Ways for your Mathematical Plays", Academic Press, New York (1982).

[3] Elwyn Berlekamp and David Wolfe: "Mathematical Go: Chilling Gets the Last Point", AK Peters (1994).

[4] Elwyn Berlekamp: "The economist's view of combinatorial games", *Games of No Chance*, Cambridge University Press, pp. 365–405 (1996).

[5] H. A. Landman: "Eyespace Values in Go", *Games of No Chance*, Cambridge University Press, pp. 227–257 (1996).

[6] Martin Müller, Elwyn Berlekamp and William Spight: "Generalized Thermography: Algorithms, Implementation, and Application to Go Endgames", International Computer Science Institute, TR–96–030, (1996).

[7] Takenobu Takizawa : "Mathematical game theory and its application to Go endgames", IPSJ SIG-GI 99–GI–1–6, pp. 39–46 (1999).

[8] Martin Müller: "Race to capture: Analyzing semeai in Go", Game Programming Workshop '99 (GPW '99), pp. 61–68 (1999).

[9] Teigo Nakamura, Elwyn Berlekamp: "Analysis of composite corridors", Proceedings of CG2002 (2002).

[10] William L. Spight: "Evaluating kos in a neutral threat environment: preliminary results", Proceedings of CG2002 (2002).

[11] Teigo Nakamura: "Counting liberties in capturing races using combinatorial game theory" (in Japanese), IPSJ SIG-GI 2003–GI–9–5, pp. 27–34 (2003).

TEIGO NAKAMURA
KYUSHU INSTITUTE OF TECHNOLOGY, FUKUOKA, JAPAN
teigo@ai.kyutech.ac.jp

Games of No Chance 3
MSRI Publications
Volume **56**, 2009

Backsliding Toads and Frogs

AARON N. SIEGEL

ABSTRACT. Backsliding Toads and Frogs is a variant of Toads and Frogs in which virtually all positions are loopy. The game is an excellent case study of Conway's theory of sides. In this paper, we completely characterize the values of all natural starting positions. We also exhibit positions with the familiar values n and 2^{-n}, as well as positions with *temperatures* n and 2^{-n}, for all n.

1. Introduction

The game of Toads and Frogs was introduced in *Winning Ways* [Berlekamp et al. 2001]. It is played on a $1 \times n$ strip, populated by some number of toads and frogs. Left plays by moving any toad one space to the right; Right by moving any frog one space to the left. If either player's move is blocked by the opponent, he may choose to leap over her, provided the next square is empty. Jumps do not result in capture. As usual, the winner is the player who makes the last move.

The variant Backsliding Toads and Frogs was also introduced in *Winning Ways*. Here both players have the additional option of retreating by one space, though reverse jumps are still prohibited. Unlike standard Toads and Frogs, the backsliding variant is loopy. As we will see, this additional rule has a monumental effect on the play of the game.

Figure 1 shows a typical position shortly after the start of the game. Each player has one advancing move and one backsliding move available, and Left has the additional option of leaping over Right's frog.

Standard Toads and Frogs was studied extensively by Erickson [1996], whose results include an analysis of certain natural starting positions, as well as the observation that other starting positions have great canonical complexity. However,

Figure 1. A typical position in Backsliding Toads and Frogs.

aside from a few very small positions analyzed in *Winning Ways*, the backsliding version has been scarcely investigated. This is likely due to the enormous difficulty in calculating its values; a direct analysis by hand is exceedingly difficult, and until very recently the tools for a machine analysis were not available.

The present research relied upon a large database of positions assembled using *CGSuite*'s implementations of the algorithms introduced in [Siegel 2009b]. (See http://www.cgsuite.org/ for *CGSuite*.) However, all of the results presented in this paper, with the exception of the museum pieces in Section 6, are fully verifiable by hand, and their proofs, as presented here, do not rely in any way on the computer's output. The transition from calculations to mathematical proofs followed a familiar pattern: a careful analysis of the database led first to a series of promising conjectures, and then ruled out many misdirections and false hypotheses, until the solutions could be isolated.

One striking result is that, in contrast to the standard version, *all* natural starting positions have simple values. Nonetheless, Backsliding Toads and Frogs is quite an interesting game if one considers arbitrary starting positions. Many typical values occur, including n, 2^{-n}, $\uparrow$ and **over**, as well as values with temperature n and 2^{-n}.

In Section 2, we introduce some notation and prove a key lemma. In Section 3, we analyze positions with just one frog, and in Section 4, those where the groups of toads and frogs are initially separated. Section 5 contains positions with the familiar values mentioned above. Finally, Section 6 lists some of the more interesting values obtained by computer search.

By the end of this paper we will be able to solve this problem:

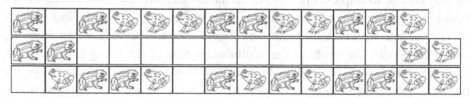

Figure 2. What is the outcome if Left plays first? If Right plays first?

2. Preliminaries

We assume familiarity with the theory of loopy games as presented in Chapter 11 of *Winning Ways*. See [Siegel 2009a] for a gentle introduction.

It is convenient to use Erickson's notation for Toads and Frogs positions. A T represents a toad, an F a frog, and an open box □ an empty space on the board. Superscripts indicate repetition, so for example,

$$T^3 \square^2 F^3 \text{ is the position TTT}\square\square\text{FFF.}$$

The Ts and Fs in a position will occasionally be subscripted, as in

$$TT_1 T_2 \square^2 F_1 F_2 F.$$

The subscripts do not affect the actual composition of the position; they are merely labels used to reference specific toads and frogs in the discussion that follows. Additionally, we will use the symbol ⊞ to represent an arbitrary sequence of zero or more empty spaces. For instance, the generality $⊞T^3⊞F^3$ would include the previous example.

Define the *configuration* of a position to be that position with all empty spaces removed. Thus the configuration depends only on the relative locations of the toads and frogs, and not on the number of spaces that separate them. For example, the configuration of the position noted above is TTTFFF. Note that sliding moves do not affect a game's configuration, while jumps change it irrevocably.

In many of the proofs that follow, the goal is to show that $X \geq 0$ for a certain position X. Elsewhere, however, we wish to show that $X = $ **on**, or that the *onside* of X is **on**. In virtually all cases, the necessary relation is established by exhibiting an explicit winning strategy for Left. However, the shapes of the strategies differ in subtle ways depending on the specific goal. The differences are worth highlighting here:

(a) To show that $X \geq 0$ for some position X, we consider X played in isolation, and show that Left, playing second, can get the last move in finite time.

(b) To show that the onside of X is **on**, we allow Right infinitely many pass moves, and show that Left can play so as never to run out of moves.

(c) To show that $X = $ **on**, we proceed as in (b), with the further restriction that Right must be forced to pass infinitely many times.

Note that (c) does *not* require Left to reach a state where Right is permanently out of moves in X. Indeed, in some cases where $X = $ **on**, it's possible for Right to make infinitely many moves in X. In such cases, Left can ensure that Right is *temporarily* out of moves infinitely often; but for Left to claim a free move, he must mobilize Right for some finite amount of time.

We close this section with a key result:

LEMMA 1 (THE DECOMPOSITION LEMMA). Let X and Y be arbitrary positions. Then:

(a) $XTTY \geq X + Y$ and $XFFY \leq X + Y$.

(b) If X contains no empty spaces, then $XTY \geq Y$ and $YFX \leq Y$.

PROOF. If Left never moves his pair of toads in $XTTY$, he can guarantee that X and Y never interact. This establishes (a), and (b) is similar: if Left never moves his extra toad in XTY, then the entire subposition X is immobilized. $\square$

3. Positions with one Frog

With just one toad and one frog, the position always has value 0 or ∗, and the value depends only on the relative position of the toad and frog (and not on the size of the board):

LEMMA 2.

$$⊞F□^k T⊞ = 0 \text{ if } k \text{ is even}, ∗ \text{ if } k \text{ is odd};$$

$$⊞T□^k F⊞ = ∗ \text{ if } k \text{ is even}, 0 \text{ if } k \text{ is odd},$$

except for the trivial case where there are no moves available to either player.

PROOF. We first show that $⊞F□^k T⊞ = 0$ if k is even. By symmetry, it suffices to show that Left can win playing second. Since no jumps are possible, every move reverses the parity of the distance k. Therefore, the distance will always be odd when Left has the move, so he can always slide toward the left end of the board. Eventually the position will reach FT⊞, and Right will be without a move. If k is initially odd, then moving to 0 is the only option available to either player, so the value is ∗.

Next we show that $⊞T□^k F⊞ = 0$ if k is odd. As before, it suffices to show that Left can win playing second. He begins by advancing until the toad and frog are adjacent. Since every sliding move reverses the parity of k, the meeting must occur immediately following an advance by Right, resulting in the position ⊞TF□⊞. At this point Left jumps, and since the toad and frog remain adjacent, the resulting position has value 0. If k is initially even, then moving to 0 is the only option available to either player, so the value is ∗ (except in the trivial case when no moves are available to either player). □

With several toads against just one frog, the position always has value **on** except in a few pathological cases:

LEMMA 3. Suppose $m \geq 2$. Then:

$$⊞FT^m = 0;$$

$$⊞F□T^m = \{\textbf{on} \mid 0\};$$

All other positions involving m toads and one frog have value **on**, except for the trivial case where there are no moves available to either player.

PROOF. *Case 1*: The frog is to the left of all toads, so that no further jumps are possible. If at least one toad has an empty space to its right, the value is **on**, as follows. On his move, Left advances his left-most toad toward the frog. If this is not possible, he moves any *other* toad arbitrarily. Eventually the left-most toad will trap the frog at the end of the board, and Left's remaining toads will still be free to move about indefinitely.

The two special cases in the statement of the lemma follow immediately.

Case 2: The frog is between two toads. We will show that Left can achieve infinitely many free moves against Right. Note that Right can jump only finitely many times; after the last jump, we are in a Case 1 position with an empty space available to Left (the one just vacated by the frog).

Left plays as follows. If there is intervening space between the frog and its adjacent toads, Left moves a surrounding toad toward the frog. Within finite time the frog will be sandwiched between two toads. Right's only move from such a position (if any) is to jump. If Left is to move from such a position, he simply makes *any* available move. This might give Right the opportunity to make an extra sliding move, but Left can reverse this by tightening the gap again. In that event Left makes two moves to Right's one, gaining a free move.

Case 3: The frog is to the right of all toads. Here Left simply advances the rightmost toad toward the frog. If the rightmost toad is adjacent to the frog, Left makes any other move (jumping permitted). Eventually Right's only move will be to jump. Any jump leads to a Case 2 position.   □

4. Natural starting positions

In this section we consider positions of the form $T^m\square^k F^n$, where the toads and frogs form two disconnected armies. These were termed $(m, n)_k$-positions in *Winning Ways*. The main result is the chart of Figure 3.

These values are, on the whole, much simpler than those for ordinary Toads and Frogs. The basic reason is that either player, if undisturbed, can assure himself infinitely many free moves by maneuvering just three of his amphibians into a "fortress":

$$T\square TTX$$

Notice that it does not matter whether X contains zero or a hundred frogs. The Decomposition Lemma implies that the value of this position is at least $T\square + X$; and since $T\square = $ **on**, the overall position must have onside **on**. This fundamental strategy accounts for the prevalence of **dud**s in the table.

The cases $m = 1$ and $n = 1$ were established in section 3 for all k. We verify the rest of the table with a series of lemmas. We study the easier limiting cases first, and then go back and fill in the gaps.

The first lemma establishes the **dud** values in the $k \geq 3$ section of the chart:

LEMMA 4. *If* $m, k \geq 3$, *then for all* n, $T^m\square^k F^n$ *has onside* **on**.

PROOF. Left, on his first two moves, advances each of the two front toads. Since $k \geq 3$, Right is powerless to interfere even if she moves first, so Left establishes a fortress:

$$T^{m-2}\square T^2 \dots$$

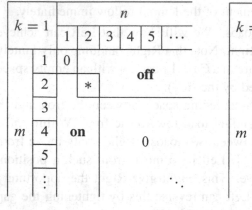

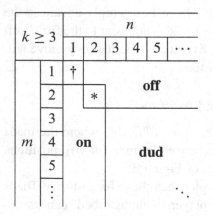

Figure 3. The value of $T^m\square^k F^n$, for all k, m, n.

Since $m \geq 3$, Left has at least one toad remaining in the rear, which he is now free to shuttle indefinitely.   □

By symmetry, we know that if $n, k \geq 3$, then the offside of $T^m\square^k F^n$ is **off**, for all m. Therefore, if $m, n, k \geq 3$, we may conclude that $T^m\square^k F^n =$ **dud**. A similar theme establishes the **dud** values for $k = 2$:

LEMMA 5. *If* $m \geq 4$, *then for all* n, $T^m\square^2 F^n$ *has onside* **on**.

PROOF. There are no problems if Left moves first: he can establish a fortress before Right can interfere. The remaining difficulty is Right's immediate move to $T^m\square F\square F^{n-1}$, which Left counters with a move to

$$T^{m-1}\square TF\square F^{n-1}.$$

If Right does anything other than jump, then Left can establish a fortress immediately. If Right jumps, Left responds by advancing his toad, leaving the position

$$T^{m-1}F_1\square T\square F^{n-1}.$$

There are now two possibilities:

- If Right takes any action other than backsliding F_1, Left jumps to the position

$$T^{m-2}\square F_1 TTX.$$

By the Decomposition Lemma, this position is $\geq T^{m-2}\square F_1 + X$. But since $m-2 \geq 2$, we know from Lemma 3 that $T^{m-2}\square F_1 = $ **on**. So the onside of the sum must be **on**.

- If Right backslides F_1, then Left can advance to

$$T^{m-2}\square T_1 F_1 T\square F^{n-1}.$$

From this position Left can shuttle his rear toads indefinitely. If Right ever jumps with F_1, Left responds by advancing T_1, establishing a virtual fortress just as before. □

When $k = 1$, all values with $m, n \geq 3$ are zero:

LEMMA 6. If $m \geq 3$, then for all n, $T^m \square F^n \geq 0$.

PROOF We can assume that $m = 3$, since Left can simply ignore any additional toads. With Left playing second, all Right moves in the following sequence are forced:

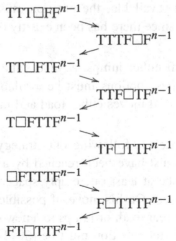

$$TTT\square FF^{n-1}$$
$$TTTF\square F^{n-1}$$
$$TT\square FTF^{n-1}$$
$$TTF\square TF^{n-1}$$
$$T\square FTTF^{n-1}$$
$$TF\square TTF^{n-1}$$
$$\square FTTTF^{n-1}$$
$$F\square TTTF^{n-1}$$
$$FT\square TTF^{n-1}$$

whereupon Right cannot move. □

Next we study those positions where Right has exactly two frogs. The following lemma verifies that $T^2 \square^k F^2 = *$ for $k \geq 2$. The remaining case $k = 1$ is analyzed in *Winning Ways* and can easily be checked by showing that the sum $* + T^2 \square F^2$ is a second-player win.

LEMMA 7. If $k \geq 1$, then $\text{TT}\square^k\text{F}\square\text{F} = 0$.

PROOF. We show that either player, moving second, can force a win in finite time. Clearly, by symmetry, it suffices to show that Left can win either $\text{TT}\square^k\text{F}\square\text{F}$ or $\text{T}\square\text{T}\square^k\text{FF}$ playing second; we will exhibit a strategy that succeeds in both cases.

Let p be the sum of the distance between the two toads and the distance between the two frogs. The following facts are apparent: p is initially odd; each sliding move changes the parity of p; and each jump maintains the parity of p. So on Left's move, p will be even just if an even number of jumps have occurred.

The remainder of the proof exhaustively describes Left's winning strategy. The strategy is broken down by configuration: at each stage, Left guarantees that the next configuration will be reached within a finite number of moves. Eventually the position will reach the configuration FTFT, whereupon we will see that Left can force a win.

Configuration TTFF: Left either jumps or advances a toad. One of these options must be available unless the position has the form ⊞TTFF⊞. Since $k > 0$, such a position cannot be reached on Right's opening move. So Left has had an opportunity to move, and therefore there is an empty space behind the pair of toads. Left backslides, jumping on the next move if Right does not.

Configuration TFTF: Left either jumps with the front toad or advances either toad. If neither option is available, the position must be ⊞TF⊞TF; but that position has even p, and since there has been exactly one jump, it cannot occur on Left's move.

Configuration TFFT: Left either jumps, advances the rear toad, or backslides the front toad. One of these options must be available unless the position is ⊞TFFT⊞, in which case Left moves either toad and jumps on the next move if Right does not.

Configuration FTTF: If Left is following our strategy, he will never jump to this configuration, so it must have been reached by a Right jump. Therefore, at the outset, there must be at least one empty space between the toads. Left plays to maintain this space: On his move, if possible, he either advances the front toad, backslides the rear toad, or jumps to a new configuration. If none of these options is available, the position must be ⊞F_1T_1⊞□T_2F_2. In this case, Left backslides T_2, reaching ⊞F_1T_1⊞T_2□F_2. Left now counters each F_1 move by backsliding T_1. Eventually Right will be forced to advance F_2 and permit a jump.

Configuration FTFT: If Left jumps into this configuration, then Right's *first* move cannot be a jump, so Left is guaranteed at least one move in this configuration. Left always backslides the rear toad if possible; otherwise he backslides the front toad. This guarantees that Right will never be allowed to jump from this configuration. The only positions from which Left cannot backslide either toad are those of the form ⊞FT⊞FT⊞; but those have even p, and so cannot occur on Left's move (as this configuration can only be reached after exactly three jumps). So play continues like this until the position reaches FTFT⊞, whereupon Right is without a move, and Left has won. □

Note: It is *not* true that all positions with two toads, two frogs, and two empty spaces have value 0 or *. For example, □F□TFT has the value $* \& \{0 \mid \textbf{off}\}$.

Three toads are sufficient to overpower Right's two frogs:

LEMMA 8. If $m \geq 3$, then $T^m \square^k F^2 = \textbf{on}$.

PROOF. It suffices to prove that $T^3 \square^k F^2 = \textbf{on}$, since Left can ignore any additional toads. We will exhibit a strategy for Left that forces Right to make infinitely many pass moves. The strategy is broken down into two major phases. In the first phase, Left ignores his rear toad completely. For each configuration K of the four remaining amphibians, our strategy will guarantee that, if Right passes only finitely many times at K, then the next configuration will eventually be reached. When the configuration reaches FTFT, the second phase of the strategy begins, and Left mobilizes his third toad.

We begin by describing the first phase, broken down by configuration.

Configuration TTFF: On his move, Left jumps if possible; otherwise he advances either toad. If neither option is available, the position must be ⊞T_1TFF⊞. Since Left is guaranteed at least one move in this configuration, there must be an empty space behind T_1, so he backslides. If Right responds by passing, Left advances T_1, returning to an earlier position with an intervening pass move. Right's only other options are to jump or to allow a jump.

Configuration TFTF: Left jumps with the *front* toad if possible (never the rear toad); otherwise he advances either toad. If neither option is available, the position must be ⊞T_1F_1⊞T_2F_2. If there is space between the toad/frog pairs, Left backslides T_2, reaching ⊞T_1F_1⊞$T_2\square F_2$. From this position Left counters each F_1 backslide by advancing T_1. Eventually Right must either jump with F_1, advance F_2, or pass. If he advances F_2, Left jumps immediately with T_2; while if he passes, Left advances T_2, having just gained a move.

Finally, if Left is ever to move from ⊞$T_1F_1T_2F_2$, he backslides T_1. If Right passes, Left advances again, gaining a move; if Right advances F_1, Left backslides T_2 to ⊞$T_1F_1T_2\square F_2$. Again Right must either jump, allow a jump with T_2, or pass, granting Left a free move.

Configuration FTTF: If Left is following our strategy, he will never jump into this configuration. So when the configuration is first reached, Right has just leapt into it, and there is at least one space between the toads. Left's moves are, in order of preference: jump; advance the front toad; backslide the rear toad; backslide the front toad. By following this strategy, Left guarantees that the position ⊞FTTF will never arise, so at least one of these options is always available. The analysis is similar to the above.

Configuration TFFT: Left jumps if possible; otherwise he moves either toad toward the frogs. If neither option is available, the position must be ⊞TFFT⊞. In this case Left slides either toad (it's possible that only one is mobile), and Right must either pass, jump, or permit a jump. If he passes, then Left returns to ⊞TFFT⊞.

Once the configuration reaches FTFT, the second phase of Left's strategy begins. The full configuration, including Left's extra toad, is $T_1F_1T_2F_2T_3$. If Right moves first in this configuration, then Left's previous move must have been a jump with T_2 or T_3. So Right's first move cannot be to jump with F_2. Likewise, since T_1 begins on the far left-hand side of the board, Right's first move cannot be to jump with F_1. Therefore Left is guaranteed at least one move in this configuration.

On his move, Left picks one of the following options, listed in order of preference.

1. Backslide T_2 or T_3, preferring T_2 *except* from the position
   $T_1F_1\square T_2\boxtimes F_2\boxtimes T_3\boxtimes$.
2. Shuttle T_1 between the two squares at the far left-hand side of the board.
3. Advance T_2.
4. Advance T_3.

There are three possible ways play might continue:

- Right never jumps again. Here a careful check of Left's strategy reveals that Right is forced to pass infinitely many times.
- Right eventually jumps with F_1. Since T_1 never leaves the two left-hand squares, the resulting position must be FTX for some X containing two toads and one frog. From Lemma 3 we know that $X = $ **on**; but by the Decomposition Lemma, FT$X \geq X$.
- Right eventually jumps with F_2. As we have observed, this cannot happen on Right's first move. Consider Left's previous move. It was not a T_2-backslide, since T_2 and F_2 must now be adjacent. So either Left was *unable* to backslide, or he *chose* not to. If he was *unable* to, then it is because F_1 and T_2 were adjacent; since they no longer are, Left must have just advanced T_2. This means T_1 is immobile and the position (before Right's jump) is exactly $Z = T_1F_1\square T_2F_2\boxtimes T_3\boxtimes$. If Left *chose* not to backslide, then again the position is exactly Z, since otherwise backsliding T_2 is top priority. So Right's jump is

to the position $T_1 F_1 F_2 T_2 \boxtimes T_3 \boxtimes$. By the Decomposition Lemma, this position has value **on**. □

This covers all cases with two frogs (or, by symmetry, two toads). All that remains now are the peculiar values along the $k = 2$, m or $n = 3$ band. The specific case $k = 2$, $m = n = 3$ can be verified computationally. A final lemma completes the analysis.

LEMMA 9. If $m \geq 4$, then $T^m \square^2 F^3 = $ **on** & **hot**.

PROOF. The onside is given by Lemma 5. If Right moves first in the offside, he can establish a fortress, so we know the offside is $\{H \mid $ **off**$\}$ for some H. Finally, a quick computation establishes that $T^3 \square T \square F^3 = $ **on**. Increasing the number of toads cannot reduce this value, so this verifies that $H = $ **on** in all cases. □

5. Some familiar values

In ordinary Toads and Frogs, it is easy to construct positions of positive integer value n: simply place a single toad at the far left of an otherwise-empty $(n + 1)$-length board. Naive constructions fail in the backsliding version, however: if $n > 0$ then such a position has value **on**.

With somewhat more effort, though, it is possible to construct positions of value n and 2^{-n} in Backsliding Toads and Frogs. Further, from these we can derive positions of temperature n and 2^{-n}.

THEOREM 10.
$$(\text{TFFT})^n \square = n;$$
$$\square(\text{TF})^n \text{TTFF} = 2^{-n}.$$

A few Lemmas are needed to prove Theorem 10:

LEMMA 11. If a Backsliding Toads and Frogs position contains just one empty space, then its value is a stopper.

PROOF. We need to show that there are no infinite alternating lines of play from any such position. Since each jump changes the position irrevocably, it suffices to show that there can be no infinite alternating sequence of sliding moves. But after any such move, the only sliding options are to return to the previous position, or to slide in the same direction as the previous move. The first is only available to the same player who just moved. So all moves in any *alternating* sequence of sliding moves must be in the same direction. Therefore any such sequence must terminate. □

Lemma 11 is fundamentally important, because it is relatively easy to compare two stoppers γ and δ. To show that $\gamma \leq \delta$, we just need to check that Left can play so as never to run out of moves in $\delta - \gamma$. (See [Berlekamp et al. 2001] for a proof of this fact.)

Our first application of this technique is the following lemma, which concerns "dead pairs" of toads and frogs. These occur in positions of the form FTX and XFT, where a toad and a frog face away from each other at the far edge of the board. The key result is that dead pairs do not change the value of positions with just one empty space.

LEMMA 12 (DEAD PAIRS LEMMA). Let X be any position with just one empty space. Then

$$\text{FT}X = X = X\text{FT}.$$

PROOF. By symmetry it suffices to prove just the first equality. Decomposition implies that FT$X \geq X$. To show $X \geq$ FTX, it suffices to show that Left, playing second, never runs out of moves in $X -$ FTX (since by Lemma 11 both games are stoppers).

Left's strategy for playing second from $X + (-X)$FT is summarized as follows. Left copies Right's move in the opposite component until Right moves the dead frog. If Right *jumps* with the dead frog, then the second component becomes $\cdots$ FT$\square$T, with no empty spaces except the one indicated. This clearly has value **on**, guaranteeing Left an infinite supply of moves. Suppose instead that Right slides the dead frog. This necessarily leaves the position

$$\square Y + (-Y)\text{F}\square\text{T}$$

for some sequence Y with no empty spaces, whereupon Left can backslide his dead toad:

$$\square Y + (-Y)\text{FT}\square.$$

Now write $Y = \text{F}^n Z$ with n maximal. Right's only possible move is to

$$\text{F}\square\text{F}^{n-1} Z + (-Y)\text{FT}\square,$$

which Left can answer by moving to

$$\text{F}\square\text{F}^{n-1} Z + (-Y)\text{F}\square\text{T}.$$

Now there are three possible options for Right:

- If Right backslides her previous move, Left does the same, returning to a prior position.
- If Right moves to

$$\text{F}\square\text{F}^{n-1} Z + (-Y)\square\text{FT},$$

  then Left simply responds with

$$\text{F}\square\text{F}^{n-1} Z + (-Z)\text{T}^{n-1}\square\text{TFT},$$

and resumes his initial strategy of mirroring Right's moves until the next time
Right activates the dead frog.

- Finally, suppose Right has another frog available:

$$FF\square F^{n-2}Z + (-Y)F\square T.$$

Then necessarily $n \geq 2$, so the second component is $\cdots TTF\square T$. By Decomposition this is $\geq TF\square T$. But Lemma 3 showed that $TF\square T = \textbf{on}$, guaranteeing
Left an infinite supply of moves. $\square$

LEMMA 13.

$$(TF)^n \square = 0 \text{ if } n \text{ is even, } * \text{ if } n \text{ is odd.}$$

PROOF. $n = 0$ is trivial. For even $n > 0$, it suffices to see that $(TF)^n \square$ is a
second-player win. By induction and the Dead Pairs Lemma, Left's only move
is to jump to a position of value $*$, which clearly loses. If Right moves first,
then Left's moves are all forced until Right chooses to jump, reaching:

$$(TF)^k F_0 T_0 \square (TF)^{k'}$$

with $k + k' = n - 1$. Left's strategy now depends on the parity of k'. If k' is
odd, then by induction (and symmetry) $\square (TF)^{k'} = *$. By Decomposition the
full position has value $\geq *$; so Left, with the move, has won. If k' is even, then
Left advances T_0 immediately, and after Right's forced response the position is

$$(TF)^k \square F_0 T_0 (TF)^{k'}.$$

Now since k' is even and $n - 1$ is odd, k must be odd. So $k \geq 1$ and Left can
respond by jumping to

$$(TF)^{k-1} \square F_1 T_1 F_0 T_0 (TF)^{k'}.$$

By induction, $(TF)^{k-1} = 0$. Henceforth Left follows his winning strategy for
$(TF)^{k-1} \square$, until Right chooses to move F_1. Then Left backslides T_1, and after
Right's forced move backslides T_0, reaching

$$XT\square (TF)^{k'}$$

for some sequence X. Since k' is even, $\square (TF)^{k'} = 0$. By Decomposition the
position has value ≥ 0, and since it is Right's move, Left has won.

When n is odd, Left's only move is to jump to a position of value 0. Right's
only move is to

$$Z = (TF)^{n-1} T\square F_0,$$

so the proof is completed by showing that Z is a second-player win. Right's
only move from Z is to return to $(TF)^n \square$, from which Left can move to 0 (as

we already observed). Finally, if Left makes his only move from Z, then Right leaps with F_0, and after a pair of forced moves the position reaches:

$$(TF)^{n-1} \square F_0 T.$$

By induction and the Dead Pairs Lemma, this is a zero position. $\square$

PROOF OF THEOREM 10.. For the first sequence, observe that in $(TFFT)^n \square$ Right has no legal move, and if Left moves first then the following three-move sequence is forced:

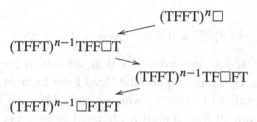

By induction and the Dead Pairs Lemma, the result has value $n - 1$. Note that it is disastrous for Right to ignore Left's opening move, since Left can then backslide for two free moves.

Next we show that $\square(TF)^n TTFF = 2^{-n}$. The proof is by induction on n. The base case is easily verified: $\square TTFF = 1$. For the general case, we show that

$$\square(TF)^n TTFF - 2^{-n}$$

is a second-player win. Suppose first that Left is playing second. If Right jumps with his only mobile frog, Left reduces to

$$FT \square (TF)^{n-1} TTFF - 2^{-n+1}.$$

By induction and the Dead Pairs Lemma, this is a zero position. If instead Right reduces -2^{-n} to 0, Left makes backsliding moves until the position

$$(TF)^n \square TTFF$$

is reached. Now Left's move depends on the parity of n:

- If n is even, Left backslides, and after a forced sequence the position $X = (TF)^{n+1} \square TF$ is reached with Right to move. But Lemma 13 showed that $(TF)^{n+2} \square = 0$, and since X occurs after Left's only response to Right's opening move, we must have $X \geq 0$.
- If n is odd, then by Decomposition $(TF)^n \square TTFF \geq (TF)^n \square = *$, so Left must have a winning move.

If Right plays second from $\square(TF)^n TTFF - 2^{-n}$, his opening strategy is similar: he counters any Left backslides with advancing moves. There are three possibilities.

- *Case 1*: The position $(TF)^n T\square TF_0 F - 2^{-n}$ is reached with Right to move. Then Right jumps with F_0; by Lemma 13 the resulting position has value -2^{-n} or $-2^{-n}*$, a win for Right.
- *Case 2*: Left jumps at some point before the above position is reached, to

$$(TF)^k \square FT(TF)^{k'} TTFF - 2^{-n}.$$

By Decomposition this is $\le (TF)^k \square - 2^{-n}$. Since $(TF)^k \square = 0$ or $*$ by Lemma 13, Right has won.

- *Case 3*: Left plays from -2^{-n} to -2^{-n+1} before either of the above occur. Then Right backslides until reaching $\square(TF)^n TTFF - 2^{-n+1}$, and jumps to

$$FT\square(TF)^{n-1} TTFF - 2^{n+1}.$$

By induction and the Dead Pairs Lemma, this position is exactly equal to 0. Note that if Left prematurely moves to -2^{n+2}, Right can return to his opening strategy, having gained appreciably.   □

Essentially as a corollary, we have:

THEOREM 14. There exist positions with arbitrarily large finite temperature, and with arbitrarily small positive temperature. In particular:

(a) For $n \ge 1$, $(TFFT)^n TTF\square F = \{n* \mid 0\}$.
(b) For $n \ge 0$, $TTF\square F(TF)^n TTFF = \{2^{-n-2}* \mid 0\}$.

PROOF. (a) We first show that $(TFFT)^n T\square FTF = n*$. Left's only move is to the position $A = (TFFT)^n \square TFTF$ and Right's is to $B = (TFFT)^n TF\square TF$; we will show that $A = B = n$. First we describe Left's strategy as second player in $A - n$. He plays just as if the position were $(TFFT)^n \square - n$ (following Theorem 10) until Right chooses to move one of the extra frogs, which necessarily gives the position $YFT\square T_0 F - m$, where $Y\square - m \ge 0$. Then Left backslides T_0, initiating a forced sequence that leads to the position $Y\square FTFT - (m - 1)$, with Left to move. Since $Y > m - 1$, the Dead Pairs Lemma implies that Left has a winning move.

If Left plays second in $B - n$, he reduces immediately to $A - n$ unless Right's first move is to jump. In that case the opening sequence forces the position $(TFFT)^n \square FTFT - n$, with Right to move, so by the Dead Pairs Lemma Left has won.

Next suppose Right plays second in $B - n$. If Left begins by backsliding, then a forcing sequence ensues ending in $(TFFT)^n \square FTFT - n$, and Right has won. If Left jumps instead, Right makes his only available move, to

$$(TFFT)^n F\square T_0 TF - n.$$

Now if Left backslides T_0, then a typical forcing sequence leads to a win for Right. His only other move is to $(\text{TFFT})^{n-1}\text{TFF}\square\text{FTTTF} - n$; but then Right can simply move in $-n$, leaving Left without a move.

Finally, suppose Right plays second in

$$A - n = (\text{TFFT})^{n-1}\text{TFF}_1 T_1 \square T_2 F_2 \text{TF} - n.$$

If Left backslides T_2, Right advances F_2, reducing the position to $B - n$. If Left advances T_1, Right backslides F_1 and follows the strategy outlined in the proof of Theorem 10 for winning $(\text{TFFT})^n - n$. This guarantees that F_1 will never move again, thereby immobilizing T_2 and ensuring that Left's extra toads do not break the strategy.

This establishes that $(\text{TFFT})^n\text{T}\square\text{FTF} = n*$. Thus $(\text{TFFT})^n\text{TTFF}\square = 0$, since Left has no move from that position, and Right's only move allows Left a response to $n*$ ($n \geq 1$). This suffices to confirm (a).

(b) As a simple corollary of Theorem 10, Left's only move is to a position of value $2^{-n-2}*$. Right's only move is to the position

$$Z = \text{TTFF}\square(\text{TF})^n\text{TTFF}.$$

The proof is completed by showing that $Z = 0$. To see that $Z \geq 0$: If Right begins with a jump, it must be to $\text{TTFFFT}\square(\text{TF})^{n-1}\text{TTFF}$. By Decomposition and Theorem 10, this position has value $\geq 2^{-(n-1)}$, so Left has won. If instead Right's first move is to backslide, then Left jumps to $\text{T}\square\text{FTF}(\text{TF})^n\text{TTFF}$. Theorem 10 implies that this position has value $2^{-n-2}*$, so again Left has won.

To see that $Z \leq 0$: Right begins by countering each backslide with an advancing move. Notice that, by symmetry, $-Z = \text{TTFF}(\text{TF})^n\square\text{TTFF}$. Since $Z \geq 0$, we know that $-Z \leq 0$, so if Left does not jump before the position reaches $-Z$, then Right has won. But Left can only jump to

$$\text{TTFF}(\text{TF})^k \square \text{FT}(\text{TF})^{k'}\text{TTFF}$$

for some k, k', and by Decomposition and Lemma 13, this position has value $\leq -2^{-k}$. Again Right has won. $\qquad\square$

6. A little museum

In this section we show some positions with particularly interesting values. All of the museum pieces were obtained by an exhaustive computer search using *CGSuite*. There are a number of surprises, including a position with offside $\uparrow$ on a board of length 14. This seems to be the smallest board on which $\uparrow$ appears in any context.

$$\text{TFF}\square\text{TTF} = \pm(1*) \qquad \text{TTFFF}\square\text{TTTFF} = \pm 1$$

Some simple switches of temperature 1.

$$\text{F}\square\text{TTTFFF}\square\text{T} = \pm(\{1 \mid *\}, \{1 \mid 0\})$$

A pure infinitesimal other than $0, *$.

$$\text{T}\square\text{F}\square\text{TTFFFT} = 1 \ \& \ \textbf{over} \qquad \text{T}\square\text{TT}\square\text{FFFFT} = \textbf{on} \ \& \ \textbf{over}$$

Some positions with offside **over**.

$$\text{F}\square\text{TTTFFTFT}\square\text{TFF} =$$

$$\{\{1 \mid *\}, \{1 \mid 0\} \mid \{* \mid -\tfrac{1}{8}*\}, \{0 \mid -\tfrac{1}{8}*\}\} \ \& \ \pm(\{\tfrac{1}{8}* \mid *\}, \{\tfrac{1}{8}* \mid 0\})$$

A pure infinitesimal with distinct sides.

$$\square\text{FTTF}\square\text{TFTFTTFF} = \tfrac{1}{8} \ \& \ \uparrow \qquad \square\text{FTTFTFTTFT}\square\text{FTFFFT} = * \ \& \ -1\Uparrow$$

$$\square\text{FTTF}\square\text{TFTFTFTTFF} = \tfrac{1}{16} \ \& \ 1/2$$

On large boards, familiar values mysteriously arise.

$$\text{TTT}\square\text{FFFFT}\square\text{TFFT}\square = \textbf{upon}* \ \& \ * \qquad \text{TF}\square\text{FT}\square\text{TF}\square\text{TTFFFT} = 1 \ \& \ \textbf{upon}$$

Some higher-order loopy infinitesimals.

$$\text{F}\square\text{TTTFTFTF}\square\text{TFF} =$$

$$\left\{\{1 \mid *\}, \{1 \mid\mid\mid \tfrac{1}{8}* \mid 0 \mid\mid -\tfrac{1}{8}\} \mid \{* \mid -\tfrac{1}{2}*\}, \{\tfrac{1}{8}* \mid 0 \mid\mid -\tfrac{1}{8} \mid\mid\mid -\tfrac{1}{2}*\}\right\}$$
$$\& \ \left\{\{\tfrac{1}{8}* \mid *\}, \{\tfrac{1}{8}* \mid\mid\mid\mid \tfrac{1}{8} \mid \uparrow \mid\mid 0 \mid\mid\mid -\tfrac{1}{8}\} \mid \{* \mid -\tfrac{1}{2}*\}, \{\tfrac{1}{8} \mid \uparrow \mid\mid 0 \mid\mid\mid -\tfrac{1}{8} \mid\mid\mid\mid -\tfrac{1}{2}*\}\right\}$$

The most complicated value known.

7. A solution

We can now solve the problem presented at the beginning of this paper. By Theorem 14 we know that the first position has value $\{0 \mid -2*\}$. The analysis of Section 4 demonstrates that the middle position has value $*$. Finally, the last position is one of the special values reported in Section 6: $\tfrac{1}{8} \ \& \ \uparrow$.

Adding these together gives a value of

$$\{\tfrac{1}{8}* \mid -\tfrac{15}{8}\} \ \& \ \{\uparrow* \mid -2\uparrow\}$$

for the overall position. The onside is fuzzy, and the offside is negative: Right can win playing first; while if Left plays first he holds the game to a draw.

References

[Berlekamp et al. 2001] E. R. Berlekamp, J. H. Conway, and R. K. Guy, *Winning Ways for Your Mathematical Plays*, Second ed., A. K. Peters, Ltd., Natick, MA, 2001.

[Erickson 1996] J. Erickson, "New Toads and Frogs results", pp. 299–310 in *Games of No Chance*, edited by R. J. Nowakowski, MSRI Publications **29**, Cambridge University Press, New York, 1996.

[Siegel 2009a] A. N. Siegel, "Coping with cycles", 2009. In this volume.

[Siegel 2009b] A. N. Siegel, "New results in loopy games", 2009. In this volume.

AARON N. SIEGEL
aaron.n.siegel@gmail.com

Games of No Chance 3
MSRI Publications
Volume **56**, 2009

New results in loopy games

AARON N. SIEGEL

ABSTRACT. We strengthen the usual notion of simplest form for stoppers
and show that under the stronger definition, equivalence coincides with graph-
isomorphism. We then show that the game graph of a canonical stopper contains no 2- or 3-cycles, but may contain n-cycles for all $n \geq 4$.

We also introduce several new methods for simplifying games γ whose
graphs contain alternating cycles. These include a generalization of dominated
and reversible moves.

1. Introduction

A *loopy game* is a combinatorial game in which repetition is permitted. The
history and basic theory of loopy games are discussed in [Siegel 2009]. In this
article we focus on two fundamental problems left unresolved by *Winning Ways*.

Long irreducible cycles. The first problem concerns the cycles that appear
in the game graph of a stopper. Conway showed that every stopper s admits
a simplest form [Conway 1978], so one would expect that certain cycles are
intrinsic to the play of s. All canonical stoppers discussed in *Winning Ways*
are *plumtrees*: their graphs contain only 1-cycles. It is therefore natural to ask
whether longer canonical cycles are possible, and to attempt to characterize the
structure of such cycles.

Conway defined the *simplest form* of s to be a representation with no dominated or reversible moves. This is not quite strong enough for our purposes, as
illustrated by the example t shown in Figure 1. Certainly t has no dominated or
reversible options, but it is easy to check that $t = \mathbf{on}$. Thus while t technically
has a 2-cycle, it is *reducible* in the sense that t has an alternate representation
with just a 1-cycle.

215

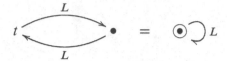

Figure 1. A 2-cycle that reduces to **on**.

In Section 3 of this paper, we introduce a stronger notion of simplicity, the *graph-canonical form* of a stopper. We show that if s and t are stoppers in graph-canonical form and $s = t$, then s and t have isomorphic game graphs. Then in Section 4, we investigate the types of cycles that can appear in a graph-canonical stopper s. We show that every such cycle of length $n > 1$ must contain at least two edges of each color. This rules out 2-cycles and 3-cycles; however, we give examples of graph-canonical stoppers with n-cycles for all $n \geq 4$.

Simplification of alternating cycles. The second problem concerns the simplification of games with alternating cycles. If γ is an arbitrary loopy game, it is desirable to know whether γ is stopper-sided, and if so to compute its sides. Previously, this problem was addressed by the technique known as *sidling* [Berlekamp et al. 2001; Conway 1978; Moews 1996], which produces a sequence of approximations to the sides of γ. If the sidling sequences converge, then they necessarily converge to the sides of γ, but there are many important cases in which they fail to converge.

In Section 5, we introduce generalizations of dominated and reversible options that apply to arbitrary loopy games. These can be used to obtain useful simplifications of γ^+ and γ^-. Often, the simplified forms are already stoppers, even in cases where sidling fails. In addition, the new methods are computationally more efficient than sidling procedures.

Finally, the Appendix (page 228) describes algorithms for comparing arbitrary games. A simplification engine can be built on these algorithms by using the techniques of Section 5. All of these algorithms and techniques have been implemented in *CGSuite* (see http://www.cgsuite.org/), with important applications to the analysis of actual games (see [Siegel 2009] for further discussion).

2. Preliminaries

We assume the reader is familiar with the theory of loopy games, as presented in *Winning Ways*. Sufficient background can be obtained from [Siegel 2009] in this volume. We briefly summarize some of the most relevant facts.

We denote loopy games by Greek letters $\gamma, \delta, \alpha, \beta, \ldots$. If every infinite play of γ is drawn, then γ is said to be *free*, and this will be the assumption when nothing is said to the contrary. When γ is free, we denote by γ^+ and γ^- the matching games with draws redefined as wins for Left and Right, respectively.

Infinite play in a sum $\alpha + \beta + \cdots + \gamma$ is assumed to be drawn unless the same player wins on every component in which play is infinite. In particular, if γ and δ are free, then the following are equivalent:

(i) $\gamma^+ \leq \delta^+$;

(ii) Left can survive $\delta^+ - \gamma^+$ playing second;

(iii) Left, playing second in $\delta - \gamma$, can guarantee that *either* he gets the last move, *or* infinitely many moves occur in the δ component.

(i) $\iff$ (ii) by the definition of $\leq$, and (ii) $\iff$ (iii) by the definition of sum (and the fact that $-\gamma^+ = (-\gamma)^-$). (iii) is a key characterization, and it will be used repeatedly in the proofs and algorithms that follow.

If s and t are *stoppers*, then the following conditions are all equivalent:

$$s \leq t; \quad s^+ \leq t^+; \quad s^- \leq t^-; \quad s^- \leq t^+; \quad \text{Left can survive } t - s \text{ playing second.}$$

Finally, throughout this paper we will assume that all games have a *finite* number of positions. Some results generalize to games with infinitely many positions; but it is usually clear when this is the case, and since the generalization will not be needed it is simpler to keep things finite.

Strategies. Often we will know that Left can survive some game γ and wish to show that he can survive a closely related game γ'. (For example, γ' might be obtained by eliminating a dominated option of γ.) In the loopfree case, this is typically handled by examining relationships between the followers of γ. However, when γ is loopy, altering the options of γ might also affect the structure of its followers. Because of this interdependence, we will usually need to take a global view of the structure of γ, and here it is useful to reason in terms of strategies.

DEFINITION 1. Let γ be a loopy game and let $\mathscr{A}$ denote the set of followers of γ. A *Left strategy for γ* is a partial mapping $S : \mathscr{A} \to \mathscr{A}$ such that, whenever $\delta \in \mathscr{A}$ has a Left option, then $S(\delta)$ is defined, and $S(\delta) = $ some δ^L.

We refer to $S(\delta)$ as the move *recommended by S*.

DEFINITION 2. Let S be a Left strategy for γ. S is a *first-player survival (winning) strategy* if Left, playing first from γ, survives (wins) every line of play in which he plays according to S.

Right strategies and second-player strategies are defined analogously. We say that Left (Right) survives (wins) γ playing first (second) if there exists an appropriate strategy.

DEFINITION 3. Let S be a Left strategy for γ. S is a *complete survival strategy* if, for each $\delta \in \mathscr{A}$ that Left can survive as first player, he can survive δ by playing according to S.

Note that a complete survival strategy recommends good moves from *every* follower of γ, even those that would never be encountered if γ itself were played according to S. Complete survival strategies always exist; this can be established by "pasting together" survival strategies.

LEMMA 4. *Let γ be any loopy game. Then there exists a complete Left survival strategy for γ.*

PROOF. First we inductively construct a sequence of strategies S_n, as follows. Let S_0 be a first-player Left survival strategy for any subposition γ_0 from which Left has a survival move. Given S_n and γ_n, let A_n be the set of positions that can be reached, with Left to move, by some line of play proceeding from γ_n, throughout which Left plays according to S_n. If $\bigcup_{i \leq n} A_i$ contains every follower of γ from which Left has a survival move, then stop. Otherwise, choose any $\gamma_{n+1} \notin \bigcup_{i \leq n} A_i$ from which Left has a survival move, and let S_{n+1} be the corresponding first-player Left survival strategy. Now define a strategy S by

$$S(\delta) = S_n(\delta) \text{ where } n \text{ is least such that } \delta \in A_n.$$

($S(\delta)$ may be chosen arbitrarily if $\delta \notin A_n$ for any n.) We claim that S is a complete Left survival strategy for γ.

To see this, let δ be some follower of γ from which Left has a survival move, and suppose Left plays δ according to S. Let $\delta = \delta_0, \delta_1, \delta_2, \ldots$ be the consecutive positions reached with Left to move (so $\delta_{i+1} = (S(\delta_i))^R$ for each i). We first show that Left has a survival move from each δ_i. This is obviously true for δ_0. For the inductive step, let n be least such that $\delta_i \in A_n$. Then $S(\delta_i) = S_n(\delta_i)$; since S_n is a survival strategy for δ_i, and $\delta_{i+1} = (S_n(\delta_i))^R$, Left has a survival move from δ_{i+1}.

If play is finite, we are done: Left must have made the last move. Otherwise, consider any δ_i, and let n be least such that $\delta_i \in A_n$. Since δ_{i+1} is reached from δ_i by play according to S_n, we also have $\delta_{i+1} \in A_n$. It follows that, for some n_0 and i_0, we have

$$S(\delta_i) = S_{n_0}(\delta_i) \text{ for all } i \geq i_0.$$

Since S_{n_0} is a survival strategy for δ_{i_0}, and the outcome does not depend on any finite initial segment of moves, Left has survived. □

Graphs. Throughout this paper, a *graph* will be a directed graph with separate Left and Right edge sets. We will use calligraphic letters $\mathcal{G}, \mathcal{H}, \ldots$ to denote graphs.

Just as every game has an associated graph, we can define games by specifying a graph and a start vertex. Given a graph $\mathcal{G}$ and a vertex v of $\mathcal{G}$, let $\mathcal{G}|v$ be the graph obtained by removing from $\mathcal{G}$ all vertices not reachable from v. Denote by $\mathcal{G}_v$ the free game whose graph is $\mathcal{G}|v$ and whose start vertex is v. Note that a game is not the same as its graph; this distinction will often be essential. Thus when we write $\mathcal{G}_u = \mathcal{G}_v$, we mean that $\mathcal{G}_u$ and $\mathcal{G}_v$ are game-theoretically equal in the sense of the usual order-relation, whereas $u = v$ means that u and v represent the exact same vertex. Clearly $u = v$ implies $\mathcal{G}_u = \mathcal{G}_v$, but the converse certainly need not be true.

DEFINITION 5. A path directed from u to v is an *alternating path* if its edges alternate colors. The path is *Left-alternating* or *Right-alternating* if the first edge out of u is blue or red, respectively. An *alternating cycle* is an alternating path of even length that starts and ends at the same vertex. We say that an edge is *cyclic* if it belongs to an alternating cycle, and a graph is *alternating cycle-free* if it contains no alternating cycles (equivalently, no cyclic edges).

Note that s is a stopper if and only if its graph is alternating cycle-free.

If u and v are vertices of a graph $\mathcal{G}$, we write $u \xrightarrow{L} v$ to indicate that $\mathcal{G}$ has a Left edge directed from u to v; likewise $u \xrightarrow{R} v$ indicates a Right edge. We sometimes write $e : u \xrightarrow{L} v$ to mean that e is the (unique) Left edge directed from u to v.

3. Fusion

Recall the *simplest form theorem* for stoppers [Berlekamp et al. 2001; Conway 1978; Siegel 2009]:

THEOREM 6 (SIMPLEST FORM THEOREM). *Let s and t be stoppers. Assume that $s = t$, and that neither s nor t has any dominated or reversible options. Then for every s^L there is a t^L with $s^L = t^L$, and vice versa; and likewise for Right options.*

If s and t satisfy this criterion along with all their followers, then they are equivalent in play. However, their graphs might still differ fundamentally. Consider the two examples s and t shown in Figure 2. $s = t = \textbf{over}$, and neither game has any dominated or reversible options, but their representations are clearly different.

A further simplification solves this problem. Suppose s is a stopper whose game graph contains two equivalent vertices, u and v, and assume that no followers of s have any dominated or reversible options. Then we can replace u

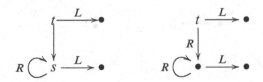

Figure 2. Two forms of **over**.

$$t \xrightarrow{L} \bullet \qquad \qquad t' \qquad \qquad$$

Figure 3. Fusion further simplifies stoppers.

and v with a single vertex, redirecting edges as appropriate, without changing the value of s or any of its followers. Repeated application of this "fusion" process ultimately produces a game with no two equivalent vertices, and this representation is unique up to graph isomorphism. In the example above, t can be reduced to s with two applications of fusion, as illustrated in Figure 3.

LEMMA 7 (FUSION LEMMA). *Let $\mathcal{G}$ be alternating cycle-free, with no dominated or reversible edges. Suppose u, v are two distinct vertices of $\mathcal{G}$ and $\mathcal{G}_u = \mathcal{G}_v$. Let $\mathcal{H}$ be the graph obtained by deleting v and replacing every edge $a \to v$ with an edge $a \to u$ of the same color. Then $\mathcal{H}$ is alternating cycle-free and $\mathcal{G}_w = \mathcal{H}_w$ for every vertex $w \neq v$.*

A cautionary note: fusion might fail when s is not a stopper, or when s is a stopper but is not in simplest form. Figure 4 gives an example: $\gamma = \delta = 2 \,\&\, 0$, but if we fuse δ to γ, then the resulting vertex has value $3 \,\&\, 0$.

PROOF OF LEMMA 7. First we show that $\mathcal{H}$ is alternating cycle-free. Assume instead (for contradiction) that $\mathcal{H}$ contains an alternating cycle. We can assume the cycle involves a redirected edge, since otherwise it would already be present in $\mathcal{G}$. So the cycle involves u, and we can assume without loss of generality that it is Left-alternating out of u. We will construct a sequence $(v_n)_{n=0}^{\infty}$ of vertices of $\mathcal{G}$ such that for all n, $\mathcal{G}_{v_n} = \mathcal{G}_{v_{n+1}}$ and there is an even-length Left-alternating path from v_n to v_{n+1}.

Let $v_0 = u$, $v_1 = v$. Since $\mathcal{H}$ contains a Left-alternating cycle out of u that involves a redirected edge, $\mathcal{G}$ must contain an even-length Left-alternating path from u to v. This establishes the base case.

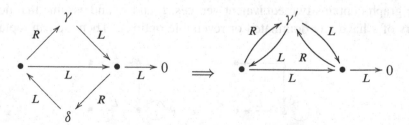

Figure 4. An example where fusion fails.

Now given v_n and v_{n+1}, we construct v_{n+2} as follows. We know that v_{n+1} is a Left-alternating follower of v_n. But $\mathcal{G}_{v_{n+1}} = \mathcal{G}_{v_n}$, so by repeated application of the Simplest Form Theorem, there is a Left-alternating follower v_{n+2} of v_{n+1} satisfying $\mathcal{G}_{v_{n+2}} = \mathcal{G}_{v_{n+1}}$. Since the path from v_n to v_{n+1} has even length, so does the path from v_{n+1} to v_{n+2}. This defines $(v_n)_{n=0}^{\infty}$.

But $\mathcal{G}$ is finite, so there must be some $m < n$ with $v_m = v_n$. It follows that there is an alternating cycle in $\mathcal{G}$ involving v_n, contradicting the assumption that $\mathcal{G}$ is alternating cycle-free. This shows that $\mathcal{H}$ is alternating cycle-free.

Next fix w, and let $s = \mathcal{G}_w$, $t = \mathcal{H}_w$. We wish to show that $s = t$. Since both are stoppers, it suffices to show that Left, playing second, never runs out of moves in $s - t$ or $t - s$. We will prove the $s - t$ case; the proof for $t - s$ is similar.

Let S be a complete Left survival strategy for $s - s$. Define the strategy S' for $s - t$ as follows: S' is equivalent to S except when S recommends a move from $\mathcal{G}_a - \mathcal{G}_b$ to $\mathcal{G}_a - \mathcal{G}_v$. In that case, S' recommends a move from $\mathcal{G}_a - \mathcal{H}_b$ to $\mathcal{G}_a - \mathcal{H}_u$. We claim that S' is a second-player Left survival strategy for $s - t$. To see this, note that whenever $\mathcal{G}_a \geq \mathcal{G}_v$, then also $\mathcal{G}_a \geq \mathcal{G}_u$. Since S is a complete survival strategy, this implies that if Left plays second from $s - t$, then any position $\mathcal{G}_a - \mathcal{H}_b$ reached according to S' will satisfy $\mathcal{G}_a \geq \mathcal{G}_b$. Therefore Left, playing according to S', will never run out of moves. This completes the proof. □

DEFINITION 8. A stopper s is said to be in *graph-canonical form* if s is in simplest form and $\mathcal{G}_u \neq \mathcal{G}_v$ for any two vertices $u \neq v$ of s.

THEOREM 9. *Suppose s, t are stoppers in graph-canonical form with $s = t$. Then the game graphs of s and t are isomorphic.*

PROOF. Let $s = \mathcal{G}_u$, $t = \mathcal{H}_v$. For every vertex a of $\mathcal{G}$, we know that there is a vertex b of $\mathcal{H}$ with $\mathcal{G}_a = \mathcal{H}_b$, and vice versa. ($b$ can be obtained by repeated application of the Simplest Form Theorem.) Since $\mathcal{G}$ and $\mathcal{H}$ contain no equivalent vertices, it follows that there is a bijection $f : V(\mathcal{G}) \to V(\mathcal{H})$ with $f(u) = v$ such that $\mathcal{G}_a = \mathcal{H}_{f(a)}$ for all vertices a of $\mathcal{G}$.

To see that f is a graph-homomorphism, suppose $\mathcal{G}$ contains a Left edge $a \xrightarrow{L} a'$. Write $b = f(a)$, so that $\mathcal{G}_a = \mathcal{H}_b$. Since $\mathcal{G}_a \leq \mathcal{H}_b$, Right has a survival response from $\mathcal{G}_{a'} - \mathcal{H}_b$. It cannot be to any $\mathcal{G}_{a'}^R$, since this would imply that

$$\mathcal{G}_a^{LR} = \mathcal{G}_{a'}^R \leq \mathcal{H}_b = \mathcal{G}_a,$$

contradicting the assumption that $\mathcal{G}$ contains no reversible moves. So $\mathcal{G}_{a'} \leq \mathcal{H}_{b'}$ for some vertex b' of $\mathcal{H}$ with $b \xrightarrow{L} b'$.

Now since $\mathcal{G}_a \geq \mathcal{H}_b$, Left has a survival response from $\mathcal{G}_a - \mathcal{H}_{b'}$. It cannot be to any $\mathcal{H}_{b'}^R$, since (as above) this would imply that $\mathcal{H}_b \geq \mathcal{H}_b^{LR}$, contradicting the assumption that $\mathcal{H}$ has no reversible moves. So $\mathcal{G}_a^L \geq \mathcal{H}_{b'}$ for some $\mathcal{G}_a^L$. Thus $\mathcal{G}_a^L \geq \mathcal{H}_{b'} \geq \mathcal{G}_{a'}$, and since $\mathcal{G}$ contains no dominated options, $\mathcal{G}_a^L = \mathcal{H}_{b'} = \mathcal{G}_{a'}$. Therefore $f(a') = b'$, so $\mathcal{H}$ contains a Left edge $f(a) \overset{L}{\longrightarrow} f(a')$. The proof for Right edges is identical. $\square$

4. Long irreducible cycles

In this section, we show that if s is a stopper in graph-canonical form, then every cycle in s of length greater than one must contain at least two edges of each color. In particular, s contains no 2- or 3-cycles. Longer cycles are possible, however: the game τ shown in Figure 5 is in graph-canonical form and has a 4-cycle. Soon we will see that there exist graph-canonical stoppers t with n-cycles for all $n \geq 4$. Such cycles are *irreducible* in the sense that any representation of t must contain at least an n-cycle.

DEFINITION 10. Let $\mathcal{G}$ be a graph. A cycle in $\mathcal{G}$ is *long* if it contains at least two edges. A cycle in $\mathcal{G}$ is *monochromatic* if all edges in the cycle are the same color; *bichromatic* otherwise.

LEMMA 11. *Let s be a stopper in graph-canonical form. Then s contains no long monochromatic cycles.*

PROOF. By symmetry, it suffices to prove the lemma for cycles consisting entirely of blue edges. So let $s_0, s_1, \ldots, s_n$ be a sequence of subpositions of s, with $s_{i+1} = s_i^L$ for $0 \leq i < n$ and $s_0 = s_n$. We will show that

$$s_0 \leq s_1 \leq s_2 \leq \cdots \leq s_n = s_0,$$

so in fact all subpositions in the sequence must be equivalent.

Left's survival strategy for $s_{i+1} - s_i$ is simple. As long as Right moves around the cycle in the $-s_i$ component, Left does the same in s_{i+1}, staying one move ahead of her. This continues until Right chooses to break the cycle. At that point the position must be either $s_{j+1}^R - s_j$ or $s_{j+1} - s_j^{L'}$ ($s_j^{L'} \neq s_{j+1}$), for some j. In the first case, we have

$$s_{j+1}^R = s_j^{LR},$$

and since s_j has no reversible options, this implies that $s_{j+1}^R \not\leq s_j$. So Left must have a winning move from $s_{j+1}^R - s_j$. Likewise, in the second case, we have

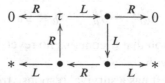

Figure 5. A stopper that is not equivalent to any plumtree.

$s_{j+1} = s_j^L$, and since s_j has no dominated options, this implies that $s_{j+1} \not\leq s_j^{L'}$. So again Left has a winning move; and we have shown that he can survive any line of play.

This shows that each $s_i \leq s_{i+1}$, and hence

$$s_0 = s_1 = s_2 = \cdots = s_n. \qquad \square$$

LEMMA 12. *Let s be a stopper in graph-canonical form. Then s contains no long cycles with just a single red edge.*

PROOF. Toward a contradiction, let s_0, s_1, ..., s_n ($n \geq 2$) be a sequence of subpositions of s, with $s_{i+1} = s_i^L$ for $0 \leq i < n$ and $s_0 = s_n^R$. We first show that

$$s_0 \leq s_1 \leq s_2 \leq \cdots \leq s_{n-1}. \tag{$\dagger$}$$

To show that $s_i \leq s_{i+1}$, we proceed just as in the previous lemma; the only difference occurs when Right has moved to the position $s_n - s_n$. Then Left responds by playing to $s_n - s_0$. If Right continues to $s_0 - s_0$, then Left plays to $s_1 - s_0$ and resumes moving around the cycle as before; while if Right makes any other move, then the absence of any dominated or reversible options hands the win to Left, as in Lemma 11.

This proves ($\dagger$), so in particular $s_0 \leq s_{n-1}$. But $s_0 = s_n^R = s_{n-1}^{LR}$, contradicting the assumption that s_{n-1} has no reversible moves. This completes the proof. $\square$

By symmetry, if s is a stopper in graph-canonical form, then s contains no long cycles with just a single blue edge. Therefore every long cycle in s must include at least two edges of each color.

Unicycles

DEFINITION 13. A stopper s is said to be a *unicycle* provided that:

(i) The graph of s has just one cycle; and
(ii) Each position on the cycle has just two options: a move to the next position on the cycle, and a move for the *other* player to a loopfree game.

We say that s is an *n-unicycle* if its cycle is an n-cycle.

For example, τ (Figure 5) is a 4-unicycle. In fact, there exist n-unicycles for all $n \geq 4$. Figure 6 gives an elegant example for all $n \geq 6$, in which 0 is the only loopfree subposition. Figure 7 is an interesting 13-unicycle: 0 and $*$ are the only loopfree subpositions; furthermore, the cycle is alternating except for the single pair of consecutive Left edges. The 13-unicycle generalizes to a $(4n+1)$-unicycle for all $n \geq 1$ (in particular, this gives an example of a 5-unicycle).

We can classify unicycles more precisely by considering the specific sequence of blue and red edges associated to each cycle. For example, τ has the pattern LLRR. Then a *P-unicycle* is a unicycle whose cycle matches the pattern P.

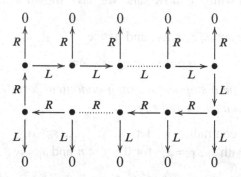

Figure 6. A particularly elegant *n*-unicycle ($n \geq 6$). It is assumed that there are *at least three* blue edges and *at least three* red edges in the cycle, though there need not be equally many of each color.

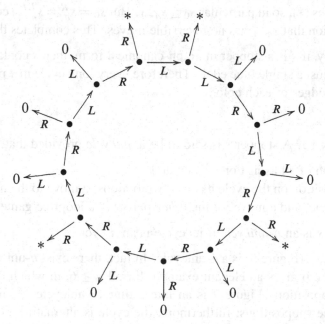

Figure 7. An "almost alternating" 13-unicycle. This generalizes to a $(4n + 1)$-unicycle by continuing the pattern: three exits to 0 followed by one exit to *.

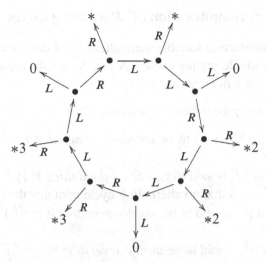

Figure 8. A 9-unicycle whose pattern cannot be realized if the exits are restricted to 0, ∗, and ∗2.

By Lemmas 11 and 12, we know that if there exists a P-unicycle, then P must have at least two edges of each color. Furthermore, P cannot be strictly alternating, since every unicycle is a stopper. As it turns out, these are the only restrictions up to length 9: if P has at most nine edges and meets both restrictions, then there exists a P-unicycle whose loopfree subpositions are all nimbers. $P = $ LLRLLRLLR is an interesting example: Figure 8 gives a P-unicycle with exits to 0, ∗, ∗2 and ∗3, but there are no P-unicycles with exits restricted to 0, ∗ and ∗2 (or any other combination of just three nimbers). All of these facts can be verified using *CGSuite*.

The same is true for patterns of length 10, with one possible exception: $Q = $ LLLRLRRRLR. It appears that there are no Q-unicycles whose exits are restricted to nimbers. However, if exits to arbitrary loopfree games are allowed, then the question remains open.

OPEN PROBLEM. Determine the patterns P for which there exists a P-unicycle. In particular, is there an LLLRLRRRLR-unicycle?

Note that the *number* of patterns of length n is equal to the number of directed binary necklaces of length n. This is sequence A000031 in Sloane's encyclopedia (http://www.research.att.com/~njas/sequences/) and is given by

$$\frac{1}{n} \sum_{d \mid n} 2^{n/d} \varphi(d),$$

where φ is the Euler phi-function.

5. Simplification of alternating cycles

This section introduces a suitable generalization of dominated and reversible moves to games with alternating cycles. All of the results are stated in terms of γ^+, but of course they dualize to γ^-.

DEFINITION 14. Let γ be a free loopy game. Then:

(a) A Left option γ^L is said to be *onside-dominated* if $(\gamma^{L'})^+ \geq (\gamma^L)^+$ for some other $\gamma^{L'}$.

(b) A Right option γ^R is said to be *onside-dominated* if $(\gamma^{R'})^+ \leq (\gamma^R)^+$ for some other $\gamma^{R'}$ such that no alternating cycle contains the edge $\gamma \xrightarrow{R} \gamma^{R'}$.

(c) A Right option γ^R is said to be *onside-reversible* if $(\gamma^{RL})^+ \geq \gamma^+$ for some γ^{RL}.

(d) A Left option γ^L is said to be *onside-reversible* if $(\gamma^{LR})^+ \leq \gamma^+$ for some γ^{LR} such that no alternating cycle contains the edges $\gamma \xrightarrow{L} \gamma^L \xrightarrow{R} \gamma^{LR}$.

The additional constraints in Definitions 14(b) and (d) are necessary, as demonstrated by examples such as Bach's Carousel [Berlekamp et al. 2001]. Of course, the point of these definitions is the following Lemma.

LEMMA 15. *Let γ be a free loopy game and let δ be any follower of γ. Suppose γ' is obtained from γ by either:*

(a) *Eliminating some onside-dominated option of δ; or*

(b) *Bypassing some onside-reversible option of δ.*

Then $\gamma^+ = (\gamma')^+$.

PROOF. We prove the lemma for onside-dominated Right options and onside-reversible Left options; the remaining cases are easier.

(a) Suppose that $(\delta^{R'})^+ \leq (\delta^R)^+$ and γ' is obtained by eliminating $\delta \xrightarrow{R} \delta^R$. Clearly $\gamma^+ \leq (\gamma')^+$, so we must show that $\gamma^+ \geq (\gamma')^+$. Let S be a complete Left survival strategy for $\gamma^+ - \gamma^+$, and define S' as follows: S' is identical to S, except that any recommendation from $-\delta^+$ to $-(\delta^R)^+$ is replaced by a recommendation to $-(\delta^{R'})^+$.

If Left plays according to S', then since S is a complete survival strategy and $(\delta^{R'})^+ \leq (\delta^R)^+$, the position $\alpha^+ - \beta^+$ reached after Left's move will always satisfy $\alpha^+ \geq \beta^+$. Therefore Left never runs out of moves. To complete the proof, we need to show that the play, if infinite, was not ultimately confined to the negative component. So assume that play was infinite. First suppose that Left was forced to deviate only finitely many times from S. Then after a finite initial sequence of moves, Left followed the survival strategy S. Therefore there must have been infinitely many plays in the positive component.

But by the assumptions of Definition 14, δ is *not* a Left-alternating follower of the dominating option $\delta^{R'}$. Thus between any two deviations from S, there must occur at least one play in the positive component. So if Left deviated infinitely many times from S, then again, infinitely many plays must have occurred in the positive component. This shows that S' is also a Left survival strategy for $\gamma^+ - \gamma^+$. Since S' never makes use of the edge $\delta \to \delta^R$, it also suffices for $\gamma^+ - (\gamma')^+$. This completes the proof.

(b) Suppose that $(\delta^{LR})^+ \leq \delta^+$ and γ' is obtained by bypassing δ^L through δ^{LR}. Let S be a complete Left survival strategy for $\gamma^+ - \gamma^+$, and consider the game $\gamma^+ - (\gamma')^+$. Note that whenever Left can survive some $\alpha^+ - \delta^+$, then $\alpha^+ \geq \delta^+ \geq (\delta^{LR})^+$, so he can also survive $\alpha^+ - (\delta^{LR})^+$. Thus he has a survival response to each $\alpha^+ - (\delta^{LRL})^+$. It follows that Left never runs out of moves if he simply plays $\gamma^+ - (\gamma')^+$ according to S. But each time Right plays from $-\delta^+$ to some $-(\delta^{LRL})^+$, the assumptions of Definition 14 guarantee a move in the positive component before the next time $-\delta^+$ is reached. By an argument similar to (a), S suffices as a Left survival strategy for $\gamma^+ - (\gamma')^+$.

To complete the proof, we must define a second-player Left survival strategy S' for $(\gamma')^+ - \gamma^+$. Let S' be identical to S, except at positions of the form $(\delta')^+ - \beta^+$, where δ' is the subposition of γ' corresponding to δ. Then there are two cases.

Case 1: If Left has a survival move from $(\delta^{LR})^+ - \beta^+$, then let

$$S'((\delta')^+ - \beta^+) = S((\delta^{LR})^+ - \beta^+).$$

That is, S' makes the same recommendation from $(\delta')^+ - \beta^+$ that S makes from $(\delta^{LR})^+ - \beta^+$. This is always valid, by definition of bypassing a reversible move.

Case 2: Otherwise, we have $(\delta^L)^+ \not\geq \beta^+$, so Left's move from $\delta^+ - \beta^+$ to $(\delta^L)^+ - \beta^+$ is losing, and therefore S does *not* recommend it (except possibly when *every* Left move from $\delta^+ - \beta^+$ is losing). In this case, S' simply follows the recommendation given by S.

If Left plays $(\gamma')^+ - \gamma^+$ according to S', then he never runs out of moves. As before, to complete the proof we must show that the play, if infinite, was not ultimately confined to the negative component. The proof is much the same as in (a): we show that each deviation from S must have been followed by a play in the positive component.

But Left only deviates from S at *Case 1* positions of the form $(\delta')^+ - \beta^+$. Until some move is made in the positive component, Left's plays in $-\beta^+$ are identical to those recommended by S from $(\delta^{LR})^+ - \beta^+$. Since *Case 1* states that Left can survive from $(\delta^{LR})^+ - \beta^+$, and since S is a complete survival strategy, this implies that some move must eventually occur in the positive component. $\square$

We can also generalize the Fusion Lemma.

LEMMA 16 (GENERALIZED FUSION LEMMA). *Let $\mathcal{G}$ be an arbitrary graph. Suppose u, v are two distinct vertices of $\mathcal{G}$ with $\mathcal{G}_u^+ = \mathcal{G}_v^+$, and assume that there is no alternating path from u to v of even length. Let $\mathcal{H}$ be the graph obtained by deleting v and replacing every edge $a \to v$ with an edge $a \to u$ of the same color. Then $\mathcal{G}_w^+ = \mathcal{H}_w^+$ for every vertex $w \neq v$.*

SKETCH OF PROOF. The proof is similar to that of Lemma 15, so we just sketch it. In playing $\mathcal{G}_a^+ - \mathcal{H}_b^+$ (or $\mathcal{H}_a^+ - \mathcal{G}_b^+$), Left follows a fixed strategy for $\mathcal{G}_a^+ - \mathcal{G}_b^+$, moving to $-\mathcal{H}_u^+$ ($\mathcal{H}_u^+$) whenever a move to $-\mathcal{G}_v^+$ ($\mathcal{G}_v^+$) is recommended. The assumptions on u and v ensure that fusion introduces no "new" alternating cycles, so two deviations in the negative component imply an intervening move in the positive one. $\square$

Appendix:
Algorithms for comparing games

The most basic computational task is the comparison of games, since comparisons form the basis for all simplifications. When G and H are loopfree, a straightforward recursion can determine whether $G \leq H$. Where loopy games are concerned, the situation is more complicated. Recall that if s and t are stoppers, then $s \leq t$ just if Left, playing second, can survive $t - s$. There is a relatively simple algorithm for testing this condition. If $s = \mathcal{G}_u$ and $t = \mathcal{H}_v$, then the basic idea is to determine those vertices of the direct sum $\mathcal{G} \oplus \mathcal{H}$ from which Right can force a win. Since this might depend on who has the move, we consider separately the pairs (A, L) and (A, R), where A is a vertex of $\mathcal{G} \oplus \mathcal{H}$; we will refer to such pairs as *states*. It is convenient to define an associated *state graph*:

DEFINITION 17. Let $\mathcal{G}$ be a game graph. Then the *state graph* S of $\mathcal{G}$ is the (monochromatic) directed graph defined as follows. The vertices of S are pairs (A, L) and (A, R), where A is a vertex of $\mathcal{G}$. Its edges are constituted as follows:

- S contains an edge $(A, L) \to (B, R)$ if and only if $\mathcal{G}$ contains a Left edge $A \to B$.
- S contains an edge $(A, R) \to (B, L)$ if and only if $\mathcal{G}$ contains a Right edge $A \to B$.
- S contains no edges $(A, L) \to (B, L)$ (or $(A, R) \to (B, R)$), for any A, B.

When we speak of predecessors, successors or outedges of a state (A, X), we mean predecessors, successors or outedges of (A, X) in the state graph.

Begin by marking as LOSING all states (A, L) with no successors. Then iteratively:

- Mark as LOSING all states (A, R) with a LOSING successor.

- Mark as LOSING all states (A, L) from which *all* successors are marked LOSING.

Stop when no further vertices can be marked.

Algorithm 1. Comparing stoppers.

The algorithm for comparing stoppers is summarized as Algorithm 1. Starting from those states (A, L) with no successors, the states of $G \oplus \mathcal{H}$ from which Right can force a win are iteratively identified. Then $s \leq t$ just if $(u \oplus v, R)$ is unmarked: if Right can win from $u \oplus v$, then he can do so in n moves, for some n; but then $(u \oplus v, R)$ will be marked on the n-th stage of the iteration.

This idea is not new. Three decades ago, Fraenkel and Perl [1975] gave a similar procedure for determining the $\mathcal{P}$- and $\mathcal{N}$-positions of an impartial loopy game. The partisan version of the algorithm was introduced several years later by Shaki [1979]. It was rediscovered independently and brought to my attention by Michael Albert (personal communication, 2004).

The algorithm can be refined to guarantee that each state is examined at most once per outedge. The improved version is summarized as Algorithm 2. A huge advantage of this refinement is that it allows substantial prunings. Traversing the states "top-down," and stopping as soon as a winner is determined, yields significant time savings when prunings are desirable. Note that wins for *both* players are determined, and not just for Right; occasionally this will quickly identify Left as the winner and permit an early pruning.

Comparing general games. If γ, δ are arbitrary loopy games, then the comparison process is substantially more difficult. Recall that $\gamma \leq \delta$ if and only if Left, playing second, can survive both $\delta^+ - \gamma^+$ and $\delta^- - \gamma^-$; see [Siegel 2009]. For clarity, and since the two cases are exactly symmetric, we consider just $\delta^+ - \gamma^+$.

Now Left survives $\delta^+ - \gamma^+$ if and only if either

(a) he gets the last move, or
(b) infinitely many plays occur in δ.

Thus if play is infinite, but is entirely confined to the $-\gamma^+$ component, then Left has lost. We can eliminate condition (a) from consideration by first applying the stopper-comparison algorithm (Algorithm 2); the remaining task is to identify those states from which Right can keep the play indefinitely in $-\gamma^+$.

The solution is to make several passes through the graph. At the start of each pass, some states will already be marked as a WIN FOR R, and the goal

---

Visit each state (A, X) *at most once* (in any order) and perform the following steps:

(1) Mark (A, X) VISITED.

(2)

- If any successor of (A, X) is already marked as a WIN FOR X, then mark (A, X) as a WIN FOR X.
- If every successor of (A, X) is already marked as a WIN FOR Y ($Y \neq X$), then mark (A, X) as a WIN FOR Y.

(3) If we just marked (A, X) as a win for either player, then examine each predecessor (B, Y) of (A, X) such that

- (B, Y) is marked VISITED; and
- the winner of (B, Y) has not been determined.

If we marked (A, X) as a WIN FOR Y, then immediately mark (B, Y) as a WIN FOR Y. If we marked (A, X) as a WIN FOR X, then rescan the successors of (B, Y), and if they are all marked as a WIN FOR X, then mark (B, Y) as a WIN FOR X.

If this determines the winner of (B, Y), repeat step 3 with (B, Y) in place of (A, X).

Algorithm 2. Comparing stoppers, refined.

is to identify new ones. Now suppose that, from some state (A, X), Right can guarantee that *either* a state marked WIN FOR R will be reached, *or* no further plays will ever occur in δ. Clearly (A, X) must be a WIN FOR R as well. Call a state BAD if it meets this test; GOOD otherwise. During each pass through the graph, we first identify all GOOD states, and then mark each BAD state as a WIN FOR R. The algorithm terminates when a pass completes with no new states identified as a WIN FOR R.

The procedure for identifying GOOD states is straightforward. For example, suppose that for some state (A, L), there exists an outedge *in δ* to a state that is not known to be a WIN FOR R. Then (A, L) can be marked GOOD immediately. The GOOD markers can then be back-propagated just as WIN markers were in the stoppers case.

In the worst case, each pass would identify just one GOOD state, so the algorithm is ostensibly $O(|V| \cdot |E|)$, where $|V|$ is the number of vertices and $|E|$ the number of edges in the state graph. In practice, however, more than a few passes are rarely necessary, and the algorithm is effectively $O(|E|)$.

The algorithm is summarized in detail as Algorithm 3.

First execute Algorithm 2 to identify states from which one of the players can force a win in finite time. Then:

(1) Visit each vertex A *at most once* and perform the following steps.

(a) If the winner of (A, L) is not yet determined, and either:

- (A, L) has an outedge *in δ* to a state whose winner is not yet determined;
  or
- (A, L) has an outedge to a state marked GOOD,

then mark (A, L) GOOD.

(b) If the winner of (A, R) is not yet determined, and every successor of (A, R) *in γ* is marked either WIN FOR L or GOOD, then mark (A, R) GOOD.

(c) If either of the previous steps caused a state (A, X) to be marked GOOD, then examine all γ-predecessors (B, Y) of (A, X) such that:

- (B, Y) is marked VISITED; and
- The winner of (B, Y) is not yet determined; and
- (B, Y) is not marked GOOD.

If $Y = L$, then immediately mark (B, Y) GOOD. If $Y = R$, then rescan the γ successors of (B, Y), and if they are all marked either WIN FOR L or GOOD, then mark (B, Y) GOOD.
If this causes (B, Y) to be marked GOOD, then repeat step 1(c) with (B, Y) in place of (A, X).

(2) Visit each state (A, X) a second time and perform the following steps:

(a) If the winner of (A, X) is not yet determined, and (A, X) is *not* marked GOOD, then mark (A, X) as a WIN FOR R.

(b) If the previous step caused a state (A, X) to be marked as a WIN FOR R, then examine all VISITED predecessors (B, Y) of (A, X) whose winner is not yet determined.
If $Y = R$, then immediately mark (B, Y) as a WIN FOR R. If $Y = L$, then rescan the successors of (B, Y), and if they are all marked as a WIN FOR R, then mark (B, Y) as a WIN FOR R.
If this determines the winner of (B, Y), then repeat step 2(b) with (B, Y) in place of (A, X).

(3) Clear all VISITED and GOOD markers. If the previous step caused any new states to be marked as a WIN FOR R, then repeat starting with step 1. Otherwise, stop.

Algorithm 3. Testing whether Left can survive $\delta^+ - \gamma^+$.

Acknowledgement

I thank Michael Albert for his considerable assistance in developing the algorithms presented in this appendix.

References

[Berlekamp et al. 2001] E. R. Berlekamp, J. H. Conway, and R. K. Guy, *Winning Ways for Your Mathematical Plays*, Second ed., A. K. Peters, Ltd., Natick, MA, 2001.

[Conway 1978] J. H. Conway, "Loopy games.", pp. 55–74 in *Advances in Graph Theory*, edited by B. Bollobás, Ann. Discrete Math. **3**, 1978.

[Fraenkel and Perl 1975] A. S. Fraenkel and Y. Perl, "Constructions in combinatorial games with cycles.", pp. 667–699 in *Infinite and Finite Sets, Vol. 2*, edited by A. Hajnal et al., Colloq. Math. Soc. János Bolyai **10**, North-Holland, 1975.

[Moews 1996] D. J. Moews, "Loopy games and Go.", pp. 259–272 in *Games of No Chance*, edited by R. J. Nowakowski, MSRI Publications **29**, Cambridge University Press, New York, 1996.

[Shaki 1979] A. Shaki, "Algebraic solutions of partizan games with cycles.", *Math. Proc. Cambridge Philos. Soc.* **85**:2 (1979), 227–246.

[Siegel 2009] A. N. Siegel, "Coping with cycles", 2009. In this volume.

AARON N. SIEGEL
aaron.n.siegel@gmail.com

Games of No Chance 3
MSRI Publications
Volume **56**, 2009

A library of eyes in Go, I:
A life-and-death definition consistent with bent-4

THOMAS WOLF

ABSTRACT. In the game of Go we develop a consistent procedural definition of the status of life-and-death problems. This computationally efficient procedure determines the number of external ko threats that are necessary and sufficient to win, and in the case of positions of the type of bent-4-in-the-corner it finds that they are unconditionally dead in agreement with common practice. A rigorous definition of the status of life-and-death problems became necessary for building a library of monolithic eyes (eyes surrounded by only one chain). It is also needed for comparisons of life-and-death programs when solving automatically thousands of problems to analyse whether different results obtained by different programs are due to different status definitions or due to bugs.

1. Introduction

1.1. Overview. In this contribution we describe a project whose aim was to built a data base of eyes together with their life-and-death status which at least reflects one aspect of ko accurately: the number of necessary external ko threats for the weaker side to win. The procedure how to determine this number is described in Section 2. After that we seem to be ready for determining the status of a life-and-death problem if there would not be the bent-4-in-the-corner positions (in the following called *bent-4*) which are characterized in Section 3 and force us in Section 4 to refine the procedure that we take as the (procedural) definition of the status of a life-and-death problem. In the appendix we discuss current limitations of the program GOTOOLS which is the implementation behind this article.

1.2. The key problem. The discussion in sections 2–4 is rather detailed and arguments are developed why procedures and rulings were designed as they are. In order not to lose sight and have an orientation when reading them we already now want to address the key problem.

In this contribution we consider the procedural definition of the status of life-and-death problems, i.e., of positions that are isolated from the rest of the board through a single solid chain of stones that has enough external safe liberties to live statically. The procedure as outlined in Section 2 is capable of classifying nearly all types of common and also strange life-and-death positions correctly within some approximations (like treating life and seki alike as listed in the appendix) and up to mastering the computational complexity. The only exception encountered so far is a class of bent-4-type positions as defined in Section 3, which includes, for example, the one on the right. Positions of this class are characterized essentially by (a) having a ko status when evaluated according to the straight forward rules of Section 2 and (b) the key property that at some stage of optimal play, the side for which this ko is unfavourable (White in Diagram 1) has as single best move only the passing move.

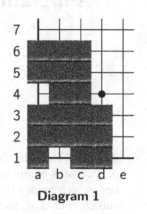

Diagram 1

The combination of these 2 properties has severe consequences. If there are only removable kos on the board (e.g. cuts, but no seki) then Black can wait until later in the game and protect all potential ko-threats (that is, 'remove' potential ko threats) at no cost before starting the ko. In that case the status would be an unconditional loss for White which also is what Go players expect from a computer Go program to find[1]. A fundamental principle of the orthodox procedure in Section 2 is that a position is alive or seki unless the attacker can prove how to kill it (unconditionally or through ko). But in bent-4-type problems the best move for the attacker is to pass, at least during the 'hot' phase of the game when playing elsewhere (*tenuki*) has some benefit. In other words, bent-4-type positions do not have a single solution for the best move, they have two: (1) if playing elsewhere is beneficial then the single best move for the attacker is to pass, (2) if playing elsewhere is not beneficial, and if the attacker is asked to prove that the position is not a seki then the best move for Black in Diagram 1 is to play on b1. Situation 2 is mastered in a straight forward manner, as described in Section 2. The challenge is situation 1: To satisfy the conflicting requests in this case (i.e., to find that the position is not alive/seki *and* that passing is the

---

[1] At times when GOTOOLS solving life and death problems online under [5] did not find bent-4 to be dead, many error reports were submitted by users.

best move for the attacker) in a consistent, local procedure, without having to check special cases separately, is the goal of Section 4. Once both situations can be handled, all that is needed is an extra boolean input parameter specifying for which of the two situations the computation shall be valid.

1.3. Notation. Throughout this paper we will call the side that builds eyes and tries to live as White and the side trying to kill as Black. The side moving next in a position will be called First and the other Second. To have finite and effective searches the right to pass is strongly regulated. In this article we will derive, modify and collect *rulings* which include statements when passing is allowed. In these rulings we will use ● and ○ instead of Black and White to get a more compact formulation. We will follow common terminology and call a position which contains an empty point that is forbidden for one side due to the ko rule as a *ko-banned* position and all other positions *regular*. When the text refers to ko threats then these are always external ko threats (ko threats outside of the problem).

2. The ko status of a life-and-death problem

The program GoTOOLS (described in more detail in [1] and with restrictions listed in the appendix) is the implementation behind the theory in this article. It performs an α/β search with only two possible outcomes: life/seki or death; it is therefore called a boolean search below. If the status is ko then it repeats the search to find the number of external ko threats needed by the weaker side to win.

2.1. Reruns with successively more ko-threats. In a first search no side is allowed to recapture a ko. If for one side, say First, all moves fail in a position in depth d then the result of this computation is not only the loss but attached is always a boolean variable *ko-chance* which, if true, means that the outcome could have been different if First would have had a ko-threat. If that is the case, the previous move by Second at depth $d - 1$ is a winning move but attached to it is ko-chance = true. This is how this information moves upwards in the search tree.

In any position at depth d ko-chance is set to true if

- the position is ko-banned and recapturing the ko would not have violated any cycle rules (see below), but was not possible due to a lack of ko threats, or
- *any one* of the winning moves of Second at depth $d + 1$ returned ko-chance = true.

For example, First has 5 possible moves in some position in depth d. The first tried move fails and has ko-chance = false. The second move of First fails too but the counterproving move of Second in depth $d + 1$ returns ko-chance = true. The third move of First happens to win, thus search in this level stops. Thus the ko-chance of First from the second try becomes irrelevant. What is relevant is whether Second has a ko-chance from any one of its losing tries at depth $d + 1$. If so, then not only will be reported to level $d - 1$ which move of First won but also that Second has a ko-chance for this move.

At the end of the first computation the program knows whether in the verification tree of the search (the minimal tree to be searched where the final winner plays only winning moves and the loser all possible moves) the loser still has a *ko-chance*, that is, a chance to win if it had one more external ko threat initially. In such a case a second run, and if necessary more runs, are performed with successively more external ko threats initially allocated to the loser. This is continued until either the loser wins or loses without having a ko-chance in the last run or until a maximum of k_m external ko threats are reached after which the status is regarded as an unconditional loss for the side that lost so far. In our calculations $k_m = 5$, which could easily be changed to an arbitrary high but fixed value.

2.2. The different ko status. We now come to the different possible status of a position. If the maximal number of allowed ko threats is k_m then $2k_m + 2$ different status may result, each characterized by a numerical value. The possible outcomes sorted from most beneficial to least beneficial for First are:

Value : Status

$k_m + 1$: an unconditional win for 1st,

k_m : a win for 1st unless 2nd has k_m external ko threats more than 1st,

...

1 : a win for 1st unless 2nd has 1 external ko threat more than 1st,

-1 : a loss for 1st unless 1st has 1 external ko threat more than 2nd,

...

$-k_m$: a loss for 1st unless 1st has k_m external ko threats more than 2nd,

$-k_m - 1$: an unconditional loss for 1st.

For example, in Diagram 2 the status is $k_m + 1$ for any side moving first, as both can win unconditionally by playing on b1. In Diagram 3 both sides would pass as the position is unconditionally dead; that is, the status is $k_m + 1$ if Black moves first and $-k_m - 1$ if White moves first.

Status values are chosen so that if the outcome is the same for both sides moving first, then the status values just differ by a sign. For example, if the outcome is that White needs one ko threat in order to win, no matter who moves

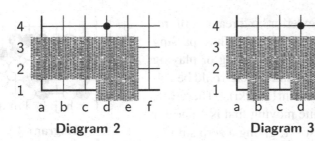

Diagram 2   **Diagram 3**

first, then the status value for White moving first is -1 and the status value for Black moving first is $+1$. In other words, if the outcome is independent of who moves first then the sum of the two status values is zero. But that is exactly the case when passing does not do any harm, i.e., passing is one of the best moves of both sides as in Diagram 3. Conversely, if the sum of the status values is nonzero then it is beneficial for both sides to move first. In Diagram 2 the sum has the maximum value $2k_m + 2$. In the context of this paper a move belongs to the best moves if no other move generates a higher numerical status value.

LEMMA. *If for a regular position (such that both sides can be considered to move first) passing belongs to the best moves of one side then passing does also belong to the best moves of the other side.*

Proof (indirect): We assume that passing belongs to the best moves of, say, White. Then, if passing would not be one of the best moves of Black, then if Black would pass then White could make a move, such that the status for Black would be worse than if Black would not pass. In other words, if passing is not one of Black's best moves then passing is also not one of White's best moves which contradicts our assumption.

If passing belongs to the best moves of both sides then we call the position *settled* if it is unconditionally dead or alive/seki, otherwise it has a ko status and we call it *calm*.

A clarification: To make clear that the terms 'best move' and 'calm' do depend on the type of computation performed, we should use *boolean-best move* and *boolean-calm* if they are determined in a boolean search but to keep the text better readable, we will continue to use simply 'best move' and 'calm' although we exclusively refer to a boolean search. It is necessary to make this remark, because not all boolean-best moves are truly best moves[2] and thus not all boolean-calm positions are truly calm positions as seen in a collection of boolean-calm positions in [4].

Diagram 4 on the next page is an example of a (boolean-) calm position. Here

---

[2] A move of White giving seki is boolean-best but not truly best if there is another move reaching life.

Black needs one exterior ko threat to kill, regardless of who moves first (White passing
or playing on m2, Black passing or playing
anywhere apart from m2 which would be followed by White on m1: seki). Therefore,
the status for White moving first is 1 and for
Black moving first -1, giving a zero sum.

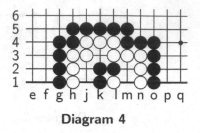

Diagram 4

LEMMA. *For a regular position the sum of status values for both sides moving first is never negative.*

(This lemma is typical for games that allow passing and have no zugzwang.)

Proof (indirect): Assume the sum of both status values is < 0. Then at least one of both status values must be negative. Let X be the side with the most negative of both status values, namely $s_X < 0$. Even if the opponent of X would have no better first move than to pass, then the achieved status value would still be $-s_X > 0$ from passing, giving a sum of at least zero in contradiction to the assumption.

3. Characterization of bent-4-type positions

The collection of plausible rules from Section 2 and the appendix describes a finite procedure with a definite result. This result is essentially identical to what one expects from real Go apart from limitations in the appendix and apart from the 4 bent-4 positions

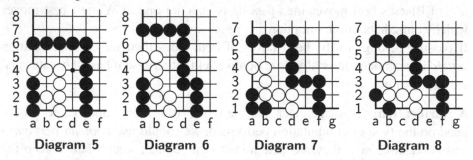

Diagram 5   **Diagram 6**   **Diagram 7**   **Diagram 8**

including those created by filling of external liberties of White, rotation, reflection and swapping colours.

When computed in accordance with the procedure definition given in the above sections, for example in Diagram 5, the first 7 moves of Black would fill White's liberties , each followed by a pass of White and further ● b1 (giving 4 bent black stones in the corner, hence the name 'bent 4 in the corner'), ○ b3, ● a2, ○ a1, ● b1 resulting in a ko where White needs an external ko threat to live, independent of who moves first.

An essential difference between bent-4 and the position in diagram 4 is that in bent-4 passing is the *only* best (internal) move for White. Hence Black has the option to wait long enough until later in the game and then protect all potential white ko threats[3], fill outside liberties of White, produce an L-shaped throw-in chain which is caught by White and then play on a2 and start a ko in the corner, which Black captures first, i.e., White needs an external ko threat. As Black had enough time to remove at least all removable ko-threats, the position on its own is commonly regarded as dead, although its value in a real game, depends on the situation on the board and the rules used.[4]

When evaluated according to Section 2 these four positions have the following more general properties.

DEFINITION 1. Any position that has the following three properties is said to be of bent-4-type:

1. The initial position does not have a ko-forbidden point.
2. The status is a ko in which side X needs at least one external ko threat (more than the other side) to win.
3. Passing is the single best first move for X (apart from playing on dame points).

Comments

- From 3 it follows that passing is also one of the best moves of the opponent of X.
- In bent-4 (diagrams 5–8) we have X = White and dame points would be external liberties of White, like a5 in diagram 5.
- There are many positions which satisfy criteria 2 and 3 for X = Black (the attacker) but not 1–3, i.e. they have a ko-forbidden point. In a search of all positions that involve a single white eye with up to 11 internal points no position showed up satisfying criteria 1–3 for X = Black (i.e. X = attacker) and only diagrams 5–8 satisfy criteria 1–3 for X = White.
- Larger bent-4-type positions are possible (Chi-Hyung Nam, personal communication). In Diagram 9 the best move for Black is to pass whereas White can always provoke a favourable ko by playing ○ e17, ● f17, ○ e19, ● f18, ○ c19 (i.e. here X = Black).

---

[3] This may not be possible, for example, in the case of an infinite source of ko threats, like a double-ko-seki somewhere else on the board, or a seki which for White is *less* costly to lose than bent-4 and for Black *more* costly to lose than bent-4.

[4] According to the Japanese 1989 rules, bent-4 positions are even unconditionally dead. More precisely, in the confirmation phase of the game the so-called 'pass-for-ko rule' implies an unconditional loss for White (see [2]). In the Japanese 2003 rules of Robert Jasiek, during the hypothetical-analysis (stage 2 of the game, following stage one, which is the alternating-sequence of moves) White would also have to ko-pass, leading to the capture of the white stones before White would be able to recapture the ko (see [3]).

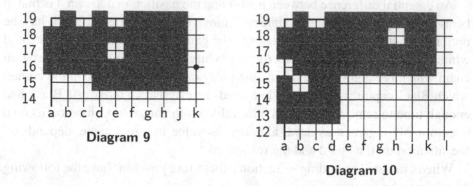

Diagram 9

Diagram 10

In Diagram 10 the best move for White is to pass whereas Black can always provoke a favourable ko by playing ▆ h18, ▆ g19, ▆ c19, ▆ h19 (i.e. here X = White).

The question to be answered in the following section is: *Can one have a consistent boolean search which on one hand evaluates anything to be alive/seki which can not be killed by a nonpassing move and on the other hand evaluates bent-4-type positions to be dead but the killing move is a pass?*

4. Modification of the status defining procedure

4.1. The passing rules so far. To define the boolean search completely one must clarify under which circumstances passing is allowed. The standard passing rules in GOTOOLS that do not yet take care of bent-4 are:

Ruling I:

1. In a regular position,

    (a) if ○ moves next then

       (i) if ○ recaptured a ko two moves earlier using one of its ko threats, and ● passed afterwards then ○ is not allowed to pass

       (ii) else ○ is allowed to pass.

    (b) ● is not allowed to pass

2. In a ko-banned position,

    (a) if First (the side moving next) has no external ko threat it may pass,

    (b) if First has an external ko threat it may not pass.

If the position contains a point that is ko-forbidden then both sides may pass (rule 2). A refinement (rule 2b) is necessary in order not to lose unnecessarily despite of having ko threats and thus unnecessarily require reruns with more and more external ko threats.

To treat *seki* as a situation in which White cannot be killed, it is necessary to allow White to pass also in regular positions (rule 1a-ii). This rule needs an exception in form of 1a-i. If White would be allowed to pass under the circumstances of 1a-i then Black could recapture the ko and a so-called *negative-value* 4 move cycle would be created in which White wasted an external ko-threat (see 'Handling cycles' on page 245).

Black, on the other hand, should not be allowed to pass in a regular position (rule 1b) because afterwards White could pass too and White could not be killed. This rule creates a problem with bent-4 positions in diagrams 5–8 for which the best move for Black is to pass. The key to get bent-4-type positions right must therefore involve a change of the passing rules.

4.2. Passing for bent-4-type positions. One way to solve the problem with bent-4-type positions would be to perform an extra computation if in the first run the status turns out to be a favourable ko for Black and if there is initially no ko-forbidden point (that is, if it is regular). In this additional search one would test whether Black can win if it passes as first move. This may work for diagrams 5–8 but a bent-4-type position could only result within one branch of a larger tree-search and the status of the larger problem may depend on the correct solution of the bent-4-type subbranch. We therefore need a 'local' rule (local in the sense of the whole search tree) about the right to pass and not a separate computation.

The key idea is to allow Black to pass if White has an external ko threat, that is, to replace rule 1b in ruling I by the rules

(1b-i) if ○ has no external ko threats then ● is not allowed to pass,
(1b-ii) if ○ has at least one external ko threat then ● is allowed to pass.

But after a type 1b-ii passing of Black, White could pass too and would not be found to be dead. It seems to be necessary to forbid White to pass after a type 1b-ii passing of Black, but this does not work either as can be seen from the problem in Diagram 11.

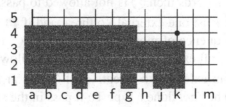

Diagram 11. Black to move first.

The solution sequence is ① f1, ② e1, ③ c1, ④ a1, ⑤ f1 such that White needs one external ko threat to live. Therefore, during the solution of this problem it comes to a second run in which White has one external ko threat. With one extra ko threat for White in the second run, the move ① f1 fails and all other alternatives have to be investigated, like ① pass. Therefore, in the second run eventually rule 1b-ii is applied yielding ① pass, ② f1, ③ pass. The resulting position is a seki, but to recognize it White must be allowed to pass with ④ (and it must be forbidden for Black to pass afterwards to have a finite algorithm). Like with bent-4 the crucial situation takes place in the second run when White has an external ko threat. The question is, what are natural rules which in a run where White has a ko threat,

- for bent-4 after ① pass *forbids* ② pass but
- for Diagram 11 after ① pass, ② f1, ③ pass, *allows* ④ pass (and forbids ⑤ pass) ?

All Black's passes are of type 1b-ii. The difference between both situations is: if there would be *no* extra White ko threat

- in bent-4 after ① pass, ② pass then Black still *wins*,
- in Diagram 11 after ① pass, ② f1, ③ pass, 4. ④ pass Black can *not win*

which in both cases is what we want. But that cannot be found out in a second run in which White *has* an extra external ko threat, *unless* one lets White pay the price of losing all external ko threats if White wants to pass *after* a Black type 1b-ii pass.

If White cannot win after losing all ko threats, then this is a situation where passing is one of Black's best moves which improves a favourable ko to an unconditional kill as Black can wait until the end of the game before starting the ko. We therefore modify rule 1a-ii and get:

Ruling II:

1. In a regular position,
   (a) if ○ moves next then
      (i) if ○ recaptured a ko two moves earlier using one of its ko threats, and ● passed afterwards then ○ is not allowed to pass, else
      (ii) if ○ has at least one external ko threat and if ● *has* done a type 1b-ii pass in the sequence of moves up to now then ○ is allowed to pass but has to *give up* all external ko threats for any following moves, else
      (iii) if ○ has no external ko threats or if ○ has at least one external ko threat and ● has *not* done a type 1b-ii pass in the sequence of moves up to now then ○ is allowed to pass *without* giving up ko threats.

(b) if ● moves next then

    (i) if ○ has no external ko threats then ● is not allowed to pass,

    (ii) if ○ has at least one external ko threat then ● is allowed to pass.

2. In a ko-banned position,

    (a) if First (the side moving next) has no external ko threat it may pass,

    (b) if First has an external ko threat it may not pass (but should instead play the ko threat and then recapture).

Comments

- These rules about passing are part of a boolean search, they make no statement whether passing is a good or bad move in a particular situation. It may very well be that one of the above rules forbids passing in a situation where passing is the only correct move. Nevertheless, the above ruling is correct, because in such a situation, nonoptimal moves have been made earlier by that side (which is forbidden to pass now) in the sequence of moves leading to this situation. The following diagrams 12–15, also known as *mannen ko*, give an example. The letter K marks a ko-forbidden point.[5]

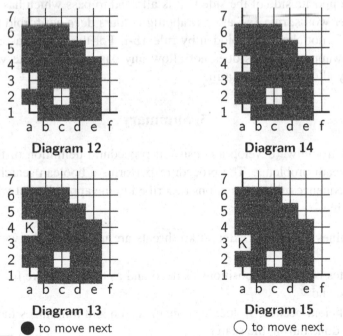

Diagram 12

Diagram 14

Diagram 13
● to move next

Diagram 15
○ to move next

[5] In all 4 positions both sides have the option to pass which leads to seki. In Diagram 12 Black can enforce a ko by playing on a1 or c2 which is about unconditional life or death and which is unfavourable for Black. In Diagram 14 White can enforce a ko, unfavourable for White, by playing on a1 or c2. In a boolean search with seki = life the single best move for White in all 4 positions is to pass and the status is an unfavourable ko for Black.

We consider Diagram 15 and a run in which White has one external ko threat available. We assume further, White (nonoptimally) uses the ko threat to recapture the ko (leading to Diagram 13) and Black passes (correctly) leading to Diagram 14. According to rule 1a-i in this situation White is not allowed to pass although the passing move is the correct move for White. If White loses in this position because it may not pass then this has only the consequence that White has to try a different move two moves earlier and do the correct move there and pass. But the fact that the proper move is forbidden has a consequence for programming. GoTools runs a hash data base in which intermediate results are stored. If White loses in a position where passing was forbidden due to rules 1a-i, 1a-ii or 2b then the status of these positions and of positions in sub trees may not enter the database.

- Rules 1a-i and 1b-ii are applicable as well after swapping attacker (Black) and defender (White) but we do not have to change our ruling because this case is already included. The symmetry between attacker and defender is broken by treating seki = life and not seki = death. Rule 1a-i after swapping Black $\leftrightarrow$ White is included in rule 1b-i because in rule 1a-i the side which is forbidden to pass had at least initially ko threats whereas in rule 1b-ii it is the opposite side of the side that is allowed to pass which has a ko threat. In other words, rule 1a-i after swapping colours does not forbid any passing which is not already forbidden by rule 1b-i. For the same reason, rule 1b-ii after swapping colours does not allow any passing for White which is not already allowed by rule 1a-iii.

5. Summary

In this article we develop a consistent procedural definition of the status of life-and-death problems. The procedure performs a boolean search which has as a consequence minor limitations described in the appendix but which, on the positive side,

- determines how many external ko threats are needed for the weaker side to win,
- evaluates bent-4-type positions as dead and as the best move for the attacker to pass, and
- it is efficient because it does a boolean search and minimizes the number of passes done during the search.

The implementation in the program GOTOOLS proved to be consistent when evaluating the status of more than $2 \cdot 10^8$ eyes with up to 11 inner points being surrounded by only one chain as reported in [4].

Appendix: Limitations

Boolean search. Because the outcome of life-and-death fights is polarized (life/seki or death) the value of such all-or-nothing fights is naturally high and it is critical to determine the status of the position and the best moves of both sides as early as possible. Thus, even with near unlimited computing power one would rather use it to solve problems earlier in the game even if one is limited to do a boolean search than optimizing the territorial value of life but being slow. The risk of losing few points by applying boolean search can be lowered by checking different first moves each in a boolean search and selecting that one of the optimal moves which in addition seems to give the most outside influence.

Seki equal life. Apart from the given position on the board we need for the definition of a life-and-death problem also one or more chains of one colour identified which are to fight for life (in this paper White tries to live) and possibly a side moving first, otherwise both sides moving first are considered. Having only two possible outcomes it is more appropriate to classify seki as life than as death. For example, if the White chain in Diagram 16 would be regarded as unconditionally alive then White would have 10 points more compared to treating it as a seki (which it is).

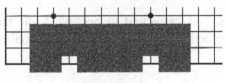

Diagram 16

The error in regarding White as dead would be larger: 22 points. Nevertheless, this is a serious limitation as a difference of, for example, 10 points is not negligible. If one has already at the starting position intruding black stones then one can *confirm* seki with a boolean search by checking whether they can be caught. This should work for the classification of monolithic eyes, although it is not done in this paper. On the other hand, an extra run will not be able to detect seki if the intruding black chains do not already exist in the initial position.

Handling cycles. For the boolean search we need a rule how to handle cycles. The side moving next, in this paper called First, is not allowed to restore a position encountered earlier in the sequence of moves that are already done

- if in the resulting cycle the opposite side Second has caught more stones than First (otherwise repeating this cycle sufficiently often would result in a loss of First exceeding the value of the life-and-death problem), or

- if First had spent external ko threats in this cycle and Second not (in the computation described in section 2, only one side is allowed to use external ko threats to recapture kos), or
- if First = Black because repeating this loop would mean that White at least reaches seki, and so wins.

With these rules for handling cycles and rules about passing discussed in Section 4.1, every search must be finite as problems of finite size can have only finitely many moves before the position must repeat.

Value of tenuki. A more serious restriction in our life-and-death computations comes from trying the passing move only as a last resort. Therefore, strictly speaking, the determined status and winning move are only correct under the assumption that playing elsewhere (passing in the life-and-death fight, or tenuki) has negligible value, as is typically the case towards the end of the game. Especially when comparing different kos, the result may depend on the value of a passing move. Since the number of available external ko threats and the value of a passing move are in general incomparable, one ideally would have to determine the status for any combination of both, i.e., for the number of external ko threats and the number of passes. Our computations cover the special case where passing has negligible value.

External ko threats. As explained in Section 2, the number of external ko threats is limited in practical computations to 5 but could easily be changed to any large value.

Acknowledgements

The author thanks Volker Wehner, Robert Jasiek and Bill Spight for comments on the manuscript.

References

[1] T. Wolf: "Forward pruning and other heuristic search techniques in tsume go", Special issue of *Information Sciences* **122**:1 (2000), 59–76.

[2] J. Davies, J. Cano, F. Hansen: "The Japanese rules of Go", http://www.cs.cmu.edu/~wjh/go/rules/Japanese.html.

[3] R. Jasiek, Japanese 2003 rules, version 35a and commentary, http://home.snafu.de/jasiek/j2003.html.

[4] T. Wolf and M. Pratola: "A library of eyes in Go, II: Monolithic Eyes", in this volume.

[5] T. Wolf: GoTools Online, http://lie.math.brocku.ca/GoTools/applet.html

THOMAS WOLF
DEPARTMENT OF MATHEMATICS, BROCK UNIVERSITY
500 GLENRIDGE AVENUE
ST. CATHARINES, ON L2S 3A1
CANADA
  twolf@brocku.ca

Games of No Chance 3
MSRI Publications
Volume **56**, 2009

A library of eyes in Go, II:
Monolithic eyes

THOMAS WOLF AND MATTHEW PRATOLA

ABSTRACT. We describe the generation of a library of eyes surrounded by only one chain which we call monolithic eyes. Apart from applying the library in the life-and-death program GoTooLS it also can be used as a source for the study of unusual positions in Go as done in the second half of the paper.

1. Introduction

In using principles of combinatorial game theory it has been discussed in the literature how in the game of Go one can assign values to eyes in order to decide whether a position lives unconditionally, simply by adding these values and checking whether or not their sum reaches the value of two (see Landman [1]). These concepts are applied in computer Go programs (as in [3]) and a computer generated library of eye shapes is available from Dave Dyer [2].

In this contribution we describe the generation of a library of eyes surrounded by only one chain which we call monolithic eyes. Compared to Dyer's library our database has a number of extensions: an evaluation of the number of ko threats needed to live or to kill the eye, the consideration of a larger number of external liberties and of an extra attached eye, larger eye sizes, the determination of all winning moves and others.

In the following section we describe a procedure to bring any set of empty or occupied points into a unique position by using shifts and symmetries of the board. The purpose is to avoid the generation of equivalent eyes as described in Section 3. The evaluation of eyes is done with the program GoTools [4] as outlined in Section 4. Computational aspects including a listing of optimizations, comments about performed consistency tests, usefulness and availability follow in Section 5.

The database of monolithic eyes is a rich source of strange positions. In Section 6 we give two examples of how to inspect it: by looking for boolean-calm eyes, i.e., eyes which are not settled but where passing belongs to the boolean-best moves[1] of both sides; and by checking for eyes where the boolean-best attacking move depends in a nonmonotonic way on the number of external liberties.

A summary concludes the paper.

2. Unique representations

If positions in the game of Go are to be stored then there is the option either to store all 8 versions obtained from rotation and reflection of a single position, or to move the position into a unique location and to store it only once. The first version is probably faster, whereas the second is definitely memory saving. Our choice follows partially from the intended use of the database of monolithic eyes which is to be applied in the program GoTools. Much of the execution time of this program is spent on tasks other than evaluating eyes. The speed of the eye database is therefore not crucial which is a reason to use the memory-saving approach. Another reason is that eyes which are not located in the corner have to be shifted to a norm position anyway, and performing a rotation and reflection in the same step does not take much more time. We therefore took the second option and bring each eye into a norm position through shift, rotation and reflection before storing or looking up the position in a database.

We have three cases: eyes in the corner, on the edge and in the centre of the board. For eyes in the corner we have no shift, the rotation is apparent as we rotate each eye into the lower left corner and only the reflection has to be decided by comparing successively two points, one on either side of the board diagonal. Eyes on the edge are rotated to the lower edge of the board. A horizontal centre of the eye is determined as the average of the extreme x-values of inner points of the eye. The middle of the eye can have an integer or a half integer value. For both cases we have a procedure to decide whether a reflection is needed. Finally, for mid-board problems we have the three cases that the centre of the problem falls onto a point of the board, between two points on a line or on the centre between 4 points. For each of these cases we have a procedure to start with points close to the centre of the position and to move gradually away in comparing points. Through the computation of dipole, quadrupole and octupole moments (in analogy to the distribution of electrostatic charges in physics) we continue with the computation until the symmetry is completely broken, i.e.

---

[1] In [4] we define 'boolean-best' moves as all the moves reaching the best possible result in a boolean search with only two outcomes: life/seki or death. In other words, moves achieving seki are regarded as good as moves achieving life.

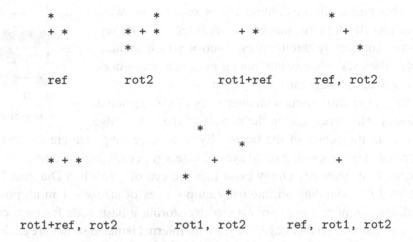

Figure 1. Examples for the possible symmetry types of positions

until a rotation around the centre of the position and a reflection are completely determined or until all points of the positions (in the application all inner points of the eye) have been considered and, as a by-product, the exact remaining symmetries of the position are determined. The 7 possible remaining symmetries can be visualized with the examples in figure 1 where + marks the centre of rotation, * a single stone, ref stands for a reflection symmetry on the x, y diagonal, rot1 for a 90 degree rotation counterclockwise around + and rot2 for a 180 degree rotation counterclockwise around +.

The procedure to *norm-locate* a position (i.e. to put it into a unique place through shift, rotation and reflection) is used repeatedly when creating all essentially different positions (to have them evaluated later). It is also used when applying the database during a life and death computation to norm-locate any encountered monolithic eye in order to look it up in the database. The norm-locating procedure does not only output the new position but also the transformation leading to it. The purpose is to use the inverse transformation on the moves stored in the database to obtain the moves that should be done in the encountered position.

3. The generation of positions

To identify a position in the database we generate a hash code according to Zobrist [7], which is the standard technique in computer Go. The hash codes have to be large enough to avoid clashes for over $2 \cdot 10^8$ eyes of size up to 11 but should be as short as possible to save memory. It turns out that a 32 bit code is too short but we confirmed that a randomly chosen 64 bit code was sufficiently large.

In this article the monolithic eye is made from white stones and Black is the attacker. Monolithic eyes are represented completely through their interior which includes empty, black and white points, not the eye enclosing stones, like ◯ on a3 in Diagram 1 to the right.

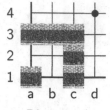

Diagram 1

The computation starts with three eyes of size 1: one in the lower left corner, one in the middle of the lower edge and one in the centre of the board. Eyes of increasing size are successively generated. The generation of all eyes of size s proceeds in four steps.

At first all internally empty eyes, like the eye of size 11 in Diagram 2, are generated by extending all internally empty eyes of size $s - 1$ in all possible directions. Duplicate eyes are avoided by storing a hash code for each empty eye in a database. At this stage all eyes with internal isolated stones are excluded. For example, an eye like in Diagram 1 is excluded when generating eyes of size three but it would come up when generating all eyes of size four.

In the second step all possible legal combinations of white stones within each eye are generated as, for example, the eye in diagram 3, with the restriction that extra white stones may not be attached to the white eye enclosure as the inner size would be reduced. Hence, white stones could not be put on points like c2 in Diagram 2. On the other hand, white stones on a1 and b2 (for example) are allowed, even if they split the monolithic eye into two or more nonmonolithic eyes. For each resulting position all possible legal combinations of additional internal black stones are generated, such as the eye in Diagram 4.

Finally, for all eyes which include one or more ko-fights the position is duplicated once for each empty potential ko point which in the duplicate position is marked as forbidden. The position in diagram 4 would therefore appear twice, once without the forbidden ko point (Diagram 4) and once with the forbidden ko point, like the point marked as K in Diagram 5. If the point a2 in Diagram 4 would be occupied by Black or White then b1 could not be a ko-forbidden point as the position could not have resulted from White making a move on c1 and capturing a single black stone on b1.

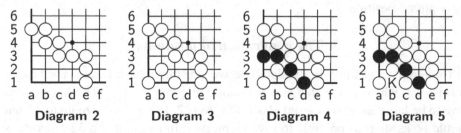

Diagram 2   Diagram 3   Diagram 4   Diagram 5

With nearly $2 \cdot 10^8$ eyes of size 11 a hash database holding all of them to exclude duplicates would be too large. Fortunately, it only needs to be big enough to hold all empty eyes of one size to exclude duplicate empty eyes of same shape. If two empty eyes are not symmetric to each other (by reflection, rotation or shift) then they cannot become symmetric by putting extra stones in. But if an empty eye has a symmetry, for example a reflection symmetry, then filling it with stones will produce positions which are pairwise equivalent with respect to this symmetry. The hash database therefore has also to be big enough to hold all eyes which are generated from a single empty eye through input stones.

Three diagrams in figure 2 illustrate the growth of the numbers of positions in dependence on the size of eyes for corner, edge and centre eyes.

Comments

- Because of $\log(0) = -\infty$ we replaced 0 on the vertical axis each time by 1 to have $\log(1) = 0$.
- Growth factors are remarkably constant and allow to estimate the number of eyes of larger sizes.
- The number of positions grows only little by adding white throw-in stones. It grows most for corner positions as for edge and centre eyes most inner points are on the edge of the eye which cannot be occupied by further white stones without decreasing the inner size.
- The number of extra positions due to an initial ko grows faster than other numbers and is highest for corner eyes, again because for corner eyes the surface of the eye (the number of inner points neighbour to the enclosing chain) is smallest.

4. The evaluation of positions

To evaluate the eyes we use the program GOTOOLS. It performs a boolean search which has, for example, the consequence that seki is treated as life. Other consequences and limitations are described in the appendix of [6] in this volume. A strength of GOTOOLS is that it is able to determine how many ko threats are needed for the weaker side to win. When evaluating eyes, beginning with size 1 and increasing the size successively, GOTOOLS already makes use of the hash database filled with data for smaller eyes.

Because GOTOOLS is a life-and-death program it cannot evaluate whether the position is at least 'worth one eye' so that it could live if it had another 'eye'.[2] We compensate this weakness by investigating the position twice, first on its

---

[2]In this article the word *eye* normally denotes a whole position consisting of a white chain with stones inside and a surrounding black enclosure. Only in this paragraph the phrase 'worth one eye' is a short form of saying 'one protected liberty which has to be the last liberty taken for capture' with the understanding that two such liberties ensure life.

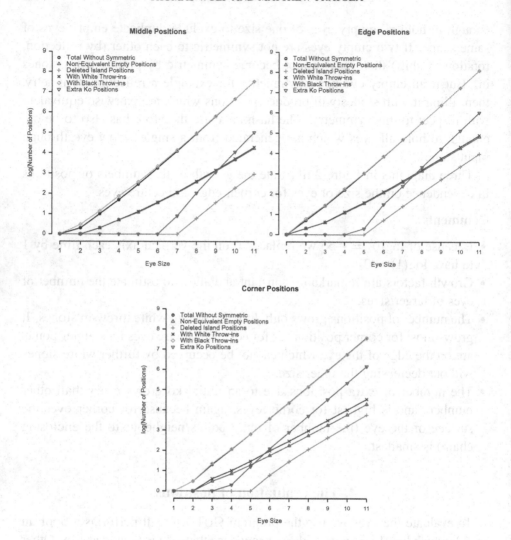

Figure 2. Numbers of generated positions.

own in order to find out whether it is 'worth two eyes' and if not then we attach externally a secure 1-point-eye and investigate it again to find out whether it is 'worth at least one eye'. The information whether a position is 'worth $\frac{1}{2}$ eye' (i.e. 'worth one eye' if White moves first and 'worth no eye' if Black moves first) or similarly 'worth $1\frac{1}{2}$ eyes' follows from evaluating it for both sides moving first.[3]

The status of the eye and/or the boolean-best moves often depend on the number of external liberties. Therefore computations are done for 0 to 31 external

---

[3] with the exception that if an initial ko forbidden point is present then only the side for which this point is forbidden may move first

s	a	b	c	d	e	f	g
1	3	3	100.0	1.000	76	< 0.1	< 0.1s
2	10	10	100.0	1.000	220	< 0.1	< 0.1s
3	56	55	98.2	1.036	868	< 0.1	0.4s
4	321	259	80,7	1.062	4,180	< 0.1	2.5s
5	1,938	809	41.7	1.116	15,004	< 0.1	7.7s
6	12,477	2,917	23.4	1.243	58,540	0.1	27.4s
7	82,808	9,955	12.0	1.464	233,572	2.5	2m 31.9s
8	565,104	41,831	7.4	1.608	1,040,548	2.5	13m 51.4s
9	3,931,849	196,402	5.0	1.659	6,601,924	5.2	1h 32m 26s
10	27,800,486	965,405	3.5	1.716	33,118,212	18.4	11h 6m 1.2s
11	199,169,127	4,990,259	2.5	1.739	215,042,444	45.1	3d 11h 19m

Table 1. Numbers of positions, sizes of final databases, evaluation times. Meaning of columns: s, size of eye; a, all positions of this size; b, positions not unconditionally alive; c, percentage of positions not unconditionally alive; d, average number of records for each not unconditionally alive eye; e, size of final read-only hash database in bytes; f, maximal time in seconds for the evaluation of one eye; g, total time for all positions of this size.

liberties for two cases: the eye on its own and one extra external 1-point eye being attached. Only in the case that the eye has no internal liberty, at least one external eye or liberty has to be present.

Table 1 gives an overview of the number of positions, the percentage of those which are not unconditionally alive, the average number of records to be stored for each not unconditionally alive position and times needed to evaluate the most expensive single eye and all eyes of each size. All times are measured on a 3GHz Pentium IV. Programs are written in FreePascal (http://www.freepascal.org).

5. Computational aspects

5.1. Optimizations. The following are observations and ideas that allow to make the computer programs more efficient.

- Because the database is generated only once, it can afterwards be converted to a read-only database saving one pointer for each record and thus memory.
- From table 1 it follows that a large proportion of all eyes are unconditionally alive. We do not need to store them. If we do not find an eye with a size that is covered by the database then we simply conclude that the eye is unconditionally alive. This works because we generated and evaluated all eyes up to the maximal considered size (currently 11).

- If we store only a small fraction of all eyes, we do not have to guarantee that there are no clashes between hash codes of eyes which are not stored. To ensure that there is no clash between any two stored eyes and not between a stored eye and an eye that is not stored, shorter hash codes are sufficient in the final read-only version.

- Column d in table 1 shows the average number of records that each (not unconditionally alive) eye contributes to the database. The larger the eye, the more likely the status and/or the boolean-best moves depend on the number of external liberties and the more records are needed for the eye. The pointers between these records have a 32 bit size but if the database has to hold eyes only of size ≤ 10 then the number of stored records is much smaller than 2^{32}, so we can use some of the bits to encode part of the hash code. Column e in table 1 shows the memory requirement for a hash database including only eyes up to this size.

- In the process of evaluating an eye of inner size n the eye either keeps its size, or its inner size shrinks. Any smaller eye that appears has already been investigated, because we evaluate eyes of successively increasing size. Therefore, either it is found in the database, then the information can already be used, or it is not found and then it is clear that it must be unconditionally alive.

  The situation is different if throw-in stones are placed which change the eye (i.e. its content and thus the eye) but in a way that its inner size is still n. Because not all eyes of size n have already been investigated, one cannot conclude that it is unconditionally alive if it is not found in the database. Nevertheless, one can look it up and use the information if the eye *is* found.

- When studying an empty eye of size n then other eyes of size n with more throw-in stones appear, but when studying an eye of size n with many throw-in stones then it is rare that eyes of same size with fewer throw-in stones turn up in this evaluation. To increase the chance that appearing eyes of that size have already been investigated, one can evaluate eyes with many interior stones first and empty ones later.

- In the worst case each of the $2 \cdot 10^8$ eyes would be evaluated 128 times ($= 2 \cdot 2 \cdot 32$ for both sides moving first, with/without extra eye and each time 0-31 external liberties) but this can always be sped up. If, for example, Black moves first and the eye has no external liberty and no extra eye is attached and Black loses then Black will always lose and White will always win.

- In each of the at most 128 computations of an eye, *all* possible first moves are checked to determine all moves giving the best status. After the first computation determining the status of the problem and one of the (boolean-)best moves is completed, any computation checking another first move can be stopped early (i.e. one can avoid ko reruns with more ko threats as described in the section on the ko status in [6]) as soon as it is clear that the status will be suboptimal.

5.2. Safety. The program GOTOOLS which evaluated the positions operates apart from the permanent monolithic eye database also a temporary hash database which is cleared before a new eye is considered. For any new entry to be stored, both databases check whether this entry is consistent with the content so far. For example, if for a position it is already known that White moving next will lose despite having one external ko threat, then a new entry saying that White even loses with two ko threats is consistent, but a new entry saying that White wins in this position without ko threat would create an inconsistency. For the eye database, further consistency checks result from the requirement that Black must not do worse when White has fewer external liberties and/or no extra external eyes compared to White having an external eye. Equivalently, White must not do worse when having more external liberties or extra eyes. These consistency checks, having been done thousands of billions of times during the generation of the database, led to the fixing of a number of bugs (mainly linked to the proper handling of ko in connection with the temporary database) and to the development of 'bent-4 compatibility' reported in [6].

5.3. Efficiency and availability. The monolithic eye database is a single permanent read-only file with a size depending on the size of eyes stored (see column *e* in table 1). To use the database, a module within the program GOTOOLS norm-transforms any monolithic eye encountered during a computation by performing a rotation, reflection and shift as necessary, generates a hash code, finds the relevant entry in the database (depending on the number of external liberties and on who moves next) and decodes the information (status and moves). The size of the database grows exponentially with the size of eyes to be stored and the frequency to encounter monolithic eyes falls quickly with increasing eye size. Being able to read the status and first moves for a big eye from a database is especially time-saving; on the other hand, loading a large database takes considerable time when starting the program, and the large amount of memory required may make cache memory less effective, slowing down the whole program. Therefore the online version of GOTOOLS on [5] uses only a database including eyes of size up to 8.

6. Unusual positions in the database

The database of monolithic eyes is a rich resource for simple but also special positions. Because of the large number of eyes a human visual inspection is out of the question. To filter out interesting positions automatically, one needs to come up with criteria that make a position interesting. These criteria depend on how positions are evaluated and which data about the positions are available. The positions in our database are computed with the program GoTools. As described in more detail in [6], the possible status are unconditional life, death or characterized by the number of external ko threats needed for the weaker side to win. The status and all first moves that result in this status are determined for both sides moving first, for 0-31 external liberties and for the two cases that an additional 1-point eye is attached or not. Given these data for all monolithic eyes in the corner, on the edge and in the centre of the board, what are criteria that make life and death positions interesting?

The filters applied in this section are not unique, they are examples and one may come up with other criteria people are interested in. Also, limitations like seki = life have their consequences on what one can find. For example, *mannen-ko* as shown in [6], Section 4.2, Diagrams 10–13, is a seki type position but both sides can start an unfavourable ko about unconditional life or death. Because the ko is unfavourable to whoever starts it, the seki is in some sense stable. In this paper we treat seki = life with the consequence that these positions are simply unfavourable ko's for Black, so passing is the boolean-best move for White.

6.1. Calm positions.

A first class of positions we want to explore has the property that passing is one of the best moves for both sides. To be more precise, passing is one of the boolean-best moves, i.e. one of the moves that achieve the best possible result in a boolean search where seki is treated as life. If such a position is not settled, i.e. if it is not unconditionally dead or alive/seki then we called it *boolean-calm* in [6] and gave an example there. Treating seki = life means that the 'double-ko seki' in Diagram 6 is regarded as settled and thus not listed as a boolean-calm position, although both sides have passing as one of their boolean-best moves.

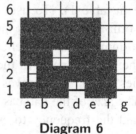

Diagram 6

A convention: *For readability, from now on we will write 'calm' instead of 'boolean-calm' and 'best' instead of 'boolean-best'.* One should keep this in mind as otherwise statements in this section become wrong. Passing is often not one of the truly best moves, for example, in Diagram 7 where ① pass, ② on h1 is boolean-best for ○, achieving a seki whereas ① on a1 achieves life.

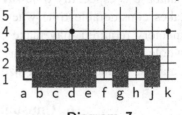

Diagram 7

If passing is one of the best moves of both sides, then the status does not depend on who moves first, i.e. such positions have only one status. Ko-banned positions (i.e. positions with a ko-forbidden point) can have only one side moving next and hence cannot be calm.

size	corner	edge	center	total
≤ 4	0	0	0	0
5	4	0	0	4
6	6	0	0	6
7	69	34	0	103
8	232	109	0	341
9	405	197	0	602
10	746	98	2	846
11	2817	470	11	3298

Table 2. Numbers of calm eyes in the corner, on the edge and in the centre

Table 2 lists the number of calm position for different eye sizes. For each size only a very small percentage is calm, each position only for one or very few specific numbers of external liberties. In this sense these positions are special, although not as exotic as bent-4. The numbers in table 2 serve merely as an illustration, as many positions are equivalent, not by rotation or reflection of the whole position but by deforming the eye, deforming the black throw-in chain(s) and rearranging internal liberties without changing anything substantially.

To construct a strange position with a large area that involves many chains, nesting eyes and multiple ko's does not seem to be too difficult. It is more of a challenge to embody the spirit of a strange position in as few stones as possible. By generating eyes beginning with minimal size, such minimal positions will be found. An example is discussed in the subsection about size 6 eyes below.

Bent-4-type positions. One way for a position to be strange is when its status depends on the global situation on the board. The simplest are straight ko's where the side with more threats wins. The next level of strangeness is reached when the status does not depend on the number of threats on the board, but on the type of threats, whether they are removable (like cuts) or unremovable (seki). This is the case when the local position is a ko, and somewhere in any optimal sequence of moves, the side for which the ko is unfavourable has the passing move as the only best/feasible move. Such positions have been called bent-4-type positions in [6], Section 3. From all the investigated eyes up to size 11, the 4 bent-4 positions in [6], Section 3, were the only ones with this property. They are commonly regarded as dead, because Black (the side for whom the ko is favourable) can wait arbitrarily long before making moves at no cost that eliminate removable ko-threats of White and then starting the favourable ko. Therefore, the program GOTOOLS had been modified as described in [6] so that bent-4-type positions are evaluated as an unconditional loss for the side with the passing move as single best move. It is with this version of GoTools that the monolithic eye database has been computed, therefore bent-4 is evaluated as unconditionally dead and will not be found below in our search for calm positions.

Eyes of size 5. Starting with eyes of small size we find that eyes need at least 5 inner points to have a status that is not unconditionally dead or alive and to allow both sides to pass without penalty. Up to rotation and reflection there are 4 such positions. One of them is shown in Diagram 8. The other three have the same black throw-in stones but the liberty at b2 is located at c1, b3 or a4. In all four positions the eye has exactly one external liberty and the status is that White needs one external ko-threat to live.[4] If a position/eye can only be calm if it has one or more external liberties then in 'real Go' (with seki $\neq$ life) they cannot be a seki because it does not harm Black to occupy these liberties. Therefore these four positions are not seki's.

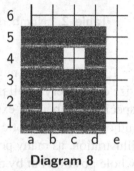

Diagram 8

The reader can find solution sequences for all positions in this paper by solving the problems in [5].

Eyes of size 6. There are two types of calm positions of size 6. One is the position in Diagram 9 where White needs one ko-threat to live.

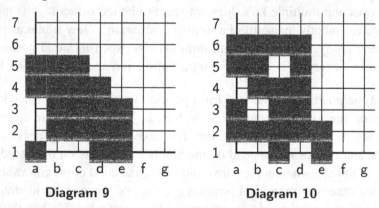

Diagram 9   **Diagram 10**

In the other 5 positions Black needs one ko-threat to kill. These are:

(i) Diagram 10 with 0, 1 or 2 external liberties for the eye, and
(ii) the same diagram but with ● added on b1 and 0, 1 or 2 external liberties, or

---

[4]If there were no external liberty of White then Black moving first can catch and if there are two external liberties, White can live moving first, so in both cases this is not a calm position. The difference to bent-4 is that White can start the Ko at b2 and does not have to pass.

(iii) ◯ added on a1 and 0 external liberties, or

(iv) ● added on a4 or b3 and 0 external liberties.

In cases 3 and 4 the only good move for White is to pass which does not make the position to be of bent-4-type because the ko is favourable for White. To be of bent-4-type the ko would have to be unfavourable for the side with passing as the single best move.

From these five positions we chose the one in Diagram 10 as the representative position — the 'mother' position — because all other ones can be reached from it. Positions in Diagram 10, cases 2 and 3 are versions of a '10,000 year ko' or, also called 'mannen ko'. Compared to the usually known and published form (see [6], Section 4.2, Diagrams 10-13) the form in Diagram 10 needs a smaller eye and fewer stones, supporting the claim that from all strange positions the database provides the more interesting minimal versions.

Eyes of size 7. For all calm eyes of size 7 Black needs exactly one external ko-threat to kill.

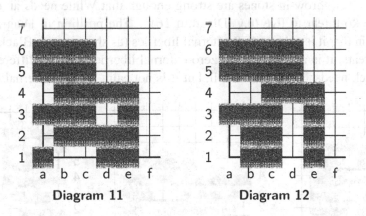

Diagram 11   Diagram 12

Most of the positions are related to the eyes in Diagrams 11 and 12. The position in Diagram 12 is a 'mother position' for a number of calm positions, including mannen ko and is itself calm for 1–3 external liberties.

Eyes of size 8. Like for size 7, in all calm eyes of size 8 Black needs exactly one external ko-threat to kill. Due to the increasing number of calm eyes we can only give examples. Many eyes are enlarged variations of mannen ko. Here are two other examples. The position in Diagram 14 is calm for 0–2 external liberties.

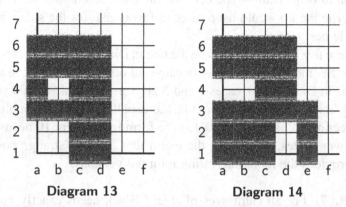

Diagram 13   Diagram 14

Eyes of size 9. Calm eyes of size 9 are large enough that for some of them Black now needs 2 ko-threats to kill (e.g. Diagram 15 with 0 or 1 external liberty), or, that the black throw-in stones are strong enough that White needs an extra eye and one ko-threat to live (e.g. Diagram 16)[5]. The position in Diagram 17 is strange in that it is calm with 2 external liberties (as shown) with Black needing 2 ko-threats, it is also calm with zero external liberties and the different status that Black needs 1 ko-threat to kill, but it is not calm for one external liberty.

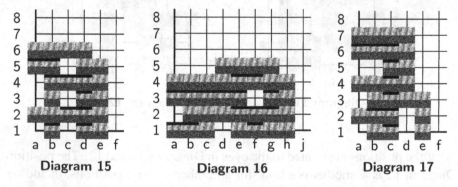

Diagram 15   Diagram 16   Diagram 17

Eyes of size 10. Calm eyes of size 10 can, similarly to size 9, have the status that Black needs 2 ko-threats (Diagrams 18, 19), Black needs one ko-threat (Diagram 20 for 0–2 external liberties), or, White needs one ko-threat (Diagram 21).

---

[5] Already an eye of size 8 is large enough to house a black living position so that even an extra white eye and many external liberties are not enough, but we are talking here only about *calm* eyes.

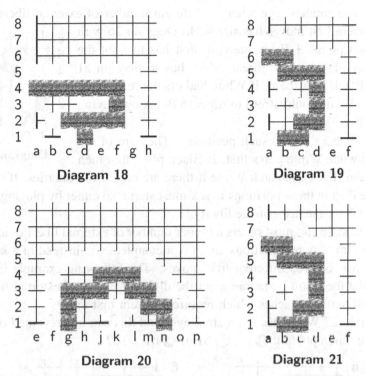

Diagram 18

Diagram 19

Diagram 20

Diagram 21

Eyes of size 11. A novelty of size 11 calm eyes is that Black may need 3 ko-threats to kill as in Diagram 22. The same diagram with 1 external liberty does not give a calm position but with zero external liberties the position is again calm with Black needing only 2 ko-threats to kill. In Diagram 23 Black needs also 2 ko-threats, in Diagram 24 with 0–2 external liberties Black needs 1 ko-threat and in Diagram 25 White needs 1 ko-threat.

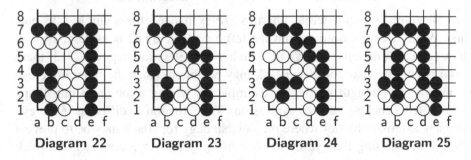

Diagram 22   Diagram 23   Diagram 24   Diagram 25

7. The influence of external liberties

In this section we search the database for eyes which change their character and/or best first moves with the number of external liberties of the eye.

7.1. Positions with a solution strategy depending on the number of external liberties.

The smallest eye where a different number of external liberties requires different first moves has size 4. In Diagram 26 with ko-forbidden point a1 White moving first has to fight the ko and after playing a ko threat White has to play on a1 otherwise Black goes there. If White had one more external liberty then White could afford to squeeze Black by playing on a3 and live.

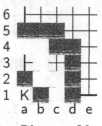

Diagram 26

For size 5 there are 10 such positions. They are of the type above when White plays first. If Black plays first then in all of them Black can catch White if there are no outside liberties. If there is at least one then in these positions Black must start a ko either by playing inside the eye or taking the last outside liberty.

In this example and most others a higher number of external liberties allowed White to go for a more ambitious aim, i.e. a complete win instead of a ko. The different aims required different first moves. The following example is more exotic in that the result is the same but the different nature of external liberties dictates different approaches which require different first moves.

In Diagram 27 White has to create 2 eyes and this can be prevented only by ● a2. Other attempts fail: ● a1, ② b1 or ● b1, ② a2.

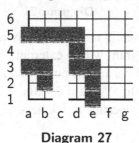

Diagram 27

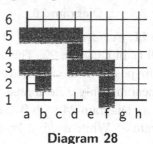

Diagram 28

In Diagram 28 two external liberties are exchanged for a single safe one. Thus, White does not need to split the left eye. Here ● a2 is useless. Instead Black can catch directly with ● b1 which was useless in Diagram 27. If there are no or few external liberties, Black may be able to catch the monolithic eye by building up strength inside, for example, by having a long and thin shape that is able to enclose an own eye inside. The situation is different if the eye has many external liberties where the only strategy for Black may be to prevent White from splitting the eye into two eyes by going for a compact and thick throw-in shape.

7.2. Nonmonotonic lists of best moves.

If the number of external liberties of a monolithic eye is decreased then the situation (i.e. the status value) should definitely not get better for the monolithic eye (i.e. for White), no matter what the shape and the internals are. Can one make a similar 'monotony' statement for the list of best moves? Let us look at the eye in the following diagrams with 2, 3 and 4 external liberties. To minimize comments underneath the diagrams, ko-threats and ko-answers are not explicitly listed as they become clear from the move sequences in the diagrams.

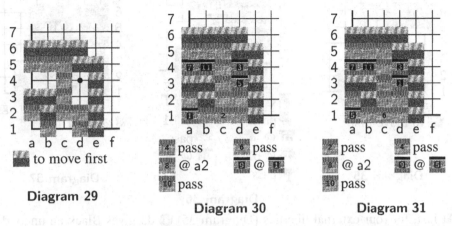

Diagram 29

Diagram 30

Diagram 31

For one and two external liberties (Diagram 29) the two moves ● a1 (Diagram 30) and ● d3 (Diagram 31) are fully equivalent. In both cases Black needs one external liberty (and White can pass or play elsewhere three times).

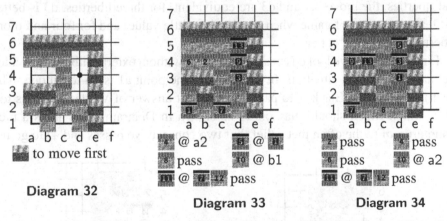

Diagram 32

Diagram 33

Diagram 34

For three external liberties (Diagram 32) the move ● d3 (Diagram 34) is the better one if we only compare the number of ko-threats, as it needs only one ko-threat compared to ● a1 (Diagram 33) which needs two ko-threats.

Diagrams 33 and 34 are also useful as an illustration that counting the number of possible passes for both sides does matter but will also complicate the characterization of positions. Depending on what the value of the passing move is in a game, either ● a1 as in Diagram 33 is the best move, costing Black two ko-threats and *two* passing moves of White, or ● d3 as in Diagram 34 which costs Black only one ko-threat but *four* passing moves of White. Passing moves could be used by White either to protect removable ko-threats or to gain points somewhere else on the board.

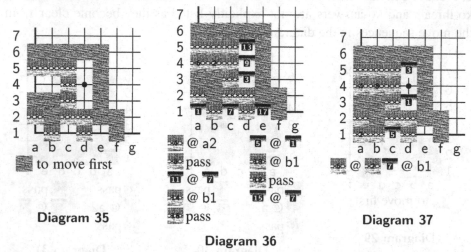

Diagram 35

Diagram 36

Diagram 37

At last, for four external liberties (Diagram 35) ❶ d3 gives Black an unconditional loss (Diagram 37) whereas ❶ a1 in Diagram 36 at least leads to a ko in which Black needs three ko-threats.

To summarize, what we call 'nonmonotonicity' is that for one and two external liberties, the moves a1 and d3 are equivalent, for three liberties, d3 is better (at least at the end of game when passing has little value) and for four and more liberties again a1 is better.

Finally, the eye we looked at earlier, but now without external liberty, provides yet another unusual situation. In Diagram 38 the point a1 is ko-forbidden. The best first move for Black is to pass and the best answer of White is to pass too after which Black should play on a1 as shown in Diagram 39. This is a nice counterproof to the often met belief that two consecutive passes end a Go game.

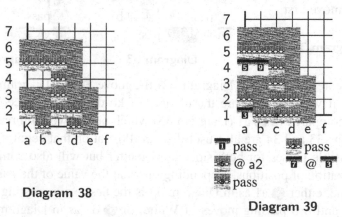

Diagram 38

Diagram 39

8. Summary

The database of monolithic eyes described in this contribution offers a number of features that other libraries of monolithic eyes do not have.

- We cover eyes of size up to 11 compared to size 7 plus incomplete groups of size 8 provided in [2].
- All possible internal configurations with stones of both colours inside are considered.
- We include eye positions that have an initial ko forbidden point.
- Each eye is evaluated for both sides moving first, with 0-31 external liberties and again with 1 external eye and in addition 0-31 external liberties.
- In each such evaluation *all* moves giving the best status are determined including the passing move if it is one of them.
- In all evaluations of eyes that result in ko, the number of external ko threats that are necessary in order to win are determined. Weaknesses of the computation are addressed in the appendix of [6]. The two main ones are 1) the number of passes is not maximized, which is correct if passing has no value, and 2) seki is treated as life.

Despite limitations in evaluating the eyes, the database is still a good source of unusual positions. We gave examples of eyes which are not settled and where nevertheless passing belongs to the best moves of both sides. In the last subsection we show an eye where the best attacking move depends on the number of external liberties of the eye in a nonmonotonic way, i.e. the best move alternates with an increasing number of external liberties.

It would be interesting to see whether positions shown in Section 6 are also special from the point of view of a thermographic analysis.

Acknowledgements

The authors thank Bill Spight for comments to the manuscript. This text and also [6] were typeset with GoL^{A}T$_E$X from Volkmar Liebscher.

References

[1] H. Landman: "Eyespace values in Go", pp. 227-257 in Richard J. Nowakowski (ed.), *Games of No Chance*, Math. Sci. Res. Inst. Publ. **29**, Cambridge University Press, New York, 1994.

[2] D. Dyer: "An eye shape library for computer Go", http://www.andromeda.com/people/ddyer/go/shape-library.html.

[3] The GnuGo project: Eyes and half eyes (2005), http://hamete.org/static/gnugo36/gnugo_8.html.

[4] T. Wolf: "Forward pruning and other heuristic search techniques in tsume go", special issue of *Information Sciences* **122**:1 (2000), 59–76.

[5] J. P. Vesinet and T. Wolf: GoTools Online, available at http://lie.math.brocku.ca/GoTools/applet.html

[6] T. Wolf: "A library of eyes in Go I: A life-and-death definition consistent with bent4", in this volume.

[7] A. L. Zobrist: "A new hashing method with application for game playing". Technical report 88, Univ. of Wisconsin, April 1970.

THOMAS WOLF
DEPARTMENT OF MATHEMATICS, BROCK UNIVERSITY
500 GLENRIDGE AVENUE
ST. CATHARINES, ON L2S 3A1
CANADA
   twolf@brocku.ca

MATTHEW PRATOLA
DEPARTMENT OF STATISTICS AND ACTUARIAL SCIENCE
SIMON FRASER UNIVERSITY
8888 UNIVERSITY DRIVE
BURNABY, BC V5A 1S6
CANADA
   mpratola@gmail.com

Complexity

Games of No Chance 3
MSRI Publications
Volume **56**, 2009

The complexity of the Dyson Telescopes puzzle

ERIK D. DEMAINE, MARTIN L. DEMAINE, RUDOLF FLEISCHER,
ROBERT A. HEARN, AND TIMO VON OERTZEN

ABSTRACT. We give a PSPACE-completeness reduction from QBF (quantified Boolean formulas) to the Dyson Telescopes puzzle where opposing telescopes can overlap in at least two spaces. The reduction does not use tail ends of telescopes or initially partially extended telescopes. If two opposing telescopes can overlap in at most one space, we can solve the puzzle in polynomial time by a reduction to graph reachability.

1. Introduction

The complexity of many motion-planning problems has been studied extensively in the literature. This work has recently focused on very simple combinatorial puzzles (one-player games) that nonetheless exhibit the theoretical difficulty of general motion planning; see, e.g., [1]. Two main examples of this pursuit are a suite of pushing-block puzzles, culminating in [2; 3], and a suite of problems involving sliding-block puzzles [4]. In pushing-block puzzles, an agent must navigate an environment and push blocks in order to reach a goal configuration, while avoiding collisions. The variations of pushing blocks began with several versions that appeared in video games (the most classic being Sokoban), and continued to consider simpler and simpler puzzles with the goal of finding a polynomially solvable puzzle. Nonetheless, all reasonable pushing-block puzzles turned out to be NP-hard, and many turned out to be PSPACE-complete, with no problems known to be in *NP*, except in one trivial case where solution paths are forced to be short. Similarly, sliding-block puzzles are usually PSPACE-complete, even in very simple models.

Fleischer's work was partially supported by a grant from the National Natural Science Fund China (grant no. 60573025).

In this paper we consider a motion-planning puzzle, the Dyson Telescopes puzzle. It takes the form of an enjoyable computer game [5], invented and developed by the Dyson company to advertise a vacuum cleaner called "Telescope" that is retractable like an astronomical telescope. The puzzle is perhaps most closely related to sliding blocks, in the sense that the agent is outside the environment. At any time, the agent can extend or retract one of several "telescopes", each of which has a specified, fixed length in extended form. Erickson [6] posed the complexity of the problem in 2003. The complexity remained open despite fairly extensive pursuit—it seemed nearly impossible to build gadgets that required multiple entrances. Thus we hoped that it would be the first "interesting" yet polynomially solvable motion-planning puzzle.

We prove that the Dyson Telescopes puzzle is indeed polynomially solvable in a fairly natural situation in which the extended forms of opposing telescopes (two telescopes on the same row or column, pointing towards each other) overlap in at most one space. However, some of Dyson's puzzles do not satisfy this restriction. We prove that this small flexibility in the general form of the problem in fact makes the problem PSPACE-complete.

The polynomial-time algorithm for the restricted form of the telescopes game is particularly interesting because such puzzles are nonetheless enjoyable for humans to play. All but a few of the hundreds of levels of the puzzle on the Dyson homepage [5] (mainly the Grandmaster levels) do not have opposing telescopes that overlap in more than one square. Therefore we expect that our algorithm can be used to design enjoyable instances of the telescope game, enumerating over puzzles within this restricted family (either by hand or by some automatic process), and automatically computing which puzzles are solvable. Our algorithm can also find the *shortest* solution, for most reasonable weighting functions, enabling the puzzle designer to find the *hardest* puzzle according to a particular difficulty measure, such as the solution requiring the longest sequence of moves or requiring a "difficult to see" sequence of moves.

1.1. Description of the problem. In the Dyson Telescopes puzzle, the goal is to maneuver a ball on a two-dimensional square grid from a starting position to a goal position, by extending and retracting telescopes on the grid; refer to Figure 1. An instance of the problem consists of an $n \times m$ grid, a number of telescopes on this grid, and the ball's starting position and goal position. Each telescope is specified by its position, its direction (up, right, down, left), and its length, i.e., the number of spaces it can be extended. Each telescope can be in either an *extended* or a *retracted* state. Initially, all telescopes are retracted. A move is made by changing the state of a telescope.

If a telescope is extended, it will expand in its direction until it is blocked (i.e., there is a telescope occupying the space where the telescope would extend

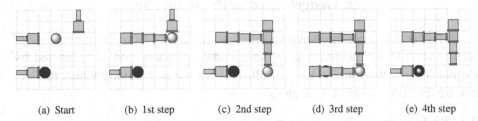

(a) Start (b) 1st step (c) 2nd step (d) 3rd step (e) 4th step

Figure 1. This example depicts a sample situation from the original game where all telescopes have length 3. We can solve this instance as follows: We extract the first telescope to push the ball to the right, where we then can push it downwards into the row of the lower telescope; when we extend and retract the lower telescope, it will finally pull the ball back to the goal position.

to next), or until it reaches its full length. If a ball blocks the extension of the telescope, the ball is pushed in the direction of the telescope, either until it is blocked by another telescope or until the pushing telescope is fully extended; see Figure 1(d). On the back side of the telescope (i.e., in the opposite direction as the telescope extends), there is a one-space tail. When the telescope is extended, this tail is retracted.

If an extended telescope is retracted, it is retracted all the way until it occupies only its base space. If the space behind the telescope is not occupied, its tail will be extended and occupy this space (and possibly push the ball). If the telescope end touches the ball when being retracted, it pulls the ball with it, so that the ball will move to the position directly in front of the retracted telescope; see again Figure 1(d).

We prove that it is PSPACE-complete to determine whether a given problem instance has a series of telescope movements that moves the ball from the starting position to the goal position (think of the goal square as a hole; the ball will fall down as soon as it is pushed across the goal square). We do this by constructing a circuit solving QBF, using gadgets of telescope configurations to simulate Boolean variables, logical gates, etc. If opposing telescopes are not allowed to overlap in more than one space, we give a polynomial time algorithm to find a solution.

Alternative versions of the game allow the telescopes to be partially extended in the initial state, or to not consider a tail end of the telescopes. We show that these modifications do not change the complexity of the problem.

2. Gadgets used in the reduction

In this section, we introduce various gadgets made from configurations of telescopes. These gadgets usually have some entrances and exits labeled by capital letters. We usually describe all the possible paths along which the ball can travel from an entrance to an exit.

2.1. Basic gadgets. We use the symbols in Figure 2 for simple tracks, simple crossings, division of the path, and union of paths, which are easy to implement. We assume that passage through one-way, split and join gadgets is possible only in the appropriate directions. Figure 3 shows the join and split gadgets.

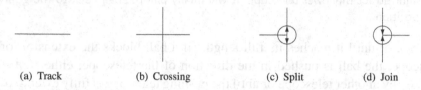

(a) Track   (b) Crossing   (c) Split   (d) Join

Figure 2. Simple gadgets.

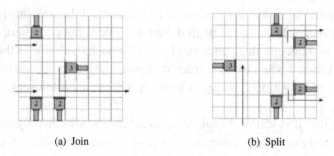

(a) Join   (b) Split

Figure 3. Details of the join and split gadgets.

Figure 4 shows a pair of *opposing telescopes* (the number on a telescope indicates its length). The pair is said to be *active* if one of the telescopes is extended with its end between the black and the white dot, and *inactive* otherwise.

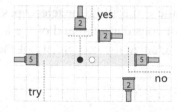

Figure 4. Opposing telescopes.

If the pair is inactive and the ball enters from *try*, it can only leave the gadget at *no*. On its way from *try* to *no*, it may activate the pair as follows. First, we retract all telescopes in the gadget. Then we extend the left telescope to full length (so that it covers the white dot square) and extend the right telescope until it is blocked by the left telescope just to the right of the white dot square (i.e., we extend it by three spaces). Then we retract the left telescope, put the ball into the gadget along the *try* path, and push it to the white dot square. Then we pull the ball to the *no* exit by retracting the right telescope. Note that this action leaves the pair in an active state.

If the pair is active and the ball enters from *try*, then the right telescope must be extended to just cover the white dot square. Then we can push the ball to the black dot square, where it can be picked up by the top telescopes so that it can leave the gadget at *yes*. We can also leave the pair at the *no* exit. In both cases, we may choose to leave the opposing pair either active or inactive.

Note that the ball can exit the gadget via *yes* and *no*, but it cannot enter the gadget at these points. We may lengthen the left and right telescopes (increasing the size of the gaps) and vertically flip the sides of the *try*, *yes*, or *no* pathways without changing the properties of the gadget, as long as we maintain the two-space overlap of the left and the right telescopes.

2.2. Variable gadget. When we move the ball from *try* to *yes* in an active opposing pair we may leave the pair active. To force it become inactive, we construct a *reset gadget*, shown schematically in Figure 5. Each grey rectangle represents an opposing pair. A single telescope extends along the lower pathway r, crossing the path of the lower telescope of β and ending in the path of the lower telescope of γ. The ball cannot enter β directly; it must first enter γ.

LEMMA 1. *The ball can always move through a reset gadget from in to out, but this forces the opposing pair α to become inactive.*

PROOF. The ball can only pass along path r if the lower telescopes of both β and γ are retracted. If an opposing telescope is also retracted, the corresponding pair will become inactive. If both upper telescopes of β and γ are extended to keep the opposing pairs active, α must be inactive. Since the ball can leave the gadget only if both β and γ are active, this is only possible if α is inactive. Note that the initial states of the opposing pairs are not important because we can activate β and γ (deactivating α). $\square$

We attach three independent reset gadgets to a single opposing pair α to construct a *variable gadget*, shown in Figure 6. Here, each pair (β_i, γ_i) corresponds to one reset gadget; the internal reset pathways are not shown. This is our workhorse gadget, forming the basis of all the following constructions. We say that the variable gadget is *open* (*closed*) if α is active (inactive).

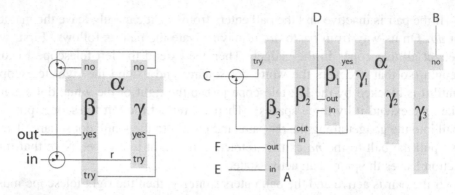

Figure 5. Reset gadget.   **Figure 6.** Variable gadget.

LEMMA 2. *In a variable gadget, traversal from either A or C to B is always possible, and may open the gadget. Traversal from C to D is possible precisely if the gadget is open, and forces it to close. Traversal from E to F is always possible, and forces the gadget to close. No other traversals are possible.*

PROOF. By properties of opposing telescopes, Lemma 1, and the pathways shown in Figure 6.   □

2.3. 3SAT gadget. Given a 3CNF formula W (a propositional formula in conjunctive normal form with three disjuncts per clause) with m clauses and n variables, we construct a *3SAT gadget*, shown in Figure 7, to test the formula. We use an $m \times 3$ array of variable gadgets. The three gadgets in row i correspond to the variables in clause i.

For each variable v and truth value $b \in \{0, 1\}$, we connect the A-B lines of all variable gadgets corresponding to $v = b$ into a chain. We also connect the E-F lines of all variable gadgets corresponding to $v = 1 - b$ into another chain. We concatenate these two chains by joining the last B line of the first chain to the first E line of the second chain. Finally we connect the first A line of the chain to an input channel $(v = b)_{in}$, and the last F line in the chain to an output channel $(v = b)_{out}$ of our 3SAT gadget.

We connect together the D lines of the three variable gadgets on row i and the C lines of the three variable gadgets on row $i + 1$, so that it is possible to go from any of the three D lines to any of the three C lines. We connect an input channel *test* to the C lines of row 1. We connect the D lines of row m to an output channel *pass*.

Thus, the 3SAT gadget has $4n + 2$ ports (in($v = b$) for each v and b, one test input, and as many outputs).

LEMMA 3. *Consider a 3SAT gadget for a formula W. If the ball enters at $(v = b)_{in}$, it can only exit the gadget at $(v = b)_{out}$. This may open all gadgets*

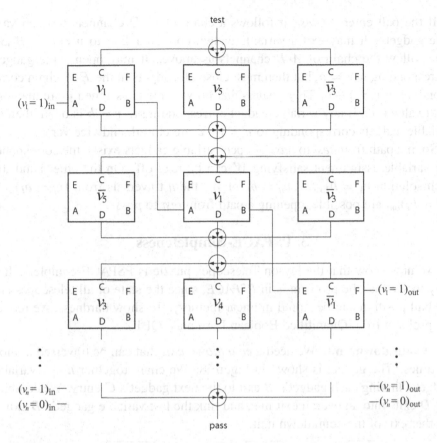

Figure 7. A 3SAT gadget. Shown are the test path and the path $(v_1 = 1)_{in}$ to $(v_1 = 1)_{out}$, where v_1 appears only in the first three clauses (twice positive, once negated).

corresponding to $v = b$ and must close all gadgets corresponding to $v = 1 - b$. The ball can also move from test to any $(v = b)_{out}$, and this must close all gadgets corresponding to $v = 1 - b$.

There exists an assignment $v_1 = b_1, \ldots, v_n = b_n$ satisfying W if and only if the ball can traverse the gadget from test to pass (after first traversing it from $(v_i = b_i)_{in}$ to $(v_i = b_i)_{out}$, for $i = 1, \ldots, n$).

PROOF. If the ball enters at $(v = b)_{in}$, it first reaches a chain of A-B channels through variable gadgets. It must follow the chain because in a variable gadget the only way from A leads to B. This may open all these gadgets. After the chain of A-B channels, the ball must traverse a chain of E-F channels which is also possible in only one way. This forces the corresponding variable channels to close.

If the ball enters at *test*, it follows a chain of C-D channels through variable gadgets. It may exit a variable gadget corresponding to $v = b$ at B and then follow the chain of A-B channels as above. It may open some gadgets corresponding to $v = b$, but then must close all gadgets in the E-F chain corresponding to $v = 1 - b$. This ensures that no variable is assigned more than one truth value (i.e., if any variable gadget corresponding to $v = b$ is open, then all variable gadgets corresponding to $v = 1 - b$ are closed, and vice versa).

So, if a path from *test* to *pass* of open variable gadgets exists, the corresponding variable assignment satisfying W can be read off. On the other hand, for each solution $v_1 = b_1, \ldots, v_n = b_n$ of W, the n traversals from $(v_i = b_i)_{in}$ to $(v_i = b_i)_{out}$ are possible, opening a path from *test* to *pass*. $\square$

3. PSPACE-completeness

We now show that the Dyson Telescopes puzzle is PSPACE-complete. It is easy to see that the problem is in *PSPACE*, since the state of all telescopes and the ball position can be stored in linear memory. To show hardness, we reduce the problem from Quanitified Boolean Formulas (QBF).

3.1. Countdown unit. We need a *countdown unit* that can be traversed at most 2^n times. The gadget is shown in Figure 8. We chain together $n + 1$ variable gadgets, linking each gadget's B exit to the next gadget's C entry. We combine the D exits into an overall exit line, and link the last variable gadget's B exit to another exit of the countdown unit.

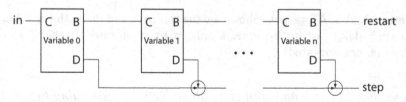

Figure 8. Countdown unit.

LEMMA 4. *When the ball enters the countdown unit for the first time, it can leave it at restart. After the gadget has been traversed from in to restart, it can be at most 2^n times traversed from in to step, before it must again be traversed from in to restart.*

PROOF. If all variable gadgets are closed, the ball can only leave them at B. After moving from *in* to *restart*, all or some of the gadgets may be open. But then the *in-step* channel can be used at most 2^n times, as can be seen by induction. $\square$

3.2. Reduction from QBF. Let

$$W = \forall v_{1,1} \exists v_{1,2} \forall v_{2,1} \exists v_{2,2} \ldots \forall v_{n,1} \exists v_{n,2} f(v_{1,1}, \ldots, v_{n,2})$$

be a quantified boolean formula with a 3CNF formula f. We build a gadget to test W using a 3SAT gadget for f, one countdown unit of size n, and a chain of n additional variable gadgets. The construction is shown in Figure 9.

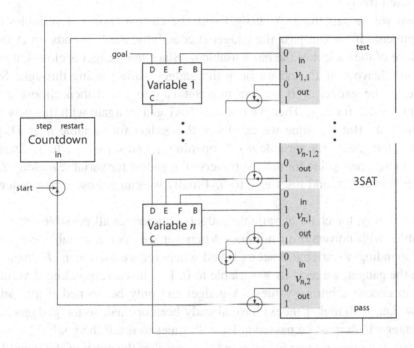

Figure 9. Reduction from QBF.

Each D exit of the variable gadgets is linked to the C entry of the previous gadget. Each F exit is linked to the E entry of the next variable gadget, however not directly but via (1) the $(v_{i,1} = 0)$ channel of the 3SAT gadget, and then (2) either the $(v_{i,2} = 0)$ or the $(v_{i,2} = 1)$ channel of the 3SAT gadget. The B exit of each variable gadget is also linked to the E entry of the next gadget, via (1) the $(v_{i,1} = 1)$ channel of the 3SAT gadget, and then (2) either the $(v_{i,2} = 0)$ or the $(v_{i,2} = 1)$ channel of the 3SAT gadget. The $(v_{n,2} = 0)$ and $(v_{n,2} = 1)$ channels of the 3SAT gadget are linked to the *in* entry of the countdown unit, whose *step* exit is linked via the test channel of the 3SAT gadget to the last variable gadget's C entry. The first gadget's D exit is linked to the goal. The starting point is also linked to the *in* entry of the countdown unit. The *restart* exit of the countdown unit is linked to the first variable gadget's E entry point.

THEOREM 5. *W is true if and only if the ball can move from start to goal.*

PROOF. We first describe how we can systematically test the formula for all possible truth assignments according to the quantifiers in W.

Initially, we must traverse the countdown unit from *in* to *restart*. Whenever the ball leaves the countdown unit at *restart*, it institutes a restart of the variable gadgets: all of them must be passed from E to F, so all of them are closed, and all variables with universal quantifiers are set to 0; all other variables can be chosen freely.

Next we can test the 3SAT gadget with the current choice of variables truth assignment. If we can pass the gadget successfully, the ball ends up at the C entrance of the gadget of the last variable n. Since this gadget is closed, the ball can only leave it at B, and we open the gadget while passing through. Since we leave the gadget at B we can now set $v_{n,1}$ to 1 and then choose a new arbitrary value for $v_{n,2}$. Then we test the 3SAT gadget again with this new truth assignment. But this time we can leave the gadget for variable n at D, pass through the gadget for variable $n-1$, opening it, and set $v_{n-1,1}$ to 1. Then we can choose a new value for $v_{n-1,2}$, traverse the gadget for variable n along E-F, thereby closing it, and reset $v_{n,1}$ to 0. Finally we can choose a new value for $v_{n,2}$.

In this way, the chain of variable gadgets enumerates all possible settings of variables with universal quantifiers. Whenever we open a variable gadget, its corresponding $\forall$-variable is set to 1, and whenever we use the E-F channel to close the gadget, we reset its $\forall$-variable to 0. For the corresponding $\exists$-variables we can choose arbitrary values. A gadget can only be opened if all gadgets below (i.e., with higher index) have already been opened, so the gadgets act as a counter which must be passed at least 2^n times to reach the goal.

Every time this counter is increased (i.e., reaches the entry of the countdown unit), it must pass the countdown unit and the 3SAT test channel. If the ball were to traverse the 3SAT unit from *test* to any $(v = b)_{out}$, then the countdown unit would have to be passed more often than the counter given by the additional variable gadgets. But these must be passed 2^n times to reach *goal*. Since the countdown unit does not allow more than 2^n traversals from A to B, it would have to be left at *restart* before we reach *goal*, which would reset the whole structure. Therefore, the ball cannot move from *test* to any other exit than *pass* if it wants to reach *goal*.

By the same argument, whenever a variable gadget is traversed from C to B, it must be opened, otherwise more than 2^n passages are required to reach *goal*, and the whole structure is reset.

If the 3SAT gadget is tested with every possible variable setting for the variables with universal quantifiers and a choice of values for the variables with existential quantifiers, W is true. If on the other hand W is true, there is such a selection for each possible setting for variables with universal quantifiers, and a path from *start* to *goal* exists.   □

4. Opposing telescopes that overlap in at most one space

In this section, we show that the Dyson Telescopes puzzle is in P if opposing telescopes cannot overlap in more than one space. Let D denote an instance of such a problem, and let $T_1, \ldots, T_n$ denote the telescopes. A *constellation* of the telescopes is an assignment of integers to the telescopes describing how far the telescopes are extended.

A *direct traversal* from T_i to T_j is a sequence of telescope extensions and retractions such that the ball is initially attached to T_i, finally attached to T_j, and in between it is not pushed or sucked by any other telescope. A *traversal* from T_1 to T_n is a sequence of direct traversals, where the ball is first attached to T_1 and ends up attached to T_n.

We first assume that D has no opposing pairs. Then we can define an induced directed graph G_D with the telescopes as vertices and an edge from T_i to T_j if

- T_i and T_j are orthogonal. Let f be the space in which they overlap.
- T_i and T_j can be extended at least up to the space before f.
- f is either the first space in front of T_i, the first space not reachable by T_i (i.e., the space to which the ball would be pushed if T_i was completely extended), or there is a third telescope T_k that can be extended to the space after f in the extension path of T_i.

LEMMA 6. *Assume D has no opposing pairs. If the ball is attached to a telescope T_i, then a direct traversal from T_i to another telescope T_j is possible precisely if there is an edge from T_i to T_j in G_D, independent of the current constellation of D.*

PROOF. If there is an edge from T_i to T_j, then obviously a direct traversal is possible.

Assume a direct traversal is possible in some constellation. Then T_i and T_j must be able to reach a common space f. Since there are no opposing pairs of telescopes, T_i and T_j must be orthogonal. Then, f is the only space reachable by both telescopes. We can transfer the ball from T_i to f if f is either the first or last reachable space of T_i, or if the space after f in the path of T_i is blocked by another telescope T_k. In any case, the edge (T_i, T_j) exists in G_D.   □

Now assume D contains an opposing pair as shown in Figure 10, where A and B overlap in at most one space, denoted by the black dot (if it exists). There may also be a third telescope C pointing to this space (or even extending beyond). There might even be a forth telescope (not shown, it can be handled

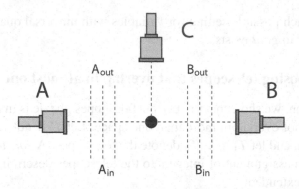

Figure 10. Opposing telescopes with at most one overlapping space.

analogously) opposing C and extending up to or beyond the black dot space. We define the graph G_D as before, but for the opposing pair we must add some additional edges as described below. For a telescope T, let T_{in} denote the set of all telescopes with an edge pointing to T, and T_{out} the set of telescopes to which T points in G_D. Note that C may or may not be in A_{in} and B_{in}, depending on the overall configuration of the at most four telescopes covering the black dot square and the initial position of the ball. Actually, for the construction below it is sufficient to assume that C is not in A_{in} and B_{in}.

LEMMA 7. *Traversal from any telescope in $A_{in} \cup B_{in}$ to C (if it exists) and any telescope in $A_{out} \cup B_{out}$ is possible in every constellation of D.*

PROOF. If C exists, we can move the ball from any telescope $T \in A_{in}$ to C as follows. First, we retract A, B, T, and C. Then we extend A completely. If we now extend B, it will be stopped just right of the black dot. Now we can retract A and move the ball from T to the line of A, which is possible since there is an edge from T to A in G_D. If we then extend A, the ball will come to rest on the black dot, where we can pick it up with C.

All other traversals are trivially possible.   □

Although the traversal from T to C in the proof above is done via A, it is impossible to traverse directly from A to C without prior preparation of the opposing pair if C extends exactly to the square above the black dot square. If the ball is initially placed in the opposing pair and the pair is not initially set up such that traversal to C is possible, C cannot be reached directly. This means we should add the following edges to G_D for each opposing pair (A, B) with one space overlap (and maybe an orthogonal telescope C pointing to the overlap space):

- edges $A \to B$ and $B \to A$;
- edges $T \to C$ for all $T \in A_{in} \cup B_{in}$, if C exists;

- edge $A \to C$, if C exists and can be extended to block A or B, or B is initially extended immediately to the right of the overlap space;
- edge $B \to C$, if C exists and can be extended to block A or B, or A is initially extended immediately to the left of the overlap space.

Note that in the second case the edges $T \to C$ are a shortcut for $T \to A \to C$ because we do not always want to add edge $A \to C$ to the graph (depending on the initial placement of the ball).

LEMMA 8. *Let D be an instance of the Dyson Telescopes puzzle with no opposing pair having more than one space overlap. Let G_D be the induced graph with edges as described above. Then, D has a solution exactly if there exists a path in G_D from a telescope that reaches the starting position of the ball to a telescope that reaches the goal position.*

PROOF. If there is a sequence of telescope movements that move the ball from start to goal, this induces a sequence of telescopes. If the start position of the ball is within an opposing pair (A, B) and both telescopes are initially retracted, the ball cannot leave the segment between A and B via C. But paths from A and B to all nodes of A_{out} and B_{out} exist in G_D, so the first telescope moves until the ball leaves the segment between A and B are reflected by edges in G_D. If the ball starts within the pair and one of the telescopes is not extended such that leaving at C would be possible, this is also reflected in G_D. Afterwards, all direct traversals of the winning strategy correspond to edges in G_D.

If on the other hand a path in G_D exists, it can easily be translated to a sequence of ball traversals (either direct or through opposing pairs) that gives a strategy to move the ball from start to goal.   □

COROLLARY 9. *The Dyson Telescopes puzzle is in P if opposing telescopes can overlap in at most one space.*

5. Summary and outlook

We showed that, in general, the problem of deciding whether the ball can move from start to goal in a setting of the Dyson Telescopes puzzle is PSPACE-complete. We also gave a polynomial-time algorithm if opposing pairs are restricted to at most one space of overlap.

Both the PSPACE-completeness proof and the algorithm for the restricted case also work if the back ends of the telescopes are taken into account and if the telescopes can initially be arbitrarily (partially) extended. Note that the PSPACE-hardness proof requires rather along telescopes. It would be interesting to investigate the complexity status of the problem with bounded-length telescopes.

Acknowledgements

We thank Jeff Erickson for posing the problem to us, and Eva M. Jungen for helpful discussions on the polynomial-time algorithm. We also thank the anonymous reviewer for his helpful suggestions.

References

[1] Erik D. Demaine. Playing games with algorithms: Algorithmic combinatorial game theory. In *Proc. 26th MFCS 2001*, pp. 18–32, LNCS 2136, 2001.

[2] Erik D. Demaine, Martin L. Demaine, Michael Hoffmann, and Joseph O'Rourke. Pushing blocks is hard. *Computational Geometry: Theory and Appl.* 26(1):21–36, 2003.

[3] Erik D. Demaine, Robert A. Hearn, Michael Hoffmann. Push-2-F is PSPACE-complete. In *Proc. of the 14th Canadian Conference on Computational Geometry (CCCG'02)*, pp. 31–35, 2002. http://www.cs.uleth.ca/~wismath/cccg/papers/31.ps.

[4] Robert A. Hearn, Erik D. Demaine. PSPACE-completeness of sliding-block puzzles and other problems through the nondeterministic constraint logic model of computation. *Theoretical Computer Science*, to appear. http://www.arXiv.org/abs/cs.CC/0205005

[5] Dyson UK's Telescope game. To be found at http://www.dyson.co.uk/game/.

[6] Jeff Erickson. Personal communication, September 2003.

ERIK D. DEMAINE
MASSACHUSETTS INSTITUTE OF TECHNOLOGY
COMPUTER SCIENCE AND ARTIFICIAL INTELLIGENCE LABORATORY
CAMBRIDGE, MA
UNITED STATES
    edemaine@mit.edu

MARTIN L. DEMAINE
MASSACHUSETTS INSTITUTE OF TECHNOLOGY
COMPUTER SCIENCE AND ARTIFICIAL INTELLIGENCE LABORATORY
CAMBRIDGE, MA
UNITED STATES
    mdemaine@mit.edu

RUDOLF FLEISCHER
FUDAN UNIVERSITY
SHANGHAI KEY LABORATORY OF INTELLIGENT INFORMATION PROCESSING
DEPARTMENT OF COMPUTER SCIENCE AND ENGINEERING
SHANGHAI
CHINA
    rudolf@fudan.edu.cn

ROBERT A. HEARN
NEUKOM INSTITUTE FOR COMPUTATIONAL SCIENCE, DARTMOUTH COLLEGE
SUDIKOFF HALL, HB 6255
HANOVER, NH 03755
UNITED STATES
  robert.a.hearn@dartmouth.edu

TIMO VON OERTZEN
MAX-PLANCK-INSTITUTE FOR HUMAN DEVELOPMENT
BERLIN
GERMANY
  vonoertzen@mpib-berlin.mpg.de

Games of No Chance 3
MSRI Publications
Volume **56**, 2009

Amazons, Konane, and Cross Purposes are PSPACE-complete

ROBERT A. HEARN

ABSTRACT. Amazons is a board game which combines elements of Chess and Go. It has become popular in recent years, and has served as a useful platform for both game-theoretic study and AI games research. Buro showed that simple Amazons endgames are NP-equivalent, leaving the complexity of the general case as an open problem.

Konane is an ancient Hawaiian game, with moves similar to peg solitaire. Konane has received some attention in the combinatorial game theory community, with game values determined for many small positions and one-dimensional positions. However, its general complexity seems not to have been previously addressed.

Cross Purposes was invented by Michael Albert, and named by Richard Guy at the Games at Dalhousie III workshop, in 2004. It is played on a Go board. Cross Purposes is a kind of two-player version of the popular puzzle Tipover: it represents stacks of crates tipping over and blocking others from tipping over.

We show that generalized versions of these games are PSPACE-complete. We give similar reductions to each game from one of the PSPACE-complete two-player formula games described by Schaefer. Our construction also provides an alternate proof that simple Amazons endgames are NP-equivalent.

1. Introduction

Combinatorial game theory is concerned with the attempt to find and analyze winning strategies for combinatorial games, or for tractable families of game positions. However, it is a curious fact that with few exceptions, any game or puzzle that is interesting to humans, and whose worst-case complexity is known,

---

During the preparation of this paper, the author learned that a group including Buro et al. was simultaneously preparing a paper showing Amazons to be PSPACE-complete [7].

is computationally as hard as possible based on very general characteristics of the game. By hardness here we mean computational complexity of determining the existence of a winning strategy for a given player, from a given position. For example, Minesweeper is a one-player game (puzzle), with a bounded number of moves; it is NP-complete [11]. Sliding-block puzzles do not have a bound on the number of moves; this raises the complexity to PSPACE-complete [10]. Two-player, bounded-move games, such as Hex, are also generally PSPACE-complete [14]. Other games can be even harder: Chess, Checkers, and Go (Japanese Rules), as two-player games with no bound on the number of moves, are EXPTIME-complete [6; 16; 15]. There are harder games still.

Amazons, Konane, and Cross Purposes are all two-player games with a poly-nomially bounded number of moves. We should therefore expect them to be PSPACE-complete, merely on the grounds that they are interesting games to play, and therefore presumably are as complex as possible given their general characteristics.

In the terminology of combinatorial game theory, all three games also follow the *normal play* convention: the first player who cannot move loses. Addition-ally, all three games are played on a square grid, with pieces that move, or are captured, or are transformed. These shared characteristics will enable us to use the same proof technique to show all of them PSPACE-hard. Only the specific gadgets differ among the three proofs. Our proof technique seems simpler than that used for many game results, and may have wider applicability. In particular, the generic crossover construction seems likely to simplify new hardness proofs.

As with most hardness results, the hardness of these games only applies di-rectly to particular configurations explicitly constructed to have computational properties. It does not say anything about the difficulty of determining the win-ner from a standard initial game configuration, or even from reasonable positions that might arise in actual play. Indeed, Hex is PSPACE-complete in general, but a simple strategy-stealing argument shows that from an empty board, it is a first-player win. Nonetheless, a hardness result for a game indicates that there are limits to the degree to which it can be theoretically analyzed.

Conway, Berlekamp, and Guy argue against a tendency to dismiss hard prob-lems as uninteresting [2, page 225]:

> Some people consider a class of problems "finished" when it has been shown to be NP-hard. Philosophically this is a viewpoint we strongly oppose. Some games which are NP-hard are very interesting!

Our view is just the reverse of that argued against: interesting games are almost of necessity hard. Showing a game to be hard is an indication that the game *is* interesting! That is the spirit in which these results are presented.

Outline. Section 2 describes the reduction to be used for each game. Sections 3, 4, and 5 detail the background, history, rules, and hardness proofs for Amazons, Konane, and Cross Purposes, respectively. Section 6 summarizes our results.

2. Reduction framework

Formula game. Schaefer [17] showed that deciding the winner of the following two-person game is PSPACE-complete: Let A be a positive CNF formula (i.e., a propositional formula in conjunctive normal form in which no negated variables occur). Each player on his move chooses a variable occurring in A which has not yet been chosen. After all variables have been chosen, player one wins if A is true when all variables chosen by player one are set to true and those chosen by player two are set to false.

We will refer to this game as the *formula game*. Our hardness reductions consist of constructing game configurations which force the two players to effectively play a given formula game.

Given a positive CNF formula A, we build logic and wiring gadgets corresponding to the variables and the formula, as shown schematically in Figure 1. (We use standard digital logic symbols for AND and OR.) If player one plays first in a variable, a signal is enabled to flow out from it; if player two plays first, that signal is blocked. When a signal arrives at or leaves from a gadget, we will speak of that input or output as *activating*. By splitting the signals, allowing them to cross, and feeding them into a network of logic gates, we may construct a particular signal line that player one may eventually activate only if A is true under the selected variable assignment. For each game, we arrange for player one to win just when he can activate that output signal.

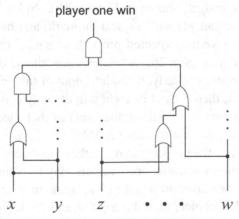

Figure 1. Reduction schematic. The circuit shown corresponds to the positive CNF formula $(x \vee y) \wedge \ldots \wedge (x \vee z \vee w)$.

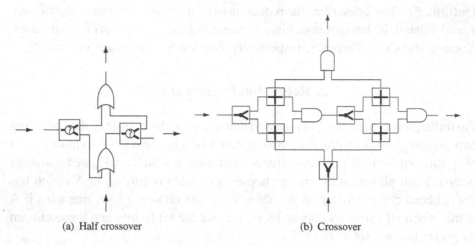

(a) Half crossover   (b) Crossover

Figure 2. Crossover gadgets.

Generic crossover. Crossover gadgets are often among the most complicated and difficult to construct in game hardness reductions. Rather than construct three separate crossover gadgets, we give a generic construction for crossing signals, given the existence of AND, OR, *split*, and *choice* gadgets. This means that in addition to the basic wiring and logic gadgets, we merely need to construct a choice gadget for each game. A choice gadget allows one, but not both, outputs to activate if the input activates. (Note that while there are traditional digital-logic methods for crossing signals in planar circuits, they require inverters, which do not fit well into our problem formalism.)

We develop the ability to cross signals in two steps. The first step is the *half-crossover* gadget, shown in Figure 2(a). Using half crossovers, we can make a full *crossover* gadget, shown in Figure 2(b). Splits are shown with a forking symbol, choice gadgets with a question mark, and half crossovers with a plus symbol. These have the expected properties: e.g., if the left input of the leftmost choice gadget in Figure 2(a) activates, then either, but not both, of its right outputs may activate; similarly, if the left input of the leftmost split gadget in Figure 2(b) activates, then both of its right outputs may activate. We note that this crossover construction is essentially the same as that used in [10]; the only difference is that in [10], the gates are reversible.

The half crossover has the property that if either input activates, either output may activate; if both inputs activate, both outputs may activate. Suppose the left input activates. Then the player propagating the signal may activate the left (i.e., upper) output of the left choice, then the top OR and the top output; or he may choose to activate the right output of the left choice, then the bottom OR, then the right output of the right choice, and the right output. Similarly, if the bottom input is activated, the signal may be directed to either output. If both inputs are activated, then by making the correct choices both outputs may be activated.

For the crossover gadget, we want the left input to be able to propagate only to the right output, and likewise vertically. First, it is clear that if one input activates, the corresponding output may also activate; simply choose the straight-through path to activate for each half crossover. The splits and ANDs then propagate the signal across the gadget. If both inputs activate, activating both half-crossover outputs allows both crossover outputs to activate.

Suppose the left split's input has not activated. Then at most one input to the left AND may activate, because the bottom-left half crossover can have at most one input, and thus output, activate. Therefore, the left AND's output may not activate. By the same reasoning, the right AND may not activate either. A similar argument shows that if the bottom split's input has not activated, the top AND may not activate. Therefore, the gadget serves to cross signals, as needed.

3. Amazons

Amazons was invented by Walter Zamkauskas in 1988. Both human and computer opponents are available for Internet play, and there have been several tournaments, both for humans and for computers.

Amazons has several properties which make it interesting for theoretical study. Like Go, its endgames naturally separate into independent subgames; these have been studied using combinatorial game theory [1; 18]. Amazons has a very large number of moves available from a typical position, even more than in Go. This makes straightforward search algorithms impractical for computer play. As a result, computer programs tend to incorporate explicit high-level knowledge of Amazons strategy [13; 12]. By showing that generalized Amazons is PSPACE-complete, we provide strong evidence that there is a practical limit to the degree of analysis possible from an arbitrary position.

As mentioned in the footnote on the first page, Furtak, Kiyomi, Uno, and Buro independently showed Amazons to be PSPACE-complete at the same time as the author [7]. (The original version of the present paper, containing only the Amazons proof, is available at [8].) Curiously, [7] already contains two different PSPACE-completeness proofs: one reduces from Hex, and the other from Generalized Geography. The paper is the result of the collaboration of two groups which had also solved the problem independently, then discovered each other. Thus, after remaining an open problem for many years, the complexity of Amazons was solved independently and virtually simultaneously by three different groups, using three completely different approaches, each of which leverages different aspects of the game to construct gadgets. This is a remarkable fact. The reduction from Generalized Geography provides the strongest result:

it shows that Amazons is PSPACE-complete even when each player only has a single Amazon. In contrast, the Hex reduction and the present formula game reduction each require a large number of Amazons.

Amazons rules. Amazons is normally played on a 10×10 board. The standard starting position, and a typical endgame position, are shown in Figure 3. (We indicate burned squares by removing them from the figures, rather than marking them with tokens.) Each player has four *amazons*, which are immortal chess queens. White plays first, and play alternates. On each turn a player must first move an amazon, like a chess queen, and then fire an *arrow* from that amazon. The arrow also moves like a chess queen. The square that the arrow lands on is burned off the board; no amazon or arrow may move onto or across a burned square. There is no capturing. The first player who cannot move loses.

Amazons is a game of mobility and control, like Chess, and of territory, like Go. The strategy involves constraining the mobility of the opponent's amazons, and attempting to secure large isolated areas for one's own amazons. In the endgame shown in Figure 3, Black has access to 23 spaces, and with proper play can make 23 moves; White can also make 23 moves. Thus from this position, the player to move will lose.

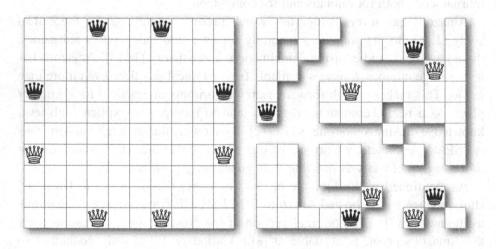

Figure 3. Amazons start position and typical endgame position.

3.1. PSPACE-completeness.
We follow the reduction framework outlined in Section 2. The game consists of a variable selection phase, during which all play occurs in variable gadgets, followed by a phase in which White attempts to activate a signal pathway leading to a large supply of extra moves, enabling him to win. Black is supplied with enough extra moves of his own to win otherwise.

Basic wiring. Signals propagate along *wires*. Figure 4(a) shows the construction of a wire. Suppose that amazon A is able to move down one square and shoot down. This enables amazon B to likewise move down one and shoot down; C may now do the same. This is the basic method of signal propagation. When an amazon moves backward (in the direction of input, away from the direction of output) and shoots backward, we will say that it has *retreated*.

Figure 4(a) illustrates two additional useful features. After C retreats, D may retreat, freeing up E. The result is that the position of the wire has been shifted by one in the horizontal direction. Also, no matter how much space is freed up feeding into the wire, D and E may still only retreat one square, because D is forced to shoot into the space vacated by C.

Figure 4(b) shows how to turn corners. Suppose A, then B may retreat. Then C may retreat, shooting up and left; D may then retreat. This gadget also has another useful property: signals may only flow through it in one direction. Suppose D has moved and shot right. C may then move down and right, and shoot right. B may then move up and right, but it can only shoot into the square it just vacated. Thus, A is not able to move up and shoot up.

By combining the horizontal parity-shifting in Figure 4(a) with turns, we may direct a signal anywhere we wish. Using the unidirectional and flow-limiting properties of these gadgets, we can ensure that signals may never back up into outputs, and that inputs may never retreat more than a single space.

Splitting a signal is a bit trickier. The *split* gadget shown in Figure 4(c) accomplishes this. A is the input; G and H are the outputs. First, observe that

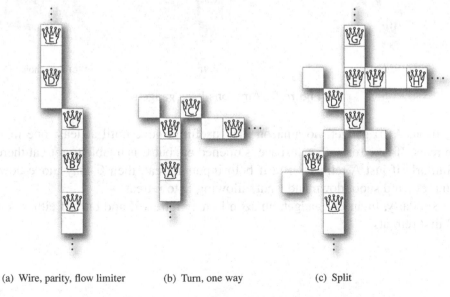

(a) Wire, parity, flow limiter   (b) Turn, one way   (c) Split

Figure 4. Amazons wiring gadgets.

until A retreats, there are no useful moves to be made. C, D, and F may not move without shooting back into the square they left. A, B, and E may move one unit and shoot two, but nothing is accomplished by this. But if A retreats, then the following sequence is enabled: B down and right, shoot down; C down and left two, shoot down and left; D up and left, shoot down and right three; E down two, shoot down and left; F down and left, shoot left. This frees up space for G and H to retreat, as required.

Logic. The *variable* gadget is shown in Figure 5(a). If White moves first in a variable, he can move A down, and shoot down, allowing B to later retreat. If Black moves first, he can move up and shoot up, preventing B from ever retreating.

The AND and OR gadgets are shown in Figures 5(b) and 5(c). In each, A and B are the inputs, and D is the output. Note that, because the inputs are protected with flow limiters — Figure 4(a) — no input may retreat more than one square; otherwise the AND might incorrectly activate.

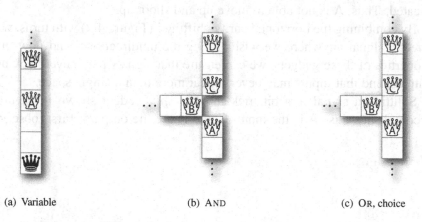

(a) Variable   (b) AND   (c) OR, choice

Figure 5. Amazons logic gadgets.

In an AND gadget, no amazon may usefully move until at least one input retreats. If B retreats, then a space is opened up, but C is unable to retreat there; similarly if just A retreats. But if both inputs retreat, then C may move down and left, and shoot down and right, allowing D to retreat.

Similarly, in an OR gadget, amazon D may retreat if and only if either A or B first retreats.

Choice. For the generic crossover construction to work, we need a choice gadget. The existing OR gadget suffices, if we reinterpret the bottom input as an output: if if B retreats, then either C or A, but not both, may retreat.

Winning. We will have an AND gadget whose output may be activated only if the formula is true under the chosen assignment. We feed this signal into a *victory* gadget, shown in Figure 6. There are two large rooms available. The sizes are equal, and such that if White can claim both of them, he will win, but if he can claim only one of them, then Black will win.

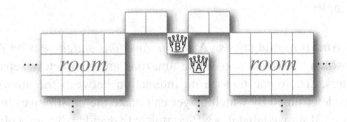

Figure 6. Amazons victory gadget.

If B moves before A has retreated, then it must shoot so as to block access to one room or the other; it may then enter and claim the accessible room. If A first retreats, then B may move up and left, and shoot down and right two, leaving the way clear to enter and claim the left room, then back out and enter and claim the right room.

THEOREM 1. *Amazons is PSPACE-complete.*

PROOF. Given a positive CNF formula A, we construct a corresponding Amazons position, as described above. The reduction may be done in polynomial time: if there are k variables and l clauses, then there need be no more than $(kl)^2$ crossover gadgets to connect each variable to each clause it occurs in; all other aspects of the reduction are equally obviously polynomial.

If the players alternate choosing variables, then when all variables have been chosen, White will be able to activate wires leading from only those variables he has chosen; these are just the variables assigned to true in the formula game. Since A contains no negated variables, White will thus be able eventually to reach both rooms of the victory gadget just if A is true under the variable assignment corresponding to the players' choices. White will then have more moves available than Black, and win; otherwise, Black's extra room will give him more moves than White, and Black will win.

Suppose a player makes a move which does not choose a variable, before all variables have been chosen. This can have no effect on the other player, apart from allowing him to choose two variables in a row, because the Black and White amazons may only interact within variable gadgets. A player who chooses two variables in a row may finish with at least the same set of variables chosen as he would otherwise. Therefore, not playing in accordance with the formula game does not allow a player to win if he could not otherwise win.

Therefore, a player may win the Amazons game if and only if he may win the corresponding formula game, and Amazons is PSPACE-hard.

Since the game must end after a polynomial number of moves, it is possible to perform a search of all possible move sequences using polynomial space, thus determining the winner. Therefore, Amazons is also in PSPACE, and thus PSPACE-complete.   □

3.2. Simple Amazons endgames. A *simple Amazons endgame* is an Amazons position in which the Black and White amazons are completely separated by burned squares. There can thus be no interaction between the amazons, and the winner is determined by which player can make the most moves in his own territory. Buro [3] showed that it is NP-complete to decide whether a player may make a given number of moves from an individual territory containing only his amazons. Buro first proved NP-completeness of the Hamilton circuit problem for cubic subgraphs of the integer grid, and then reduced from that problem. As a result, deciding the outcome of a simple Amazons endgame is NP-equivalent (that is, it can be decided with a polynomial number of calls to an algorithm for an NP-complete problem, and vice versa). Our gadgets provide a simple alternate proof.

THEOREM 2. *Deciding the outcome of a simple Amazons endgame is NP-equivalent.*

PROOF. We reduce SAT to a single-color Amazons position. Given a propositional formula A, we construct the same position as in Theorem 1, with the following modifications. We remove the Black amazons, then connect each variable output to the input of a choice gadget. We connect one choice output path to the non-negated occurrences of the corresponding variable in the formula, and the other output path to the negated occurrences.

Then, White may reach both rooms of the victory gadget if and only if A is satisfiable, by choosing the correct set of choice output paths. Therefore, it is NP-hard to decide whether a player may make a given number of moves from a position containing only his amazons. We may nondeterministically guess a satisfying move sequence and verify it in polynomial time; therefore, the problem is NP-complete. As in [3], it follows automatically that deciding the winner of a simple Amazons endgame is NP-equivalent.   □

4. Konane

Konane is an ancient Hawaiian game, with a long history. Captain Cook documented the game in 1778, noting that at the time it was played on a 14×17 board. Other sizes were also used, ranging from 8×8 to 13×20. The game was usually played with pieces of basalt and coral, on stone boards with indentations to hold the pieces. King Kamehameha the Great was said to be an expert player; the game was also popular among all classes of Hawaiians.

More recently, Konane has been the subject of combinatorial game-theoretic analysis [5; 4]. Like Amazons, its endgames break into independent games whose values may be computed and summed. However, as of this writing, even $1 \times n$ Konane has not been completely solved, so it is no surprise that complicated positions can arise.

Konane rules. Konane is played on a rectangular board, which is initially filled with black and white stones in a checkerboard pattern. To begin the game, two adjacent stones in the middle of the board or in a corner are removed. Then, the players take turns making moves. Moves are made as in peg solitaire – indeed, Konane may be thought of as a kind of two-player peg solitaire. A player moves a stone of his color by jumping it over a horizontally or vertically adjacent stone of the opposite color, into an empty space. Stones so jumped are captured, and removed from play. A stone may make multiple successive jumps in a single move, as long as they are in a straight line; no turns are allowed within a single move. The first player unable to move wins.

4.1. PSPACE-completeness. The Konane reduction is similar to the Amazons reduction; the Konane gadgets are somewhat simpler. As before, the game consists of a variable selection phase, during which all play occurs in variable gadgets, followed by a phase in which White attempts to activate a signal pathway leading to a large supply of extra moves, enabling him to win. Black is supplied with enough extra moves of his own to win otherwise.

Basic wiring. A Konane wire is simply a string of alternating black stones and empty spaces. By capturing the black stones, a white stone traverses the wire. Note that in Konane, in contrast with the Amazons reduction, signals propagate by stones moving forwards, capturing opposing stones.

Turns are enabled by adjoining wires as shown in Figure 7; at the end of one wire, the white stone comes to rest at the beginning of another, protected from capture by being interposed be-

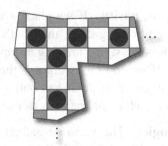

Figure 7. Konane wire, turn.

tween two black stones. If the white stone tried to traverse the turn in the other direction, it would not be so protected, and Black could capture it. Thus, as in the Amazons reduction, the turn is also a one-way device, and we assume that gadget entrances and exits are protected by turns to ensure that signals can only flow in the proper directions.

Conditional gadget. A single gadget serves the purpose of AND, split, and positional parity adjustment. It has two input / output pathways, with the property that the second one may only be used if the first one has already been used. This *conditional gadget* is shown in Figure 8; the individual uses are outlined below.

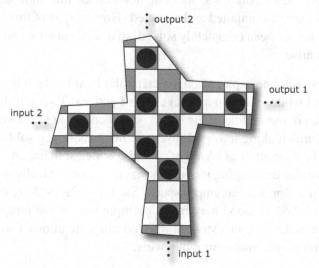

Figure 8. Konane wiring: conditional.

Observe that a white stone arriving at input 1 may only leave via output 1, and likewise for input 2 and output 2. However, if White attempts to use pathway 2 before pathway 1 has been used, Black can capture him in the middle of the turn. But if pathway 1 has been used, the stone Black needs to make this capture is no longer there, and pathway 2 opens up.

Split, parity. If we place a white stone within the wire feeding input 2 of a conditional gadget, then both outputs may activate if input 1 activates. This splits the signal arriving at input 1.

If we don't use output 1, then this split configuration also serves to propagate a signal from input 1 to output 2, with altered positional parity. This enables us to match signal parities as needed at the gadget inputs and outputs.

Logic. The *variable* gadget consists of a white stone at the end of a wire, as in Figure 9(a). If White moves first in a variable, he can traverse the wire, landing safely at an adjoining turn. If Black moves first, he can capture the white stone and prevent White from ever traversing the wire.

The AND gadget is a conditional gadget with output 1 unused. By the properties of a conditional gadget, a white stone may exit output 2 only if white stones have arrived at both inputs. The OR gadget is shown in Figure 9(b). The inputs are on the bottom and left; the output is on the top. Clearly, a white stone arriving via either input may leave via the output.

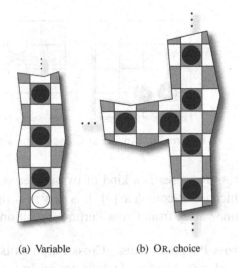

(a) Variable   (b) OR, choice

Figure 9. Konane logic gadgets.

Choice. For the generic crossover to work, we need a choice gadget. As was the case with Amazons, the OR gadget suffices, if we relabel the bottom input as an output: a white stone arriving along the left input may exit via either the top or the bottom. (For Konane, it turns out that crossover is a trivial gadget to make in any case.)

Winning. We will have an AND gadget whose output may be activated only if the formula is true under the chosen assignment. We feed this signal into a long series of turns, providing White with enough extra moves to win if he can reach them. Black is provided with his own series of turns, made of white wires, with a single black stone protected at the end of one of them, enabling Black to win if White cannot activate the final AND.

THEOREM 3. *Konane is PSPACE-complete.*

PROOF. Given a positive CNF formula A, we construct a corresponding Konane position, as described above. As in the Amazons construction, the reduction is clearly polynomial. Also as in Amazons, White may reach his supply of extra moves just when he can win the formula game on A.

Therefore, a player may win the Konane game if and only if he may win the corresponding formula game, and Konane is PSPACE-hard. As before, Konane is clearly also in PSPACE, and therefore PSPACE-complete. □

5. Cross Purposes

Cross Purposes was invented by Michael Albert, and named by Richard Guy, at the Games at Dalhousie III workshop, in 2004. It was introduced to the author by Michael Albert at the 2005 BIRS Combinatorial Game Theory Workshop.

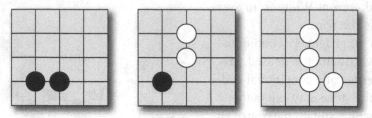

Figure 10. An initial Cross Purposes configuration, and two moves.

Cross Purposes is a kind of two-player version of the popular puzzle Tipover, which is NP-complete [9]. It is easy to construct many interesting combinatorial game values from Cross Purposes positions.

Cross Purposes rules. Cross Purposes is played on the intersections of a Go board, with black and white stones. In the initial configuration, there are some black stones already on the board. A move consists of replacing a black stone with a pair of white stones, placed in a row either directly above, below, to the left, or to the right of the black stone; the spaces so occupied must be vacant for the move to be made. See Figure 10. The idea is that a stack of crates, represented by a black stone, has been tipped over to lie flat. Using this idea, we describe a move as *tipping* a black stone in a given direction.

The players are called *Vertical* and *Horizontal*. Vertical moves first, and play alternates. Vertical may only move vertically, up or down; Horizontal may only move horizontally, left or right. All the black stones are available to each player to be tipped, subject to the availability of empty space. The first player unable to move loses.

5.1. PSPACE-completeness. The Cross Purposes construction largely follows those used for Amazons and Konane; we build the necessary gadgets to force the two players to effectively play a formula game.

One new challenge in constructing the gadgets is that each player may only directly move either horizontally or vertically, but not both. Yet, for formula game gadgets to work, one player must be able to direct signals two dimensionally. We solve this problem by restricting the moves of Horizontal so that, after the variable selection phase, his possible moves are constrained so as to force him to cooperate in Vertical's signal propagation. (We assume that the number of variables is even, so that it will be Vertical's move after the variable selection phase.) An additional challenge is that a single move can only empty a single square, enabling at most one more move to be made, so it is not obviously possible to split a signal. Again, we use the interaction of the two players to solve this problem.

We do not need a supply of extra moves at the end, as used for Amazons and Konane; instead, if Vertical can win the formula game, and correspondingly activate the final AND gadget, then Horizontal will have no move available, and lose. Otherwise, Vertical will run out of moves first, and lose.

Basic wiring. Signals flow diagonally, within surrounding corridors of white stones. A *wire* is shown in Figure 11(a). Suppose that Vertical tips stone A down, and suppose that Horizontal has no other moves available on the board. Then his only move is to tip B left. This then enables Vertical to tip C down. The result of this sequence is shown in Figure 11(b).

The turn gadget is shown in Figure 11(c); its operation is self-evident. Also

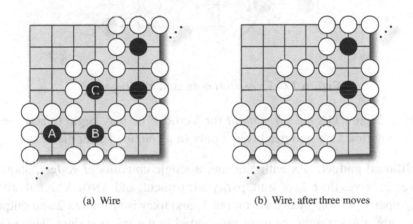

(a) Wire   (b) Wire, after three moves

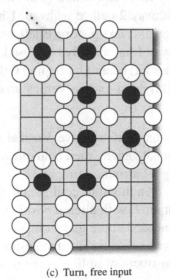

(c) Turn, free input

Figure 11. Cross Purposes wiring.

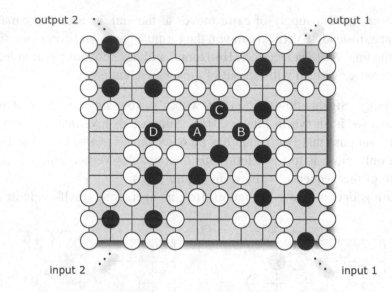

output 2   output 1

input 2   input 1

Figure 12. Cross Purposes conditional gadget.

shown in Figure 11(c) is a *free input* for Vertical: he may begin to activate this
wire at any time. We will need free inputs in a couple of later gadgets.

Conditional gadget. As with Konane, a single *conditional gadget*, shown in
Figure 12, serves the role of split, parity adjustment, and AND. A signal arriving
along input 1 may only leave via output 1, and likewise for input 2 and output 2;
these pathways are ordinary turns embedded in the larger gadget. However, if
Vertical attempts to use pathway 2 before pathway 1 has been used, then after
he tips stone A down, Horizontal can tip stone B left, and Vertical will then have
no local move. But if pathway 1 has already been used, stone B is blocked from
this move by the white stones left behind by tipping C down, and Horizontal has
no choice but to tip stone D right, allowing Vertical to continue propagating the
signal along pathway 2.

Split, parity. As with Konane, if we give Vertical a free input to the wire
feeding input 2 of a conditional gadget, then both outputs may activate if input 1
activates. This splits the signal arriving at input 1.

   If we don't use output 1, then this split configuration also serves to propagate
a signal from input 1 to output 2, with altered positional parity. This enables us
to match signal parities as needed at the gadget inputs and outputs. We must be
careful with not using outputs, since we need to ensure that Vertical has no free
moves anywhere in the construction; unlike in the previous two constructions,
in Cross Purposes, there is no extra pool of moves at the end, and every available
move within the layout counts. However, blocking an output is easy to arrange;

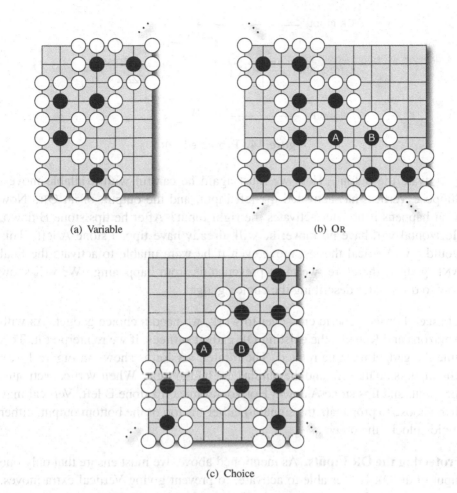

(a) Variable   (b) OR

(c) Choice

Figure 13. Cross Purposes logic gadgets.

we just terminate the wire so that Horizontal has the last move in it. Then Vertical gains nothing by using that output.

Logic. The *variable* gadget is shown in Figure 13(a). If Vertical moves first in a variable, he can begin to propagate a signal along the output wire. If Horizontal moves first, he will tip the bottom stone to block Vertical from activating the signal.

The AND gadget is a conditional gadget with output 1 unused. By the properties of the conditional gadget, output 2 may activate only if both inputs have activated.

The OR gadget is shown in Figure 13(b). The inputs are on the bottom; the output is on the top. Whether Vertical activates the left or the right input, Horizontal will be forced to tip stone A either left or right, allowing Vertical

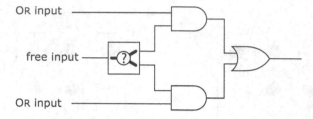

Figure 14. Protected OR.

to activate the output. Here we must again be careful with available moves. Suppose Vertical has activated the left input, and the output, of an OR. Now what happens if he later activates the right input? After he tips stone B down, Horizontal will have no move; he will already have tipped stone A left. This would give Vertical the last move even if he were unable to activate the final AND gadget; therefore, we must prevent this from happening. We will show how to do so after describing the choice gadget.

Choice. For the generic crossover to work, we need a choice gadget. As with Amazons and Konane, the existing OR gadget suffices, if we reinterpret it. This time the gadget must be rotated. The rotated version is shown in Figure 13(c). The input is on the left, and the outputs are on the right. When Vertical activates the input, and tips stone A down, Horizontal must tip stone B left. Vertical may then choose to propagate the signal to either the top or the bottom output; either choice blocks the other.

Protecting the OR Inputs. As mentioned above, we must ensure that only one input of an OR is ever able to activate, to prevent giving Vertical extra moves. We do so with the circuit shown in Figure 14. Vertical is given a free input to a choice gadget, whose output combines with one of the two OR input signals in an AND gadget. Since only one choice output can activate, only one AND output, and thus one OR input, can activate. Inspection of the relevant gadgets shows that Vertical has no extra moves in this construction; for every move he can make, Horizontal has a response.

Winning. We will have an AND gadget whose output may be activated only if the formula is true under the chosen assignment. We terminate its output wire with Vertical having the final move. If he can reach this output, Horizontal will have no moves left, and lose. If he cannot, then since Horizontal has a move in reply to every Vertical move within all of the gadgets, Vertical will eventually run out of moves, and lose.

THEOREM 4. *Cross Purposes is PSPACE-complete.*

PROOF. Given a positive CNF formula A, we construct a corresponding Cross Purposes position, as described above. As before, the reduction is clearly polynomial. Also as before, Vertical may activate a particular AND output, and thus gain the last move, just when he can win the formula game on A.

Therefore, a player may win the Cross Purposes game if and only if he may win the corresponding formula game, and Cross Purposes is PSPACE-hard. As before, Cross Purposes is clearly also in PSPACE, and therefore PSPACE-complete. □

6. Conclusion

We have shown that generalized versions of Amazons, Konane, and Cross Purposes are PSPACE-complete, indicating that it is highly unlikely that an efficient algorithm for optimal play exists for any of them. Their hardness is also additional evidence, if any were needed, that the games are interesting – they are sufficiently rich games to represent abstract computations.

Additionally, we have demonstrated a simple proof technique for showing planar, two-player, bounded move games hard. The generic crossover, in particular, seems likely to make further proofs along these lines easier. It would be interesting to revisit some classic game hardness results, to see whether the proofs can be simplified with these techniques.

Acknowledgments

We thank Michael Albert for introducing us to Cross Purposes, and for his invaluable assistance in constructing the Cross Purposes signal propagation mechanism.

References

[1] Elwyn R. Berlekamp. Sums of $N \times 2$ Amazons. In *Lecture Notes - Monograph Series*, volume 35, pages 1–34. Institute of Mathematical Statistics, 2000.

[2] Elwyn R. Berlekamp, John H. Conway, and Richard K. Guy. *Winning Ways, 2nd edition*. A. K. Peters, Ltd., Wellesley, MA, 2001–2004.

[3] Michael Buro. Simple Amazons endgames and their connection to Hamilton circuits in cubic subgrid graphs. In *Proceedings of the 2nd International Conference on Computers and Games*, Lecture Notes in Computer Science, Hamamatsu, Japan, October 2000.

[4] Alice Chan and Alice Tsai. $1 \times n$ konane: a summary of results. In R. J. Nowakowski, editor, *More Games of No Chance*, pages 331–339, 2002.

[5] Michael D. Ernst. Playing Konane mathematically: A combinatorial game-theoretic analysis. *UMAP Journal*, 16(2):95–121, Spring 1995.

[6] Aviezri S. Fraenkel and David Lichtenstein. Computing a perfect strategy for $n \times n$ chess requires time exponential in n. *Journal of Combinatorial Theory*, Series A, 31:199–214, 1981.

[7] Timothy Furtak, Masashi Kiyomi, Takeaki Uno, and Michael Buro. Generalized amazons is PSPACE-complete. In Leslie Pack Kaelbling and Alessandro Saffiotti, editors, *IJCAI*, pages 132–137. Professional Book Center, 2005.

[8] Robert A. Hearn. Amazons is PSPACE-complete. Manuscript, February 2005. http://www.arXiv.org/abs/cs.CC/0008025.

[9] Robert A. Hearn. Tipover is NP-complete. *Mathematical Intelligencer*, 28(3):10–14, 2006.

[10] Robert A. Hearn and Erik D. Demaine. PSPACE-completeness of sliding-block puzzles and other problems through the nondeterministic constraint logic model of computation. *Theoretical Computer Science*, 343(1-2):72–96, October 2005.

[11] Richard Kaye. Minesweeper is NP-complete. *Mathematical Intelligencer*, 22(2): 9–15, 2000.

[12] Jens Lieberum. An evaluation function for the game of Amazons. *Theoretical Computer Science*, 349(2):230–244, December 2005.

[13] Martin Müller and Theodore Tegos. Experiments in computer Amazons. In R. J. Nowakowski, editor, *More Games of No Chance*, pages 243–257. Cambridge University Press, 2002.

[14] Stefan Reisch. Hex ist PSPACE-vollständig. *Acta Informatica*, 15:167–191, 1981.

[15] J. M. Robson. The complexity of Go. In *Proceedings of the IFIP 9th World Computer Congress on Information Processing*, pages 413–417, 1983.

[16] J. M. Robson. N by N Checkers is EXPTIME complete. *SIAM Journal on Computing*, 13(2):252–267, May 1984.

[17] Thomas J. Schaefer. On the complexity of some two-person perfect-information games. *Journal of Computer and System Sciences*, 16:185–225, 1978.

[18] Raymond Georg Snatzke. New results of exhaustive search in the game Amazons. *Theor. Comput. Sci.*, 313(3):499–509, 2004.

ROBERT A. HEARN
NEUKOM INSTITUTE FOR COMPUTATIONAL SCIENCE, DARTMOUTH COLLEGE
SUDIKOFF HALL, HB 6255
HANOVER, NH 03755
UNITED STATES
robert.a.hearn@dartmouth.edu

Impartial

Games of No Chance 3
MSRI Publications
Volume **56**, 2009

Monotonic sequence games

M. H. ALBERT, R. E. L. ALDRED, M. D. ATKINSON, C. C. HANDLEY,
D. A. HOLTON, D. J. MCCAUGHAN, AND B. E. SAGAN

ABSTRACT. In a monotonic sequence game, two players alternately choose
elements of a sequence from some fixed ordered set. The game ends when the
resulting sequence contains either an ascending subsequence of length a or a
descending one of length d. We investigate the behaviour of this game when
played on finite linear orders or $\mathbb{Q}$ and provide some general observations for
play on arbitrary ordered sets.

1. Introduction

Monotonic sequence games were introduced by Harary, Sagan and West in
[6]. We paraphrase the description of the rules as follows:

From a deck of cards labelled with the integers from 1 through n, two
players take turns choosing a card and adding it to the right hand end of
a row of cards. The game ends when there is a subsequence of a cards in
the row whose values form an ascending sequence, or of d cards whose
values form a descending sequence.

The parameters a, d, and n are set before the game begins. There are two
possible methods for determining the winner of the game. In the *normal* form
of the game, the winner is the player who places the last card (which forms an
ascending or descending sequence of the required length). In the *misère* form
of the game, that player is the loser. In [6] these are called the *achievement* and
avoidance forms of the game respectively.

As a consequence of the Erdős–Szekeres theorem [5], the game cannot end
in a draw if $n > (a-1)(d-1)$. It is therefore natural to attempt to classify the
parameters (a, d, n) according to whether the first player can force a win, the
second player can force a win, or either player can ensure at least a draw. Some
results towards such a classification were presented in [6] and the problem of
extending and generalising these results was posed there and by Sagan in [8].

In this paper we will report on some progress on this and related problems. As regards the original game, we have been able to extend the computer-assisted analysis to decide many instances which were left open in [6]. We also provide some general results concerning the long run behaviour of these games (that is, for fixed a and d but large n). However, most of the work reported here deals with variations of the original game. In particular, we consider the case where the deck of cards is $\mathbb{Q}$ rather than a finite linear order. Finally, we examine some other variations of the game obtained either by relaxing the rules, or by playing with a deck of cards that is partially ordered. We list some open problems in the final section of the paper.

We adopt, and in some cases adapt, the notation and terminology of *Winning Ways* [1; 2; 3; 4] in discussing our results. This differs somewhat from that used in [6] so, where necessary, we will also provide translations of the results from that paper.

2. The general framework

Any version of the monotonic sequence game specifies at the outside, a *deck* D which is simply some partially ordered set, and two positive integer parameters a and d which we call the *critical lengths* of ascending and descending sequences respectively. There are two players, A and B (for convenience in assigning pronouns, A is assumed to be male and B female), who alternately choose an element which has not previously been chosen from the deck and add it to a sequence whose elements consist of the cards chosen up to this point. This sequence will be called the *board*. Conventionally, A plays first while B plays second. In the *basic* form of the game the board is constructed from left to right. That is, if the current board is $bc \cdots v$ and the next player chooses a value $w \in D$ then the new board is $bc \cdots vw$. An ascending subsequence of length a or a descending subsequence of length d of the board is called a *critical sequence*. As soon as the board contains a critical sequence the game ends. In *normal* play, the winner is the player whose move terminated the game. In *misère* play that player is the loser. We henceforth assume that $a, d \geq 2$ since the cases $a = 1$ or $d = 1$ are completely trivial. If the deck is exhausted without creating a critical sequence, then the game is considered drawn. If the deck is infinite then the game is also considered drawn if play proceeds without termination. By default we assume that normal play is being considered unless otherwise noted.

PROPOSITION 1. *If D is finite and contains a chain of length greater than $(a-1)(d-1)$, or D is infinite and contains no infinite antichain then no draws are possible in either normal or misère play.*

PROOF. In the first case any supposedly drawn board would contain all the elements of the specified chain. However, by the Erdős–Szekeres theorem any such sequence contains a critical sequence. In the latter case a similar result follows from the well-known observation that, as a consequence of Ramsey's theorem, any infinite sequence of elements from a partially ordered set contains an infinite subsequence which is either ascending, descending, or an antichain. Since the last possibility is ruled out by hypothesis, one of the former two must apply, and the play producing that sequence could not have been drawn. □

OBSERVATION 2. If D has a fixed-point-free order-preserving involution then the second player can force at least a draw.

B's strategy is to play the image of A's move under the involution, *unless* she has an immediate win available. Since no chain can involve both a point and its image she thereby never plays a suicidal move, that is one which makes it possible for Alexander to win the game on his next turn, and hence she cannot lose.

OBSERVATION 3. If $a = d$ and D has a fixed point free order reversing involution i with the property that whenever x and x^i are comparable, one is minimal and the other maximal, then the second player can force at least a draw.

Again the strategy for B is to play a winning move if one exists, and otherwise the image of A's previous move. The minimality/maximality criterion guarantees that in the resulting sequence of plays no chain can arise using both x and x^i unless $a = d = 2$ which is trivially a second player win.

This observation applies to play on the cube 2^n or equivalently on the lattice of subsets of a set. In particular it is easy to check that for $a = d = 3$ play on 2^3 is a second player win though cooperatively the two players can play to a draw.

Since D, a and d are fixed parameters of any particular game, all the relevant information about a position is contained in its board. A board which could arise in play may not have a proper prefix containing either an ascending sequence of length a or a descending sequence of length d. Subject to this condition we may define the *type* of a board to be one of $\mathcal{N}$, $\mathcal{P}$ or $\mathcal{D}$. We say that the type is $\mathcal{N}$ (next) if the player whose turn it is to move (that is, the next player) has a winning strategy. The type is $\mathcal{P}$ (previous) if the previous player (that is, the player who is not next) has a winning strategy. Finally, the type is $\mathcal{D}$ (drawn) if each player has a strategy that guarantees her or him at least a draw.

A board which contains the entire deck or which contains a critical sequence is called a *terminal board*. A terminal board containing a critical sequence is of type $\mathcal{P}$ in normal play and $\mathcal{N}$ in misère play, while a terminal board that does not contain a critical sequence is of type $\mathcal{D}$. Otherwise, the type of a board, X, is determined by the set of types of the boards that can be obtained in one further move. We call these boards the *children* of X. If this set contains any board of type $\mathcal{P}$ then the type of X is $\mathcal{N}$. If *all* the boards in this set are of type $\mathcal{N}$ then the type of X is $\mathcal{P}$. Otherwise, the type of X is $\mathcal{D}$.

These rules may not be immediately sufficient for determining the type of an arbitrary board when arbitrarily long plays or even draws with infinite play are possible. However, even in this case the boards are partitioned into the three types above. The algorithm for performing the partitioning is to begin by labelling all the terminal boards according to the winning conditions. Then inductively any currently unlabelled boards which either have a child of type $\mathcal{P}$, or all of whose children have type $\mathcal{N}$, are labelled appropriately. After completing this induction, any boards remaining unlabelled are of type $\mathcal{D}$.

Our principal goal will be to determine the type of the empty board – that is, to determine whether the first player has a winning strategy, or failing that, whether he can force a draw. We denote this type by $W_{\text{nor}}(a, d, D)$ for normal play, or $W_{\text{mis}}(a, d, D)$ for misère play.

3. Double bumping

Given a sequence of distinct elements $\mathbf{v} = v_1 v_2 \cdots$ from a linearly ordered set C, a well known algorithm due to Schensted [9] determines (explicitly) the length of the longest increasing subsequence of any prefix $v_1 v_2 \cdots v_k$ and (implicitly) the elements of such a sequence. This is sometimes called the "bumping" algorithm. An increasing sequence $\mathbf{w} = w_1 w_2 \cdots w_m$ is maintained as the elements of $\mathbf{v}$ are processed in order. When v_i is processed, $\mathbf{w}$ is modified as follows: if $w_m < v_i$ then v_i is appended to $\mathbf{w}$; otherwise v_i bumps (that is, replaces) the smallest element of w that is larger than v_i.

It is easy to check that, after processing $v_1 v_2 \cdots v_k$ the element w_j of $\mathbf{w}$ is the least maximum element of an ascending subsequence of $v_1 v_2 \cdots v_k$ of length j. In particular, the length of $\mathbf{w}$ is equal to the length of the longest ascending subsequence obtained to that time.

Of course there is a dual algorithm that allows one to keep track of the length of the longest descending subsequence. In this version an element is either prepended to the sequence being maintained (if smaller than all the elements of the sequence), or it bumps the immediately smaller element.

For the purposes of analysing some forms of the monotonic sequence game it will be useful to be able to combine these two algorithms into a single one. However, in doing so, we need to keep track of whether the elements in the single ordered sequence which we are maintaining represent elements of the ascending or descending type – that is, whether an element takes part in the sequence $\mathbf{w}$ of the original algorithm, the corresponding sequence $\mathbf{m}$ in the dual algorithm, or both.

Initially we will do this by marking the elements with overlines (if they belong to **w**), underlines (if they belong to **m**) or both (if both). Thus we maintain a single marked sequence which we shall call the *recording sequence*. The double bumping form of the combined algorithm can then be described as follows.

- Initially set the recording sequence to be empty, and process the elements of the permutation in order from left to right.
- Repeatedly, until the permutation is exhausted:
  - insert the next element of the permutation into the recording sequence with both an underline and an overline (maintaining the increasing order of the recording sequence);
  - delete the first overline if any to its right and the first underline if any to its left;
  - remove any naked elements (ones which no longer have an underline or an overline).

For example, when we process the permutation 514263 in this way we obtain

$$\underline{\overline{5}} \rightarrow \overline{1}\underline{5} \rightarrow \overline{1}\overline{4}\underline{5} \rightarrow \overline{1}\overline{2}\overline{4}\underline{5} \rightarrow \overline{1}\overline{2}\overline{4}\overline{6} \rightarrow \overline{1}\overline{2}\overline{3}\overline{4}\overline{6}$$

Frequently the precise identity of the elements of the recording sequences will not be important, but only their type (that is, what decoration they have). This remark will be exactly true when we deal with monotonic sequence games on $\mathbb{Q}$, and is still of some relevance in the case of monotonic sequence games on finite chains. For typographical purposes it is easier to record type sequences as colours rather than bars, and so we will also call them *colour sequences*. Specifically we associate the colour Blue with an underline, and Red with an overline. Elements having both underlines and overlines will be called Purple. An element is reddish if it is Red or Purple, and bluish if it is Blue or Purple. The process of the double bumping algorithm on the permutation above, purely in terms of colours is

$$P \rightarrow PB \rightarrow RPB \rightarrow RPBB \rightarrow RPBP \rightarrow RRPBB.$$

Of course the length of the colour sequence corresponding to a permutation is not more than the length of the permutation itself. Different permutations can easily have the same colour sequence (e.g. 231 and 213 both have colour sequence PP) and indeed permutations of different sizes can have the same colour sequence (e.g. 312 and 2143 both have colour sequence RPB).

It is clear that only some sequences of colours can occur as a result of applying the double bumping algorithm. We call such colour sequences *admissible*. In terms of colour, when we add a new element, we insert a Purple somewhere in the sequence and remove the red tinge from the first reddish element to the right (deleting it entirely if it were Red) and the blue tinge from the first bluish element to the left. In particular, a colour sequence can never begin with Blue nor end with Red. In fact we can completely characterise the admissible colour sequences. Recall that a *factor* of a sequence is a subword consisting of a block of consecutive elements from the sequence.

PROPOSITION 4. *The language of admissible colour sequences consists precisely of the empty sequence, together with those sequences which contain at least one P, do not begin with B nor end with R, and do not contain RB as a factor.*

PROOF. Necessity is relatively straightforward. Each insertion leaves a P so a nonempty admissible sequence must contain a P. Of the remaining conditions, the first two conditions are obviously preserved by any legitimate insertion. To see that the final condition is preserved as well consider an insertion which supposedly creates an RB factor. It could not create both the R and the B since only an insertion between those two elements could do that. Suppose, without loss of generality, that the newly created element was the B. Then previously that element was represented by a P. But in order to eliminate its reddish tinge, the insertion would have had to be after any preceding R, so we could not get the RB factor as claimed.

The proof of sufficiency is by induction. We show that if w is a nonempty sequence of the form described, then there is some parent word v also of the form described such that w can be obtained from v by the bumping algorithm. That this suffices is based on the observation that for *any* starting word u (admissible or not), after $ad + 1$ bumps the resulting word must contain at least $d + 1$ bluish or $a + 1$ reddish (red or purple elements). Thus the backwards chain of parents from w is bounded in length by the product of the number of bluish elements and the number of reddish elements in w, and can only terminate in the empty sequence which is admissible.

If $w = P$ the result is clear, so we may assume that the length of w is at least two. Suppose first that $w = Pu$. If $u = Bu'$ let $v = Pu'$ (which still has the form required) and note that v produces w by an insertion on the left hand side. If u begins with a P or an R let $v = Ru$ which is admissible and produces w by an insertion on the left hand side.

Now suppose that $w = R^i Pu$ with $i > 0$ and let $w' = Pu$. Then w' is admissible, and by the case just proven we can find v' which produces w' by an insertion into the first position. Let $v = R^i v'$. Then v produces w by insertion after the first block of R's.   □

The number of nonempty admissible words is enumerated by the sequence of alternate Fibonacci numbers

$$1, 3, 8, 21, 55, 144, \ldots$$

This is easily established by standard transfer matrix approaches or by the observation that the association

$$R \rightarrow 01 \quad B \rightarrow 10 \quad P \rightarrow 00$$

almost provides a bijection between admissible colour sequences and binary sequences of even length which contain no consecutive 1's.

4. Finite chains

In this section we assume throughout that the deck is a finite chain which, for convenience, we take to be

$$[n] = \{1, 2, \ldots, n\}$$

with the usual ordering. This was the basic situation investigated by Harary, Sagan and West in [6]. On the theoretical front we have relatively little to add to their results in this area, however, we have extended their computational results considerably.

PROPOSITION 5. *For fixed a and d both sequences*

$$W_{\text{nor}}(a, d, [n]) \quad and \quad W_{\text{mis}}(a, d, [n])$$

for $n = 1, 2, 3, \ldots$ are eventually constant.

PROOF. Since we know that a play of the game with parameters a and d cannot last more than $(a-1)(d-1) + 1$ moves, the existence of a winning strategy for either player, in either termination condition, can be expressed as a first order sentence in the language of linear orders. Consider, for example, the case of a first player win in normal play. In this case this sentence begins with an existential quantifier, followed by a long alternation of quantifiers representing the moves which might be chosen by the two players. These quantifiers are followed by a quantifier free formula expressing the condition "the first ascending sequence of length a or descending sequence of length d arising in this play occurred after a move made by the first player". The other cases are all similar.

However, it is well known that the theory of finite linear orders admits quantifier elimination (see [7], specifically sections 2.7 and A.6 and their exercises). In particular, any sentence in this language is either true in $([n], <)$ for all sufficiently large n or false in $([n], <)$ for all sufficiently large n. Since one of the statements "the game is of type $\mathcal{N}$" and "the game is of type $\mathcal{P}$" must be true for every $n > (a-1)(d-1)$, it must be the case that the same one is true for all sufficiently large n. $\square$

The proof above is a little unsatisfying from the standpoint of attempting to understand the structure of the monotonic sequence game played with a finite deck. By essentially recreating the quantifier elimination for the theory of finite linear orders but tailoring it to the situation at hand we can make it somewhat more concrete. As a side-effect we obtain improved bounds for the onset of the "long term behaviour" of such games.

Specifically, consider boards that arise in the play of the monotonic sequence game. Suppose that the colour sequence at this point $c_1 c_2 \cdots c_k$. There is an associated sequence of gap lengths $g_0, g_1, \cdots g_k$ where g_i is the number of cards remaining in the deck between the elements representing c_{i-1} and c_i. Note that this is not necessarily the same as the difference between these elements minus one, as some of the intervening elements may have been played earlier but no longer form part of the colour sequence.

The basic idea of the argument is to divide gaps into two categories *large* and *small*. All gaps whose length is larger than a certain number (which may depend on the colour sequence and the position of the gap relative to that sequence) will be considered large. We aim to show that if two boards have the same colour sequences and corresponding gaps are either both large or both small and of equal length then we can emulate the following play in one game within the other game and vice versa. This *Tweedledum–Tweedledee argument* then establishes that the two games have the same outcome type (and in fact the same nim-value or Grundy number). The first part of the argument must establish just what the bounds are for large gaps.

Imagine for the moment that the next play of the game will be a card from the deck that lies in some particular gap. Among the values in the board below this card there will be some maximal increasing sequence whose length, r, is the number of reddish elements lying below the gap. Likewise there is some maximal decreasing sequence on the board whose length, b is the number of bluish elements lying above the gap. Within this particular gap, the game will certainly end if we create an increasing sequence of length $a - r$ or a decreasing one of length $d - b$. That is, within the gap we are essentially playing a game with parameters $a - r$ and $d - b$ (the play within this gap may influence plays in other gaps, but only by reducing their associated parameters). Suppose that we temporarily let $B(x, y)$ denote some value which is "big enough" to define a large gap for parameters x and y. A play into such a gap leaves two gaps, a lower one with parameters x and $y - 1$ and an upper one with parameters $x - 1$ and y. Since we must ensure that we can match small gaps exactly and create corresponding large gaps it will be sufficient to have

$$B(x, y) \geq B(x - 1, y) + B(x, y - 1) + 1.$$

If we choose equality and note that we may take $B(x, 1) = B(1, y) = 1$ then simple algebraic manipulation shows that we may choose

$$B(x, y) = 2\binom{x + y - 2}{x - 1} - 1.$$

Henceforth we take this as the definition of $B(x, y)$ and hence of what constitutes a large gap.

PROPOSITION 6. *For fixed a and d, any two boards having the same colour sequence with the property that corresponding gaps are either both large, or otherwise equal have the same outcome type.*

PROOF. As promised, the proof is what is known as a Tweedledum–Tweedledee argument [1] or a back and forth argument [7]. The idea is that any move made in either position has one or more matching moves on the other position which preserve the equality of colour sequences and corresponding gaps. Specifically, a move in a small gap is mirrored by the obvious corresponding move of the other position. A move in a large gap leaves either large gaps on either side or one small gap and one large gap. In either case there is a corresponding move in the other position leaving two large gaps, or one small gap (of the same size) and a large gap.

Suppose, for the sake of argument, that the first position has a second player winning strategy. We devise a second player winning strategy in the second position as follows. Given a move in the second game to which we must reply, we consider a matching move in the first game. Our strategy there will dictate a certain response to this move. We make the matching response in the second game. Proceeding in this way, we cannot fail to win in the second game (in fact we will win in precisely the same number of moves as we win the matched sequence of plays in the first game). All the other cases are very similar. □

In particular, any two games beginning with an empty board and having decks of size $2\binom{a+d-2}{a-2} - 1$ or larger must have the same outcome type. As indicated by the computations below, this bound appears to be somewhat extravagant, though not as much so as the naïve bound arising from a direct translation of the quantifier elimination for the theory of finite linear orders which would be $2^{(a-1)(d-1)} - 1$.

4.1. Computational results: normal play.

We will assume throughout that $a \geq d$ because the outcome type for parameters (a, d) is the same as that for parameters (d, a). We begin by recapitulating results from [6] recast into our notation.

If $d = 2$ then any move other than the smallest remaining element at that time gives your opponent a "win in one". So the outcome type is determined by the parity of a and we have

$$W_{nor}(a, 2, n) = \begin{cases} \mathcal{D} & \text{if } n < a, \\ \mathcal{N} & \text{if } a \leq n \text{ is odd,} \\ \mathcal{P} & \text{if } a \leq n \text{ is even.} \end{cases}$$

For $d = 3$ then, depending on parity, the first player can choose to play either the largest or second largest element of the deck as his first move. This more or less reduces the game to the $d = 2$ case, and provided that $n > a$ and a is even, or $n > a + 1$ and a is odd $W_{nor}(a, 3, n) = \mathcal{N}$, with the remaining cases being drawn.

Finally, [6] showed that $W_{nor}(4, 4, n) = \mathcal{N}$ for $n \geq 9$. A winning strategy is to play near the middle, and to ensure after your second move that all remaining moves must be the smallest or largest remaining element.

We implemented a straightforward game tree traversal algorithm to determine the outcome type of the empty board for various combinations of the parameters (a, d, n). Although the observations made in the proof of Proposition 6 could improve the efficiency of this algorithm (by storage and reuse of previously computed outcomes for equal or equivalent colour and gap sequences) such time improvement would come at significant cost in space, and complexity of the underlying code. Since we could extend the results of [6] considerably using just the raw improvement in computing power between 1983 and now, we did not choose to pursue these improvements. Our program permitted computations with deck sizes up to 20 in a few minutes on a standard desktop machine. Note that whenever a type $\mathcal{P}$ position is found, two other positions are immediately known to be of type $\mathcal{N}$, namely

$$W_{nor}(a, d, n) = \mathcal{P} \Rightarrow$$
$$W_{nor}(a + 1, d, n + 1) = \mathcal{N} \text{ and } W_{nor}(a, d + 1, n + 1) = \mathcal{N},$$

since the first player can reduce the game to the preceding case by playing the smallest (respectively largest) element as his first move.

We give our new computational results in the following form: first we specify the smallest nondrawn game of that type and its winner; then a sequence of values until we (appear) to reach an eventually constant block. Thus, the first line below means that $W_{nor}(5, 4, n) = \mathcal{D}$ for $n \leq 10$, and $W_{nor}(5, 4, n) = \mathcal{N}$ for $11 \leq n \leq 20$.

$$(5, 4, 11) \in \mathcal{N}$$
$$(6, 4, 14), (6, 4, 15) \in \mathcal{P}, \ (6, 4, 16) \in \mathcal{N}$$
$$(5, 5, 15) \in \mathcal{N}$$
$$(7, 4, 15), (7, 4, 16) \in \mathcal{N}, \ (7, 4, 17) \in \mathcal{P}, \ (7, 4, 18) \in \mathcal{N}.$$

4.2. Computational results: misère play. We also computed results for misère play. In this case it appears to be true that the game is drawn much less frequently, and so the data include some more interesting observations. In this case, the table below lists the sequence of outcome types for the various combinations of parameters a and d with n ranging from 1 through 20.

a	d	Misère winner			
3	3	$DDDNN$	$NNNNN$	$NNNNN$	$NNNNN$
4	3	$DDDDP$	$NNNNN$	$NNNNN$	$NNNNN$
5	3	$DDDDD$	$NPPPN$	$NNNNN$	$NNNNN$
6	3	$DDDDD$	$DDNNN$	$NNNNN$	$NNNNN$
7	3	$DDDDD$	$DDDPN$	$NNNNN$	$NNNNN$
8	3	$DDDDD$	$DDDDN$	$PPNNN$	$NNNNN$
9	3	$DDDDD$	$DDDDD$	$DNNNN$	$NNNNN$
4	4	$DDDDD$	$NPNNN$	$NNNNN$	$NNNNN$
5	4	$DDDDD$	$DDNNN$	$PNNNN$	$NNNNN$
6	4	$DDDDD$	$DDDDN$	$PNPPN$	$NNNNN$
7	4	$DDDDD$	$DDDDN$	$PNNNN$	$NNNNN$
8	4	$DDDDD$	$DDDDD$	$DNNND$	$NNNNP$
9	4	$DDDDD$	$DDDDD$	$DDDNP$	$NPNND$
5	5	$DDDDD$	$DDDDD$	$DNPNP$	$NNNNN$
6	5	$DDDDD$	$DDDDD$	$DNPNN$	$NPNNN$
7	5	$DDDDD$	$DDDDD$	$DNDNP$	$NPNNP$

Most of the blocks of trailing N's do seem to represent long run behaviour. The evidence supporting this is that the smallest winning first move is also constant across these blocks.

The cases $a = 8, 9$, $d = 4$ seem particularly interesting. First of all, with $a = 8$ there is the interposed D at $n = 15$. Thus, with a 14 or 16 card deck the first player can force the second player to make an ascending sequence of size 8 or a descending one of size 4 but with a 15 card deck he cannot! A further oddity of this sequence concerns the fact that for $a = 9$, $d = 4$, the *second* player wins $n = 15$. This means that the second player can force the first to create an ascending sequence of length 9 or a descending one of length 4 in a 15 card deck, but can't force an ascending sequence of length 8 or a descending sequence of length 4 in the same deck. Why can't the first player simply follow an "at least draw" strategy from the latter case to get the same result in the former case? Because there is a hidden assumption in this strategy – that the second player will never create an ascending sequence of length 8 or a descending sequence of length 4 either.

4.3. Computation: further remarks. As noted above the program used to obtain these results was exceedingly straightforward. Essentially, every response to every move was examined from lowest to highest. Only when a response of type $\mathcal{P}$ was found (permitting the current board to be labelled as $\mathcal{N}$) was any pruning done. Likewise, no heuristic choices of responses were considered. This alone would probably improve the efficiency of the program considerably since it was observed that in many cases if y was a good countermove to first move x (and was quite different from x) then it was also a good countermove to $x + 1$. Secondly, storage and reuse of previously computed results, or some form of "orderly" generation based on Proposition 6 would permit even more pruning. For example, the first three moves $10, 5, 20$ and $10, 20, 5$ result in identical colour sequences and gaps, so have the same outcome type.

However, beyond some obvious observations and conjectures which we propose in the final section, our opinion is that the data (particularly for the misère version) suggest rather "noisy" behaviour for small values of n. So, the benefits of pursuing these optimisations seems rather limited.

5. Dense linear order

We now consider playing the monotonic sequence game with $\mathbb{Q}$ (or any other dense linear order without endpoints) as the deck.

PROPOSITION 7. *For any $a, d \geq 1$, $W_{\text{nor}}(a, d, \mathbb{Q}) = W_{\text{mis}}(a - 1, d - 1, \mathbb{Q})$.*

PROOF. In order to win the normal game, you cannot ever create an ascending chain of length $a - 1$ or a descending chain of length $d - 1$ since your opponent would then have the opportunity to win immediately. Conversely, if your opponent creates such a sequence on the board then you can win immediately. So the outcome of the misère $(a - 1, d - 1, \mathbb{Q})$ game is the same as that of the normal $(a, d, \mathbb{Q})$. □

We note that the proposition above requires only that the deck not have a maximal or minimal element. Owing to this proposition we restrict our attention to the normal form of the game.

The outcome type of a particular board depends only on the relative ordering among the elements currently on the board. This is clear, since with two boards having the same relative ordering among their elements, there is an order preserving bijection from $\mathbb{Q}$ to itself which maps one board to the other. Any strategy which applies to the first board, then also applies to the second by taking its image under this bijection. However, in fact all that we need to know in order to determine the outcome of a game is the colour sequence of the board. As noted previously, different boards and even boards of different sizes can have the same colour sequence.

PROPOSITION 8. *For any $a, d \geq 1$, and playing with $\mathbb{Q}$ as a deck, the outcome type of a particular board is determined by the colour sequence of that board.*

PROOF. First we observe that the colour sequence of the board is sufficient to determine whether or not the game has ended since the length of the longest ascending (descending) sequence on the board is equal to the number of reddish (bluish) elements of its colour sequence.

Next we note that given the colour sequence of a board, the possible colour sequences which can be obtained by making a single move are determined. Any move involves the insertion of a P somewhere in the existing colour sequence, and then "first higher red reduction" and "first lower blue reduction". Moreover, because the deck is dense, any such insertion *can* be made.

So, in terms of determining the outcome, we need only know the colour sequence of the current board, exactly as claimed. □

In considering the basic form of the monotonic sequence game with parameters $(a, d, \mathbb{Q})$ we will work almost exclusively with the colour sequences. We define the *children* of a colour sequence to be all those sequences that can be obtained from it in a single move. A colour sequence is *terminal* if it contains a reddish, or d bluish elements.

As before, we will assume that $a \geq d$ and for a few values of d we are able to determine the type of the general game with parameters $(a, d, \mathbb{Q})$.

THEOREM 9. *For $d \leq 5$ the types of the monotonic sequence games with parameters $(a, d, \mathbb{Q})$ are as follows:*

(i) *For $a \geq 2$, $W_{nor}(a, 2, \mathbb{Q}) = \mathcal{P}$.*
(ii) *For $a \geq 3$, $W_{nor}(a, 3, \mathbb{Q}) = \mathcal{N}$ precisely when a is odd.*
(iii) *For $a \geq 4$, $W_{nor}(a, 4, \mathbb{Q}) = \mathcal{N}$.*
(iv) *For $a \geq 5$, $W_{nor}(a, 5, \mathbb{Q}) = \mathcal{N}$.*

PROOF. Throughout the argument we consider an equivalent version of the monotonic sequence game with parameters $(a, d, \mathbb{Q})$. In this version, a suicidal move i.e. one which creates an ascending sequence of length $a - 1$ or a descending sequence of length $d - 1$ on the board is forbidden, unless forced. Since the player with a winning strategy in the original game will never make a suicidal move, and the other player may choose not to so until forced, the outcome type of the modified game is the same as that of the original.

For parameters $(a, 2, \mathbb{Q})$ the game is truly trivial, since the very first move is suicidal.

For the parameters $(a, 3, \mathbb{Q})$, any move below an element already played is suicidal. So, in the modified form, the two players alternately add to an increasing sequence, and clearly the first player wins only if a is odd.

Now consider parameter sequences of the form $(a, 4, \mathbb{Q})$. We will show that the set of colour sequences representing nonterminal $\mathcal{P}$-positions in this game is

$$P_4 = \{P, R^{a-3} PB\} \cup \{R^i P^2 \mid 0 \le i \le a - 5\}.$$

To establish this result we must show that for any position which can arise in the play of $(a, 4, \mathbb{Q})$, if it is not in P_4 then it has a child which is in P_4 or a terminal position, and if it is in P_4 then there is no such child. The second part is easily checked.

Suppose that we have a colour sequence w which is not terminal and not in P_4. If it has three or more bluish elements, then it has a terminal child. Suppose that w has exactly one bluish element. Then it is of the form $R^i P$ for some $0 < i \le a - 2$. If $i = a - 2$ it has a terminal child. If $i < a - 2$ then an insertion just before the last R yields $R^{i-1} P^2$ which is in P_4 unless $i = a - 3$. In that case, inserting before the P yields $R^{a-3} PB$.

Next consider the case where w has two bluish elements, both purple. Ignoring positions with terminal children, it must be of the form $R^i PR^j P$ where either $j > 0$ or $i = a - 4$. If $j > 0$ inserting before the last R yields $R^{i+j} P^2$ while inserting before the last P yields $R^{i+j+1} PB$ and one of these two is in P_4. If $j = 0$ and $i = a - 4$ then inserting between the two P's yields $R^{a-3} PB$.

Finally consider the case of one purple and one blue element. Then w is $R^i PB$ for some i. If $i \le a - 5$ then moving at the right hand end produces $R^i P^2$, while if $i = a - 4$, moving just after the P produces $R^{a-3} PB$.

Thus for the parameters $(a, 4, \mathbb{Q})$ we have established that $P \in \mathcal{P}$ and hence the initial position is in $\mathcal{N}$.

We give a similar argument for the parameter sequences of the form $(a, 5, \mathbb{Q})$. In this case though we do not provide an exhaustive listing of the type $\mathcal{P}$ nonterminal colour sequences, but only a sufficient set of these. By this we mean that we provide a set P_5 of colour sequences, and an argument that the following conditions hold:

- $P \in P_5$;
- if $w \in P_5$ and v is a child of w, then v has a child which is either terminal or in P_5;
- no $w \in P_5$ has a terminal child.

This establishes that $P \in \mathcal{P}$, since from any position not in P_5 the player whose turn it is to move can simply take either the immediate win, or the move guaranteed by the second of the conditions above. We take

$$P_5 = \{P, RPB\} \cup \{R^i PRPB : 0 \le i \le a - 6\}$$
$$\cup \{R^i RPBP : 0 \le i \le a - 6\} \cup \{R^{a-5} P^3, R^{a-3} PB^2\}.$$

The first part of the verification is routine. From the initial position P the second player can ensure that after her play, the resulting code will be RPB by always replying "in the second position". Likewise from RPB she can always guarantee that her opponent's next move will be from one of $PRPB$ or $RPBP$.

Now suppose that $0 \leq i \leq a - 7$ and that a single move has been made from $R^i PRPB$ or $R^i RPBP$. If this move occurs below the first P it creates a descent of length 4 and can be countered by an immediate win (i.e. it is suicidal). In all of the remaining cases there is a counter move to one of $R^{i+1} PRPB$ or $R^{i+1} RPBP$.

If a single move has been made from $R^{a-6} PRPB$ or $R^{a-6} RPBP$ which is nonsuicidal, then again there are only a few positions near the end of the colour sequence that need to be examined, and each of these allows a response to either $R^{a-5} P^3$ or $R^{a-3} PB^2$.

The final cases to consider are moves from $R^{a-5} P^3$. There are only two nonsuicidal moves and they both permit replies to $R^{a-3} PB^2$. $\qquad \square$

We have strong experimental evidence that the monotonic sequence game with parameters $(a, d, \mathbb{Q})$ and $a, d \geq 4$ *always* has type $\mathcal{N}$. Computation has established this result for $4 \leq d \leq 8$ and any a with $d \leq a \leq 16$. We can establish this result for the symmetrical form of the game:

THEOREM 10. *Let* $a \geq 4$. *The monotonic sequence game with parameters* $(a, a, \mathbb{Q})$ *has type* $\mathcal{N}$.

PROOF. The argument we provide uses a form of *strategy stealing* together with symmetry. That is, we show that if the second player had a winning strategy then the first player could appropriate it for his own use. This contradiction implies that it must be the first player who has a winning strategy.

If the result were false then the type of the colour sequence P would have to be $\mathcal{N}$. As the moves from P to RP and PB are symmetrical (under order reversal) both these positions would have to be of type $\mathcal{P}$.

In particular the two children PP and RPB of RP would both lie in $\mathcal{N}$. The children of PP are PBP, RPB and PRP. By assumption, $RPB \in \mathcal{N}$. By symmetry PBP and PRP have the same type, so these two positions would have type $\mathcal{P}$. The children of PRP and PBP would all be of type $\mathcal{N}$. These include the positions

$$RPP, \ RRPB, \ RPB^2, \ PPB.$$

However, these are *all* the children of RPB, so RPB must be of type $\mathcal{P}$, contradicting our assumption. $\qquad \square$

Finally, for this section, we consider an extended form of the monotonic sequence game when the deck is $\mathbb{Q}$. In this extension, a chosen card can be inserted anywhere in the board, in other words you are allowed to choose the position as well as the value of the next element to insert in the sequence. A useful model of this game is that the players alternately choose points in the open unit square (or the plane, but using the square saves paper) subject to the condition that no two chosen points can lie on a vertical or horizontal line. The game ends when there are either a points such that the segments connecting them all have positive slope, or d such that the segments connecting them all have negative slope. We refer to such sequences of points as increasing or decreasing respectively.

This extra power reduces the analysis of the game to a simple parity argument owing to the following lemma:

LEMMA 11. *Let a set of fewer than rs points in the open unit square be given no two of which lie on a horizontal or vertical line. If the longest increasing sequence of points has length at most r and the longest decreasing sequence of points has length at most s then it is possible to add an additional point without creating a sequence of $r + 1$ increasing or $s + 1$ decreasing points.*

PROOF. View the points as a permutation. To avoid trivialities, suppose that there is indeed an increasing subsequence of length r and a decreasing subsequence of length s. Under these conditions, it is well known that the permutation has a decomposition into s disjoint increasing subsequences, I_1 through I_s, each of length at most r which can be obtained by a simple greedy algorithm. Since the number of elements of the permutation is less than rs, one of these subsequences, without loss of generality I_1, will contain at most $r - 1$ points. Now consider a decomposition of the permutation into r disjoint decreasing subsequences D_1 through D_r each of length at most s (which can also be obtained by a greedy algorithm). Since for each i and j, $|D_i \cap I_j| \le 1$ any of the D_i of size s must intersect each I_j. However, some D_i has empty intersection with I_1 (since there are r D's and at most $r - 1$ points in I_1). Without loss of generality, suppose it is D_1 and note that necessarily $|D_1| < s$.

Now return to thinking of the elements of the permutation as points in the square. It is possible to find a point (x, y) whose addition to D_1 forms a decreasing sequence, and whose addition to I_1 forms an increasing sequence. Such a point can be obtained by "connecting the dots" for D_1, and connecting the ends horizontally to the sides of the square. Do likewise for I_1 only connect the ends vertically. The resulting two paths have a point P in common. Suppose that P lies in a vertical or horizontal line determined by any of the finitely many points in the set. In that case, it is possible to perturb P slightly, so that this is no longer true and so that P's addition to D_1 forms a decreasing sequence, and its addition to I_1 forms an increasing one, without otherwise changing P's relative horizontal or vertical position with respect to the elements of the set. The point P thus satisfies the lemma since its addition still permits the partitioning of the set of points into r decreasing sequences of size at most s, and s increasing sequences of size at most r.   □

In terms of the extended monotonic sequence game with parameters $(a, d, \mathbb{Q})$ the lemma above implies that for the first $(a-2)(d-2)$ moves neither player can be forced to play suicidally. However, at this point, by the Erdős–Szekeres theorem the next move is necessarily suicidal. Since the parity of ad is the same as that of $(a-2)(d-2)$ we obtain:

THEOREM 12. *The extended monotonic sequence game with parameters*

$$(a, d, \mathbb{Q})$$

has type $\mathcal{N}$ if ad is odd, and type $\mathcal{P}$ if ad is even.

6. Observations and open problems

It appears that the monotonic sequence game, particularly with normal termination criteria, has a fairly strong bias towards the first player. Specifically, our computational results suggest the following pair of conjectures:

- For any $a \geq d \geq 3$ and all sufficiently large n, $W_{\text{nor}}(a, d, n) = W_{\text{mis}}(a, d, n) = \mathcal{N}$.
- For any $a \geq d \geq 3$, $W_{\text{nor}}(a, d, \mathbb{Q}) = \mathcal{N}$.

We would be surprised (assuming the correctness of these conjectures) if similar results did not also hold for other infinite linear orders (not models of the theory of almost all finite partial orders) such as $\mathbb{N}$ or $\mathbb{Z}$.

In the finite form of the game it appears that the last $\mathcal{D}$ occurring in the sequence $W_{\text{mis}}(a, d, n)$ is generally closer to position $a + d$ than to position $(a-1)(d-1)$. It would be of interest to determine a good upper bound for the position of this last $\mathcal{D}$ (the same observation and question applies to the sequence $W_{\text{nor}}(a, d, n)$ though the computational evidence is less compelling). Likewise, the "long run behaviour" of these games seems to become established well before the bound obtained using the argument of Proposition 6. That the trailing sequences of $\mathcal{N}$'s observed in the computational results do generally represent long run behaviour is supported by a more detailed examination of these positions showing that there is a large central block of equivalent moves, which extends by a single element each time the deck size is increased (extensions to CGSUITE [10] were used for some of these computations).

Another area of interest to investigate would be the behaviour of the extended form of the game played with a finite deck. In this form, players take turn naming pairs (i, π_i) subject to the constraint that the chosen values form part of the graph of some permutation of $\{1, 2, \ldots, n\}$ (and with termination based on increasing or decreasing sequences as normally). An equivalent formulation has the players placing nonattacking rooks on a (generalised) chessboard.

References

[1] Elwyn R. Berlekamp, John H. Conway, and Richard K. Guy. *Winning ways for your mathematical plays. Vol. 1.* A K Peters Ltd., Natick, MA, second edition, 2001.

[2] Elwyn R. Berlekamp, John H. Conway, and Richard K. Guy. *Winning ways for your mathematical plays. Vol. 2.* A K Peters Ltd., Natick, MA, second edition, 2003.

[3] Elwyn R. Berlekamp, John H. Conway, and Richard K. Guy. *Winning ways for your mathematical plays. Vol. 3.* A K Peters Ltd., Natick, MA, second edition, 2003.

[4] Elwyn R. Berlekamp, John H. Conway, and Richard K. Guy. *Winning ways for your mathematical plays. Vol. 4.* A K Peters Ltd., Wellesley, MA, second edition, 2004.

[5] Paul Erdős and George Szekeres. A combinatorial problem in geometry. *Compositio Mathematica*, 2:464–470, 1935.

[6] Frank Harary, Bruce Sagan, and David West. Computer-aided analysis of monotonic sequence games. *Atti Accad. Peloritana Pericolanti Cl. Sci. Fis. Mat. Natur.*, 61:67–78, 1983.

[7] Wilfrid Hodges. *Model theory*, volume 42 of *Encyclopedia of Mathematics and its Applications*. Cambridge University Press, Cambridge, 1993.

[8] Bruce E. Sagan. *The symmetric group*, volume 203 of *Graduate Texts in Mathematics*. Springer, New York, second edition, 2001. Representations, combinatorial algorithms, and symmetric functions.

[9] C. Schensted. Longest increasing and decreasing subsequences. *Canad. J. Math.*, 13:179–191, 1961.

[10] Aaron Siegel. Combinatorial game suite. http://cgsuite.sourceforge.net/.

M. H. ALBERT
DEPARTMENT OF COMPUTER SCIENCE
UNIVERSITY OF OTAGO
NEW ZEALAND
    malbert@cs.otago.ac.nz

R. E. L. ALDRED
DEPARTMENT OF MATHEMATICS AND STATISTICS
UNIVERSITY OF OTAGO
NEW ZEALAND

M. D. ATKINSON
DEPARTMENT OF COMPUTER SCIENCE
UNIVERSITY OF OTAGO
NEW ZEALAND

C. C. HANDLEY
DEPARTMENT OF COMPUTER SCIENCE
UNIVERSITY OF OTAGO
NEW ZEALAND

D. A. HOLTON
DEPARTMENT OF MATHEMATICS AND STATISTICS
UNIVERSITY OF OTAGO
NEW ZEALAND

D. J. MCCAUGHAN
DEPARTMENT OF MATHEMATICS AND STATISTICS
UNIVERSITY OF OTAGO
NEW ZEALAND

B. E. SAGAN
DEPARTMENT OF MATHEMATICS
MICHIGAN STATE UNIVERSITY
UNITED STATES

Games of No Chance 3
MSRI Publications
Volume **56**, 2009

The game of End-Wythoff

AVIEZRI S. FRAENKEL AND ELNATAN REISNER

ABSTRACT. Given a vector of finitely many piles of finitely many tokens. In End-Wythoff, two players alternate in taking a positive number of tokens from either end-pile, or taking the *same* positive number of tokens from both ends. The player first unable to move loses and the opponent wins. We characterize the P-positions (a_i, K, b_i) of the game for any vector K of middle piles, where a_i, b_i denote the sizes of the end-piles. A more succinct characterization can be made in the special case where K is a vector such that, for some $n \in \mathbb{Z}_{\geq 0}$, (K, n) and (n, K) are both P-positions. For this case the (noisy) initial behavior of the P-positions is described precisely. Beyond the initial behavior, we have $b_i - a_i = i$, as in the normal 2-pile Wythoff game.

1. Introduction

A position in the (impartial) game *End-Nim* is a vector of finitely many piles of finitely many tokens. Two players alternate in taking a positive number of tokens from either end-pile ("burning-the-candle-at-both-ends"). The player first unable to move loses and the opponent wins. Albert and Nowakowski [1] gave a winning strategy for End-Nim, by describing the P-positions of the game. (Their paper also includes a winning strategy for the partizan version of End-Nim.)

Wythoff's game [8] is played on two piles of finitely many tokens. Two players alternate in taking a positive number of tokens from a *single* pile, or taking the *same* positive number of tokens from both piles. The player first unable to move loses and the opponent wins. From among the many papers on this game, we mention just three: [2], [7], [3]. The P-positions (a'_i, b'_i) with $a'_i \leq b'_i$ of Wythoff's game have the property: $b'_i - a'_i = i$ for all $i \geq 0$.

Keywords: combinatorial games, Wythoff's game, End-Wythoff's game, P-positions.

Richard Nowakowski suggested to one of us (F) the game of *End-Wythoff*, whose positions are the same as those of End-Nim but with Wythoff-like moves allowed. Two players alternate in taking a positive number of tokens from either end-pile, or taking the *same* positive number of tokens from both ends. The player first unable to move loses and the opponent wins.

In this paper we characterize the P-positions of End-Wythoff. Specifically, in Theorem 1 the P-positions (a_i, K, b_i) are given recursively for any vector of piles K.

The rest of the paper deals with values of K, deemed *special*, such that (n, K) and (K, n) are both P-positions for some $n \in \mathbb{Z}_{\geq 0}$. Theorem 3 gives a slightly cleaner recursive characterization than in the general case. In Theorems 4 and 5, the (noisy) initial behavior of the P-positions is described, and Theorem 6 shows that after the initial noisy behavior, we have $b_i - a_i = i$ as in the normal Wythoff game. Before all of that we show in Theorem 2 that if K is a P-position of End-Wythoff, then (a, K, b) is a P-position if and only if (a, b) is a P-position of Wythoff.

Finally, in Section 4, a polynomial algorithm is given for finding the P-positions (a_i, K, b_i) for any given vector of piles K.

2. P-positions for general End-Wythoff games

DEFINITION 1. A *position* in the game of End-Wythoff is the empty game, which we denote by (0), or an element of $\bigcup_{i=1}^{\infty} \mathbb{Z}_{\geq 1}^i$, where we consider mirror images identical; that is, $(n_1, n_2, \ldots, n_k)$ and $(n_k, n_{k-1}, \ldots, n_1)$ are the same position.

NOTATION 1. For convenience of notation, we allow ourselves to insert extraneous 0s when writing a position. For example, $(0, K)$, $(K, 0)$, and $(0, K, 0)$ are all equivalent to K.

LEMMA 1. *Given any position K, there exist unique $l_K, r_K \in \mathbb{Z}_{\geq 0}$ such that (l_K, K) and (K, r_K) are P-positions.*

PROOF. We phrase the proof for l_K, but the arguments hold symmetrically for r_K.

Uniqueness is fairly obvious: if (n, K) is a P-position and $m \neq n$, then (m, K) is not a P-position because we can move from one to the other.

For existence, if $K = (0)$, then $l_K = r_K = 0$, since the empty game is a P-position. Otherwise, let t be the size of the rightmost pile of K. If any of $(0, K), (1, K), \ldots, (2t, K)$ are P-positions, we are done. Otherwise, they are all N-positions. In this latter case, the moves that take $(1, K), (2, K), \ldots, (2t, K)$ to P-positions must all involve the rightmost pile. (That is, none of these moves take tokens only from the leftmost pile. Note that we cannot make this guarantee for $(0, K)$ because, for example, $(0, 2, 2) = (2, 2)$ can reach the P-position $(1, 2)$ by taking only from the leftmost pile.)

In general, if L is a position and $m < n$, then it cannot be that the same move takes both (m, L) and (n, L) to P-positions: if a move takes (m, L) to a P-position (m', L'), then that move takes (n, L) to $(m' + n - m, L')$, which is an N-position because we can move to (m', L').

In our case, however, there are only $2t$ possible moves that involve the right-most pile: for $1 \leq i \leq t$, take i from the rightmost pile, or take i from both end-piles. We conclude that each of these moves takes one of $(1, K)$, $(2, K)$, $\ldots$, $(2t, K)$ to a P-position, so no move involving the rightmost pile can take $(2t + 1, K)$ to a P-position. But also, no move that takes only from the left-most pile takes $(2t + 1, K)$ to a P-position because (n, K) is an N-position for $n < 2t + 1$. Thus $(2t + 1, K)$ cannot reach any P-position in one move, so it is a P-position, and $l_K = 2t + 1$. $\qquad \square$

We now state some definitions which will enable us to characterize P-positions as pairs at the 2 ends of a given vector K. For any subset $S \subset \mathbb{Z}_{\geq 0}$, $S \neq \mathbb{Z}_{\geq 0}$, let mex $S = \min(\mathbb{Z}_{\geq 0} \setminus S) =$ least nonnegative integer not in S.

DEFINITION 2. Let K be a position of End-Wythoff, and let $l = l_K$ and $r = r_K$ be as in Lemma 1. For $n \in \mathbb{Z}_{\geq 1}$, define

$$d_n = b_n - a_n$$
$$A_n = \{0, l\} \cup \{a_i : 1 \leq i \leq n - 1\}$$
$$B_n = \{0, r\} \cup \{b_i : 1 \leq i \leq n - 1\}$$
$$D_n = \{-l, r\} \cup \{d_i : 1 \leq i \leq n - 1\},$$

where

$$a_n = \text{mex } A_n \tag{1}$$

and b_n is the smallest number $x \in \mathbb{Z}_{\geq 1}$ satisfying both

$$x \notin B_n, \tag{2}$$
$$x - a_n \notin D_n. \tag{3}$$

Finally, let

$$A = \bigcup_{i=1}^{\infty} a_i \quad \text{and} \quad B = \bigcup_{i=1}^{\infty} b_i.$$

Note that the definitions of A and B ultimately depend only on the values of l and r. Thus, Theorem 1 below shows that if K and L are positions with $l_K = l_L$ and $r_K = r_L$, then the pairs (a_i, b_i) that form P-positions when placed as end-piles around them will be the same.

THEOREM 1.

$$P_K = \bigcup_{i=1}^{\infty} (a_i, K, b_i)$$

is the set of P-positions of the form (a, K, b) with $a, b \in \mathbb{Z}_{\geq 1}$.

i	$K = (1,2)$ $l = 0$ | 0 $r = 0$ | 0			$K = (1,3)$ $l = 4$ | -4 $r = 1$ | 1			$K = (2,3)$ $l = 5$ | -5 $r = 3$ | 3			$K = (1,2,2)$ $l = 1$ | -1 $r = 1$ | 1		
	a_i	b_i	d_i	a_i	b_i	d_i	a_i	b_i	d_i	a_i	b_i	d_i
1	1	2	1	1	3	2	1	1	0	2	2	0
2	2	1	-1	2	2	0	2	4	2	3	5	2
3	3	5	2	3	6	3	3	2	-1	4	7	3
4	4	7	3	5	4	-1	4	5	1	5	3	-2
5	5	3	-2	6	10	4	6	10	4	6	10	4
6	6	10	4	7	5	-2	7	12	5	7	4	-3
7	7	4	-3	8	13	5	8	6	-2	8	13	5
8	8	13	5	9	15	6	9	15	6	9	15	6
9	9	15	6	10	7	-3	10	7	-3	10	6	-4
10	10	6	-4	11	18	7	11	18	7	11	18	7
11	11	18	7	12	20	8	12	8	-4	12	20	8
12	12	20	8	13	8	-5	13	21	8	13	8	-5
13	13	8	-5	14	23	9	14	23	9	14	23	9
14	14	23	9	15	9	-6	15	9	-6	15	9	-6
15	15	9	-6	16	26	10	16	26	10	16	26	10

Table 1. The first 15 outer piles of P-positions for some values of K.

PROOF. Since moves are not allowed to alter the central piles of a position, any move from (a, K, b) with $a, b > 0$ will result in (c, K, d) with $c, d \geq 0$. Since $(l, K) = (l, K, 0)$ and $(K, r) = (0, K, r)$ are P-positions, they are the only P-positions with $c = 0$ or $d = 0$. Thus, to prove that P_K is the set of P-positions of the desired form, we must show that, from a position in P_K, one cannot reach (l, K), (K, r), or any position in P_K in a single move, and we must also show that from any $(a, K, b) \notin P_K$ with $a, b > 0$ there is a single move to at least one of these positions.

We begin by noting several facts about the sequences A and B.

(a) We see from (1) that $a_{n+1} = \begin{cases} a_n + 1, & \text{if } a_n + 1 \neq l \\ a_n + 2, & \text{if } a_n + 1 = l \end{cases}$ for $n \geq 1$, so A is strictly increasing.

(b) We can also conclude from (1) that $A = \mathbb{Z}_{\geq 1} \setminus \{l\}$.

(c) It follows from (2) that all elements in B are distinct. The same conclusion holds for A from (1).

We show first that from $(a_m, K, b_m) \in P_K$ one cannot reach any element of P_K in one move:

(i) $(a_m - t, K, b_m) = (a_n, K, b_n) \in P_K$ for some $0 < t \leq a_m$. Then $m \neq n$ but $b_m = b_n$, contradicting (c).

(ii) $(a_m, K, b_m - t) = (a_n, K, b_n) \in P_K$ for some $0 < t \leq b_m$. This implies that $a_m = a_n$, again contradicting (c).

(iii) $(a_m - t, K, b_m - t) = (a_n, K, b_n) \in P_K$ for some $0 < t \leq a_m$. Then $b_n - a_n = b_m - a_m$, contradicting (3).

It is a simple exercise to check that $(a_m, K, b_m) \in P_K$ cannot reach (l, K) or (K, r).

Now we prove that from $(a, K, b) \notin P_K$ with $a, b > 0$, there is a single move to (l, K), (K, r), or some $(a_n, K, b_n) \in P_K$.

If $a = l$, we can take all of the right-hand pile and reach (l, K). Similarly, if $b = r$, we can move to (K, r) by taking the left-hand pile.

Now assume $a \neq l$ and $b \neq r$. We know from (b) that $a \in A$, so let $a = a_n$. If $b > b_n$, then we can move to (a_n, K, b_n). Otherwise, $b < b_n$, so b must violate either (2) or (3).

If $b \in B_n$, then $b = b_m$ with $m < n$ (because $b \neq r$ and $b > 0$). Since $a_m < a_n$ by (a), we can move to (a_m, K, b_m) by drawing from the left pile.

If, on the other hand, $b - a_n \in D_n$, then there are three possibilities: if $b - a_n = b_m - a_m$ for some $m < n$, then we can move to (a_m, K, b_m) by taking $b - b_m = a_n - a_m > 0$ from both end-piles; if $b - a_n = -l$, then drawing $b = a_n - l$ from both sides puts us in (l, K); and if $b - a_n = r$, then taking $a_n = b - r$ from both sides leaves us with (K, r). $\qquad\square$

3. P-positions for special positions

Examining Table 1 reveals a peculiarity that occurs when $l = r$.

DEFINITION 3. A position K is *special* if $l_K = r_K$.

In such cases, if (a_i, b_i) occurs in a column, then (b_i, a_i) also appears in that column. Examples of special K are P-positions, where $l = r = 0$, and palindromes, where (l, K) is the unique P-position of the form (a, K), but $(K, r) = (r, K)$ is also a P-position, so $l = r$. However, other values of K can also be special. We saw $(1, 2)$—a P-position—and $(1, 2, 2)$ in Table 1; other examples are $(4, 1, 13)$, $(7, 5, 15)$, and $(3, 1, 4, 10)$, to name a few.

We begin with the special case $l_K = r_K = 0$.

THEOREM 2. *Let K be a P-position of End-Wythoff. Then (a, K, b) is a P-position of End-Wythoff if and only if (a, b) is a P-position of Wythoff.*

PROOF. Induction on $a + b$, where the base $a = b = 0$ is obvious. Suppose the assertion holds for $a + b < t$, where $t \in \mathbb{Z}_{\geq 1}$. Let $a + b = t$. If (a, b) is an N-position of Wythoff, then there is a move $(a, b) \to (a', b')$ to a P-position of Wythoff, so by induction (a', K, b') is a P-position hence (a, K, b) is an N-position. If, on the other hand, (a, b) is a P-position of Wythoff, then every follower (a', b') of (a, b) is an N-position of Wythoff, hence every follower (a', K, b') of (a, K, b) is an N-position, so (a, K, b) is a P-position. $\qquad\square$

The remainder of this paper deals with other cases of special K. We will see that this phenomenon allows us to ignore the distinction between the left and the right side of K, which will simplify our characterization of the P-positions. We start this discussion by redefining our main terms accordingly. (Some of these definitions are not changed, but repeated for ease of reference.)

DEFINITION 4. Let $r = r_K$, as above. For $n \in \mathbb{Z}_{\geq 1}$, define

$$d_n = b_n - a_n,$$
$$A_n = \{0, r\} \cup \{a_i : 1 \leq i \leq n - 1\},$$
$$B_n = \{0, r\} \cup \{b_i : 1 \leq i \leq n - 1\},$$
$$V_n = A_n \cup B_n,$$
$$D_n = \{r\} \cup \{d_i : 1 \leq i \leq n - 1\},$$

where

$$a_n = \operatorname{mex} V_n \qquad (4)$$

and b_n is the smallest number $x \in \mathbb{Z}_{\geq 1}$ satisfying both

$$x \notin V_n, \qquad (5)$$
$$x - a_n \notin D_n. \qquad (6)$$

As before, $A = \bigcup_{i=1}^{\infty} a_i$ and $B = \bigcup_{i=1}^{\infty} b_i$.

With these definitions, our facts about the sequences A and B are somewhat different:

(A) The sequence A is strictly increasing because $1 \leq m < n \Longrightarrow a_n = \operatorname{mex} V_n > a_m$, since $a_m \in V_n$.
(B) It follows from (5) that all elements in B are distinct.
(C) Condition (5) also implies that $b_n \geq a_n = \operatorname{mex} V_n$ for all $n \geq 1$.
(D) $A \cup B = \mathbb{Z}_{\geq 1} \setminus \{r\}$ due to (4).

(E) $A \cap B$ is either empty or equal to $\{a_1\} = \{b_1\}$. First, note that $a_n \neq b_m$ for $n \neq m$, because $m < n$ implies that a_n is the mex of a set containing b_m by (4), and if $n < m$, then the same conclusion holds by (5). If $r = 0$, then $b_i - a_i \neq 0$ for all i, so $A \cap B = \varnothing$. Otherwise $r > 0$, and for $n = 1$, the minimum value satisfying (5) is $\text{mex}\{0, r\} = a_1$, and in this case a_1 also satisfies (6); that is, $0 = a_1 - a_1 \notin \{r\}$. Therefore, $b_1 = a_1$, and $b_i - a_i \neq 0$ for $i > 1$, by (6).

THEOREM 3. *If K is special, then*

$$P_K = \bigcup_{i=1}^{\infty} (a_i, K, b_i) \cup (b_i, K, a_i)$$

is the set of P-positions of the form (a, K, b) with $a, b \in \mathbb{Z}_{\geq 1}$.

Table 2 lists the first few such (a_i, b_i) pairs for several special values of K. Note that the case $K = (0)$ corresponds to Wythoff's game.

PROOF. As in the proof for general K, we need to show two things: from a position in P_K one cannot reach (r, K), (K, r), or any position in P_K in a single move, and from any $(a, K, b) \notin P_K$ with $a, b > 0$ there is a single move to at least one of these positions.

It is a simple exercise to see that one can reach neither (r, K) nor (K, r) from $(a_m, K, b_m) \in P_K$, so we show that it is impossible to reach any position in P_K in one move:

(i) $(a_m - t, K, b_m) \in P_K$ for some $0 < t \leq a_m$. We cannot have $(a_m - t, K, b_m) = (a_n, K, b_n)$ because it contradicts (B). If $(a_m - t, K, b_m) = (b_n, K, a_n)$, then $a_n = b_m$, so $m = n = 1$ by (E). But then $a_m - t = b_n = a_m$, a contradiction.

(ii) $(a_m, K, b_m - t) \in P_K$ for some $0 < t \leq b_m$. This case is symmetric to (i).

(iii) $(a_m - t, K, b_m - t) \in P_K$ for some $0 < t \leq a_m$. We cannot have $(a_m - t, K, b_m - t) = (a_n, K, b_n)$ because it contradicts (6). If $(a_m - t, K, b_m - t) = (b_n, K, a_n)$, then $b_m - a_m = -(b_n - a_n)$. But (C) tells us that $b_m - a_m \geq 0$ and $b_n - a_n \geq 0$, so $b_m - a_m = b_n - a_n = 0$, contradicting (6).

Similar reasoning holds if one were starting from $(b_m, K, a_m) \in P_K$.

Now we prove that from $(a, K, b) \notin P_K$ with $a, b > 0$ there is a single move to (r, K), to (K, r), to some $(a_n, K, b_n) \in P_K$, or to some $(b_n, K, a_n) \in P_K$. We assume that $a \leq b$, but the arguments hold symmetrically for $b \leq a$.

If $a = r$, we can move to (r, K) by taking the entire right-hand pile. Otherwise, by (D), a is in either A or B. If $a = b_n$ for some n, then $b \geq a = b_n \geq a_n$. Since $(a, K, b) \notin P_K$, we have $b > a_n$, so we can move b to a_n, thereby reaching $(b_n, K, a_n) \in P_K$.

	$K = (0)$ $r = 0$			$K = (10)$ $r = 6$			$K = (15, 15)$ $r = 10$			$K = (8, 6, 23)$ $r = 14$		
i	a_i	b_i	d_i	a_i	b_i	d_i	a_i	b_i	d_i	a_i	b_i	d_i
1	1	2	1	1	1	0	1	1	0	1	1	0
2	3	5	2	2	3	1	2	3	1	2	3	1
3	4	7	3	4	7	3	4	6	2	4	6	2
4	6	10	4	5	9	4	5	8	3	5	8	3
5	8	13	5	8	10	2	7	11	4	7	11	4
6	9	15	6	11	16	5	9	14	5	9	15	6
7	11	18	7	12	19	7	12	18	6	10	17	7
8	12	20	8	13	21	8	13	20	7	12	20	8
9	14	23	9	14	23	9	15	23	8	13	18	5
10	16	26	10	15	25	10	16	25	9	16	25	9
11	17	28	11	17	28	11	17	28	11	19	29	10
12	19	31	12	18	30	12	19	31	12	21	32	11
13	21	34	13	20	33	13	21	34	13	22	34	12
14	22	36	14	22	36	14	22	36	14	23	36	13
15	24	39	15	24	39	15	24	39	15	24	39	15
16	25	41	16	26	42	16	26	42	16	26	42	16
17	27	44	17	27	44	17	27	44	17	27	44	17
18	29	47	18	29	47	18	29	47	18	28	46	18
19	30	49	19	31	50	19	30	49	19	30	49	19
20	32	52	20	32	52	20	32	52	20	31	51	20

Table 2. The first 20 outer piles of P-positions for some values of K. Note that B, while usually strictly increasing, need not always be, as illustrated at $K = (8, 6, 23)$, $i = 9$.

If $a = a_n$ for some n, then if $b > b_n$, we can move to $(a_n, K, b_n) \in P_K$. Otherwise we have, $a = a_n \le b < b_n$. We consider 2 cases.

I. $b - a_n \in D_n$. If $b - a_n = r$, then we can take $b - r = a_n$ from both ends to reach (K, r). Otherwise, $b - a_n = b_m - a_m$ for some $m < n$, and $b - b_m = a_n - a_m > 0$ since $a_n > a_m$ by (A). Thus we can move to $(a_m, K, b_m) \in P_K$ by taking $a_n - a_m = b - b_m$ from both a_n and b.

II. $b - a_n \notin D_n$. This shows that b satisfies (6). Since $b < b_n$ and b_n is the smallest value satisfying both (5) and (6), we must have $b \in V_n$. By assumption, $b > 0$. If $b = r$, then we can move to (K, r) by taking the entire left-hand pile. Otherwise, since $b \ge a_n > a_m$ for all $m < n$, it must be that $b = b_m$ with $m < n$. We now see from (A) that $a_m < a_n$, so we can draw from the left-hand pile to obtain $(a_m, K, b_m) \in P_K$. $\square$

LEMMA 2. *For* $m, n \in \mathbb{Z}_{\geq 1}$, *if* $\{0, \ldots, m-1\} \subseteq D_n$, $m \notin D_n$ *and* $a_n + m \notin V_n$, *then* $b_n = a_n + m$ *and* $\{0, \ldots, m\} \subseteq D_{n+1}$.

PROOF. We have $x < a_n \implies x \in V_n$, and $a_n \leq x < a_n + m \implies x - a_n \in D_n$, so no number smaller than $a_n + m$ satisfies both (5) and (6). The number $a_n + m$, however, satisfies both since, by hypothesis, $a_n + m \notin V_n$ and $m \notin D_n$, so $b_n = a_n + m$. Since $b_n - a_n = m$, $\{0, \ldots, m\} \subseteq D_{n+1}$. □

LEMMA 3. *For* $m \in \mathbb{Z}_{\geq 1}$, *if* $D_m = \{0, \ldots, m-1\}$, *then* $b_n = a_n + n$ *for all* $n \geq m$.

PROOF. We see that $m \notin D_m$. Also, $a_m + m \notin V_m$: it cannot be in A_m because A is strictly increasing, and it cannot be in B_m because if it were, we would get $m = b_i - a_m < b_i - a_i \in D_m$, a contradiction. So Lemma 2 applies, and $b_m = a_m + m$.

This shows that $D_{m+1} = \{0, \ldots, m\}$, so the result follows by induction. □

LEMMA 4. *If* $1 \leq m \leq r < a_m + m - 1$ *and* $D_m = \{r, 0, 1, \ldots, m-2\}$, *then* $d_m = m - 1$. *Thus, for* $m \leq n \leq r$, $d_n = n - 1$.

PROOF. For $0 < i < m$ we have $a_i < a_m$ by (A) and $d_i < m - 1$ since we cannot have $d_i = r$. Hence $a_m + m - 1 > a_i + d_i = b_i$. Also by hypothesis, $a_m + m - 1 > r$, so $a_m + m - 1 \notin V_m$. Since $m - 1 \notin D_m$, Lemma 2 (with $n = m$ and $m = m - 1$) implies $d_m = m - 1$.

For $m \leq n \leq r$, the condition in the lemma holds inductively, so the conclusion holds, as well. □

We will now begin to note further connections between the P-positions in End-Wythoff and those in standard Wythoff's Game, to which end we introduce some useful notation.

NOTATION 2. The P-positions of Wythoff's game—i.e., the 2-pile P-positions of End-Wythoff, along with $(0, 0) = (0)$—are denoted by $\bigcup_{i=0}^{\infty}(a_i', b_i')$, where $a_n' = \lfloor n\phi \rfloor$ and $b_n' = \lfloor n\phi^2 \rfloor$ for all $n \in \mathbb{Z}_{\geq 0}$, and $\phi = (1 + \sqrt{5})/2$ is the golden ratio. We write $A' = \bigcup_{i=0}^{\infty} a_i'$ and $B' = \bigcup_{i=0}^{\infty} b_i'$.

An important equivalent definition of A' and B' is, for all $n \in \mathbb{Z}_{\geq 0}$ (see [3]),

$$a_n' = \text{mex}\{a_i', b_i' : 0 \leq i \leq n - 1\},$$
$$b_n' = a_n' + n.$$

The following is our main lemma for the proof of Theorem 4.

LEMMA 5. *Let* $n \in \mathbb{Z}_{\geq 0}$. *If* $a_n' + 1 < r$, *then* $a_{n+1} = a_n' + 1$. *If* $b_n' + 1 < r$, *then* $b_{n+1} = b_n' + 1$.

PROOF. Note that $a'_0 + 1 = b'_0 + 1 = 1$. If $1 < r$, then $a_1 = \text{mex}\{0, r\} = 1$, and 1 satisfies both (5) and (6), so $b_1 = 1$. So the result is true for $n = 0$.

Assume that the lemma's statement is true for $0 \le i \le n - 1$ $(n \ge 1)$, and assume further that $a'_n + 1 < r$. Then $a_0 = 0 < a'_n + 1$. Also, $a'_i + 1 < a'_n + 1 < r$ for $0 \le i \le n - 1$ because A' is strictly increasing. But, $a_{i+1} = a'_i + 1$ for $0 \le i \le n - 1$ by the induction hypothesis, so $a_i < a'_n + 1$ for $1 \le i \le n$. Thus, we have shown that $a'_n + 1 \notin A_{n+1}$.

Let m be the least index such that $b'_m + 1 \ge r$, and let $j = \min\{m, n\}$. Then $b'_{i-1} + 1 < r$ for $1 \le i \le j$, so $b_i = b'_{i-1} + 1$ by the induction hypothesis. We know that $b'_r = a'_s \Longrightarrow r = s = 0$, so $b'_{i-1} \ne a'_n$ because $n \ge 1$. Therefore $b_i = b'_{i-1} + 1 \ne a'_n + 1$, so $a'_n + 1 \notin B_{j+1}$.

If $j = n$, then we have shown that $a'_n + 1 \notin V_{n+1}$. Otherwise, $j = m$. For $k \ge m+1$ we have $d_k \ge m$ by (6), since $d_i = b_i - a_i = b'_{i-1} + 1 - (a'_{i-1} + 1) = i - 1$ for $1 \le i \le m$ by our induction hypothesis. Also, $a_i \ge a_{m+1}$ for $i \ge m + 1$, by (A). Therefore, for $i \ge m + 1$, $b_i = a_i + d_i \ge a_{m+1} + m = (a'_m + 1) + m = b'_m + 1 \ge r > a'_n + 1$. Thus we see that $a'_n + 1 \notin \{b_i : i \ge m + 1\}$, and we have shown that $a'_n + 1 \notin V_{n+1}$.

Now, $0 \in V_{n+1}$, and if $1 \le x < a'_n + 1$, then $0 \le x - 1 < a'_n$, so $x - 1 \in \{a'_i, b'_i : 0 \le i < n\}$. Thus, for some i with $0 \le i < n$, either $x = a'_i + 1 = a_{i+1}$ or $x = b'_i + 1 = b_{i+1}$ by the induction hypothesis, so $x \in V_{n+1}$. Hence $a'_n + 1 = \text{mex } V_{n+1}$. This proves the first statement of the lemma: $a_{n+1} = \text{mex } V_{n+1} = a'_n + 1$.

Note that if $b'_i + 1 < r$ for some $i \in \mathbb{Z}_{\ge 0}$, then a fortiori $a'_i + 1 < r$. Hence by the first part of the proof, $a_{i+1} = a'_i + 1$. Thus,

$$b'_i + 1 < r \implies a_{i+1} = a'_i + 1. \tag{7}$$

For the second statement of the lemma, assume that the result is true for $0 \le i \le n - 1$ $(n \ge 1)$, and that $b'_n + 1 < r$. Then, for $0 \le i \le n - 1$, we know $a_{i+1} = a'_i + 1$ by (7), and $b_{i+1} = b'_i + 1$ by the induction assumption. Therefore $d_i = i - 1$ for $1 \le i \le n$, so b_{n+1} cannot be smaller than $a_{n+1} + n$. Also $a_{n+1} = a'_n + 1$ by (7).

Consider $a_{n+1} + n = a'_n + 1 + n = b'_n + 1$. We have $0 < b'_n + 1 < r$, and for $1 \le i \le n$, $a_i \le b_i = b'_{i-1} + 1 < b'_n + 1$. This implies that $b'_n + 1 \notin V_{n+1}$, and we conclude that $b_{n+1} = b'_n + 1$. $\qquad \square$

COROLLARY 1. *Let* $n \in \mathbb{Z}_{\ge 0}$. *If* $a_{n+1} < r$, *then* $a_{n+1} = a'_n + 1$. *If* $b_{n+1} < r$, *then* $b_{n+1} = b'_n + 1$.

PROOF. Note that $a_{n+1} > 0$. Since $A' \cup B' = \mathbb{Z}_{\ge 0}$, either $a_{n+1} = a'_i + 1$ or $a_{n+1} = b'_i + 1$. If $a'_i + 1 = a_{n+1} < r$, then $a_{n+1} = a'_i + 1 = a_{i+1}$ by Lemma 5, so $i = n$. If $b'_i + 1 = a_{n+1} < r$, then $a_{n+1} = b'_i + 1 = b_{i+1}$ by Lemma 5, so $i = n = 0$ by (E), and $a_1 = b_1 = b'_0 + 1 = a'_0 + 1$. The same argument holds for b_{n+1}. $\qquad \square$

COROLLARY 2. *For* $1 \leq n \leq r-1$, $n \in A$ *if and only if* $n-1 \in A'$ *and* $n \in B$ *if and only if* $n-1 \in B'$.

PROOF. This follows from Lemma 5 and Corollary 1. $\qquad\square$

THEOREM 4. *If* $r = a'_n + 1$, *then for* $1 \leq i \leq r$, $d_i = i - 1$. *Furthermore, for* $1 \leq i \leq n$, $a_i = a'_{i-1} + 1$ *and* $b_i = b'_{i-1} + 1$.

PROOF. If $r = 1$, then $n = 0$. In this case, note that $a_1 = b_1 = 2$, so $d_1 = 0$, and that the second assertion of the theorem is vacuously true.

Otherwise, $r \geq 2$, and we again let m be the least index such that $b'_m + 1 \geq r$. Note that $m \geq 1$ because $b'_0 + 1 = 1 < r$. Thus $b'_m \neq a'_n = r - 1$, so in fact $b'_m + 1 > r$. For $1 \leq i \leq m$, since $a'_{i-1} \leq b'_{i-1}$ and B' is increasing, we have $a'_{i-1} + 1 \leq b'_{i-1} + 1 \leq b'_{m-1} + 1 < r$, so $a_i = a'_{i-1} + 1$ and $b_i = b'_{i-1} + 1$ by Lemma 5. We see that $d_i = i - 1$ for $1 \leq i \leq m$, so $D_{m+1} = \{r, 0, \ldots, m-1\}$.

Notice that $a_{m+1} + m > r$ because either $a_{m+1} > r$ and the fact is clear, or $a_{m+1} < r$, so $a_{m+1} = a'_m + 1$ by Corollary 1, which implies that $a_{m+1} + m = a'_m + 1 + m = b'_m + 1 > r$. Also, $m + 1 \leq b'_{m-1} + 2$ (because $1 + 1 = b'_0 + 2$ and B' is strictly increasing) and $b'_{m-1} + 1 < r$, so $m + 1 \leq b'_{m-1} + 2 \leq r$. We can now invoke Lemma 4 to see that $d_i = i - 1$ for $m + 1 \leq i \leq r$, so we have $d_i = i - 1$ for $1 \leq i \leq r$.

Since $n \leq a'_n < r$, in particular $d_i = i - 1$ for $1 \leq i \leq n$. With i in this range, we know $a'_{i-1} + 1 < a'_n + 1 = r$, so we get $a_i = a'_{i-1} + 1$ by Lemma 5, and since $d_i = i - 1$, $b_i = a_i + i - 1 = a'_{i-1} + 1 + i - 1 = b'_{i-1} + 1$. $\qquad\square$

THEOREM 5. *If* $r = b'_n + 1$, *then for* $1 \leq i \leq r$, $d_i = i - 1$ *except as follows:*

- *If* $n = 0$, *there are no exceptions.*
- *If* $a'_n + 1 \in B'$, *then* $d_{n+1} = n + 1$ *and* $d_{n+2} = n$.
- *If* $n = 2$, *then* $d_3 = 3$, $d_4 = 4$ *and* $d_5 = 2$.
- *Otherwise,* $d_{n+1} = n + 1$, $d_{n+2} = n + 2$, $d_{n+3} = n + 3$, *and* $d_{n+4} = n$.

PROOF. One can easily verify the theorem for $0 \leq n \leq 2$—that is, when $r = 1$ (first bullet), 3 (second bullet), or 6 (third bullet). So we assume $n \geq 3$.

Lemma 5 tells us that for $1 \leq i \leq n$, $a_i = a'_{i-1} + 1$ and $b_i = b'_{i-1} + 1$ because $a'_{i-1} + 1 \leq b'_{i-1} + 1 < b'_n + 1 = r$. This implies that $d_i = i - 1$ for $1 \leq i \leq n$. This is not the case for d_{n+1}: $a'_n + 1 < b'_n + 1 = r$, so $a_{n+1} = a'_n + 1$, but $a_{n+1} + n = b'_n + 1 = r$, which cannot be b_{n+1}. We must have $b_{n+1} \geq a_{n+1} + n$, however, and $a_i \leq b_i < r = a_{n+1} + n$ for $1 \leq i \leq n$, so we see that $a_{n+1} + n + 1 \notin V_{n+1}$; thus $b_{n+1} = a_{n+1} + n + 1 = a'_n + n + 2 = b'_n + 2$, and $d_{n+1} = n + 1$.

If $a'_n + 1 \in B'$, then $a'_{n+1} = a'_n + 2$ (because B' does not contain consecutive numbers) and $a'_{n+1} + 1 = a'_n + 3 \leq a'_n + n = b'_n = r - 1$, so Lemma 5 tells us that $a_{n+2} = a'_{n+1} + 1 = a'_n + 3$. Now, $a_{n+2} + n = a'_n + n + 3 > a'_n + n + 2 = b_{n+1} \geq b_i$

for all $i \leq n+1$, so $b_{n+2} = a_{n+2} + n$, and we see $d_{n+2} = n$. That is, we have $a_{n+1}, b_i, a_{n+2}, \ldots, r, b_{n+1}, b_{n+2}$.

This gives us $D_{n+3} = \{r, 0, \ldots, n+1\}$. Also, $5 < b'_2 + 1 = 6$ so, since B' is strictly increasing and $n + 3 \geq 5$, we know $n + 3 < b'_n + 1 = r$. Furthermore, $r < b_{n+1} = a_{n+1} + n + 1 < a_{n+3} + n + 2$. Therefore, we can cite Lemma 4 to assert that $d_i = i - 1$ for $n + 3 \leq i \leq r$.

If, on the other hand, $a'_n + 1 \notin B'$, then $a'_n + 1 = a'_{n+1}$. Note that $a'_3 + 1 = 5 = b'_2$ and $a'_4 + 1 = 7 = b'_3$, so we can assume $n \geq 5$. We have $a'_{n+1} + 1 = a'_n + 2 < a'_n + n = b'_n < r$, so $a_{n+2} = a'_{n+1} + 1 = a'_n + 2$, and we find that $a_{n+2} + n = a'_n + n + 2 = b_{n+1} \in V_{n+2}$. Also, a difference of $n + 1$ already exists, but $a_{n+2} + n + 2$ is not in V_{n+2}, as it is greater than all of the previous B values. So we get $b_{n+2} = a_{n+2} + n + 2$, and $d_{n+2} = n + 2$. We have the following picture: $a_{n+1}, a_{n+2}, \ldots, r, b_{n+1}, -, b_{n+2}$.

Now, since $a'_n + 1 \in A'$, $a'_n + 2$ must be in B' because A' does not contain three consecutive values. Because $a'_n + 3 \leq a'_n + n = b'_n = r - 1$, we have $a_{n+2} + 1 = a'_n + 3 \in B$ by Corollary 2. Also, $a'_n + 3 \in A'$ because B' does not contain consecutive values, and $a'_n + 4 \leq r - 1$, so $a'_n + 4 \in A$. We therefore have $a_{n+1}, a_{n+2}, b_j, a_{n+3}, \ldots, r, b_{n+1}, -, b_{n+2}$. Since $a_{n+3} + n = b_{n+2}$ and differences of $n + 1$ and $n + 2$ already occurred, we get $b_{n+3} = a_{n+3} + n + 3$, and $d_{n+3} = n + 3$, and the configuration is

$$a_{n+1}, a_{n+2}, b_j, a_{n+3}, \ldots, r, b_{n+1}, -, b_{n+2}, -, -, b_{n+3}.$$

If $n = 5$, then $r = b'_5 + 1 = 14$, and one can check that $a_{n+3} = a_8 = 12$ and $a_{n+4} = a_9 = 13 = a_{n+3} + 1$. If $n \geq 6$, then $a_{n+3} + 2 = a_{n+1} + 5 \leq a_{n+1} + n - 1 = r - 1$. The sequence B' does not contain consecutive values, so either $a_{n+3} \in A'$ or $a_{n+3} + 1 \in A'$, and therefore either $a_{n+3} + 1 \in A$ or $a_{n+3} + 2 \in A$. So regardless of the circumstances, either $a_{n+4} = a_{n+3} + 1$ or $a_{n+4} = a_{n+3} + 2$.

This means that either $a_{n+4} + n = a_{n+3} + n + 1 = b_{n+2} + 1$ or $a_{n+4} + n = a_{n+3} + n + 2 = b_{n+2} + 2$. In either case, this spot is not taken by an earlier b_i, so $b_{n+4} = a_{n+4} + n$, and $d_{n+4} = n$.

A few moments of reflection reveal that $4 \leq a_3$. Since A is strictly increasing, this gives us that $5 \leq a_4$ and, in general, $n + 5 \leq a_{n+4}$. We now have $n + 5 \leq a_{n+4} < r < b_{n+1} = a_{n+1} + n + 1 < a_{n+5} + n + 4$, and $D_{n+5} = \{r, 0, \ldots, n+3\}$, so Lemma 4 completes the proof. $\square$

THEOREM 6. *If $n \geq r + 1$, then $d_n = n$.*

PROOF. The smallest n which fall under each of the bullets of Theorem 5 are $n = 0$, $n = 1$, $n = 2$, and $n = 5$, respectively. ($n = 3$ and $n = 4$ fall under the second bullet.) Notice that $n + 2 \leq b'_n + 1$ when $n \geq 1$ since $1 + 2 \leq b'_1 + 1 = 3$ and B' is strictly increasing. Similarly, $n + 3 \leq b'_n + 1$ when $n \geq 2$ since

$2 + 3 \leq b_2' + 1 = 6$, and $n + 4 \leq b_n' + 1$ when $n \geq 3$ because $3 + 4 \leq b_3' + 1 = 8$. Therefore, we see that all of the exceptions mentioned in Theorem 5 occur before index $r + 1 = b_n' + 2$.

Theorems 4 and 5, combined with this observation, reveal that $D_{r+1} = \{0, \ldots, r\}$, whether $r = a_n' + 1$ or $r = b_n' + 1$. Thus, by Lemma 3, $d_n = n$ for $n \geq r + 1$. $\qquad\square$

4. Generating P-positions in polynomial time

Any position of End-Wythoff is specified by a vector whose components are the pile sizes. We consider K to be a constant. The input size of a position (a, K, b) is thus $O(\log a + \log b)$. We seek an algorithm polynomial in this size.

Theorem 6 shows that we can express A and B beyond r as

$$a_n = \operatorname{mex}(X \cup \{a_i, b_i : r + 1 \leq i < n\}), \quad n \geq r + 1,$$
$$b_n = a_n + n, \qquad\qquad\qquad\qquad\qquad n \geq r + 1,$$

where $X = V_{r+1}$. This characterization demonstrates that the sequences generated from special End-Wythoff positions are a special case of those studied in [4], [5], [6]. In [4] it is proved that $a_n' - a_n$ is eventually constant except for certain "subsequences of irregular shifts", each of which obeys a Fibonacci recurrence. That is, if i and j are consecutive indices within one of these subsequences of irregular shifts, then the next index in the subsequence is $i + j$. This is demonstrated in Figure 1.

Relating our sequences to those of [4] is useful because that paper's proofs give rise to a polynomial algorithm for computing the values of the A and B sequences in the general case dealt with there. For the sake of self-containment, we begin by introducing some of the notation used there and mention some of the important theorems and lemmas.

DEFINITION 5. Let $c \in \mathbb{Z}_{\geq 1}$. (For Wythoff's game, $c = 1$.)

$$a_n' = \operatorname{mex}\{a_i', b_i' : 1 \leq i < n\}, \quad n \geq 1;$$
$$b_n' = a_n' + cn, \quad n \geq 1;$$
$$m_0 = \min\{m : a_m > \max(X)\};$$
$$s_n = a_n' - a_n, \quad n \geq m_0;$$
$$\alpha_n = a_{n+1} - a_n, \quad n \geq m_0;$$
$$\alpha_n' = a_{n+1}' - a_n', \quad n \geq 1;$$
$$W = \{\alpha_n\}_{n=m_0}^{\infty};$$
$$W' = \{\alpha_n'\}_{n=1}^{\infty}.$$

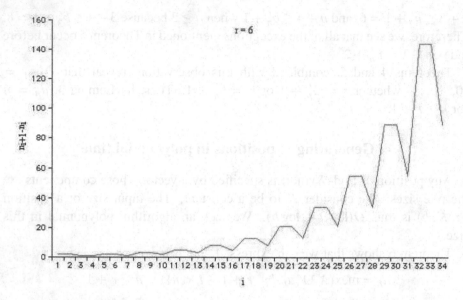

Figure 1. With $r = 6$, the distance between consecutive indices of P-positions which differ from Wythoff's game's P-positions. (That is, n_i is the subsequence of indices where $(a_n, b_n) \neq (a'_n, b'_n)$.) Note that every third point can be connected to form a Fibonacci sequence.

$F : \{1, 2\}^* \to \{1, 2\}^*$ is the nonerasing morphism

$$F : \begin{array}{l} 2 \to 1^c 2, \\ 1 \to 1^{c-1} 2. \end{array}$$

A *generator* for W is a word of the form $u = \alpha_t \cdots \alpha_{n-1}$, where $a_n = b_t + 1$; similarly, a generator for W' is a word $u' = \alpha'_r \cdots \alpha'_{m-1}$, where $a'_m = b'_r + 1$. We say that W, W' are generated *synchronously* if there exist generators u, u', such that $u = \alpha_t \cdots \alpha_{n-1}, u' = \alpha'_t \cdots \alpha'_{n-1}$ (same indices t, n), and

$$\forall k \geq 0, F^k(u) = \alpha_g \cdots \alpha_{h-1} \iff F^k(u') = \alpha'_g \cdots \alpha'_{h-1},$$

where $a_h = b_g + 1$.

A *well-formed string of parentheses* is a string $\vartheta = t_1 \cdots t_n$ over some alphabet which includes the letters '(', ')', such that for every prefix μ of ϑ, $|\mu|_(\geq |\mu|_)$ (never close more parentheses than were opened), and $|\vartheta|_(= |\vartheta|_)$ (don't leave opened parentheses).

The *nesting level* $N(\vartheta)$ of such a string is the maximal number of opened parentheses. More formally, let $p_1, \ldots, p_n$ satisfy $p_i = 1$ if $t_i = ($, $p_i = -1$ if $t_i =)$, and $p_i = 0$ otherwise. Then

$$N(\vartheta) = \max_{1 \leq k \leq n} \left\{ \sum_{i=1}^{k} p_i \right\}.$$

With these definitions in mind, we cite the theorems, lemmas, and corollaries necessary to explain our polynomial algorithm.

THEOREM 7. *There exist $p \in \mathbb{Z}_{\geq 1}, \gamma \in \mathbb{Z}$, such that, either for all $n \geq p$, $s_n = \gamma$; or else, for all $n \geq p$, $s_n \in \{\gamma - 1, \gamma, \gamma + 1\}$. If the second case holds, then:*

1. *s_n assumes each of the three values infinitely often.*
2. *If $s_n \neq \gamma$ then $s_{n-1} = s_{n+1} = \gamma$.*
3. *There exists $M \in \mathbb{Z}_{\geq 1}$, such that the indices $n \geq p$ with $s_n \neq \gamma$ can be partitioned into M disjoint sequences, $\{n_j^{(i)}\}_{j=1}^{\infty}, i = 1, \ldots, M$. For each of these sequences, the shift value alternates between $\gamma - 1$ and $\gamma + 1$:*

$$s_{n_j^{(i)}} = \gamma + 1 \Longrightarrow s_{n_{j+1}^{(i)}} = \gamma - 1;$$
$$s_{n_j^{(i)}} = \gamma - 1 \Longrightarrow s_{n_{j+1}^{(i)}} = \gamma + 1.$$

THEOREM 8. *Let $\{n_j\}_{j=1}^{\infty}$ be one of these subsequences of irregular shifts. Then it satisfies the following recurrence:*

$$\forall j \geq 3, n_j = c n_{j-1} + n_{j-2}.$$

COROLLARY 3. *If for some $t \geq m_0$, $b_t + 1 = a_n$ and $b_t' + 1 = a_n'$, then the words*

$$u = \alpha_t \cdots \alpha_{n-1},$$
$$u' = \alpha_t' \cdots \alpha_{n-1}',$$

are permutations of each other.

LEMMA 6 (SYNCHRONIZATION LEMMA). *Let m_1 be such that $a_{m_1} = b_{m_0} + 1$. Then there exists an integer $t \in [m_0, m_1]$, such that $b_t + 1 = a_n$ and $b_t' + 1 = a_n'$.*

COROLLARY 4. *If for some $t \geq m_0$, $b_t + 1 = a_n$ and $b_t' + 1 = a_n'$, then W, W' are generated synchronously by u, u', respectively.*

In comparing u and u', it will be useful to write them in the following form:

$$\begin{bmatrix} u \\ u' \end{bmatrix} = \begin{bmatrix} \alpha_t \cdots \alpha_{n-1} \\ \alpha_t' \cdots \alpha_{n-1}' \end{bmatrix},$$

and we will apply F to these pairs:

$$F\left(\begin{bmatrix} u \\ u' \end{bmatrix} \right) := \begin{bmatrix} F(u) \\ F(u') \end{bmatrix}.$$

Since u, u' are permutations of each other by Corollary 3, if we write them out in this form, then the columns $\begin{bmatrix} 1 \\ 2 \end{bmatrix}$ and $\begin{bmatrix} 2 \\ 1 \end{bmatrix}$ occur the same number of times.

Thus we can regard $\left[\begin{smallmatrix} u \\ u' \end{smallmatrix}\right]$ as a well-formed string of parentheses: put '$\bullet$' for $\left[\begin{smallmatrix} 1 \\ 1 \end{smallmatrix}\right]$ or $\left[\begin{smallmatrix} 2 \\ 2 \end{smallmatrix}\right]$, and put '(', ')' for $\left[\begin{smallmatrix} 1 \\ 2 \end{smallmatrix}\right]$, $\left[\begin{smallmatrix} 2 \\ 1 \end{smallmatrix}\right]$ alternately such that the string remains well-formed. That is, if the first nonequal pair we encounter is $\left[\begin{smallmatrix} 1 \\ 2 \end{smallmatrix}\right]$, then '(' stands for $\left[\begin{smallmatrix} 1 \\ 2 \end{smallmatrix}\right]$ and ')' stands for $\left[\begin{smallmatrix} 2 \\ 1 \end{smallmatrix}\right]$ until all opened parentheses are closed. Then we start again, by placing '(' for the first occurrence different from $\left[\begin{smallmatrix} 1 \\ 1 \end{smallmatrix}\right]$, $\left[\begin{smallmatrix} 2 \\ 2 \end{smallmatrix}\right]$.

EXAMPLE 1.

$$\begin{bmatrix} 122 \\ 221 \end{bmatrix} \longrightarrow (\bullet), \quad \begin{bmatrix} 1221 \\ 2112 \end{bmatrix} \longrightarrow ()(), \quad \begin{bmatrix} 22211211 \\ 21112122 \end{bmatrix} \longrightarrow \bullet((\bullet)()).$$

LEMMA 7 (NESTING LEMMA). *Let $u(0) \in \{1, 2\}^*$, and let $u'(0)$ be a permutation of $u(0)$. If $c = 1$ and $u(0)$ or $u'(0)$ contains 11, put $u := F(u(0))$, $u' := F(u'(0))$. Otherwise, put $u := u(0)$, $u' := u'(0)$. Let $\vartheta \in \{\bullet, (,)\}^*$ be the parentheses string of $\left[\begin{smallmatrix} u \\ u' \end{smallmatrix}\right]$. Then successive applications of F decrease the nesting level to 1. Specifically,*

(I) *If $c > 1$, then $N(\vartheta) > 1 \Longrightarrow N(F(\vartheta)) < N(\vartheta)$.*
(II) *If $c = 1$,*

    (a) *$N(\vartheta) > 2 \Longrightarrow N(F(\vartheta)) < N(\vartheta)$;*
    (b) *$N(\vartheta) = 2 \Longrightarrow N(F^2(\vartheta)) = 1$.*

LEMMA 8. *Under the hypotheses of the previous lemma, if $N(\vartheta) = 1$, then $F^2(\vartheta)$ has the form*

$$\cdots () \cdots () \cdots () \cdots, \tag{8}$$

where the dot strings consist of '$\bullet$' letters and might be empty. Further applications of F preserve this form, with the same number of parentheses pairs; the only change is that the dot strings grow longer.

We now have the machinery necessary to sketch the polynomial algorithm for generating the sequences A and B. There is a significant amount of initial computation, but then we can use the Fibonacci recurrences from Theorem 8 to obtain any later values for s_n, and thus for a_n and b_n as well. Here are the initial computations:

- Compute the values of A and B until $a_n = b_t + 1$ and $a'_n = b'_t + 1$. The Synchronization Lemma assures us that we can find such values with $m_0 \le t \le m_1$, where m_1 is the index such that $a_{m_1} = b_{m_0} + 1$. Corollary 4 tells us that W, W' are generated synchronously by $u = \alpha_t \ldots \alpha_{n-1}$, $u' = \alpha'_t \ldots \alpha'_{n-1}$.
- Iteratively apply F to u and u' until the parentheses string of $w = F^k(u)$ and $w' = F^k(u')$ is of the form (8). We know this will eventually happen because of Lemmas 7 and 8.
- Let p and q be the indices such that $w = \alpha_p \ldots \alpha_q$ and $w' = \alpha'_p \ldots \alpha'_q$. Compute A up to index p, and let $\gamma = a'_p - a_p$. At this point, noting the

differences between w and w' gives us the initial indices for the subsequences of irregular shifts. Specifically, if letters $i, i+1$ of the parentheses string of $\left[\begin{smallmatrix} w \\ w' \end{smallmatrix}\right]$ are '()', then $i+1$ is an index of irregular shift. Label these indices of irregular shifts $n_1^{(1)}, \ldots, n_1^{(M)}$ and, for $1 \le i \le M$, let

$$o_i = s_{n_1^{(i)}} - \gamma \in \{-1, 1\}.$$

(The o_i indicate whether the i-th subsequence of irregular shifts begins offset by $+1$ or by -1 from the regular shift, γ.)

- Apply F once more to w and w'. The resulting sequences will again have M pairs of indices at which $w' \ne w$; label the indices of irregular shifts $n_2^{(1)}, \ldots, n_2^{(M)}$.

With this initial computation done, we can determine a_n and b_n for $n \ge n_1^{(1)}$ as follows: for each of the M subsequences of irregular shifts, compute successive terms of the subsequence according to Theorem 8 until reaching or exceeding n. That is, for $1 \le i \le M$, compute $n_1^{(i)}, n_2^{(i)}, \ldots$ until $n_j^{(i)} \ge n$. Since the $n_j^{(i)}$ are Fibonacci-like sequences, they grow exponentially, so they will reach or exceed the value n in time polynomial in $\log n$. If we obtain $n = n_j^{(i)}$ for some i, j, then

$$s_n = \begin{cases} \gamma + o_i & \text{if } j \text{ is odd,} \\ \gamma - o_i & \text{if } j \text{ is even,} \end{cases}$$

since each subsequence alternates being offset by $+1$ and by -1, by Theorem 7. If, on the other hand, every subsequence of irregular shifts passes n without having a term equal n, then $s_n = \gamma$. Once we know s_n, we have $a_n = a'_n - s_n$ and $b_n = a_n + n$. This implies $b_{n+1} - b_n \in \{2, 3\}$, hence the mex function implies $a_{n+1} - a_n \in \{1, 2\}$. Therefore $a_n \le 2n$, and the algorithm is polynomial.

In the case of sequences deriving from special positions of End-Wythoff, we must compute the value of r before we can begin computing A and B. After that, the initial computation can be slightly shorter than in the general case, as we are about to see.

The only fact about m_0 that is needed in [4] is that $a_{n+1} - a_n \in \{1, 2\}$ for all $n \ge m_0$. For the A and B sequences arising from special positions of End-Wythoff, this condition holds well before m_0, as the following proposition illustrates.

PROPOSITION 1. *For all $n \ge r + 1$, $1 \le a_{n+1} - a_n \le 2$.*

PROOF. $n \ge r + 1$ implies that $b_{n+1} - b_n = a_{n+1} + n + 1 - a_n - n = a_{n+1} - a_n + 1 \ge 2$. That is, from index $r + 1$ onward, B contains no consecutive values. Therefore, since we know that $r + 1 \le a_{r+1} \le a_n$, (D) tells us that if $a_n + 1 \notin A$, then $a_n + 1 \in B$, so $a_n + 2 \notin B$, so $a_n + 2 \in A$, again by (D). This shows that $a_{n+1} - a_n \le 2$. $\qquad\square$

Therefore, in the first step of the initial computation, we are guaranteed to reach synchronization with $r + 1 \leq t \leq m$, where m is the index such that $a_m = b_{r+1} + 1$. Now, $a_{m_0} > b_r$ because $b_r \in V_{r+1} = X$. Also, $b_r \geq a_r$, so $a_{m_0} > a_r$ and $a_{m_0} \geq a_{r+1}$, which implies by (A) that $m_0 \geq r + 1$. Thus, this is an improvement over the bounds in the general case. Furthermore, note that as r grows larger, this shortcut becomes increasingly valuable.

5. Conclusion

We have exposed the structure of the P-positions of End-Wythoff, which is but a first study of this game. Many tasks remain to be done. For example, it would be useful to have an efficient method for computing l_K and r_K. The only method apparent from this analysis is unpleasantly recursive: if $K = (n_1, \ldots, n_k)$, then to find l_K, compute the P-positions for $(n_1, \ldots, n_{k-1})$ until reaching $(l_K, n_1, \ldots, n_{k-1}, n_k)$, and to find r_K, compute P-positions for $(n_2, \ldots, n_k)$ until reaching $(n_1, n_2, \ldots, n_k, r_K)$.

Additionally, there are two observations that one can quickly make if one studies special End-Wythoff positions for different values of r. Proving these conjectures would be a suitable continuation of this work:

- For $r \in \mathbb{Z}_{\geq 0}$, $\gamma = 0$.
- If $r \in \{0, 1\}$, then M, the number of subsequences of irregular shifts, equals 0. If $r = b'_n + 1$ and $a'_n + 1 \in B'$, then $M = 1$. Otherwise, $M = 3$.

Furthermore, evidence suggests that, with the appropriate bounds, Theorem 6 can be applied to any position of End-Wythoff rather than only special positions. In general, it seems that $b_n - a_n = n$ for $n > \max\{l_K, r_K\}$, if we enumerate only those P-positions with the leftmost pile smaller than or equal to the rightmost pile. This is another result that would be worth proving.

6. Acknowledgment

Elnatan Reisner is indebted to the Karyn Kupcinet International Science School for Overseas Students, http://www.weizmann.ac.il/acadaff/kkiss.html, and to the Jerome A. Schiff Undergraduate Fellows Program at Brandeis University. These programs enabled him to spend the summer of 2005 at the Weizmann Institute working with Aviezri Fraenkel.

References

[1] M. H. Albert and R. J. Nowakowski [2001], The game of End-Nim, *Electr. J. Combin.* **8(2)**, #R1, 12pp. http://www.combinatorics.org/

[2] H. S. M. Coxeter [1953], The golden section, phyllotaxis and Wythoff's game, *Scripta Math.* **19**, 135–143.

[3] A. S. Fraenkel [1982], How to beat your Wythoff games' opponent on three fronts, *Amer. Math. Monthly* **89**, 353–361.

[4] A. S. Fraenkel and D. Krieger [2004], The structure of complementary sets of integers: a 3-shift theorem, *Internat. J. Pure and Appl. Math.* **10**, 1–49.

[5] X. Sun [2005], Wythoff's sequence and N-heap Wythoff's conjectures, *Discrete Math.* **300**, 180–195.

[6] X. Sun and D. Zeilberger [2004], On Fraenkel's N-heap Wythoff's conjectures, *Ann. Comb.* **8**, 225–238.

[7] A. M. Yaglom and I. M. Yaglom [1967], *Challenging Mathematical Problems with Elementary Solutions*, Vol. II, Holden-Day, San Francisco, translated by J. McCawley, Jr., revised and edited by B. Gordon.

[8] W. A. Wythoff [1907], A modification of the game of Nim, *Nieuw Arch. Wisk.* **7**, 199–202.

AVIEZRI S. FRAENKEL
DEPARTMENT OF COMPUTER SCIENCE AND APPLIED MATHEMATICS
WEIZMANN INSTITUTE OF SCIENCE
REHOVOT 76100
ISRAEL
    fraenkel@wisdom.weizmann.ac.il

ELNATAN REISNER
DEPARTMENT OF COMPUTER SCIENCE
UNIVERSITY OF MARYLAND
COLLEGE PARK, MD 20742
UNITED STATES
    elnatan@cs.umd.edu

Games of No Chance 3
MSRI Publications
Volume **56**, 2009

On the geometry of combinatorial games:
A renormalization approach

ERIC J. FRIEDMAN AND ADAM S. LANDSBERG

ABSTRACT. We describe the application of a physics-inspired renormalization technique to combinatorial games. Although this approach is not rigorous, it allows one to calculate detailed, probabilistic properties of the geometry of the P-positions in a game. The resulting geometric insights provide explanations for a number of numerical and theoretical observations about various games that have appeared in the literature. This methodology also provides a natural framework for several new avenues of research in combinatorial games, including notions of "universality," "sensitivity-to-initial-conditions," and "crystal-like growth," and suggests surprising connections between combinatorial games, nonlinear dynamics, and physics. We demonstrate the utility of this approach for a variety of games — three-row Chomp, 3-D Wythoff's game, Sprague–Grundy values for 2-D Wythoff's game, and Nim and its generalizations — and show how it explains existing results, addresses longstanding questions, and generates new predictions and insights.

1. Introduction

In this paper we introduce a method for analyzing combinatorial games based on renormalization. As a mathematical tool, renormalization has enjoyed great success in virtually all branches of modern physics, from statistical mechanics [Goldenfeld 1992] to particle physics [Rivasseau 2003] to chaos theory [Feigenbaum 1980], where it is used to calculate properties of physical systems or objects that exhibit so-called 'scaling' behavior (i.e., geometric similarity on different spatial scales). In the present context we adapt this methodology to the study of combinatorial games. Here, the main "object" we study is the set of

This material is based in part upon work supported by the National Science Foundation under Grant No. CCF-0325453.

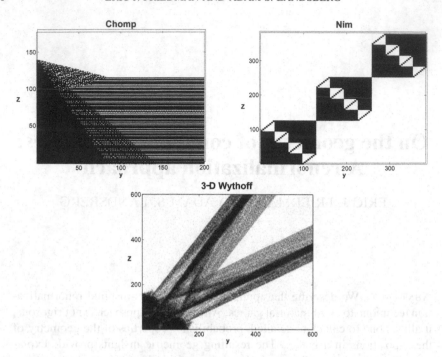

Figure 1. The underlying geometries of combinatorial games. Shown are the IN-sheet structures for Chomp, Nim, and 3-D Wythoff's game.

P-positions[1] of the game, viewed as a geometric entity in the abstract position space of the game (see, e.g., Figure 1, which will be explained later). As we will show, this geometric object exhibits a strong scaling property, and hence can be analyzed via a suitably adapted renormalization technique. Since all critical information about a game is encoded in this geometry, as a methodology renormalization has broad explanatory powers and impressive (numerically verifiable) predictive capabilities, as will be demonstrated through examples.

When we compare renormalization to other traditional analytical techniques for analyzing combinatorial games, such as Sprague–Grundy theory, nimbers, algebraic approaches, and so on [Berlekamp et al. 1979; Bouton 1902; Sprague 1936; Grundy 1939; Conway 1976], several significant distinctions, advantages, and disadvantages emerge:

I: As a mathematical technique, the renormalization procedure described here does not, at present, possess the strict level of rigor needed for formal mathematical proof. In this respect, this renormalization procedure for games 'suffers' the same defect as the renormalization of modern physics: even though renormalization is a highly successful, well-established technique in physics

---

[1] In games without ties or draws, every game position is either of type N (Next player to move wins) or P (Previous player wins).

that is routinely used to correctly predict physical phenomena with (sometimes unwarranted!) accuracy, there are very few cases where it has been rigorously proven to be correct. In this same spirit, we hope that the reader will find the insights provided by the renormalization analysis of games to be sufficiently compelling so as to warrant its serious consideration as a practical and powerful method of analysis for combinatorial games, despite its non-rigorous status at present.

II: Unlike Sprague–Grundy theory [Bouton 1902; Sprague 1936; Grundy 1939] and its extensions to numbers and nimbers [Berlekamp et al. 1979] which have proven extremely successful for analyzing games that can be decomposed (i.e., expressed as a disjunctive sum of simpler games) such as Dots-and-Boxes [Berlekamp 2000] and Go endgames [Berlekamp 1994], the renormalization approach to games works equally well for decomposable and non-decomposable games. Indeed, many interesting games, such as the early play in chess and Go, have resisted analysis using traditional methods due to their intrinsic non-decomposability. Very little is in fact understood about the optimal strategies in non-decomposable games, even such "elementary" ones as Chomp. We will demonstrate how renormalization can readily handle a non-decomposable game such as Chomp and raise the possibility that such an approach could be extended to more complex games such as go, chess or checkers.

III: One of the interesting features of renormalization is that it results in probabilistic information about the game, despite the fact that we consider only purely deterministic games (i.e., games of no chance). In particular, rather than providing a description of the exact locations (in position space) of the P-positions of a game, renormalization only specifies the probability that a given position is P. However, there are often relatively sharp boundaries associated with these probabilities. So even though renormalization cannot give us the precise (point-by-point) geometry of the game, it will allow us to calculate its broad, overall geometric features, which in fact provides significant information about the game. Indeed, we believe that this inherent "imprecision" in the methodology, rather than being a shortcoming, is in fact what allows renormalization to proceed and what gives it its power. We conjecture that for many combinatorial games there do not exist any simple formulas or polynomial-time algorithms for efficiently computing the exact location of the P-positions, but that probabilistic information is possible. By sacrificing exact geometric information about the game for probabilistic information, significant insights into "hard" combinatorial games can be obtained. This is reminiscent of chaotic dynamical systems and strange attractors in which there does not exist formulas for specific trajectories [Cvitanovic 1989], but global information about the overall structure of attractor does exist. We discuss this further in later sections, but comment here that this view is supported by numerical evidence on the "sensitive dependence on initial conditions" displayed by the game Chomp and discussed in Section 3.2.

IV: The renormalization analysis brings to light several previously unexplored features of combinatorial games, and indeed in certain respects we consider these new lines of inquiry to be one of the highlights of this new approach. In Section 3.1 we introduce the notion of universality, and describe how renormalization provides a natural classification scheme for combinatorial games, wherein games can be grouped into "universality classes" such that all members of a class share key features in common. In Section 3.2 we show how it is possible to discuss the "sensitivity" of a game to certain types of perturbations (i.e., rule changes) within the renormalization framework. And in Section 3.3 we describe how this method reveals unexpected similarities between the geometric structures seen in games and various crystal-growth models and aggregation processes in physics.

V: As a final comment, we emphasize that as a new approach to combinatorial games, renormalization is still very much in its infancy; its limitations, shortcomings, and scope of applicability are not fully understood at present. Hence, in what follows we will simply give a number of worked examples of this method applied to specific games, so that the reader might develop a working feel for how the procedure is actually implemented, and perhaps appreciate its potential utility.

2. Renormalization framework

We begin with a schematic overview of the general renormalization procedure.

The first step is to create a natural geometry for the game. Towards this end, consider the abstract "space" of all positions of a game. Typically, this space can be realized by mapping game positions to a subset of the integer lattice $\mathbb{Z}^d$ for some dimension $d > 1$. The set of all P-positions in this d-dimensional position space, which we call the "P-set", is the key geometric 'object' for study.

To proceed, we next define various sets of $(d - 1)$-dimensional hyperplanes ("sheets") that foliate position space. Here, we will let $x \in \mathbb{Z}$ specify the index of a sheet, and $y \in \mathbb{Z}^{d-1}$ the coordinates on the sheet. As we will see, there exist various types of recursion relations and nonlinear operators that relate the different sheets to one another. These sheets will prove instrumental for determining the overall geometric structure of the game's P-set.

There are several basic types of foliating sheets to consider. The first are the P-sheets. These simply mark the location of the game's P-positions within each hyperplane. More precisely, define P_x, the P-sheet at level x, to be a $(d - 1)$-

dimensional, semi-infinite matrix consisting of zeros and ones, with ones marking the locations of the P-positions (i.e., $P_x(y) = 1$ if game position $[x, y] \in \mathbb{Z}^d$ is a P-position and 0 otherwise). We will be interested in the geometric patterns (of the ones and zeros) on these sheets, since, taken together, they capture the full geometry of the P-set in d-dimensional position space.

A second type of foliating sheets are the instant-N sheets (IN-sheets for short). They are constructed as follows: Following [Zeilberger 2004], we declare an N-position in the game, $[x, y]$, to be an IN if there exists a legal move from that position to some P-position $[x', y']$ on a lower sheet, $x' < x$. The IN-sheets are simply hyperplanes through position space that mark the locations of the IN's (i.e., defining matrix W_x to be the IN-sheet at level x, set $W_x(y) = 1$ if position $[x, y]$ is an IN and 0 otherwise). As we will see, their key significance lies in the fact that the P-sheets (and, ultimately, the P-set itself) can be computed directly from the IN-sheets via the relation $P_x = \mathcal{M} W_x$, where $\mathcal{M}$ denotes a "supermex" operator (a generalization of the standard Mex operator). Hence, we can think of the IN-sheets as effectively encoding the critical information about the game. Moreover, they will prove useful for visualizing a game's geometric features.

Now, in many examples (e.g., the first three discussed in this paper), it is possible to write down a recursion relation on the IN-sheets:

$$W_x = \mathcal{R} W_{x-1}.$$

As will become clear, this is the key step in the renormalization analysis, since it allows the IN-sheet at level x to be generated directly from its immediate predecessor.[2]

We note, however, that in general it is not always possible to construct a recursive formulation on the IN-sheets themselves, as shown in our 4th example. In such cases, we show how to construct auxiliary sheets $V_x^1, \ldots, V_x^k$ for which a (vector) recursion relation of the form $V_x = \mathcal{R} V_{x-1}$ does exist. (The W_x can then be computed from the vector of sheets V_x.) For ease of presentation we will assume for the remainder of this section that there exists a direct recursion relation for the IN-sheets themselves (making auxiliary sheets unnecessary); however, we will demonstrate the alternate case in an example.

Thus far, the overall scheme is as follows (see Figure 2): We first recursively generate the IN-sheets using the recursion operator $\mathcal{R}$, then use the supermex operator $\mathcal{M}$ to construct the P-sheets. The final key to the renormalization

---

[2] We remark that for all the games considered here, a judicious choice of position-space coordinates allows one to recursively compute the P-sheet at level x, P_x, from all the preceding sheets, $P_0, P_1, \ldots P_{x-1}$. However, this type of 'infinite'-dimensional recursive formulation is not directly useful for renormalization purposes, since one has to know *all* preceding sheets just to compute the current one; to apply renormalization effectively we require a 'finite'-dimensional recursion relation, like that for the IN-sheets. (Nonetheless, the assumption that P_x can be expressed in terms of all preceding sheets is useful for other parts of the analysis, and we will assume that this is always the case.)

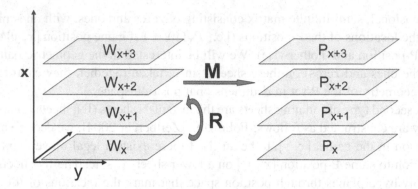

Figure 2. Foliating sheets and associated operator relations.

scheme is the observation that the IN-sheets exhibit a form of 'geometrical invariance'. Loosely speaking, the geometric patterns on sheets at different x levels all look similar to one another (i.e., they exhibit "scaling" — here, shape-preserving growth with increasing x). This allows for a compact description of W_x for large x. In the examples considered we will see that, in some sense, the sheets W_x converge to a specific geometry. However, understanding and defining this 'convergence' relies on two key observations:

The first is that since the geometric structures (i.e. patterns of 0's and 1's) on these IN-sheets 'grow' with increasing x (but maintain their overall shape), we must re-scale (shrink) the sheets with larger x values to see the convergence. We will introduce a rescaling operator $\mathcal{S}$ for this purpose. Second, while the precise structure (point by point) of the IN-sheets does not converge, the overall probabilistic structure (i.e., densities of points) on the IN-sheets does converge. Formally, we must define a probability measure on the sheets for this purpose.

Hence, the asymptotic behavior of the IN-sheets (i.e., their 'invariant geometry') is described by the limiting probability measure $\mathbf{W} = \lim_{x \to \infty} (\mathcal{S}\mathcal{R})^x W_0$. Here, we think of the operator $\mathcal{R}$ as 'growing' an IN-sheet at a given level to the next higher level (since $\mathcal{R}W_x = W_{x+1}$), and we think of the operator $\mathcal{S}$ as inducing a simple geometrical rescaling of the grown sheet back to the original size. Repeated application of these growing and rescaling operators, starting at the initial sheet W_0, yields the desired limiting probability measure $\mathbf{W}$. However, since this limit is independent of the initial sheet W_0 for most interesting cases, we can alternatively express $\mathbf{W}$ as a fixed point of the equation

$$\mathbf{W} = \mathcal{S}\mathcal{R}\mathbf{W}.$$

This "renormalization" equation (with $\mathcal{S}\mathcal{R}$ the "renormalization operator") is our key equation. It states that the invariant measure on a sheet is unchanged if you grow the sheet and then rescale it, thereby expressing the invariance of the

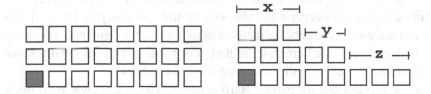

Figure 3. The game of Chomp. Left: the starting configuration of three-
-row ($M = 3$) Chomp. Right: a sample game configuration after play has
begun (describable by coordinates $[x, y, z]$).

geometry on the different IN-sheets. The solution to this equation will provide
a complete probabilistic description of the game (including the geometry of its
P-positions), and will allow us to understand much about the game. As will be
illustrated, in practice the above renormalization equation is most easily solved
by deriving a series of related algebraic self-consistency conditions.

We now present several examples. All the games we have chosen share
certain features in common. First, they are all impartial games, and do not
allow draws or ties. They are also all poset games for which it is possible
to define a complete ordering on position space such that the position values
strictly decrease during play. Whether these conditions are inherent limitations
on the scope of applicability of the renormalization methodology remains to be
seen, although we strongly suspect that they are not. For convenience, we have
chosen all of our examples to have three-dimensional position spaces, so as to
make the visualization of the resulting patterns (and the analytic calculations)
more transparent.

2.1. Chomp. We focus first on the game of Chomp, which is, in some sense,
among the simplest of the "unsolved" games. Its history is marked by some
significant theoretical advances [Gale 1974; Schuh 1952; Zeilberger 2001; Zeilberger 2004; Sun 2002; Byrnes 2003], but it has yet to succumb to a complete
analysis in the 30 years since its introduction by Gale [1974] and Schuh [1952].
(Chomp is an example of a non-decomposable game, where traditional methods
so far have not proven to be especially effective.)

The rules of Chomp are easily explained. Play begins with an $M \times N$ array
of tokens, with the (dark) token in the southwest corner considered "poison"
(Figure 3, left). On each turn a player selects a token and removes ("chomps")
it along with all tokens to the north and east of it. (Figure 3, right, shows a
sample token configuration after two chomps.) Players alternate turns until one
player takes the poison token, thereby losing the game.

For simplicity, we consider here the case of three-row ($M = 3$) Chomp, a
subject of recent study [Zeilberger 2001; Zeilberger 2004; Sun 2002; Byrnes

2003]. To start, note that the token configuration at any stage of play can be described (using Zeilberger's coordinates) by the position $p = [x, y, z]$, where x, y, z specify the number of columns of height three, two, and one, respectively (Figure 3, right). (Note here that the first coordinate x will eventually serve as our sheet index, and $[y, z]$ the coordinates within a sheet.)

In these coordinates, the game's starting position is $[x, 0, 0]$, while the opening move must be to a position of the form $[x-r, r, 0]$, $[x-s, 0, s]$ or $[x-t, 0, 0]$ (these are the "children" of the starting position). Every position may be classified as either an N-position — if a player starting from that position can guarantee a win under perfect play — or as a P-position otherwise (draws and ties not being possible). The computation of N- and P-positions rests on the standard observation that all children of a P-position must be N-positions, and at least one child of an N-position must be a P-position. For example, position $[0, 0, 1]$ (where only the poison token remains) is a P-position by definition, so $[0, 1, 0]$ must be an N-position since its child is $[0, 0, 1]$. (Note that a *winning move* in the game is always from an N-position to a P-position.)

An intriguing feature of Chomp, as shown by Gale [1974], is that the player who moves first can always win under optimal play (i.e., $[x, 0, 0]$ is an N-position). The proof uses an elegant strategy-stealing argument: Consider the "nibble" move to $[x - 1, 1, 0]$. If this is a winning move, then we are done. If it is not a winning move, then the second player must have a winning response, in which case the first player could have chosen this as the opening move instead of the nibble, leading to a win. Observe that this argument provides no information as to what the desired opening move for the first player should be (or even whether it is unique), only that it exists — a longstanding question that the renormalization analysis will address.

In previous numerical studies of the game by Brouwer [2004] and others, several linear scaling relations were noticed. For example, for every x (under $\approx 80,000$) there is a P-position of the form $[x, 0, z]$ where $z = 0.7x \pm 1.75$; other sequences with similar linear scaling behavior were also observed. Zeilberger [2001], Sun [2002] and Byrnes [2003] also find more complex patterns in the P-positions, including periodic orbits and intimations of possible chaotic-like behavior. The existence of these numerically observed scaling behaviors provides the first hint that some type of renormalization approach may be possible, as we now describe.

To begin, we introduce the foliating sheets, indexed by their x values, as in [Zeilberger 2004]. (As noted previously, sheet index x here corresponds to the first coordinate of Chomp's three-dimensional position space $[x, y, z]$.) For any x, recall that the P-sheet P_x is a two-dimensional, semi-infinite matrix that marks the location of all P-positions at the specified x value. The $(y, z)^{\text{th}}$

element of this matrix is denoted $P_x(y, z)$. (We note for future reference the easily proven fact that for every x there exists at most one z, call it $z^*(x)$, such that $[x, 0, z^*(x)]$ is a P-position, i.e. $P_x(0, z^*(x)) = 1$.) The IN-sheets are defined as in the previous section: $W_x(y, z) = 1$ if $[x, y, z]$ is an IN, and 0 otherwise[3].

As noted earlier, the IN-sheet W_x contains all the necessary information for computing the corresponding P-sheet P_x, and, moreover, one can calculate W_{x+1} directly from W_x. To see this, we define the following operators:

Identity I: for any sheet A, let $(IA)(y, z) = A(y, z)$.

Left shift $\mathcal{L}$: for any sheet A, let $(\mathcal{L}A)(y, z) = A(y+1, z)$.

Diagonal $\mathcal{D}$: for any x the action of $\mathcal{D}$ on the P-sheet P_x is given by

$$(\mathcal{D}P_x)(z^*(x) - t, t) = 1 \text{ for } 0 \leq t \leq z^*(x),$$
$$(\mathcal{D}P_x)(y, z) = P_x(y, z) \text{ otherwise.}$$

Supermex $\mathcal{M}$: for any x the action of $\mathcal{M}$ on W_x is defined via the following algorithm:

(1) Set $\mathcal{M}W_x = 0$, $T_x = W_x$, $y = 0$.
(2) Let z_s be the smallest z such that $T_x(y, z) = 0$ and set $(\mathcal{M}W_x)(y, z_s) = 1$, $T_x(y + t, z_s - t) = 1$ for all $0 \leq t \leq z_s$.
(3) If $z_s = 0$ stop; else let $y \to y + 1$ and go to step 2.

A direct computation shows (see [Friedman and Landsberg 2007] for details):

$$P_x = \mathcal{M}W_x, \quad W_{x+1} = \mathcal{L}(I + \mathcal{D}\mathcal{M})W_x.$$

Thus, defining $\mathcal{R} \equiv \mathcal{L}(I + \mathcal{D}\mathcal{M})$ yields $W_{x+1} = \mathcal{R}W_x$. (These relations all follow simply from a careful application of the game rules.) This provides the setting for a renormalization analysis. (For future reference, we also mention one additional relation which is sometimes of use: $W_x = \sum_{t=1}^{x} \mathcal{L}^t \mathcal{D} P_{x-t}$, where all sums are binary and interpreted as logical OR's. This relation follows from the observation that the IN positions at level x are generated from the parents[4] of P-positions at all lower levels.)

Numerical solution of the recursion equation $W_{x+1} = \mathcal{R}W_x$ reveals an interesting structure for the IN-sheets (W_x), characterized by several distinct regions (Figure 4a). Most crucially, the IN-sheets at different x levels 'scale' (see, e.g., Figure 4b): their overall geometric structures are identical (in a probabilistic sense) up to a linear scale factor. In particular, the boundary-line slopes and densities of points in the interior regions of the sheets W_x are the same for all

---

[3] This definition differs in a small but important way from the "instant-winner" sheets introduced in [Zeilberger 2004].

[4] A *parent* of position p is defined as any position from which it is possible to reach p in a single move.

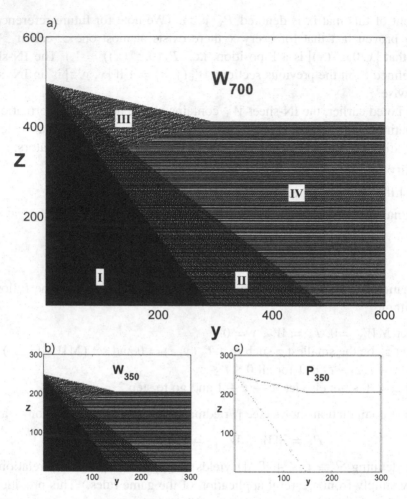

Figure 4. The geometry of Chomp. (a) the IN-sheet geometry W_x for three-row Chomp, shown for $x = 700$. Here, IN locations in the y-z plane (i.e., the 1's in the matrix) are shown in black. (b) The IN-sheet W_x for $x = 350$. Comparison with W_{700} illustrates the geometric invariance of the sheets. (c) The geometry of P_x, shown for $x = 350$. The P-sheets also exhibit geometrical invariance, i.e., the P_x for all x exhibit identical structure (in the probabilistic sense) up to an overall scale factor.

x (though the actual point-by-point locations of the instant-N positions within the sheets will differ).

Thus, the invariant geometry W of the sheets satisfies the renormalization fixed-point equation $W = SRW$, with operators S and R defined above. We now show how to analyze this equation, and ultimately determine the structure of Chomp's P-set. We point out that a direct, formal assault on the renormalization equation would require that the renormalization operator be carefully defined on the space of probability distributions over IN-sheets, which proves somewhat delicate. In practice, the desired result can be more easily obtained through a somewhat less formal procedure, as we now describe.

We begin by considering the structure of a typical P-sheet P_x (Figure 4c). Numerically, it is found to consist of three (diffuse) lines (heretofore called P-lines) that may be characterized by six fundamental geometric parameters: a lower P-line of slope m_L and density of points λ_L, an upper line of slope m_U and density λ_U, and a flat line extending to infinity. The upper and lower P-lines originate from a point whose height (i.e., z-value) is αx. The flat line (with density one) is only present with probability γ in randomly selected P-sheets. Our goal is to determine analytical values for these six geometric parameters that characterize the P-set. (Recall that the IN-sheet geometry can be directly linked with this P-sheet geometry via $W_x = \sum_{t=1}^{x} \mathcal{L}^t \mathcal{D} P_{x-t}$.) Hence, a determination of the parameters $m_L, \lambda_L, m_U, \lambda_U, \alpha, \gamma$ will provide a complete probabilistic description of the entire geometric structure of the game[5]

To get at this geometry, we will derive a set of algebraic self-consistency equations relating the six geometric parameters. Intuitively, these equations arise from the demand that as an IN-sheet at level x (W_x) 'grows' to W_{x+1} under the action of the recursion operator $\mathcal{R}$, its overall geometry is preserved. The key to actually implementing this analysis is to observe that the P-positions in sheet P_x (i.e., the 1's of the matrix; see Figure 4c) are constrained to lie along certain boundaries in W_x (Figure 4b); the various interior regions of W_x remain "forbidden" to P-positions. Geometric invariance of the sheets demands that these forbidden regions be preserved as an IN-sheet grows under the recursion operator. Each such forbidden region yields a constraint on the allowable geometry of the W_x's, and may be formulated as an algebraic equation relating the hitherto unknown parameters $m_L, \lambda_L, m_U, \lambda_U, \alpha, \gamma$ that define the P-sheets. In all, we find six independent geometric constraints:

$$\frac{\lambda_U}{1 + m_U} = 1, \tag{2-1}$$

$$\frac{1}{1 + \alpha} - \frac{\lambda_L}{1 + m_L} = 1, \tag{2-2}$$

---

[5]Here, we will not be addressing the interesting issue of the small 'scatter' of points around the P-lines. Numerical simulations show that the range of scatter is in fact extremely narrow (e.g., the distance from a P-position to the idealized P-line is always less than 5 for $x < 1,000$, and does not appear to increase with x). Despite its smallness, the scatter is not at all irrelevant, and indeed, we believe it is largely the scatter that makes a purely deterministic analysis of Chomp hard. Our probabilistic description provides a means of bypassing much of this difficulty while still extracting useful information about the game. We will revisit this notion briefly in Section 4.

$$(\gamma - 1)\frac{m_L}{\alpha - m_L} + \frac{1}{1 + \alpha} = 1, \qquad (2\text{-}3)$$

$$\lambda_U + \lambda_L = 1, \qquad (2\text{-}4)$$

$$\frac{\alpha \lambda_L}{\alpha - m_L}\left(\frac{m_U - m_L}{(m_U - m_L)\alpha + m_L\gamma}\right) + \frac{1}{1 + \alpha} = 1, \qquad (2\text{-}5)$$

$$\frac{\lambda_L}{\alpha - m_L} - \frac{\alpha}{\alpha + 1}\left(1 - \frac{\lambda_U}{\alpha - m_U}\right) = 0. \qquad (2\text{-}6)$$

These six constraints arise as follows (see Figure 4a): (1) arises from forbidden region III; (2) from region II; (3) from the bottom row of region I; (4) from operator $\mathcal{M}$ in regions I,II,III; (5) from the lower part of region I; and (6) from the upper part of region I. To illustrate we derive constraint (3) here. (For detailed derivations of the others see [Friedman and Landsberg 2007].)

Recall that $W_x = \sum_{t=1}^{x} \mathcal{L}^t \mathcal{D} P_{x-t}$, so that the IN-sheet at level x is 'built up' from a series of earlier P-lines (coming from lower-level sheets) and diagonal lines (associated with operator $\mathcal{D}$). Constraint (3) arises because the lower P-lines and diagonal lines each contribute points (i.e., instant-N's) to the bottommost row of region I and completely fill it up, thereby rendering it forbidden. Now, the density of the diagonal lines along the bottom row of W_x can be computed from elementary geometry to be $(1 + \alpha)^{-1}$. The density of the lower P-lines is $-m_L/(\alpha - m_L)$. However, each lower P-line will only contribute a point to the bottom row (at $z = 0$) with probability $(1 - \gamma)$, since this equals the probability that the flat P-line doesn't exist (by step 3 in the supermex algorithm). Hence, the actual density of points contributed by the lower P-lines to the bottom row becomes $-(1 - \gamma)m_L/(\alpha - m_L)$. Summing this density with that of the diagonals and equating to unity yields constraint (3) (i.e., this is the condition that the bottom row of the IN-sheets always remains forbidden to P-positions even as the sheets grow.)

Taken together, the above six constraints may be solved exactly, yielding precise values for the geometric parameters of the game. These are:

$$\alpha = \tfrac{1}{\sqrt{2}}, \ \lambda_L = 1 - \tfrac{1}{\sqrt{2}}, \ \lambda_U = \tfrac{1}{\sqrt{2}},$$

$$m_L = -1 - \tfrac{1}{\sqrt{2}}, \ m_U = -1 + \tfrac{1}{\sqrt{2}}, \ \gamma = \sqrt{2} - 1.$$

Thus, we have found the renormalization fixed point, and hence have a complete (probabilistic) characterization of the game of Chomp—i.e., the global geometric structure of its P-set.

With this geometric insight, it becomes straightforward to explain virtually all numerical properties of the game previously reported in the literature, including various numerical conjectures by Brouwer [2004], along with a variety of new results. As an illustration of its utility, we show how this geometric result lets us decide (in a probabilistic sense) the optimal first move of the game, which has been a longstanding open question.

To start, recall that the possible opening moves from the starting position $[x, 0, 0]$ are to positions of the form $[x-r, r, 0]$, $[x-s, 0, s]$ or $[x-t, 0, 0]$ (bearing in mind that the desired (winning) opening move will be to a P-position). The last of these, $[x - t, 0, 0]$, can never be a P-position, by Gale's strategy-stealing argument. Next consider $[x - r, r, 0]$, which we will refer to as an "r"-position. From the geometric structure of the P-sheets, a simple calculation shows that the only accessible P-position of this form is for $r(x) \approx \alpha x / (\alpha - m_L)$. Likewise, the only possible P-position of the form $[x - s, 0, s]$ (i.e., an "s"-position) is for $s(x) \approx \alpha x / (\alpha + 1)$. (Note that we use $\approx$ here since these are asymptotic values; for any finite x there are small deviations owing to the slight scatter of P-positions around the P-lines in Figure 4c.) Thus, the P-set geometry has allowed us to identify the asymptotic locations of the only two possible winning opening moves in the game! Moreover, since $r(x) < s(x)$, the "s"-position is a child of the "r"-position, so only one of these two positions can be an actual P-position (for a given x). Hence the winning opening move is unique — a result which was previously only known numerically [Brouwer 2004] for x values up to a certain level. Taking this further, we can also compute the probabilities that this unique winning opening move will be to the "r"-position or to the "s"-position, as follows: For each starting position $[x, 0, 0]$ from $x = 1 \ldots x_{\max}$ there is an associated "r"-position $[x - r(x), r(x), 0]$, which may or may not be a P-position. The total number of actual "r"-type P-positions with an x-value less than or equal to $x_{\max} - r(x_{\max})$ is just $\gamma(x_{\max} - r(x_{\max}))$. So the fraction of "r"-positions which are actually P-positions is $\gamma(x_{\max} - r(x_{\max})) / x_{\max} = \sqrt{2} - 1$. Thus, the winning opening move is to the "r"-position with probability $\sqrt{2} - 1$, and to the "s"-position with probability $2 - \sqrt{2}$.

The ease with which the above results were obtained (once the P-set geometry was determined) illustrates the utility of this geometrically based, renormalization approach as a potentially powerful tool for analyzing games.

2.2. Nim. We next consider three-heap Nim [Bouton 1902]. Note that this simple game is decomposable and "solvable" (in the sense that a simple criterion exists for deciding if a given position is N or P). In this game we let $[x, y, z]$ represent the heights of the three heaps. An allowed move in Nim consists of reducing a single position coordinate by an arbitrary amount. As in Chomp, we will let the various hyperplanes be indexed by x, and the coordinates within each plane by $[y, z]$. (Note that while these coordinates break the natural permutation symmetry of the heaps, the resulting analysis is not affected.)

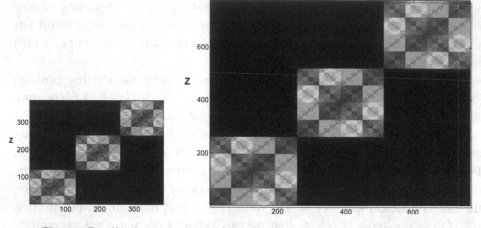

Figure 5. IN-sheet geometry for ordinary Nim: W_{128} (left) and W_{256} (right). (Note here that the instant-N positions have been color-coded based on the order in which they were recursively generated; the black background corresponds to non-instant-N's.)

A straightforward calculation based on the game rules shows that the IN-sheets are related to the P-sheets by

$$W_x = \sum_{x'=0}^{x-1} P_{x'}, \qquad (*)$$

where addition denotes the logical OR operation. This relation simply reflects the fact that the IN's at level x are, by definition, determined by the parents of the P-positions at the lower levels.

We define the action of the supermex operator $\mathcal{M}$ on W_x via the following algorithm:

(1) Set $\mathcal{M}W_x = 0$, $T_x = W_x$, $y = 0$.
(2) Let z_s be the smallest z such that $T_x(y, z) = 0$ and set $(\mathcal{M}W_x)(y, z_s) = 1$, $T_x(y + t, z_s) = 1$ for all $0 \le t$.
(3) Let $y \to y + 1$ and go to step 2.

This yields the relation,

$$P_x = \mathcal{M}W_x, \qquad (**)$$

as the reader may verify. Combining $(*)$ and $(**)$ yields the desired recursion formula,

$$W_{x+1} = \mathcal{R}W_x,$$

with $\mathcal{R} = I + \mathcal{M}$ and I the identity operator.

The IN-sheets are readily constructed using this recursion operator. Figure 5 displays the sheet geometry W_x for $x = 128$ and $x = 256$. Again, we observe that the geometry of the game's IN-sheets scale (linearly with x), just as in the preceding case of Chomp. As before, one could construct algebraic self-consistency conditions that exploit this scaling, and thereby develop a geometric characterization of the sheets. This is unnecessary here since the game of Nim is easily solvable, and all sheets can be directly constructed using nimbers instead. So for the moment we will content ourselves with having set up the basic renormalization framework for the game. However, we will revisit this issue when we discuss a modified (nontrivial) version of Nim later in this paper.

Before moving to our next example, we remark briefly on one unique feature of Nim (not seen in the earlier Chomp example). In Chomp, all W_x's (regardless of x) look geometrically similar up to linear rescaling. In Nim, the sheets exhibit linear scaling, but also display a periodicity (in x) in powers of 2. For instance, the sequence $W_{128}, W_{256}, W_{512}, \ldots$ exhibits geometric invariance (up to rescaling), as does the sequence $W_{100}, W_{200}, W_{400}, \ldots$. However, the basic patterns for these two sequences will differ somewhat (e.g., Figure 1b illustrates the geometry for this second sequence.) The existence of this periodicity means that the original renormalization equation will not have a true fixed point; however, in practice this can be handled by using a slightly modified renormalization equation which exploits this periodicity (loosely speaking, $W_{2x} = (\mathcal{SR})^x W_x$), but we do not pursue this further. (Nim is the only one of our examples to display this feature.)

2.3. 2-D Wythoff's game and Sprague–Grundy values.

In our next example we show how renormalization can be used to compute the Sprague–Grundy values of a game. We illustrate here with Wythoff's game [Wythoff 1907], whose Grundy values have been the subject of a recent study by Nivasch [2004]. Wythoff's game is equivalent to two-heap Nim, where in addition to removing an arbitrary number from either heap a player can also remove the same number from both heaps. Thus if coordinates $[y, z]$ represent the heights of the two heaps then a legal move reduces either of the coordinates by an arbitrary amount, or both by the same amount.

It is well known that the P-positions of Wythoff's game are all of the form $(\lfloor \phi k \rfloor, \lfloor \phi^2 k \rfloor)$ and $(\lfloor \phi^2 k \rfloor, \lfloor \phi k \rfloor)$ for all positive integers $k > 0$, where $\phi = (\sqrt{5} + 1)/2$ is the golden ratio [Wythoff 1907]. Thus, they lie near the lines through the origin with slopes ϕ and ϕ^{-1}. However, the characterization of the Sprague–Grundy values for Wythoff's game is significantly more difficult. Nivasch [2004] has shown that these Grundy values also lie 'close' to these lines; specifically, that a position with Grundy value g is bounded within a distance $O(g)$ of these lines. We show that this result follows directly from a straightforward renormalization analysis — although our proof is not rigorous, whereas Nivasch's is.

The general procedure for computing the Sprague–Grundy values of a game via renormalization is straightforward: Add a single Nim heap to be played in conjunction (i.e., disjunctive sum) with the game of interest, and then do ordinary renormalization on the combined game. In the present case we represent the position space of the combined game by coordinates $[x, y, z]$, where a player can either move in Wythoff's game, $[y, z]$, or play on the Nim pile by reducing x by an arbitrary amount. Then, a standard argument shows that Wythoff position $[y, z]$ has Grundy value x if $[x, y, z]$ is a P-position of the combined game. Thus the sheets P_x correspond to the set of all positions in Wythoff's game with Grundy value x.

As in Nim, we use the IN-sheets and note that[6]

$$W_x = \sum_{x'=0}^{x-1} P_{x'}.$$

One can compute $\mathcal{M}$ from the properties of Wythoff's game:

(1) Set $\mathcal{M}W_x = 0$, $T_x = W_x$, $y = 0$.
(2) Let z_s be the smallest z such that $T_x(y, z) = 0$ and set $(\mathcal{M}W_x)(y, z_s) = 1$, $T_x(y + t, z_s) = 1$, $T_x(y + t, z_s + t) = 1$ for all $0 \le t$.
(3) Let $y \to y + 1$ and go to step 2.

Combining this supermex algorithm with the preceding expression yields the recursion operator

$$\mathcal{R} = I + \mathcal{M}.$$

We now analyze the invariant geometry of the game, and show how it explains Nivasch's result.

A representative IN-sheet and P-sheet are shown in Figure 6. We will focus on the outer regions of these graphs (i.e., the large $[y, z]$ regime), avoiding the more complicated structures near the origin. Here, the IN-sheet consists of two (thick) lines, an upper and lower one, whose slopes we denote by m_U, m_L. The P-sheet also exhibits two related lines (P-lines) of the same slopes as in the IN-sheet (we neglect the fact that closer inspection shows that each P-line is a double line, as this will not affect the calculation). We let λ_U, λ_L denote the density of points (i.e., P-positions) along these lines (per unit horizontal).

The renormalization analysis of the invariant geometry is simplified by the observation that the regions of Figure 6 labeled I, II, III, IV are entirely devoid of P-positions. (More precisely, if the P-sheet and IN-sheet plots are superimposed, the P-positions do not appear in the four labeled regions.) These four regions are thus "forbidden." The fact that they contain no P-positions provides

---

[6]This holds true for a sum of any game with a single Nim pile, which arises whenever one analyzes the Sprague–Grundy values.

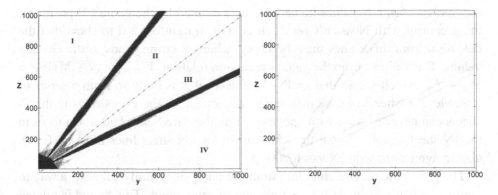

Figure 6. IN-sheet and P-sheet associated with the Sprague–Grundy values of Wythoff's game: W_{100} (left) and P_{100} (right). Note that a 45°-line has been artificially added to the W_{100} plot so as to demarcate regions II and III.

constraints that allow us to compute analytical values for the four parameters $m_U, m_L, \lambda_U, \lambda_L$ that characterize the invariant geometry of the P-set.

The absence of P-positions in the forbidden regions is due to the fact that these empty regions get completely filled up (during the supermex operation $\Gamma_x = \mathcal{M} W_x$) by parents of the P-positions. (Note that these parents cannot themselves be P-positions.) Within any given sheet, the parents of a position $[y, z]$ lie along three lines (one vertical, one horizontal, and one diagonal): $V = \{[y+k, z] \mid k > 0\}$, $H = \{[y, z+k] \mid k > 0\}$, and $D = \{[y+k, z+k] \mid k > 0\}$. Forbidden region (I) gets completely covered by the vertical lines V arising from (parents of) P-positions along the lower and upper P-lines. The density of these (per unit y) is given by $\lambda_L + \lambda_U$ which must equal 1, since they can't overlap and must completely fill the region. Likewise, Forbidden region (IV) is completely filled by horizontal lines arising from the upper and lower P-lines. Since their densities (per unit z) are λ_i / m_i, $i = L, U$, it follows that $\lambda_L / m_L + \lambda_U / m_U = 1$. Forbidden region (II) is filled by the diagonal lines emerging from the upper P-line. Elementary geometry shows the density of these lines to be $\lambda_U / (m_U - 1)$, yielding $\lambda_U / (m_U - 1) = 1$. (We note that horizontal lines from the upper P-line and vertical lines from the lower P-line also contribute to region (II), but since neither of these — either alone or in combination — is sufficient to completely fill the region, and since they are not well correlated with the diagonal line, it must be the case that the diagonal lines alone are sufficient to fill the region.) A similar argument for region (III) shows that the diagonals emerging from the lower P-line completely fill the region, yielding $\lambda_L / (1 - m_L) = 1$.

Solving these four constraints we find that $m_U = \phi = m_L^{-1}$ and $\lambda_U = \phi^{-1}$, $\lambda_L = 1 - \phi^{-1}$, which agree with numerical observations. Thus, we see that the Sprague–Grundy values lie near the rays defined by the game's P-positions, in agreement with Nivasch's result. It is also straightforward to show that the deviation from these lines must be $O(x)$ (where x corresponds to the Grundy value). This follows from the game's recursion relation, $W_{x+1} = (I + \mathcal{M})W_x = W_x + P_x$, which shows that an IN-sheet at level x is built up from a series of lower-level P-sheets (whose total number is x). Since the P-positions in the P-sheets can never overlap with one another as they are being laid down to form the IN-sheet, it follows that the width of the two N-sheet lines must be $O(x)$, also in agreement with [Nivasch 2004].

The geometric picture emerging from our analysis actually suggests a way to compute a crude estimate for the tightness of this bound. This bound is related to the width of the two (thick) lines in the sheet W_x, which we can calculate as follows: Consider a section of horizontal extent S of one of these lines. The area occupied by this section of line is just that of a rectangle with length $S\sqrt{1+m^2}$ and thickness w (where m denotes the slope of the line, either m_U or m_L). The total number of points making up this area is just $\lambda S x$ (since the thick line is built from x P-lines, each one contributing λS points, with $\lambda = \lambda_U$ or λ_L). Since this area is completely filled, it must have density 1. Solving, we find that the line thickness is $w = x/(\phi\sqrt{1+\phi^2})$. Thus, our probabilistic estimate is that, asymptotically speaking, a game position with Grundy value g will roughly lie within a distance of $g/(\phi\sqrt{1+\phi^2})$ of the known P-lines. (Here, asymptotic refers to game positions with suitably large values of $[y, z]$, so that we are far from the complex structure located near the origin of Figure 6.)

In summary, this analysis illustrates how the renormalization method can be used to rather easily (albeit nonrigorously) obtain results that are difficult to obtain by more traditional methods, including Sprague–Grundy results. Moreover, this type of geometric analysis reveals insights that are less apparent by other means. In the present case, for instance, we find a complex structure near the origin of the IN- and P-sheets, which (to our knowledge) has not been recognized before. Separate treatments of the local and asymptotic structures in this game would presumably allow one to derive even tighter analytical bounds than were obtained in [Nivasch 2004].

2.4. Three-dimensional Wythoff's game.

As our last example, we consider a 3-D generalization of the ordinary (2-D) Wythoff's game, for which, as far as we know, relatively little is known. Here, we will not carry out the complete renormalization analysis, but will derive the necessary analytical operators and recursion relations (which in this case will require the use of auxiliary sheets in addition to the IN-sheets), and also numerically illustrate the geometric scaling property of these sheets.

The rules of the game are as follows: Letting $[x, y, z]$ denote the three heap sizes, one can remove one or more tokens from a single heap, or the same number from any pair of heaps. Other versions of the game are also possible: For instance, one could replace the rule about removing tokens from any pair of heaps with one allowing removal of an equal number of tokens from all three heaps, or keep all the original rules and supplement them with this additional one. In any case, the derivation of the recursion operators for these alternate versions will be entirely analogous to the game version we will illustrate here.

We note that in this example there does not exist a recursion relation among the IN-sheets. This is related to the set of legal moves in the game. In this case, there are three distinct types of legal moves from a higher sheet at level x to a lower sheet at level x', with $x' < x$. These are the 'straight' move $[x, y, z] \rightarrow [x', y, z]$ and the two 'diagonal' moves $[x, y, z] \rightarrow [x', y-(x-x'), z]$ and $[x, y, z] \rightarrow [x', y, z - (x - x')]$. We will require one auxiliary sheet for each of these moves, V^1, V^2, V^3. The first sheet, associated with the straight move, is constructed as a sum (logical OR's): $V_x^1 = \sum_{x'=0}^{x-1} P_{x'}$. The second sheet is constructed via right-shifted sums (i.e., shifts along the y-axis in $[y, z]$ space) $V_x^2 = \sum_{x'=0}^{x-1} y^{x-x'} P_{x'}$ and the third via 'upward'-shifts along the z-axis, $V_x^3 = \sum_{x'=0}^{x-1} z^{x-x'} P_{x'}$.[7] We also note that the IN-sheets can be expressed as sums (logical ORs) of the auxiliary sheets, $W_x = V_x^1 + V_x^2 + V_x^3$, and that the supermex operator for this game is the same as that for our previous example, the Sprague–Grundy values for 2-dimensional Wythoff's game. Thus $P_x = \mathcal{M}W_x = \mathcal{M}(V_x^1 + V_x^2 + V_x^3)$.

The key observation is that the auxiliary sheets have been constructed so as to obey a recursion relation:
$$V_{x+1}^1 = V_x^1 + \mathcal{M}(V_x^1 + V_x^2 + V_x^3),$$
$$V_{x+1}^2 = yV_x^2 + y\mathcal{M}(V_x^1 + V_x^2 + V_x^3),$$
$$V_{x+1}^3 = zV_x^3 + z\mathcal{M}(V_x^1 + V_x^2 + V_x^3),$$
and hence can be recursively generated from one another. The IN-sheets and the P-sheets can in turn be derived from these.

Plots of the IN-sheets and P-sheets for this game are given in Figure 7. They display complex probabilistic geometrical structures (which, as in our other examples, exhibit scaling behavior). In theory one should be able to compute the fixed points of the renormalization operator for this game, although this is clearly a complicated calculation that we leave for the future.

Lastly we remark that the IN-sheets for the other versions of 3-D Wythoff mentioned above can be constructed straightforwardly, and display similar (but

---

[7]Note that when right-shifting (resp. up-shifting), we fill in new columns (resp. rows) with 0's.

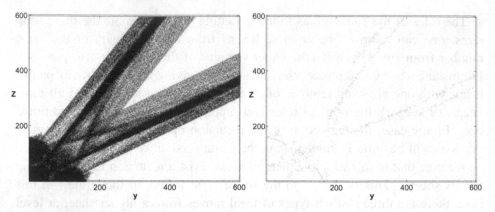

Figure 7. IN-sheet and P-sheet for three-dimensional Wythoff's game: W_{100} (left) and P_{100} (right).

not identical) complex geometrical structure and obey analogous scaling relations.

3. Implications and new directions

Apart from being a practical tool for garnering new insight into specific games, the renormalization methodology opens up several interesting new lines of inquiry into combinatorial games in general, as we now discuss.

3.1. Perturbations, structural stability, and universality in combinatorial games.
Having established a renormalization framework, it is natural to inquire about the stability of the associated renormalization fixed point. In other words, is the underlying geometry of a game stable to perturbations? In the present context, we perturb a game by adding one (or more) new points to one of its IN-sheets. We then repeatedly operate on the modified sheet with the recursion operator $\mathcal{R}$, and examine the perturbation's effect on the asymptotic geometry. We can think of such a perturbation to an IN-sheet as creating a variant of the original game with slightly modified rules: In these variant games, one or more of the P-positions of the original game have been arbitrarily "declared" (by the perturbation) to now be N-positions. How does the geometry of these variant games compare to the original?

In the game of three-row Chomp, a numerical analysis shows that for a wide range of perturbations the system quickly returns to the same renormalization fixed point (in the probabilistic sense) as in the original game, i.e., the overall geometric structure seen in Figure 4 re-emerges. Thus, adopting terminology from physics, we would say that these variant games lie in the same "universality class" as ordinary Chomp. In this manner, renormalization provides a natural classification scheme for combinatorial games: games can be grouped

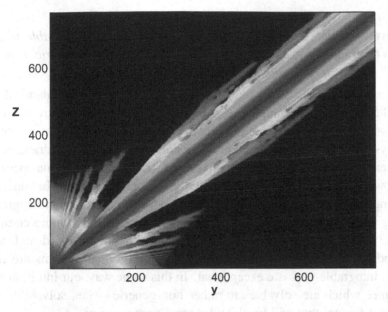

Figure 8. The geometry of generic Nim, illustrated for W_{256}.

into universality classes based on the nature of their associated renormalization fixed point. (We note that, like Chomp, the three-dimensional Wythoff's game discussed in the preceding section also appears to be structurally stable.)

Interestingly, not all games are stable to perturbations (i.e., the perturbation may create a game in a different universality class). For the game of ordinary Nim [Bouton 1902] considered earlier, we find that its IN-sheet geometry is structurally unstable to perturbations (i.e., the renormalization fixed point is unstable), resulting in a radically different geometry. Figure 8 shows the geometry[8] of a typical variant of Nim. We emphasize that this new geometry is stable and reproducible — it is the *typical* geometry that one observes if one makes a random perturbation to ordinary Nim. Hence we think of these variants of Nim as forming their own universality class, with Nim an outlier. In this manner we can see that Nim has a highly delicate (and non-generic) underlying geometric structure.

Why do some games like Chomp and 3-D Wythoff's game possess stable underlying geometries, while a game like Nim does not? We observe that Nim, unlike Chomp, is a solvable[9], decomposable game, and we believe that its inherent instability in the renormalization setting says something deep about the computational complexity of the game. Thus, we are led to this conjecture:

---

[8] We note that the W_x sheets display a weak periodicity in x in powers of 2, as was the case for ordinary Nim.

[9] i.e., a simple algorithm exists for determining if a given position in Nim is N or P

Conjecture: *Solvable combinatorial games are structurally unstable to perturbations (and hence have unstable sheet geometries), while generic, complex games will be structurally stable.*

— a suggestion which, if true, would relate geometric structure, dynamical stability, and computational complexity! We note an analogous feature from dynamical systems theory — that of *integrability* in Hamiltonian systems. For integrable systems, the solution to the problem can be reduced to quadratures and is characterized by simple behaviors: fixed points, periodic and quasiperiodic orbits. However, integrable systems are highly susceptible to perturbations, and adding a random perturbation will typically render the system non-integrable, destroying (some of) its simple structures and leading to much more complex dynamics, as described by the Poincaré–Birkhoff theorem[Arnold and Avez 1968]. Indeed, most many-degree-of-freedom Hamiltonian systems are non-integrable; integrable ones are exceptional. In this same way, our intuition here is that games which are solvable are rather non-generic — i.e., solvability is a delicate, rare feature that will break under most perturbations.

3.2. Sensitivity to initial conditions. One of the hallmarks of the modern understanding of dynamical systems and chaos theory is the concept of "sensitivity to initial conditions." Colloquially, this is the idea that a butterfly flapping its wings in New York can alter the weather in Chicago a few days later. More formally, it implies that one cannot predict the long term behavior of a dynamical system due to the rapid growth of small uncertainties. (See [Devaney 1986] for an elementary introduction.)

In this section, we will show that games can exhibit a related behavior. To do this, we will view the game's asymptotic distribution (i.e., the IN-sheet or P-sheet geometry) as a type of attractor.

We start, as in the preceding section, by perturbing an IN-sheet W_x in a game, and then iterate (with $\mathcal{R}$). Here, we explicitly assume the game to be structurally stable. We then examine how the precise locations of P-positions in sheets $P_{x'}$, $x' \geq x$ are affected by the initial perturbation. (Recall that by the structurally stability assumption, the same probabilistic structure will emerge in the perturbed and unperturbed cases, but the actual point-by-point locations of the P-positions in the P-set will differ.)

We illustrate this idea with the game of Chomp. Consider Figure 9. The blue data demonstrates that Chomp's attractor appears to exhibit a form of sensitivity to initial conditions. It was generated by changing a single IN on sheet W_{100} and then plotting (as a function of iteration number) the fractional discrepancy between the locations of the P-positions for the perturbed and unperturbed initial conditions (restricting here to P-positions on the lower and upper P-lines in P_x).

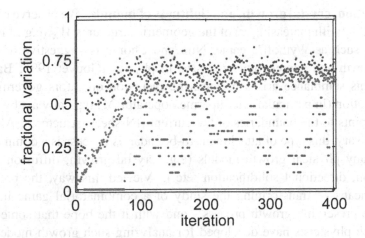

Figure 9. Dependence on initial conditions. The figure shows the fraction of P-positions affected by a small initial perturbation to an IN-sheet, as a function of iteration number.

Remarkably, after only 25 iterations, over half the losing positions have shifted their locations, while still remaining on the attractor. (The red data is similarly computed for an initial perturbation to P_{400}, while the green data shows a rolling average of the corresponding effect for P-positions lying on the flat line of P_x.) Note that despite the strong sensitivity on initial conditions, it is somewhat surprising that the growth of a perturbation appears to be roughly linear, rather than exponential. The resolution of this remains an open problem (as does the formal definition and analysis of Lyapunov exponents in this setting).

3.2.1. Renormalization and correlations. This sensitivity to initial conditions provides some justification for the renormalization procedure. We note that the main (unproven) assumption used in the renormalization analysis is that the various P-lines (at different x levels sufficiently far apart) were essentially uncorrelated with one another. (This was used implicitly, for instance, in the derivation of a few of the algebraic constraints given in Sections 2.1 and 2.3.) In the limit of large x we believe that this is justified because these lines are determined by sheets with large differences in x values, and since the system displays sensitive dependence on initial conditions, the precise point-by-point details of distant sheets should be uncorrelated in the limit.[10] Thus, we see that the renormalization analysis and assumptions about correlations are self-consistent.

---

[10]We remark that this lack of correlation is not a universal trait of all renormalization analyses. For example, in one of the most famous uses of renormalization, the phase transition in Ising Models (and many other phase transitions), in the 'frozen' case, correlation lengths in fact become infinite. [Ising 1925; Cipra 1987].

3.3. Accretion, crystal growth, and tightness of bounds. We observe here that the "growth" (with increasing x) of the geometric structures W_x (e.g., Figure 1) for games such as Wythoff's game, Nim and Chomp is suggestive of certain crystal growth and aggregation processes in physics [Gouyet 1995; Bar-Yam 1997]. This semblance arises because the recursion operators governing the game evolution (in particular, the supermex operator $\mathcal{M}$) typically act by attaching new points to the boundaries of the current (IN-sheet) structures. Although the details vary, this type of attachment-to-boundaries process is a common feature of many physical growth models (e.g., crystal growth, diffusion-limited aggregation, directional solidification, etc.). Viewed this way, the procedure offers a means of transforming the study of a combinatorial game into that of a shape-preserving growth process - and with it the hope that some of the tools which physicists have developed for analyzing such growth models may be brought to bear on combinatorial games. Most promising in this context would be a PDE description of the evolving boundaries in the game geometry, or a non-markovian diffusion formulation.

4. Open questions

Clearly this work raises many open questions and research problems. We provide a list of some of the key ones below:

1. **Making renormalization rigorous:** Despite its apparent practical capabilities as a tool for analyzing combinatorial games, it would be extremely valuable to make this renormalization approach mathematically rigorous. A first step would be to prove that, for stable games, the fixed point of the renormalization procedure is globally attracting, i.e., all initial conditions converge to the fixed point, or as is more likely, almost all converge. The local version of this stability problem is far more tractable, as it reduces to the computation of the spectrum of the linearization of the renormalization operator at the fixed point and standard techniques should suffice.

2. **3-D Wythoff's Game and Generic ('perturbed') Nim:** Solve analytically for the invariant geometry in these games, which have interesting and complex IN-sheets.

3. **Four-row Chomp:** Solve for the renormalization fixed point in four-row Chomp. Two approaches seem promising and both could be combined with automated procedures described in Item 5 below.

   (i) Consider three-dimensional sheets and apply the analogous renormalization procedure as used for three-row Chomp. (This can be done in a straightforward manner.)

(ii) Given a four-row position (w, x, y, z), fix w and then apply this analysis of Chomp to the subgame with the last three coordinates. Note that in this case, the renormalization equations become inhomogeneous, of the form $\mathbf{W} = \mathcal{R}\mathbf{W} + \mathcal{B}$, where $\mathcal{B}$ comes from the solutions with smaller w.

4. **Sprague–Grundy values for (2-D) Wythoff's game** Analyze the complex structure of the sheets found near the origin (i.e., for small position values).

5. **Automated Renormalization:** Design an algorithm for analytically computing the renormalization fixed point, in the spirit of Zeilberger's automated analysis of Chomp. (An example of an automated renormalization procedure, in a very different setting is given in [Friedman and Landsberg 2001].)

6. **NP-Hardness of Combinatorial Games:** Prove that some game which can be solved by renormalization techniques is NP-hard.

7. **Hardness of Perturbed Games:** Provide a class of solvable games such that 'most' perturbations lead to games that are 'difficult' to solve.

8. **Partisan Games:** Apply renormalization techniques to partisan games.

9. **Accretion and Partial Differential Equations:** Apply modern tools from accretion theory to a combinatorial game. In particular, find a PDE approximation to the renormalization operator to compute the fixed points.

10. **Lyapunov Exponents:** Formally define and calculate Lyapunov exponents to describe the sensitivity to initial conditions of an interesting game.

11. Our analysis of Chomp appears to suggest that one can answer the following two new and fundamental questions in complexity theory in the affirmative.

   (i) **Probabilistic Solutions of Hard Problems ("betting on NP"):** Our results suggest that the computation of P-positions in 3-rowed Chomp is not polynomial. (Note that it is not clear whether the problem is in NP or even NP-hard.) Thus we do not expect to find simple formulas or fast algorithms. Nonetheless, this analysis implies that we can compute probabilistic estimates. This raises the question of whether such estimates are possible in complexity theory and raises the following challenging (but fundamental) problem: For an NP-complete (or NP-hard) problem, find a polynomial time algorithm which can accurately estimate the probability that a word is in the language. This would not allow one to solve NP-hard problems (which is not possible if $P \neq NP$), but would allow one to "bet effectively" on such problems. Clearly one needs a more precise formulation to allow one to sensibly evaluate the notion of 'probability'. One possibility is a computational formulation of the notion of 'calibration' from Bayesian analysis [Dawid 1982].

   (ii) **Stochastic NP-Hard Problems:** A dual to the previous problem is to consider a set of NP-complete (or NP-hard) languages, generated stochastically and ask whether there exists a polynomial time algorithm which, given a word, can estimate the fraction of languages that it is a member of. For example, one could take a traveling salesman problem on a computer network and assume that each link exists with probability $p \in [0, 1]$. Then one could ask for the probability that a given graph has an expected tour less than some fixed length.

12. **Difficult Combinatorial Games:** Clearly the proof of this new approach is in the pudding. What other combinatorial games can be analyzed using these methods?

5. On the application of renormalization to games

First, we want to emphasize that (at least) some games of no chance have interesting and revealing underlying geometric structures. This suggests that simply computing the geometric structure in a game's position space could, in and of itself, lead to new and potentially powerful insights into a game (even in the absence of a full-blown renormalization analysis). For example, as we saw, the plot of Sprague–Grundy values for 2-D Wythoff's game reveals an interesting structure near the origin.

Second, we wish to reiterate that the renormalization approach to games is still very much in its infancy, with much unexplored terrain — its scope of applicability and limitations are not fully understood. Its primary limitation at present is that, like many renormalization procedures, making it fully rigorous is likely to prove challenging, and most renormalization results do not constitute formal mathematical proofs. Nonetheless, at a minimum, one can view the renormalization results for a game as representing strong conjectures, and then seek independent formal proofs of these conjectures. An alternative is that one can ignore rigor and simply compute — as is done in modern physics — to help understand the complex structure of non-decomposable combinatorial games. Given our lack of knowledge about the solutions of such games, we suggest that this last approach might be extremely valuable.

References

[Arnold and Avez 1968] V. I. Arnold and A. Avez, *Ergodic problems of classical mechanics*, Benjamin, New York, 1968.

[Bar-Yam 1997] Y. Bar-Yam, *Dynamics of complex systems*, Addison-Wesley, Reading, MA, 1997.

[Berlekamp 1994] E. R. Berlekamp, *Mathematical Go: chilling gets the last point*, A K Peters, Nattick, MA, 1994.

[Berlekamp 2000] E. R. Berlekamp, *The game of Dots and Boxes: sophisticated child's play*, A K Peters, Nattick, MA, 2000.

[Berlekamp et al. 1979] E. R. Berlekamp, J. H. Conway, and R. K. Guy, *Winning ways for your mathematical plays*, Academic Press, London, 1979. Second ed., Vol. 1, A K Peters, Wellesley, MA 2001.

[Bouton 1902] C. L. Bouton, "Nim, a game with a complete mathematical theory", *Ann. of Math.* (2) **3** (1902), 35–39.

[Brouwer 2004] A. E. Brouwer, The game of Chomp, 2004. http://www.win.tue.nl/ ~aeb/games/chomp.html.

[Byrnes 2003] S. Byrnes, "Poset games periodicity", *Integers* **2** (2003), G3.

[Cipra 1987] B. A. Cipra, "An introduction to the Ising model", *Amer. Math. Monthly* **94** (1987), 937–959.

[Conway 1976] J. H. Conway, *On numbers and games*, Academic Press, Boston, MA, 1976. Second ed., A K Peters, Wellesley, MA 2001.

[Cvitanovic 1989] P. Cvitanovic (editor), *Universality in chaos: a reprint selection*, 2nd ed., Adam Hilger, Bristol, 1989.

[Dawid 1982] P. Dawid, "The well-calibrated Bayesian", *J. Amer. Statist. Ass.* **77** (1982), 604–613.

[Devaney 1986] R. L. Devaney, *An introduction to chaotic dynamical systems*, Benjamin/Cummings, Menlo Park, CA, 1986.

[Feigenbaum 1980] M. Feigenbaum, "Universal behavior in nonlinear systems", *Los Alamos Science* **1** (1980), 4–27.

[Friedman and Landsberg 2001] E. J. Friedman and A. S. Landsberg, "Large-scale synchrony in weakly interacting systems", *Physical Review E.* **63** (2001).

[Friedman and Landsberg 2007] E. Friedman and A. S. Landsberg, "Nonlinear dynamics in combinatorial games: Renormalizing Chomp", *CHAOS* **17** (2007), 023117.

[Gale 1974] D. Gale, "A curious Nim-type game", *Amer. Math. Monthly* **81** (1974), 876–879.

[Goldenfeld 1992] N. Goldenfeld, *Lectures on phase transitions and the renormalization group*, Frontiers in Physics **85**, Westview Press, 1992.

[Gouyet 1995] J.-F. Gouyet, *Physics of fractal structures*, Springer, New York, 1995.

[Grundy 1939] P. Grundy, "Mathematics and games", *Eureka* **2** (1939), 6–8.

[Ising 1925] E. Ising, "Beitrag zur Theorie des Ferromagnetismus.", *Zeitschr. f. Physik* **31** (1925), 253–258.

[Nivasch 2004] G. Nivasch, *The Sprague–Grundy function for Wythoff's game: on the location of the g-values*, M.Sc. Thesis, Weizmann Institute of Science, 2004.

[Rivasseau 2003] V. Rivasseau, *An introduction to renormalization: Poincaré Seminar 2002*, Progress in Mathematical Physics **30**, Birkhäuser, 2003.

[Schuh 1952] F. Schuh, "Spel van delers", *Nieuw Tijdschrift voor Wiskunde* **39** (1952), 299–304.

[Sprague 1936] R. Sprague, "Uber mathematische Kampfspiele", *Tohoku Mathematical Journal* **41** (1936), 438–444.

[Sun 2002] X. Sun, "Improvements on Chomp", *Integers* **2** (2002), G1.

[Wythoff 1907] W. A. Wythoff, "A modification of the game of Nim", *Nieuw Arch. Wiskunde* **8** (1907), 199–202.

[Zeilberger 2001] D. Zeilberger, "Three-rowed Chomp", *Adv. in Appl. Math* **26** (2001), 168–179.

[Zeilberger 2004] D. Zeilberger, "Chomp, recurrences, and chaos", *J. Diff. Equ. and its Apps.* **10** (2004), 1281–1293.

ERIC J. FRIEDMAN
SCHOOL OF ORIE
CORNELL UNIVERSITY
ITHACA, NY 14853
UNITED STATES
ejf27@cornell.edu

ADAM S. LANDSBERG
JOINT SCIENCE DEPARTMENT
CLAREMONT MCKENNA, PITZER AND SCRIPPS COLLEGES
CLAREMONT, CA 91711
UNITED STATES
landsberg@jsd.claremont.edu

Games of No Chance 3
MSRI Publications
Volume **56**, 2009

More on the Sprague–Grundy function for Wythoff's game

GABRIEL NIVASCH

ABSTRACT. We present two new results on Wythoff's Grundy function $\mathcal{G}$. The first one is a proof that for every integer $g \geq 0$, the g-values of $\mathcal{G}$ are within a bounded distance to their corresponding 0-values. Since the 0-values are located roughly along two diagonals, of slopes ϕ and ϕ^{-1}, the g-values are contained within two strips of bounded width around those diagonals. This is a generalization of a previous result by Blass and Fraenkel regarding the 1-values.

Our second result is a *convergence* conjecture and an accompanying recursive algorithm. We show that for every g for which a certain conjecture is true, there exists a recursive algorithm for finding the n-th g-value in $O(\log n)$ arithmetic operations. Our algorithm and conjecture are modifications of a similar result by Blass and Fraenkel for the 1-values. We also present experimental evidence for our conjecture for small g.

1. Introduction

The game of Wythoff [10] is a two-player impartial game played with two piles of tokens. On each turn, a player removes either an arbitrary number of tokens from one pile (between one token and the entire pile), or the same number of tokens from both piles. The game ends when both piles become empty. The last player to move is the winner.

Wythoff's game can be represented graphically with a quarter-infinite chessboard, extending to infinity upwards and to the right (Figure 1). We number the rows and columns sequentially $0, 1, 2, \ldots$. A chess queen is placed in some cell

This work was done when the author was at the Weizmann Institute of Science in Rehovot, Israel. The author was partially supported by a grant from the German-Israeli Foundation for Scientific Research and Development.

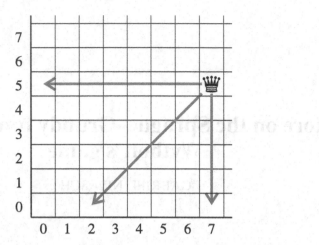

Figure 1. Graphic representation of Wythoff's game.

of the board. On each turn, a player moves the queen to some other cell, except that the queen can only move left, down, or diagonally down-left. The player who takes the queen to the corner wins.

Wythoff found a simple, closed formula for the P-positions of his game. Let $\phi = (1 + \sqrt{5})/2$ be the golden ratio. Then:

THEOREM 1.1 (WYTHOFF [10]). *The P-positions of Wythoff's game are given by*

$$(\lfloor \phi n \rfloor, \lfloor \phi^2 n \rfloor) \quad \text{and} \quad (\lfloor \phi^2 n \rfloor, \lfloor \phi n \rfloor), \tag{1-1}$$

for $n = 0, 1, 2, \ldots$.

1.1. Impartial games and the Sprague–Grundy function. We briefly review the Sprague–Grundy theory of impartial games [1].

An impartial game can be represented by a directed acyclic graph $G = (V, E)$. Each position in the game corresponds to a vertex in G, and edges join vertices according to the game's legal moves. A token is initially placed on some vertex $v \in V$. Two players take turns moving the token from its current vertex to one of its direct followers. The player who moves the token into a sink wins.

Given two games G_1 and G_2, their *sum* $G_1 + G_2$ is played as follows: On each turn, a player chooses one of G_1, G_2, and moves on it, leaving the other game untouched. The game ends when no moves are possible on G_1 nor on G_2.

Let $\mathbb{N} = \{0, 1, 2, \ldots\}$ be the set of natural numbers. Given a finite subset $S \subset \mathbb{N}$, let mex $S = \min(\mathbb{N} \setminus S)$ denote the smallest natural number not in S. Then, given a game $G = (V, E)$, its *Sprague–Grundy function* (or just *Grundy function*) $\mathcal{G} : V \to \mathbb{N}$ is defined recursively by

$$\mathcal{G}(u) = \text{mex}\{\mathcal{G}(v) \mid (u, v) \in E\}, \qquad \text{for } u \in V. \tag{1-2}$$

This recursion starts by assigning sinks the value 0.

The Grundy function $\mathcal{G}$ satisfies the following two important properties:

1. A vertex $v \subset V$ is a P-position if and only if $\mathcal{G}(v) = 0$.
2. If $G = G_1 + G_2$ and $v = (v_1, v_2) \in V_1 \times V_2$, then $\mathcal{G}(v)$ is the bitwise XOR, or *nim-sum*, of the binary representations of $\mathcal{G}(v_1)$ and $\mathcal{G}(v_2)$.

Clearly, the sum of games is an associative operation, as is the nim-sum operation. Therefore, knowledge of the Grundy function provides a winning strategy for the sum of any number of games.

The following lemma gives some basic bounds on the Grundy function.

LEMMA 1.2. *Given a vertex v, let n_v be the number of direct followers of v, and let p_v be the number of edges in the longest path from v to a leaf. Then $\mathcal{G}(v) \leq n_v$ and $\mathcal{G}(v) \leq p_v$.*

PROOF. The first bound follows trivially from equation (1-2). The second bound follows by induction.   ☐

The following lemma shows that the Grundy function of a game-graph can be calculated up to a certain value g using the mex property.

LEMMA 1.3. *Given a game-graph $G = (V, E)$ and an integer $g \geq 0$, let $\mathcal{H}$ be a function $\mathcal{H} : V \rightarrow \{0, \ldots, g, \infty\}$ such that for all $v \in V$,*

1. *if $\mathcal{H}(v) \leq g$ then*

$$\mathcal{H}(v) = \mathrm{mex}\,\{\mathcal{H}(w) \mid (v, w) \in E\};$$

2. *if $\mathcal{H}(v) = \infty$ then*

$$\mathrm{mex}\,\{\mathcal{H}(w) \mid (v, w) \in E\} > g.$$

Then $\mathcal{G}(v) = \mathcal{H}(v)$ whenever $\mathcal{H}(v) \leq g$, and $\mathcal{G}(v) > g$ whenever $\mathcal{H}(v) = \infty$.

In the function $\mathcal{H}$, the labels ∞ are placeholders that indicate values larger than g. Theorem 1.3 is easily proven by induction; see [6].

1.2. Notation and terminology. From now on, $\mathcal{G}$ will denote specifically the Grundy function of Wythoff's game. Thus, $\mathcal{G} : \mathbb{N} \times \mathbb{N} \rightarrow \mathbb{N}$ is given by

$$\mathcal{G}(x, y) = \mathrm{mex}\,(\{\mathcal{G}(x', y) \mid 0 \leq x' < x\} \cup$$
$$\{\mathcal{G}(x, y') \mid 0 \leq y' < y\} \cup \qquad (1\text{-}3)$$
$$\{\mathcal{G}(x - k, y - k) \mid 1 \leq k \leq \min(x, y)\}).$$

Table 1 shows the value of $\mathcal{G}$ for small x and y. This matrix looks quite chaotic at first glance, as has been pointed out before [1]. And indeed, it is still an open problem to compute $\mathcal{G}$ in time polynomial in the size of the input (meaning, to compute $\mathcal{G}(x, y)$ in time $O((\log x + \log y)^c)$ for some constant c).

15	15	16	17	18	10	13	12	19	14	0	3	21	22	8	23	20
14	14	12	13	16	15	17	18	10	9	1	2	20	21	7	11	23
13	13	14	12	11	16	15	17	2	0	5	6	19	20	9	7	8
12	12	13	14	15	11	9	16	17	18	19	7	8	10	20	21	22
11	11	9	10	7	12	14	2	13	17	6	18	15	8	19	20	21
10	10	11	9	8	13	12	0	15	16	17	14	18	7	6	2	3
9	9	10	11	12	8	7	13	14	15	16	17	6	19	5	1	0
8	8	6	7	10	1	2	5	3	4	15	16	17	18	0	9	14
7	7	8	6	9	0	1	4	5	3	14	15	13	17	2	10	19
6	6	7	8	1	9	10	3	4	5	13	0	2	16	17	18	12
5	5	3	4	0	6	8	10	1	2	7	12	14	9	15	17	13
4	4	5	3	2	7	6	9	0	1	8	13	12	11	16	15	10
3	3	4	5	6	2	0	1	9	10	12	8	7	15	11	16	18
2	2	0	1	5	3	4	8	6	7	11	9	10	14	12	13	17
1	1	2	0	4	5	3	7	8	6	10	11	9	13	14	12	16
0	0	1	2	3	4	5	6	7	8	9	10	11	12	13	14	15
	0	1	2	3	4	5	6	7	8	9	10	11	12	13	14	15

Table 1. The Grundy function of Wythoff's game.

Nevertheless, as we will see now, several results on $\mathcal{G}$ have been established. But let us first introduce some notation.

A pair (x, y) is also called a *point* or a *cell*. If $\mathcal{G}(x, y) = g$, we call (x, y) a *g-point* or a *g-value*.

Note that by symmetry, $\mathcal{G}(x, y) = \mathcal{G}(y, x)$ for all x, y. We refer to this property as *diagonal symmetry*.

We will consistently use the following graphical representation of $\mathcal{G}$: The first coordinate of $\mathcal{G}$ is plotted vertically, increasing upwards, and the second coordinate is plotted horizontally, increasing to the right.

Thus, we call *row r* the set of points (r, x) for all $x \geq 0$, and *column c* the set of points (x, c) for all $x \geq 0$. Also, *diagonal e* is the set of points $(x, x + e)$ for all x if $e \geq 0$, or the set of points $(x - e, x)$ for all x if $e < 0$. (Note that we only consider diagonals parallel to the movement of the queen.)

We also define, for every integer $g \geq 0$, the sequence of g-values that lie on or to the right of the main diagonal, sorted by increasing row. Formally, we let

$$T_g = ((a_0^g, b_0^g), (a_1^g, b_1^g), (a_2^g, b_2^g), \ldots) \qquad (1\text{-}4)$$

be the sequence of g-values having $a_n^g \leq b_n^g$, ordered by increasing a_n^g. We also

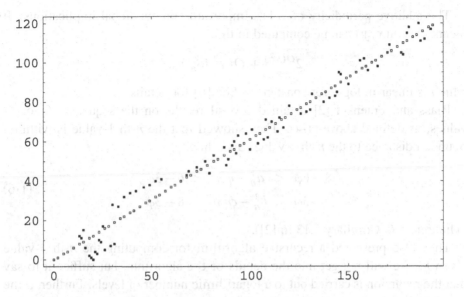

Figure 2. First terms of the sequences T_0 (framed squares) and T_{20} (filled squares).

let $p_n^g = (a_n^g, b_n^g)$. Note that, by Theorem 1.1, we have

$$p_n^0 = (\lfloor \phi n \rfloor, \lfloor \phi^2 n \rfloor).$$

For example, Figure 2 plots the first points in T_0 and T_{20}. A pattern is immediately evident: The 20-values seem to lie within a strip of constant width around the 0-values. This is true in general. In fact, we will prove something stronger as one of the main results of this paper.

We also let $d_n^g = b_n^g - a_n^g$ be the diagonal occupied by the point p_n^g.

Finally, we place the a, b, and d coordinates of g-points into sets, as follows:

$$A_g = \{a_i^g \mid i \geq 0\}, \quad B_g = \{b_i^g \mid i \geq 0\}, \quad D_g = \{d_i^g \mid i \geq 0\}. \quad (1\text{-}5)$$

1.3. Previous results on Wythoff's Grundy function. We now give a brief overview of previous results on the Grundy function of Wythoff's game.

It follows directly from formula (1-3) that no row, column, or diagonal of $\mathcal{G}$ contains any g-value more than once. In fact, it is not hard to show that every row and column of $\mathcal{G}$ contains every g-value exactly once [2; 4]. Furthermore, every diagonal contains every g-value exactly once [2], although this is somewhat harder to show. We will rederive these results in this paper.

It has also been shown that every row of $\mathcal{G}$ is *additively periodic*. This result was first proven by Norbert Pink in his doctoral thesis [8] (published in [3]), and Landman [4] later found a simpler proof. Both [3] and [4] derive an upper bound of $2^{O(x^2)}$ for the preperiod and the period of row x.

The additive periodicity of $\mathcal{G}$ has important computational implications: It means that $\mathcal{G}(x, y)$ can be computed in time

$$2^{O(x^2)} + O(x^2 \log y),$$

which is linear in log y for constant x. See [6] for details.

Blass and Fraenkel [2] obtained several results on the sequence T_1 of 1-values, as defined above (1-4). They showed that the n-th 1-value is within a bounded distance to the n-th 0-value. Specifically,

$$
\begin{aligned}
8 - 6\phi &< a_n^1 - \phi n < 6 - 3\phi, \\
-3\phi &< b_n^1 - \phi^2 n < 8 - 3\phi
\end{aligned}
\tag{1-6}
$$

(Theorem 5.6, Corollary 5.13 in [2]).

They also presented a recursive algorithm for computing the n-th 1-value given n. We will not get into the details of the algorithm, but suffice it to say that the recursion is carried out to a logarithmic number of levels. Further, if the computation done at each level were shown to be constant, then the algorithm would run in $O(\log n)$ steps altogether. For the computation at each level to be constant, certain arrangements in the sequence of 0- and 1-values must occur infinitely many times with at least constant regularity. The authors did not manage to prove this latter property, so they left the polynomiality of their algorithm as a conjecture.

1.4. Our results. In this paper we make two main contributions on the function $\mathcal{G}$. The first one is a generalization of the result for the 1-values described above. We will prove that for every g, the point p_n^g is within a bounded distance to p_n^0, where the bound depends only on g, not on n. Our theoretical bound turns out to be much worse than the actual distances seen in practice. We present experimental data and compare them to our theoretical result.

Our second contribution is a modification and generalization of Blass and Fraenkel's recursive algorithm. We present a conjecture, called the *Convergence Conjecture*, which claims a certain property about the sequences T_0 through T_g. We show that for every g for which the conjecture is true, there exists an algorithm that computes p_n^g in $O(\log n)$ arithmetic operations, where the factor implicit in the O notation depends on g.

We present experimental results that seem to support the conjecture for small g. We finally use our recursive algorithm to predict the value of several points p_n^g for small g and very large n.

1.5. Significance of our results. Suppose we are playing the sum of Wythoff's game with some other game, like a Nim pile. Our winning strategy, then, is to make the Grundy values of the two games equal. Suppose that the position in

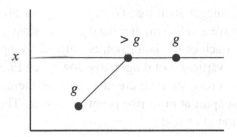

Figure 3. A supporter from a row lower than x.

Wythoff has Grundy value m, and the Nim pile is of size n. Then, if $m < n$, we should reduce the Nim pile to size m, and if $m > n$, we should move in Wythoff to a position with Grundy value n.

How do the results in this paper help us in this scenario? The first result, regarding the location of the g-values, is of no practical help: It gives us only the approximate location of the g-values, not their precise position.

The recursive algorithm, on the other hand, has much more practical significance. If the conjecture is true for small values of g, then we can play on sums where the Nim pile is of size $\leq g$. And even if there are sporadic counterexamples to the conjecture, the recursive algorithm will probably give the correct answer in most cases, so it is a good heuristic.

1.6. Organization of this paper. The rest of this paper is organized as follows. In Section 2 we prove the closeness of the g-values to the 0-values, and in Section 3 we present the Convergence Conjecture and its accompanying recursive algorithm. The Appendix (page 407) contains some proofs omitted from Section 2.5.

2. Closeness of the g-values to the 0-values

In this Section we will prove that for every g, the point p_n^g is within a bounded distance to p_n^0, where the bound depends only on g. We begin with some basic results on $\mathcal{G}$.

LEMMA 2.1 (LANDMAN [4]). *Given x and g, there exists a unique y such that $\mathcal{G}(x, y) = g$. Moreover,*

$$g - x \leq y \leq g + 2x. \tag{2-1}$$

PROOF. Uniqueness follows trivially from the definition of $\mathcal{G}$.

As for existence, for any x, y, the longest path from cell (x, y) to the corner $(0, 0)$ has length $x + y$, so by Lemma 1.2, $g \leq x + y$, implying the lower bound in (2-1).

Next, let y be an integer such that $G(x, y') \neq g$ for all $y' < y$. At most g such points (x, y') have a value smaller than g, so at least $y - g$ of them have a value larger than g. Each of the latter points must be "supported" by a g-point in a lower row, either vertically of diagonally down (see Figure 3). No two such supporters can share a row, so there are at most x of them. On the other hand, each supporter can support at most two points in row x. Therefore, $y - g \leq 2x$, yielding the upper bound of (2-1). $\square$

The following result was also known already to Landman [5]:

LEMMA 2.2. *Given* $g \geq 0$, *there exists a unique integer* x *such that* $G(x, x) = g$. *Moreover,*

$$g/2 \leq x \leq 2g. \tag{2-2}$$

PROOF. As before, uniqueness follows from the definition of G. And the lower bound follows from equation (2-1).

For the upper bound, suppose x is such that $G(y, y) \neq g$ for all $y < x$. At most g of such points (y, y) have value smaller than g, so at least $x - g$ of them have value larger than g. By diagonal symmetry, for each of the latter points there exist two g-points, one to the left of (y, y) and the other one below it.

Therefore, there are at least $2(x - g)$ g-points below and to the left of point (x, x). No two of them can share a column, so $2(x - g) \leq x$, or $x \leq 2g$. $\square$

Recall the definition of the sets A_g, B_g, and D_g (1-5). By Lemmas 2.1 and 2.2, together with diagonal symmetry, we have $|A_g \cap B_g| = 1$ and $A_g \cup B_g = \mathbb{N}$ for all g. Therefore, we could say that the sequences $\{a_i^g\}$ and $\{b_i^g\}$ are "almost complementary". We will show later on that $D_g = \mathbb{N}$ for all g.

2.1. Algorithm WSG for computing T_g.

On page 385 we show a greedy algorithm that computes, given an integer $g \geq 0$, the sequences T_h and the sets A_h, B_h, D_h for $0 \leq h \leq g$. This algorithm was first described in [2].

The idea behind the algorithm is simple: We traverse the rows in increasing order, and for each row, we go through the values $h = 0, \ldots, g$ in increasing order. If the current row needs an h-point on or to the right of the main diagonal (because it contains no h-point to the left of the main diagonal), then we greedily find the first legal place for an h-point, and insert the h-point there. We also reflect the h-point with respect to the main diagonal, into a higher row (so that higher row will not receive an h-point to the right of the main diagonal).

Of course, since this algorithm works row by row, it takes exponential time.

For simplicity, we do not specify when the algorithm halts, although we could make it halt after, say, computing the first n terms of T_g.

We will rely heavily on this algorithm later on, in our analysis of T_g.

The correctness of Algorithm WSG follows easily from Lemma 1.3; see [6] for the details.

Algorithm WSG (Wythoff Sprague–Grundy)

1. Initialize the sets A_h, B_h, D_h, and the sequences T_h, to $\varnothing$, for $0 \le h \le g$.

2. For $r = 0, 1, 2, \ldots$ do:

3.   For $h = 0, \ldots, g$ do:

4.   If $r \notin B_h$ then:

5.   • find the smallest $d = 0, 1, 2, \ldots$ for which:

6.   ◦ $(r, r + d) \notin T_k$ for all $0 \le k < h$,

7.   ◦ $r + d \notin B_h$, and

8.   ◦ $d \notin D_h$;

9.   • append $(r, r + d)$ to T_h;

10.   • insert r into A_h;

11.   • insert $r + d$ into B_h;

12.   • insert d into D_h.

2.2. Statement of the main Theorem. Recall from Theorem 1.1 that the 0-values of Wythoff's game are given by

$$(a_n^0, b_n^0) = (\lfloor \phi n \rfloor, \lfloor \phi^2 n \rfloor).$$

Graphically, the 0-values lie close to a straight line of slope ϕ^{-1} that starts at the origin.

Our main result for this Section is the following:

THEOREM 2.3. *For every Grundy value $g \ge 0$ and every diagonal $e \ge 0$, there exists an n such that*

$$d_n^g = e$$

(i.e., every diagonal e contains a g-value).

Further, for every $g \ge 0$ there exist constants α_g, β_g, such that

$$|a_n^g - a_n^0| \le \alpha_g, \quad |b_n^g - b_n^0| \le \beta_g, \quad \text{for all } n$$

(i.e., the n-th g-value is close to the n-th 0-value).

Our strategy for proving Theorem 2.3 is as follows. We first show that for every g there is a g-value in every diagonal, and furthermore, for every g there exists a constant δ_g such that

$$|d_n^g - n| \le \delta_g \quad \text{for all } n. \tag{2-3}$$

In other words, the g-values occupy the diagonals in roughly sequential order.

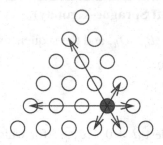

Figure 4. Queen in a triangular lattice.

Then we show how equation (2-3), together with the almost-complementarity of the sequences $\{a_n^g\}$ and $\{b_n^g\}$, implies that $|a_n^g - \phi n|$ and $|b_n^g - \phi^2 n|$ are bounded.

Note that for $g = 0$, Theorem 1.1 gives us

$$d_n^0 = b_n^0 - a_n^0 = (\lfloor \phi n \rfloor + n) - \lfloor \phi n \rfloor = n;$$

in other words, the 0-values occupy the diagonals in sequential order. This can be confirmed easily by following Algorithm WSG with $g = 0$.

2.3. Nonattacking queens on a triangle. The following is a variation on the well-known "eight queens problem". We will use its solution in proving bound (2-3).

We are given a triangular lattice of side n, as shown in Figure 4. A queen on the lattice can move along a straight line parallel to any of the sides of the triangle. How many queens can be placed on the lattice, without any two queens attacking each other?

LEMMA 2.4. *The maximum number of nonattacking queens that can be placed on a triangular lattice of side n is exactly*

$$q(n) = \left\lfloor \frac{2n + 1}{3} \right\rfloor.$$

A simple proof of this fact is given by Vaderlind et al. in [9, Problem 252]. Another proof is given in [7].

2.4. d_n^g is close to n. We will now prove bound (2-3). For convenience, in this subsection we fix g, and we write $a_n = a_n^g$, $b_n = b_n^g$, $d_n = d_n^g$, $p_n = p_n^g$. Whenever we refer to h-points, $h < g$, we will say so explicitly.

Recall that the points p_n are ordered by increasing row a_n, so that $a_n > a_m$ if and only if $n > m$.

In this subsection we will make extensive use of Algorithm WSG.

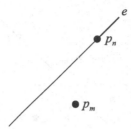

Figure 5. Point p_m skips diagonal e.

Observe that Algorithm WSG does not place a g-point on certain rows r, because it skips row r on line 4. Then such an r is not added to A_g (line 10). We call such an r a *skipped row*.

Similarly, sometimes a certain column c is never inserted into B_g (line 11), because no point p_m falls on that column. In that case, we call column c a *skipped column*.

Let us define the notion of a g-point skipping a diagonal. Intuitively, point p_m skips diagonal e if Algorithm WSG places point p_m to the right of diagonal e, while diagonal e does not yet contain a g-point (see Figure 5). Formally:

DEFINITION 2.5. Diagonal e is *empty up to row* r if there is no point p_n with $d_n = e$ and $a_n \le r$.

DEFINITION 2.6. Point p_m is said to *skip diagonal* e if $d_m > e$ and e is empty up to row a_m.

Our goal in this subsection is to derive a bound on $|d_n - n|$. We will derive separately upper bounds for $d_n - n$ and $n - d_n$. We do this by bounding the number of diagonals that a given point can skip, and the number of points that can skip a given diagonal:

LEMMA 2.7. *If no point p_n skips more than k diagonals, then $d_n - n \le k$ for all n. If no diagonal is skipped by more than k points, then $n - d_n \le k$ for all n.*

PROOF. For the first claim, suppose by contradiction that $d_n - n > k$ for some n. Then, of the d_n diagonals $0, \ldots, d_n - 1$, only n can be occupied by points $p_0, \ldots, p_{n-1}$. Therefore, point p_n skips at least $k + 1$ diagonals.

For the second claim, suppose by contradiction that $n - d_n > k$ for some n. Then, of the n points $p_0, \ldots, p_{n-1}$, only d_n can fall on diagonals $0, \ldots, d_n - 1$. Therefore, diagonal d_n is skipped by at least $k + 1$ points. □

Let us inspect why a g-point skips a diagonal according to Algorithm WSG. Suppose point p_m skips diagonal e, and let $C = (a_m, a_m + e)$ be the cell on diagonal e on the row in which p_m was inserted. Then, point p_m skipped diagonal e, either because cell C was already assigned some value $k < g$ (WSG

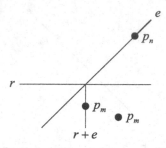

Figure 6. Points p_m active with respect to diagonal e and row r.

line 6), or because there was already a point $p_{m'}$ directly below cell C (WSG line 7).

We need a further definition: Let e be a diagonal and r be a row, such that diagonal e is empty up to row $r - 1$. Draw a line from the intersection of row r and diagonal e vertically down. If a point p_m is strictly below row r, and on or to the right of the said vertical line, then we say that p_m is *active with respect to diagonal e and row r* (see Figure 6). In other words:

DEFINITION 2.8. *If diagonal e is empty up to row $r - 1$, then point p_m is active with respect to diagonal e and row r if $a_m < r$ and $b_m \geq r + e$.*

We can bound the number of active g-points:

LEMMA 2.9. *The number of g-points active with respect to any diagonal e and any row r is at most g.*

PROOF. By assumption, diagonal e is empty up to row $r - 1$.

We will show that for every $r' \leq r$, if diagonal e contains k h-points, $h < g$, below row r', then there can be at most k active g-points with respect to diagonal e and row r'. This implies our Lemma, since there are at most g h-points, $h < g$, on diagonal e.

We prove the above claim by induction on r'. If $r' = 0$ then clearly $k = 0$ and there are no active g-points with respect to e and r'.

Suppose our claim is true up to row r', and let us examine Algorithm WSG on row r' itself. If no point p_m is inserted on row r', then the number of active g-points does not increase when we go from row r' to row $r' + 1$. And if point p_m is inserted on row r' and it skips diagonal e, it must be for one of the two reasons mentioned above. If there is an h-point, $h < g$, on the intersection of row r' with diagonal e, then the number k of our claim increases by 1 when we go to row $r' + 1$. And if there is no such h-point, then there must be an earlier point $p_{m'}$ directly below the intersection of e and r'. But then $p_{m'}$ is active with respect to e and r', but not with respect to e and $r' + 1$, so the number of active g-points stays the same when we go from row r' to row $r' + 1$.

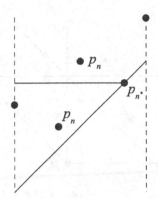

Figure 7. p_{n*} is the point p_n with maximum d_n.

So in either case, the inductive claim is also true for row $r' + 1$.   □

We can now bound the number of diagonals a given g-point can skip:

LEMMA 2.10. *A point p_m can skip at most $2g$ diagonals.*

PROOF. Let e_0 be the first diagonal skipped by point p_m. For every diagonal e skipped by p_m, there must be either an active g-point with respect to diagonal e_0 and row a_m, or an h-point, $h < g$, on cell $(a_m, a_m + e)$. There can be at most g of the latter, and by Lemma 2.9, at most g of the former.   □

We proceed to bound the number of g-points that can skip a given diagonal. For this we need a lower bound on the number of skipped columns in an interval of consecutive columns:

LEMMA 2.11. *An interval of k consecutive columns contains at least $k/3 - 2g$ skipped columns.*

PROOF. Consider the points p_n that lie within the given interval of columns. Let p_{n*} be the point p_n with maximum d_n (see Figure 7). The number of points p_n with $n > n^*$ is at most $2g$ by Lemma 2.10. And the points p_n with $n < n^*$ are confined to a triangular lattice; but this situation is isomorphic to the nonattacking queens of subsection 2.3!

The triangular lattice has side at most $k - 2$ (since the lattice cannot reach the column of p_{n*} nor the preceding column). Therefore, the number of points p_n with $n < n^*$ is at most

$$\left\lfloor \frac{2(k-2)+1}{3} \right\rfloor \leq 2k/3 - 1.$$

Thus, the total number of points p_n is at most $2k/3 - 1 + 2g + 1 = 2k/3 + 2g$, so the number of skipped columns is at least $k/3 - 2g$.   □

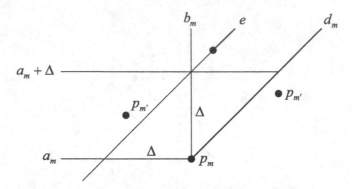

Figure 8. Points $p_{m'}$ between rows a_m and $a_m + \Delta$.

COROLLARY 2.12. *An interval of k consecutive rows contains at least $k/3 - 2g$ points p_n.*

PROOF. By diagonal symmetry: If column c is a skipped column, then row c contains a point p_n. □

LEMMA 2.13. *Suppose point p_m skips diagonal e, and let $\Delta = d_m - e$. Suppose diagonal e is empty up to row $a_m + \Delta$ (see Figure 8). Then $\Delta \leq 15g$.*

PROOF. Let us bound the number of points $p_{m'}$ in the interval between row $a_m + 1$ and row $a_m + \Delta$. For every such $p_{m'}$, either $d_{m'} < d_m$, or $b_{m'} > b_m$, or both (see Figure 8). In the former case, p_m skips diagonal $d_{m'}$, and in the latter case, $p_{m'}$ is active with respect to diagonal e and row $a_m + \Delta + 1$. So by Lemmas 2.9 and 2.10, there are at most $3g$ such points $p_{m'}$.

Therefore, by Corollary 2.12, we must have

$$\Delta/3 - 2g \leq 3g,$$

so $\Delta \leq 15g$. □

COROLLARY 2.14. *The number of points p_n that can skip a given diagonal e is at most $16g$.*

PROOF. Every point that skips diagonal e must lie on a different diagonal. Therefore, Lemma 2.13 already implies that diagonal e must eventually be occupied by a point p_n.

Now, consider the points p_m, $m < n$, that skip diagonal $e = d_n$. Partition these points into two sets: those having $b_m > b_n$ and those having $b_m < b_n$.

By Lemma 2.9, there are at most g points in the first set, since each such point is active with respect to the diagonal and row of p_n. And every point in the second set satisfies the assumptions of Lemma 2.13, so there are at most $15g$ such points. Therefore, diagonal e is skipped by no more than $16g$ points. □

We used somewhat messy arguments, but we have finally proven:

THEOREM 2.15. *For every Grundy value g and every diagonal e, there exists a g-point with* $d_n^g = e$. *Furthermore,*

$$-16g \leq d_n^g - n \leq 2g. \qquad \square$$

2.5. The g-values are close to the 0-values. We proceed to show the existence of the constants α_g and β_g of Theorem 2.3. In order to understand the idea behind our proof, it is helpful to look first at the following proof that the ratio between consecutive Fibonacci numbers tends to ϕ:

CLAIM 2.16. *Let F_n be the n-th Fibonacci number. Then $F_n/F_{n-1} \to \phi$.*

PROOF. Let $x_n = F_n - \phi F_{n-1}$. Then,

$$x_{n+1} = F_{n+1} - \phi F_n = (F_n + F_{n-1}) - \phi F_n$$
$$= -\phi^{-1}(F_n - \phi F_{n-1}) = -\phi^{-1} x_n. \qquad (2\text{-}4)$$

Therefore, $x_n \to 0$, since $|-\phi^{-1}| < 1$. Therefore,

$$\frac{F_n}{F_{n-1}} = \frac{x_n}{F_{n-1}} + \phi \to \phi. \qquad \square$$

Next, we introduce the following notation, which will help make our arguments clearer:

DEFINITION 2.17. Given sequences $\{f_n\}$ and $\{g_n\}$, we write

$$f_n \sim g_n$$

if, for some k, $|f_n - g_n| \leq k$ for all n.

Note that the relation $\sim$ is transitive: If $f_n \sim g_n$ and $g_n \sim h_n$, then $f_n \sim h_n$.

In this subsection we make a few claims that are intuitively obvious. We therefore decided to defer their proofs to the Appendix, in order not to interrupt the main flow of the arguments. Our first intuitive claim is the following:

LEMMA 2.18. *Let $\{x_n\}$ be a sequence that satisfies $x_{n+1} \sim c x_n$ for some $|c| < 1$. Then $\{x_n\}$ is bounded as a sequence.*

The main result of this subsection is the following somewhat general theorem:

THEOREM 2.19. *Let $a_0 < a_1 < a_2 < \cdots$ be a sequence of increasing natural numbers, and let $b_0, b_1, b_2, \ldots$ be a sequence of distinct natural numbers. Let $A = \{a_n \mid n \geq 0\}$, $B = \{b_n \mid n \geq 0\}$. Suppose the following conditions hold:*

1. $|A \cap B|$ *is finite;*
2. $A \cup B = \mathbb{N}$;
3. $b_n - a_n \sim n$.

Then $a_n \sim \phi n$ and $b_n \sim \phi^2 n$.

Note that, in particular, our Wythoff sequences $\{a_n^g\}$ and $\{b_n^g\}$ satisfy all of the above requirements, so the above theorem yields Theorem 2.3, as desired.

PROOF OF THEOREM 2.19.

We start with the following claim, which we prove in the Appendix:

LEMMA 2.20. *Regarding the sequences $\{a_n\}$ and $\{b_n\}$ in the statement of the Theorem:*

(a) *There is a constant k such that for all n, the number of $b_m > b_n$, $m < n$, is at most k.*

(b) *There is a constant k' such that for all n, the number of $b_m < b_n$, $m > n$, is at most k'.*

(c) $a_n \sim a_{n-1}$ *and* $b_n \sim b_{n-1}$.

(Note that, for our sequences $\{a_n^g\}$ and $\{b_n^g\}$, the lemmas of Section 2.4 already give bounds on the number of m's in Lemma 2.20(a,b). But we still want to prove Theorem 2.19 in general.)

Now, for $n \geq 0$, define

$$x_n = \phi n - a_n,$$

and let

$$f(n) = \left| A \cap \{0, 1, 2, \ldots, b_n - 1\} \right|$$

be the number of a's smaller than b_n.

By Lemma 2.20(a,b), the number of b's smaller than b_n is $\sim n$, so by conditions 1 and 2 of our Theorem, the number of a's smaller than b_n is $\sim b_n - n$. And by condition 3 we have $b_n - n \sim a_n$. Therefore,

$$f(n) \sim a_n. \tag{2-5}$$

Further, $a_{f(n)}$ is the first a that is $\geq b_n$ (by the definition of $f(n)$), so $a_{f(n)} \sim b_n$ by Lemma 2.20(c). Therefore (compare with (2-4)),

$$x_{f(n)} = \phi f(n) - a_{f(n)} \sim \phi a_n - b_n \sim \phi a_n - a_n - n$$
$$= \phi^{-1} a_n - n = -\phi^{-1} x_n. \tag{2-6}$$

The following lemma is proven in the Appendix:

LEMMA 2.21. *There exists an integer n_1 such that $f(n) > n$ for all $n \geq n_1$.*

Now, choose n_1 as in Lemma 2.21, and recursively define the integer sequence $n_1, n_2, n_3, \ldots$, by $n_{i+1} = f(n_i)$. This sequence, therefore, is strictly increasing. Also let $n_0 = 0$.

Define the sequence $\{y_j\}_{j=0}^{\infty}$ by

$$y_j = \max_{n_j \le i \le n_{j+1}} |x_i|, \quad \text{for } j \ge 0. \tag{2-7}$$

We want to show that $\{y_j\}$ is bounded as a sequence, which would imply that $\{|x_n|\}$ is bounded as a sequence. But it follows from equation (2-6) that:

LEMMA 2.22. *The sequence $\{y_j\}$ satisfies $y_{j+1} \sim \phi^{-1} y_j$.*

The full proof of Lemma 2.22 is given in the Appendix.

Lemmas 2.22 and 2.18 together imply that $\{y_j\}$ is bounded as a sequence, so $\{|x_n|\}$ is bounded as a sequence, as desired. Therefore, $a_n \sim \phi n$, and by condition 3 of our Theorem, $b_n \sim \phi^2 n$.   □

This completes the proof of the existence of the constants α_g and β_g of Theorem 2.3.

2.6. Experimental results. In this subsection we present experimental results on a few aspects of the function $\mathcal{G}$.

Experimental bounds on $d_n^g - n$. Let us compare the rigorous bound of $-16g \le d_n^g - n \le 2g$ given by Theorem 2.15 with data obtained experimentally.

Figure 9 shows a histogram of $d_n^g - n$ for $g = 30$, counting points up to row $5 \cdot 10^6$.

Table 2 shows the extreme values of $d_n^g - n$ achieved for different g by points lying in rows up to $5 \cdot 10^6$. In each case we show the earliest appearance of the extremal value.

We notice an interesting phenomenon: For $g \ge 7$ the maximum is achieved by the zeroth point $p_0^g = (0, g)$. This phenomenon is due to the fact that the sequence T_g tends to start with an anomalous behavior that "smooths out" over time.

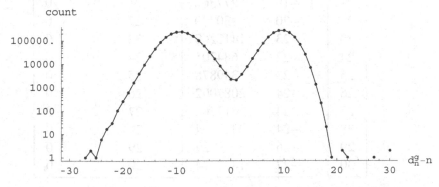

Figure 9. Histogram of $d_n^g - n$ for $g = 30$, counting points up to row $5 \cdot 10^6$.

Table 3 shows the maximum $d_n^g - n$ achieved by points having $n \geq 100$, for $g \geq 7$. We see that as g grows, the maximum achieved decreases substantially from Table 2.

g	min $d_n^g - n$	n	max $d_n^g - n$	n
0	0	0	0	0
1	−4	57	2	282
2	−6	35745	3	38814
3	−8	149804	4	2335
4	−10	569350	5	15486
5	−11	1245820	6	2638
6	−11	30165	7	1974933
7	−11	75459	7	0
8	−12	701260	8	0
9	−13	17972	9	0
10	−13	516328	10	0
11	−14	722842	11	0
12	−16	2853838	12	0
13	−17	2860809	13	0
14	−18	2814039	14	0
15	−18	2597774	15	0
16	−18	1027151	16	0
17	−18	2979529	17	0
18	−19	789978	18	0
19	−20	22347	19	0
20	−21	2548028	20	0
21	−19	277362	21	0
22	−20	30200	22	0
23	−23	1412268	23	0
24	−22	684205	24	0
25	−23	349878	25	0
26	−24	2087092	26	0
27	−24	617166	27	0
28	−24	2343474	28	0
29	−26	27	29	0
30	−27	1872274	30	0

Table 2. Extreme values of $d_n^g - n$ for given g, for points p_n^g having $a_n^g \leq 5 \cdot 10^6$.

g	max $d_n^g - n$	n	g	max $d_n^g - n$	n
7	7	131307	19	14	594141
8	8	20735	20	14	2482469
9	9	1056831	21	14	90130
10	9	258676	22	15	347510
11	10	987102	23	15	323425
12	10	1295870	24	16	129240
13	10	90426	25	17	1880006
14	11	453415	26	17	36662
15	11	61780	27	18	332552
16	12	509772	28	18	370321
17	12	86093	29	19	2425182
18	13	32439	30	18	444272

Table 3. Maximum value of $d_n^g - n$ for given g, for points p_n^g having $n \geq 100$ and $a_n^g \leq 5 \cdot 10^6$.

The conclusion from these observations is the following: The bounds observed experimentally for $d_n^g - n$ are much tighter than those given by Theorem 2.15. Therefore, either the theoretical bound is much looser than necessary, or it is a worst-case bound that is achieved very rarely in practice.

The converse of Theorem 2.3. Theorem 2.3 implies that if $\mathcal{G}(a, b)$ is bounded with $a \leq b$, then $|b - \phi a|$ is bounded. Is the converse also true? Namely, does a bound on $|b - \phi a|$ imply a bound on $\mathcal{G}(a, b)$? Or, on the contrary, are there arbitrarily large values very close to 0-values? We do not know the answer, but we explored this question experimentally.

We looked for the cells with the largest Grundy value lying at a given Manhattan distance from the closest 0-value. (The *Manhattan distance* between (a_1, b_1) and (a_2, b_2) is defined as $|a_2 - a_1| + |b_2 - b_1|$.) We looked up to row 10^6, calculating points up to $g = 199$. Our results are shown in Table 4.

The first column in Table 4 lists the cell with the largest Grundy value at a given Manhattan distance from the closest 0-value, for cells in rows $\leq 10^6$. Some cells are labelled "≥ 200" because they were not assigned any Grundy value ≤ 199.

The second column in Table 4 lists the cell with the largest Grundy value at a given Manhattan distance from the closest 0-value, restricted to cells between rows 600 and 10^6. (Only entries differing from the first column are shown.)

We see a very significant difference in the Grundy values between the first and second columns. This phenomenon is due to the fact that the sequences T_g

distance	cell	$\mathcal{G}$ value	row ≥ 600 cell	$\mathcal{G}$ value
1	(283432, 458601)	82		
2	(944634, 1528447)	96		
3	(44, 67)	89	(82399, 133320)	81
4	(49, 86)	115	(665224, 1076349)	82
5	(58, 86)	116	(402997, 652071)	103
6	(62, 110)	147	(538568, 871413)	99
7	(97, 168)	≥ 200	(839162, 1357804)	108
8	(95, 167)	≥ 200	(182922, 295987)	115
9	(87, 155)	≥ 200	(319656, 517229)	122
10	(85, 154)	≥ 200	(927492, 1500730)	125

Table 4. Large Grundy values at a given Manhattan distance from the closest 0-value.

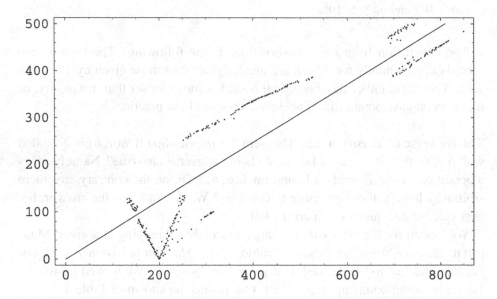

Figure 10. First terms of the sequence T_{200}.

tend to start by passing very close to the 0-points, before "smoothing out". This can be seen in Figure 10, which plots the beginning of T_{200}.

3. A recursive algorithm for the n-th g-value

In this Section we will present our so-called *Convergence Conjecture* and show how, if the conjecture is true, it leads to a recursive algorithm for finding the n-th g-point in $O(f(g)\log n)$ arithmetic operations, where f is some function on g.

To do this, we will first show how Algorithm WSG (page 385) can be considered, in a certain sense, as a *finite-state automaton* that receives input symbols and jumps from state to state as it calculates rows one by one.

Consider Algorithm WSG as it is about to start inserting points in row r. For each h, $0 \le h \le g$, there are cells on row r that cannot take an h-point: some because they have an h-point below them, some because they have an h-point diagonally below, and others because they have already been assigned a value $k < h$.

We will show how we can represent this information about row r, which we will call its *state*, in such a way that there is a finite number of states among all rows. We will further consider the set of h-points to be inserted in row r as a single *symbol* out of a finite alphabet of symbols. Then, knowing the state of row r and the said symbol, we can correctly place along row r the relevant h-points, and compute the state of row $r + 1$.

(Here we ignore the fact that the points to be inserted at row r are determined by the points inserted in lower rows.)

Let us develop these ideas formally.

3.1. A finite-state algorithm.

Let us fix g. The variable h will always take the values $0 \le h \le g$.

In this Section, whenever we refer to an h-point, we mean a point (x, y) with $\mathcal{G}(x, y) = h$ and $x \le y$. Points with $x \ge y$ will be referred to as *reflected* h-points. A point with $x = y$ is both reflected and unreflected.

DEFINITION 3.1.

1. $index_h(r) = |A_h \cap \{0, \ldots, r - 1\}|$ is the number of h-points strictly below row r. It is also the index of the first h-point on a row $\ge r$.
2. $D_h(r) = \{d_n^h \mid 0 \le n < index_h(r)\}$ is the set of diagonals occupied by h-points on rows $< r$.
3. $firstd_h(r) = \mathrm{mex}\, D_h(r)$ denotes the first empty diagonal on row r.
4. $ocdiag_h(r) = \{d \in D_h(r) \mid d > firstd_h(r)\}$ is the set of occupied diagonals after the first empty diagonal on row r.

    Note that

    $$index_h(r) = firstd_h(r) + |ocdiag_h(r)|. \qquad (3\text{-}1)$$

    This is because there are $index_h(r)$ occupied diagonals on row r: diagonals 0 through $firstd_h(r) - 1$, and $|ocdiag_h(r)|$ additional ones.
5. Similarly, $B_h(r) = \{b_n^h \mid 0 \le n < index_h(r)\}$ is the set of columns occupied by h-points on rows $< r$.

6. We let $occol_h(r) = \{b - r \mid b \in B_h(r), b - r \geq firstd_h(r)\}$. This set has the following interpretation: Identify each cell on row r by the diagonal it lies on, i.e., cell (r, b) is identified by $b - r$. Then $occol_h(r)$ represents the set of cells on row r on a diagonal $\geq firstd_h(r)$ that lie on a column occupied by a lower h-point.

LEMMA 3.2. *The expression*

$$\big(\max ocdiag_h(r)\big) - firstd_h(r) \tag{3-2}$$

is bounded for all r, and so is $|ocdiag_h(r)|$.

PROOF. Expression (3-2) corresponds to the maximum distance between a free diagonal and a subsequent occupied diagonal on row r; and this is bounded by Theorem 2.3. And since $ocdiag_h(r)$ only contains integers $> firstd_h(r)$, its size is also bounded. □

LEMMA 3.3. $\max occol_h(r) < \max ocdiag_h(r)$ *for all h, r.*

PROOF. Suppose a cell on row r lies on a column occupied by a lower h-point. Then a cell on row r further to the right lies on the diagonal occupied by that h-point. □

LEMMA 3.4. *For $g = 0$ we can compute explicitly the quantities of Definition 3.1. In particular,*

$$index_0(r) = firstd_0(r) = \lceil \phi^{-1} r \rceil, \tag{3-3}$$
$$ocdiag_0(r) = occol_0(r) = \varnothing. \tag{3-4}$$

PROOF. $index_0(r)$ is the index of the first 0-point on a row $\geq r$. Since $a_n^0 = \lfloor \phi n \rfloor$, the first n that gives $a_n^0 \geq r$ is $n = \lceil \phi^{-1} r \rceil$. (3-3) follows from the fact that the 0-points fill the diagonals in sequential order. And the second part of (3-3) follows from (3-1). □

DEFINITION 3.5. We define the following "normalized" quantities by subtracting $index_0(r)$:

$$n\_index_h(r) = index_h(r) - index_0(r),$$
$$n\_firstd_h(r) = firstd_h(r) - index_0(r),$$
$$n\_ocdiag_h(r) = \{d - index_0(r) \mid d \in ocdiag_h(r)\},$$
$$n\_occol_h(r) = \{c - index_0(r) \mid c \in occol_h(r)\}.$$

LEMMA 3.6. *The quantities $n\_index_h(r)$ and $n\_firstd_h(r)$, as well as the elements of $n\_ocdiag_h(r)$ and $n\_occol_h(r)$, are bounded in absolute value for all r.*

PROOF. This follows from Theorem 2.3. □

DEFINITION 3.7. The *state* of a given row r consists of

$$n\_index_h(r), \ n\_firstd_h(r), \ n\_ocdiag_h(r), \ n\_occol_h(r),$$

for $0 \le h \le g$.

By Lemma 3.6 we have:

COROLLARY 3.8. *There is a finite number of distinct states among all rows* $r \ge 0$.   □

Note that the state of a row r always satisfies

$$n\_index_h(r) = n\_firstd_h(r) + |n\_ocdiag_h(r)|. \qquad (3\text{-}5)$$

Note also that if $g = 0$ there is a single state for all rows:

$$n\_index_0(r) = n\_firstd_0(r) = 0, \quad n\_ocdiag_0(r) = n\_occol_0(r) = \varnothing.$$

DEFINITION 3.9. Given row r, we denote by

$$insert(r) = \{h \mid r \in A_h, 0 \le h \le g\}$$

the set of h-points that Algorithm WSG must insert in this row.

Definitions 3.7 and 3.9 enable us to reformulate Algorithm WSG as a finite-state automaton that jumps from state to state as it reads symbols from a finite alphabet Σ. The automaton is in the state of row r when it begins to calculate row r, and after reading the symbol $insert(r)$, it goes to the state of row $r + 1$. The input alphabet is $\Sigma = 2^{\{0,\dots,g\}}$, the set of all possible values of $insert(r)$.

Algorithm FSW (page 400) spells out in detail this finite-state formulation. (This algorithm is not equivalent to Algorithm WSG because it does not calculate the sets $insert(r)$ from previous rows, but only receives them as input.)

3.2. Convergence of states. Suppose we run Algorithm FSW starting from some row r_1, giving it as input the correct values of $insert(r_1)$, $insert(r_1 + 1)$, ..., but with a different initial state

$$n\_index'_h, n\_firstd'_h, n\_ocdiag'_h, n\_occol'_h, \quad 0 \le h \le g, \qquad (3\text{-}7)$$

instead of (3-6). Then the algorithm will output 4-tuples $(h, n, a_n'^h, b_n'^h)$, where the b-coordinates of the points will not necessarily be correct.

Could it happen that at some row $r > r_1$ the algorithm reaches the correct state for row r? If that happens, then for all subsequent rows the algorithm will be in the correct state, since the state of a row depends only on the state of the previous row. Therefore, for all rows $\ge r$ the algorithm will output the correct 4-tuples (h, n, a_n^h, b_n^h).

Algorithm FSW (Finite-State Wythoff)

Input: Integer g; integers r_1, r_2 (initial and final rows); state at row r_1, given by the variables

$$n\_index_h, n\_firstd_h, n\_ocdiag_h, n\_occol_h, \quad \text{for } 0 \le h \le g; \qquad (3\text{-}6)$$

sets $insert(r_1), \ldots, insert(r_2)$ (points to insert in rows $r_1, \ldots, r_2$).

Output: h-points in rows r_1 through r_2, given as 4-tuples

$$(h, n, a_n^h, b_n^h) \quad \text{for } 0 \le h \le g, \, r_1 \le a_n^h \le r_2.$$

1. For $r = r_1, \ldots, r_2$ do:

2.   • Let $S \leftarrow \varnothing$ [location of points inserted in this row].

3.   • For $h = 0, \ldots, g$ do:

4.   ○ If $h \in insert(r)$ then:

5.   * find the smallest $d \ge n\_firstd_h$ which is in none of the sets n_ocdiag_h, n_occol_h, and S;

6.   * output the 4-tuple

$$(h, \, n\_index_h + index_0(r), \, r, \, r + d + index_0(r))$$

[note that $index_0(r)$ can be calculated by Lemma 3.4];

7.   * let $n\_index_h \leftarrow n\_index_h + 1$;

8.   * insert d into n_ocdiag_h, n_occol_h, and S;

9.   * while $n\_firstd_h \in n\_ocdiag_h$ do $n\_firstd_h \leftarrow n\_firstd_h + 1$.

10.   ○ Subtract 1 from each element of n_occol_h [since r increases by 1: see Definition 3.1–6].

11.   ○ Remove from n_ocdiag_h and n_occol_h all elements $< n\_firstd_h$.

12.   • If $n\_index_0 = 1$ then [renormalize]:

13.   ○ subtract 1 from n_index_h and n_firstd_h for all h;

14.   ○ subtract 1 from each element of n_ocdiag_h and n_occol_h for all h.

Denote by

$$n\_index_h'(r), n\_firstd_h'(r), n\_ocdiag_h'(r), n\_occol_h'(r), \quad 0 \le h \le g, \quad r \ge r_1,$$

the state of the algorithm at row r when run with the initial state (3-7). We assume that the initial state (3-7) is consistent with property (3-5).

Observe that if $n\_index_h'(r_1) \ne n\_index_h(r_1)$, this difference will persist in all subsequent rows, since changes to n_index_h (at lines 7 and 13 of Algorithm

FSW) depend only on the input symbol $insert(r)$. Therefore, convergence can only occur if the initial state contains the correct values of $n\_index_h(r_1)$ for all h.

Now, we make the following conjecture:

CONJECTURE 3.10 (CONVERGENCE CONJECTURE). *For every g there exists a constant R_g such that for every row r_1, if Algorithm FSW is run starting from row r_1 with the initial "dummy" state*

$$
\begin{aligned}
n\_index'_h = n\_firstd'_h = n\_index_h(r_1), \\
n\_ocdiag'_h = n\_occol'_h = \varnothing,
\end{aligned}
\qquad \text{for } 0 \le h \le g, \qquad (3\text{-}8)
$$

and with the correct values of $insert(r_1)$, $insert(r_1 + 1), \ldots$, then the algorithm will converge to the correct state within at most R_g rows.

3.3. Experimental evidence for convergence.

We tested Conjecture 3.10 experimentally as follows: For some constant r_{max} we precalculated the state of row r and the value of $insert(r)$ for all r between 0 and r_{max}. We then ran Algorithm FSW starting from each row r, $0 \le r \le r_{max}$, with the dummy initial state (3-8), comparing at each step whether the algorithm's internal state converged to the correct state of the current row. We carried out this experiment for different values of g.

Our results are as follows: For $g = 0$ convergence always occurs after 0 rows; i.e., convergence is immediate.

For $g = 1$ the maximum time to convergence found was 45 rows. In fact, up to row 10^7 there are 3019 instances of convergence taking 45 rows.

For $g = 2$ the maximum found was 72 rows. Below row 10^7 there are 91 instances of convergence taking 72 rows.

For $g = 3$ the maximum of 140 rows to convergence is achieved only once below row 10^7.

For larger values of g we ran our experiment until row 10^6. Table 5 shows our findings. In each case we indicate the largest number of rows to convergence, and the starting row that achieves the maximum (or the first such starting row in case there are several).

Finally, Figure 11 shows a histogram of the time to convergence for $g = 10$ up to row 10^6. The shape of the curve suggests that there might be instances of higher convergence times that occur very rarely. However, we still find it plausible that a theoretical maximum R_g exists.

Note that Conjecture 3.10 could also be true only up to a certain value of g.

g	rows to convergence	starting row
0	0	0
1	45	2201
2	72	72058
3	140	804421
4	180	862429
5	235	732494
6	395	685531
7	395	685531
8	461	827469
9	630	59948
10	909	443109
⋮		
15	2041	8662
⋮		
20	4136	896721

Table 5. Maximum number of rows to convergence for different g up to row 10^6, and first starting row that achieves the maximum.

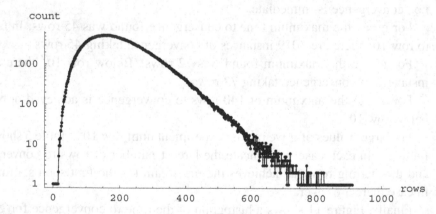

Figure 11. Histogram of the number of rows to convergence for $g = 10$, up to row 10^6.

3.4. The recursive algorithm.

We now show how Conjecture 3.10 leads to an algorithm for computing point p_n^g in $O(f(g) \log n)$ arithmetic operations, where f is some function on the constant R_g of Conjecture 3.10 and the constants α_g, β_g of Theorem 2.3.

Algorithm RW below is a recursive algorithm that receives as input an integer g and an interval $[r_1, r_2]$ of rows, and calculates all h-points, $0 \leq h \leq g$, in that interval.

Algorithm RW (Recursive Wythoff)

Input: Integer g; integers r_1, r_2 (initial and final rows).

Output: $index_h(r_1)$ for $0 \leq h \leq g$; set S of h-points in rows r_1 through r_2, given as 4-tuples (h, n, a_n^h, b_n^h) for $0 \leq h \leq g$, $r_1 \leq a_n^h \leq r_2$.

1. Let $r_0 \leftarrow r_1 - R_g$.

2. Let L and H be lower and upper bounds for $a_n^h - \phi^{-1} b_n^h$ for $0 \leq h \leq g$ and all n, according to Theorem 2.3. [Note that $a_n^h < \phi^{-1} r + L$ implies $b_n^h < r$, and $a_n^h > \phi^{-1} r + H$ implies $b_n^h > r$.]

3. Let $r_1' \leftarrow \lceil \phi^{-1} r_0 + L \rceil$, $r_2' \leftarrow \lfloor \phi^{-1} r_2 + H \rfloor$.

4. If $r_2' \geq r_0$ or $r_1' \leq 2g$ then:

5. • calculate and return the desired points by starting from row 0 using Algorithm WSG;

6. else:

7. • call Algorithm RW recursively and get $index_h(r_1')$ and the set S' of h-points in rows r_1' through r_2' for $0 \leq h \leq g$;

8. • calculate $insert(r_0), \ldots, insert(r_2)$ as
    $$insert(r) = \{h \mid \nexists n \text{ for which } (h, n, a_n^h, r) \in S'\}$$
    for $r_0 \leq r \leq r_2$;

9. • let t_h be the number of h-points in S' with $b_n^h < r_0$, for $0 \leq h \leq g$;

10. • calculate $index_h(r_0)$ as $index_h(r_0) = r_0 + 1 - index_h(r_1') - t_h$, for $0 \leq h \leq g$ [see explanation];

11. • calculate $n\_index_h(r_0)$ as $n\_index_h(r_0) = index_h(r_0) - index_0(r_0)$, for $0 \leq h \leq g$;

12. • run Algorithm FSW from rows r_0 to r_2 starting from the dummy state
    $$n\_index_h' = n\_firstd_h' = n\_index_h(r_0),$$
    $$n\_ocdiag_h' = n\_occol_h' = \varnothing, \qquad 0 \leq h \leq g,$$
    using $insert(r_0), \ldots, insert(r_2)$; get set T of 4-tuples $(h, n, a_n^h, b_n'^h)$ for $r_0 \leq a_n^h \leq r_2$;

13. • return $index_h(r_1)$ for $0 \leq h \leq g$, and the 4-tuples in T with $r_1 \leq a_n^h \leq r_2$.

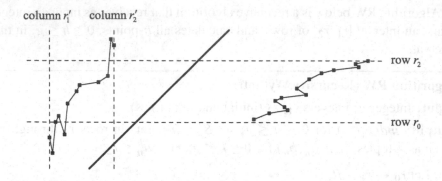

Figure 12. Points and reflected points in rows r_0 through r_2.

The idea behind Algorithm RW is the following: To calculate the h-points between rows r_1 and r_2, we run Algorithm FSW starting from row $r_0 = r_1 - R_g$ and the dummy initial state (3-8). Then, by Conjecture 3.10, the 4-tuples obtained from row r_1 on will be the correct ones (h, n, a_n^h, b_n^h).

We face two problems, however:

- To run Algorithm FSW we also need to know $insert(r_0), \ldots, insert(r_2)$, i.e., which h-points to insert between rows r_0 and r_2.
- For the dummy initial state (3-8) we need to know $index_h(r_0)$ for $0 \leq h \leq g$, i.e., how many h-points there are below r_0.

We solve both problems with a recursive call, in which we calculate all the *reflected h-points* that lie between rows r_0 and r_2, to the left of the main diagonal (see Figure 12). By the definition of L and H (line 2 of Algorithm RW), all these reflected h-points lie between columns r_1' and r_2' as computed in line 3. Of course, finding these reflected h-points is equivalent to finding the unreflected originals.

Once we have the reflected h-points, constructing $insert(r_0), \ldots, insert(r_2)$ is simple, since every row r, $r_0 \leq r \leq r_2$ that does not contain a reflected h-point must contain an h-point, and vice versa.

And computing $index_h(r_0)$ is also no problem, once we know $index_h(r_1')$ from the recursive call. Recall that $index_h(r_0)$ is the number of h-points on rows $0, \ldots, r_0 - 1$. Let k_h be the number of *reflected h-points* on rows $0, \ldots, r_0 - 1$. Then

$$index_h(r_0) + k_h = r_0 + 1,$$

since there is one h-point on the main diagonal, which is counted twice.

To calculate k_h, note that all reflected h-points before column r_1' lie below row r_0, and there are $index_h(r_1')$ such reflected h-points. Therefore,

$$k_h = index_h(r_1') + t_h,$$

where t_h is the number of reflected h-points below row r_0 that lie on or after column r_1', as in line 9. Putting all this together, we get

$$index_h(r_0) = r_0 + 1 - index_h(r_1') - t_h,$$

as in line 10.

The above calculation is only valid if the h-point on the main diagonal lies before column r_1'. That is why we check for the case $r_1' \leq 2g$ at line 4 (recall Lemma 2.2).

Finally, the check $r_2' \geq r_0$ at line 4 prevents making a recursive call if the new interval $[r_1', r_2']$ is not strictly below the old interval $[r_0, r_2]$.

If we cannot make a recursive call (for either of the two possible reasons), we calculate the h-points in the standard way, using Algorithm WSG starting from row 0.

3.5. Algorithm RW's running time. If we want to use Algorithm RW to calculate a single point p_n^g, we must first estimate its row number a_n^g. By Theorem 2.3, we can bound it between $r_1 = \lceil \phi n + L' \rceil$ and $r_2 = \lfloor \phi n + H' \rfloor$ for some constants L', H' that depend on g.

Whenever Algorithm RW makes a recursive call, it goes from an interval of length $\Delta r = r_2 - r_1$ to an interval of length $\Delta r' = r_2' - r_1'$, where

$$\Delta r' = \phi^{-1} \Delta r + (H - L + \phi^{-1} R_g)$$

(ignoring the rounding to integers). After repeated application of this transformation, the interval length converges to the constant

$$\Delta r^* = \phi^2 (H - L) + \phi R_g.$$

The number of recursive calls is $O(\log n)$, since each interval is ϕ times closer to the origin than its predecessor. And in the base case of the recursion, Algorithm WSG runs for at most a bounded number of rows, taking constant time.

Therefore, altogether Algorithm RW runs in $O(f(g) \log n)$ steps, for some function f that depends on the constants R_g, α_g, and β_g, as claimed.

3.6. Application of Algorithm RW. Let us discuss how to apply Algorithm RW to the problem raised in the Introduction, namely playing the sum of Wythoff's game with a Nim pile.

Suppose we are given the sum of a game of Wythoff in position (a, b), $a \leq b$, with a Nim pile of size g, where a and b are very large and g is relatively small. Suppose Conjecture 3.10 is true for this value of g, and we know the value of R_g.

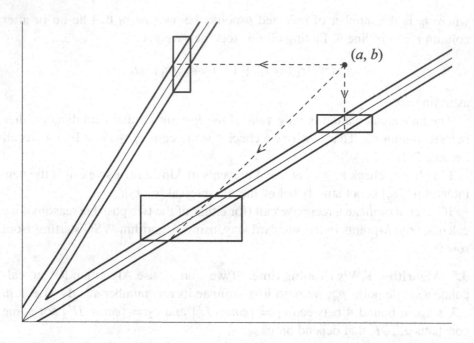

Figure 13. Intervals in which to look for a successor with Grundy value g to position (a, b).

We have to determine whether $\mathcal{G}(a, b)$ is larger than, smaller than, or equal to g. By Theorem 2.3, we can only have $\mathcal{G}(a, b) \leq g$ if $|b - \phi a| \leq k_g$ for some constant k_g that depends on α_g and β_g.

Therefore, if $|b - \phi a| > k_g$, we know right away that $\mathcal{G}(a, b) > g$. If, on the other hand, $|b - \phi a| \leq k_g$, then we use Algorithm RW to find all the h-points, $h \leq g$, in the vicinity of (a, b), and we check whether (a, b) is one of them.

If we find that $\mathcal{G}(a, b) = g$, then the overall game is in a P-position, so there is no winning move. If $\mathcal{G}(a, b) = h < g$, then our winning move is to reduce the Nim pile to size h. And if $\mathcal{G}(a, b) > g$, then our winning move consists of moving in Wythoff's game to a position with Grundy value g. There are at most three alternatives to check — moving horizontally, vertically, or diagonally. Therefore, the winning move can be found by making at most three calls to Algorithm RW with bounded-size intervals, as shown schematically in Figure 13.

g	$p_{10^{12}}^{g}$	g	$p_{10^{12}}^{g}$	g	$p_{10^{12}}^{g}$
0	$(a+49, b+49)$	7	$(a+50, b+46)$	14	$(a+49, b+54)$
1	$(a+50, b+50)$	8	$(a+51, b+51)$	15	$(a+47, b+51)$
2	$(a+49, b+50)$	9	$(a+51, b+56)$	16	$(a+49, b+43)$
3	$(a+50, b+49)$	10	$(a+49, b+52)$	17	$(a+53, b+51)$
4	$(a+50, b+51)$	11	$(a+51, b+49)$	18	$(a+48, b+52)$
5	$(a+50, b+52)$	12	$(a+49, b+53)$	19	$(a+52, b+61)$
6	$(a+49, b+51)$	13	$(a+50, b+55)$	20	$(a+49, b+39)$

where $a = 1\,618\,033\,988\,700$ and $b = 2\,618\,033\,988\,700$.

Table 6. Predicted value of $p_{10^{12}}^{g}$ for $0 \le g \le 20$.

3.7. Algorithm RW in practice.
We wrote a C++ implementation of Algorithm RW. For the constants L and H we used experimental lower and upper bounds for $a_n^g - \phi^{-1} b_n^g$, to which we added safety margins. For the constant R_g we added a safety margin to the values shown in Table 5.

We checked our program's results against those produced by the nonrecursive Algorithm WSG. The results were in complete agreement as far as we tested.

We also used our recursive program to predict the trillionth g-values for g between 0 and 20. We used $L = 15.0$, $H = 15.0$ (which are a safe distance away from the experimental bounds of -12.4 and 11.3), and $R_{20} = 8000$ (almost twice the value in Table 5).

Our predictions are shown in Table 6. The program actually performed this calculation in just twenty seconds. These predictions might be verified one day with a powerful computer.

To conclude, note that if there are only sporadic counterexamples to Conjecture 3.10 for a certain value of R_g, then Algorithm RW is still likely to give correct results in most cases. Algorithm RW will only fail if one of the rows r_0 in the different recursion levels constitutes the initial row of a counterexample to Conjecture 3.10. But, as we showed earlier, the number of recursion levels is logarithmic in the magnitude of the initial parameters.

Appendix: Lemmas of Section 2.5

We relegated some proofs in Section 2.5 to this Appendix.

LEMMA 2.18. *Let $\{x_n\}$ be a sequence that satisfies $x_{n+1} \sim c x_n$ for some $|c| < 1$. Then $\{x_n\}$ is bounded as a sequence.*

PROOF. Let k be the constant such that $|x_{n+1} - c x_n| \le k$, as in Definition 2.17; and let $d = k/(1 - |c|)$. Let I be the real interval $I = [-d, d]$. It can be verified that if $x_n \in I$, then $x_{n+1} \in I$, and if $x_n \notin I$, then $|x_{n+1}| - d \le |c|\,(|x_n| - d)$; in other words, the distance between x_n and I is multiplied by a factor of at most $|c|$. Therefore, since $|c| < 1$, the sequence $\{x_n\}$ is "attracted" towards I.   □

LEMMA 2.20. *Regarding the sequences $\{a_n\}$ and $\{b_n\}$ in the statement of Theorem 2.19:*

(a) *There is a constant k such that for all n, the number of $b_m > b_n$, $m < n$, is at most k.*

(b) *There is a constant k' such that for all n, the number of $b_m < b_n$, $m > n$, is at most k'.*

(c) $a_n \sim a_{n-1}$ *and* $b_n \sim b_{n-1}$.

PROOF. According to condition 3 in the Theorem, let L and H be such that

$$n + L \le b_n - a_n \le n + H \quad \text{for all } n.$$

Suppose $b_m > b_n$, with $m < n$. Then

$$b_n \ge a_n + n + L, \quad b_m \le a_m + m + H.$$

But $a_n - a_m \ge n - m$, so $0 < b_m - b_n \le 2(m - n) + H - L$, so

$$m > n - \frac{H - L}{2}.$$

Therefore, for every n there are at most $(H - L)/2$ possible values for m. This proves claim (a). Claim (b) follows analogously.

For claim (c), let $k = a_n - a_{n-1}$. Then, by condition 2 in the Theorem, the interval

$$I = \{a_{n-1} + 1, \ldots, a_n - 1\},$$

whose size is $k - 1$, is a subset of B. Let i be the smallest index and j the largest index such that both b_i and b_j are in I. Then $b_j - b_i \le k - 2$ and $j - i \ge k - 2$. But we have

$$b_j \ge a_j + j + L, \quad b_i \le a_i + i + H, \quad a_j - a_i \ge j - i,$$

so

$$k - 2 \ge b_j - b_i \ge 2(j - i) + L - H \ge 2k - 4 + L - H,$$

so

$$k \le H - L + 2,$$

proving that $a_n \sim a_{n-1}$. Moreover

$$b_n - b_{n-1} \le (a_n + n + H) - (a_{n-1} + n - 1 + L) \le 2(H - L) + 3$$

and

$$b_n - b_{n-1} \ge (a_n + n + L) - (a_{n-1} + n - 1 + H) \ge L - H + 2,$$

since $a_n - a_{n-1} \geq 1$. Therefore, $b_n \sim b_{n-1}$. $\qquad\square$

LEMMA 2.21. *There exists an integer n_1 such that $f(n) > n$ for all $n \geq n_1$.*

PROOF. The number of a's smaller than a_n is exactly n, so the number of b's smaller than a_n is $\sim a_n - n$. On the other hand, $b_n \sim b_{n-1}$, so the number of b's smaller than a_n goes to infinity as $n \to \infty$. Therefore, $a_n - n \to \infty$. But $a_n \sim f(n)$ (equation (2-5) in Section 2.5). Therefore, $f(n) - n \to \infty$, which is even stronger than our claim. $\qquad\square$

LEMMA 2.22. *The sequence $\{y_j\}$ defined in (2-7) satisfies $y_{j+1} \sim \phi^{-1} y_j$.*

PROOF. First note that if $\{c_n\}$ and $\{d_n\}$ are sequences such that $c_n \sim d_n$, then by equation (2-5) and Lemma 2.20(c) we have $f(c_n) \sim f(d_n)$. We also have $x_{c_n} \sim x_{d_n}$.

For each $j \geq 0$, let $m(j)$, $n_j \leq m(j) \leq n_{j+1}$, be the index for which the maximum $y_j = |x_{m(j)}|$ is achieved.

We claim that for each $j \geq 1$ there exists an integer $p(j)$ in the range $n_j \leq p_j \leq n_{j+1}$ such that

$$f(p(j)) \sim m(j+1). \qquad (A\text{-}9)$$

Indeed, at the one extreme we have $f(n_j) = n_{j+1} \leq m(j+1)$, while at the other we have $f(n_{j+1}) = n_{j+2} \geq m(j+1)$. Thus, there exists some intermediate value $p(j)$ such that $f(p_j) \leq m(j+1)$ and $f(p_j+1) \geq m(j+1)$. This choice of $p(j)$ satisfies (A-9).

Therefore, using equation (2-6),

$$y_{j+1} = |x_{m(j+1)}| \sim |x_{f(p(j))}| \sim \phi^{-1} |x_{p(j)}| \leq \phi^{-1} |x_{m(j)}| = \phi^{-1} y_j.$$

So

$$y_{j+1} \sim h_j \leq \phi^{-1} y_j \qquad (A\text{-}10)$$

for some sequence $\{h_j\}$.

Similarly, for each $j \geq 1$ there exists an integer $q(j)$ in the range $n_j \leq q(j) \leq n_{j+1}$ such that

$$f(m(j)) \sim q(j+1)$$

for all $j \geq 1$. Specifically, let

$$q(j+1) = \min\{\max\{f(m(j)), n_{j+1}\}, n_{j+2}\}.$$

(It is not hard to show that if $f(n) > f(n'), n < n'$, then $f(n) - f(n')$ is bounded.)

Therefore, using again equation (2-6),

$$y_{j+1} \geq |x_{q(j+1)}| \sim |x_{f(m(j))}| \sim \phi^{-1} |x_{m(j)}| = \phi^{-1} y_j;$$

so $y_{j+1} \geq h'_j \sim \phi^{-1} y_j$ for some sequence $\{h'_j\}$. This, together with (A-10), implies that $y_{j+1} \sim \phi^{-1} y_j$. $\qquad\square$

Acknowledgements

This paper is based on the author's M.Sc. thesis work [6]. I would like to thank my advisor, Aviezri Fraenkel, for suggesting the topic for this work, and for many useful discussions we had together.

I also owe thanks to Achim Flammenkamp, who gave me extensive comments on a draft of my thesis, and who also checked independently most of the experimental data presented here.

References

[1] E. R. Berlekamp, J. H. Conway and R. K. Guy, *Winning Ways for your Mathematical Plays*, vol. 1, second edition, A K Peters, Natick, MA, 2001. (First edition: Academic Press, New York, 1982.)

[2] U. Blass and A. S. Fraenkel, The Sprague–Grundy function for Wythoff's game, *Theoret. Comput. Sci.*, vol. 75 (1990), pp. 311–333.

[3] A. Dress, A. Flammenkamp and N. Pink, Additive periodicity of the Sprague–Grundy function of certain Nim games, *Adv. in Appl. Math.*, vol. 22 (1999), pp. 249–270.

[4] H. Landman, A simple FSM-based proof of the additive periodicity of the Sprague–Grundy function of Wythoff's game, in: *More Games of No Chance*, MSRI Publications, vol. 42, Cambridge University Press, Cambridge, 2002, pp. 383–386.

[5] H. Landman, Personal communication.

[6] G. Nivasch, *The Sprague–Grundy function for Wythoff's game: On the location of the g-values*, M.Sc. thesis, Weizmann Institute of Science, 2004.

[7] G. Nivasch and E. Lev, Nonattacking queens on a triangle, *Mathematics Magazine*, vol. 78 (2005), pp. 399–403.

[8] N. Pink, *Über die Grundyfunktionen des Wythoffspiels und verwandter Spiele*, dissertation, Universität Heidelberg, 1993.

[9] P. Vaderlind, R. K. Guy and L. C. Larson, *The Inquisitive Problem Solver*, The Mathematical Association of America, 2002.

[10] W. A. Wythoff, A modification of the game of Nim, *Niew Archief voor Wiskunde*, vol. 7 (1907), pp. 199–202.

GABRIEL NIVASCH
SCHOOL OF COMPUTER SCIENCE
TEL AVIV UNIVERSITY
TEL AVIV 69978
ISRAEL
gabriel.nivasch@cs.tau.ac.il

Theory of the small

Yellow-Brown Hackenbush

ELWYN BERLEKAMP

ABSTRACT. This game is played on a sum of strings. In its "restricted" form, Left, at her turn, picks a bichromatic string and removes its highest yeLLow branch. Right, at his turn, picks a bichromatic string and removes its highest bRown branch. As in the well-known game of bLue-Red Hackenbush, all higher branches, being disconnected, also disappear. But in yellow-brown Hackenbush, unlike blue-red Hackenbush, all moves on monochromatic strings are *illegal*. This makes all values of yellow-brown Hackenbush all-small.

This paper presents an explicit solution of restricted yellow-brown Hackenbush. The values are sums of basic infinitesimals that have appeared in many other games found in *Winning Ways*.

Yellow-Brown (YB) Hackenbush is played on sums of strings. Each mixed string is played analogously to LR Hackenbush. Left can remove a yeLLow branch, and all other branches (if any) above it; Right can remove a bRown branch and all other branches (if any) above it. But yellow-brown strings differ from blue-red strings in an important respect:

> Neither player is allowed to move on any monochromatic string.

That rule ensures that all stopping positions are 0, and that all YB values are infinitesimal.

There are (at least) two variations of YB Hackenbush: restricted and unrestricted. In the restricted variation, each player is allowed at most a single option on any string, namely, his branch which is highest above the ground. Although this restriction would have no effect on the values of LR Hackenbush, it has a major effect on YB Hackenbush. In the unrestricted YB Hackenbush, either player can move to 0 by playing his lowest branch, so all nonzero values are confused with 0.

We now present a complete solution for *restricted YB strings*.

To each YB string we may associate a number x which is the value of the corresponding LR Hackenbush string. It is not hard to see, recursively, that the value of the YB string is

$$v = \int_0^0 x$$

Although this operator, $\int_0^0$, is nonlinear, it can be used to compute the value v. This value can be expressed in terms of a basis of standard infinitesimals. The key to accomplishing this is another number, namely

$$y = \lceil x \rceil + \tfrac{1}{2} - x.$$

Here are the values corresponding to some short YB strings:

String	x	y	v
Y	1	.1	0
YYBBB	1.001	.011	⇑*
YYBB	1.01	.01	↑
YYBBYB	1.0101	.0011	$2.\uparrow[2]$
YYBBY	1.011	.001	$\uparrow[2]$
YYBBYY	1.0111	.0001	$\uparrow[3]$
YYB	1.1	0	*

where the basis infinitesimals in the v column are defined by

$$\uparrow[1] = 0 \mid *$$
$$\uparrow[2] = \uparrow[1] \mid *$$
$$\uparrow[3] = \uparrow[2] \mid *$$

$$\cdots$$

$$\uparrow[n+1] = \uparrow[n] \mid *$$

In general, if $y = \sum_{i=1}^n Y_i 2^{-i}$, where each Y_i equals 0 or 1, with $Y_1 = 0$ and $Y_n = 1$, then our asserted value is

$$v(y) = * + \sum_{i=1}^n Y_i \left(\uparrow \left[\sum_{j=1}^i (1 - Y_i) \right] * \right),$$

where $\uparrow[n]$ is defined above. So, for example, consider the string

$$\text{Y Y B B B Y B Y Y B Y B B B Y}$$

Using the well-known rule for Blue-Red Hackenbush strings (see *Winning Ways*, vol. 1), we have the binary expansion of x, namely

$$\begin{array}{c}\text{Y Y B B B Y B Y Y B Y B B B Y}\\ x = \;1\;\overbrace{}\;0\;0\;1\;0\;1\;1\;0\;1\;0\;0\;0\;1\;1\end{array}$$

from which we proceed as follows:

$$\begin{array}{llllllllllllllllll}
x = & 1 & \cdot & 0 & 0 & 1 & 0 & 1 & 1 & 0 & 1 & 0 & 0 & 0 & 1 & 1 \\
y = & & \cdot & 0 & 1 & 0 & 1 & 0 & 0 & 1 & 0 & 1 & 1 & 1 & 0 & 1 \\
v = & * & + & \uparrow[1]* & & & & & & & & & & & & \\
& & & + & & \uparrow[2]* & & & & & & & & & & \\
& & & & & + & & \uparrow[4]* & & & & & & & & \\
& & & & & & + & & 3.\uparrow[5]* & & & & & & & \\
& & & & & & & + & & \uparrow[6]* & & & & & &
\end{array}$$

SKETCHED PROOF. The negatives of the Left incentives of the basis infinitesimals are positive infinitesimals of increasingly higher orders:

$$\uparrow \;=\; -\Delta^L\uparrow[1] \;\gg\; -\Delta^L\uparrow[2] \;\gg\; \cdots \;\gg\; \Delta^L\uparrow[j] \;>\; 0$$

In real analysis, one expects equations such as

$$\varepsilon \gg \varepsilon^2 \gg \cdots \gg \varepsilon^k > 0,$$

so, by analogy, it is common to denote $-\Delta^L\uparrow[j]$ by $\uparrow^j$, even though no multiplication is defined nor intended.

Then

$$\uparrow[j] = \uparrow[j-1] + \uparrow^j$$

$$= \sum_{i=1}^{j} \uparrow^i .$$

For every j,

$$* \not\gtreqless \sum_{i=1}^{j} \uparrow^i$$

but

$$* \;<\; \sum_{i=1}^{j-1} \uparrow^i + \uparrow^j + \uparrow^j .$$

Both sequences of incentives increase

$$\Delta^L(\uparrow[j]) < \Delta^L(\uparrow[j+1])$$
$$\Delta^R(\uparrow[j]) < \Delta^R(\uparrow[j+1]).$$

Therefore, in any sum of nonnegative integer multiples of $\uparrow[j]$, each player's dominant move is on a term with maximum superscript.

The cases

$$y = .1, \quad v = \uparrow[0] = 0$$
$$\text{and} \quad y = 0, \quad v = *$$

are degenerate. In the general nondegenerate case, $0 < y < \frac{1}{2}$,

$$y = \sum_{i=1}^{n} Y_i 2^{-i} \quad \text{for each } Y_i = 0 \text{ or } 1.$$

Let m be the integer for which $Y_m = 0$, but $Y_i = 1$ for $m < i \le n$.

Let

$$k = \sum_{i=1}^{n}(1 - Y_i) = \sum_{i=1}^{m}(1 - Y_i).$$

Then

$$v(y) = * + \sum_{i=1}^{m} Y_i \left(\uparrow\left[\sum_{j=1}^{i}(1 - Y_j)\right]* \right) + (n - m).(\uparrow[k]*)$$

We now explore properties of this asserted value, leading to a sketched proof that it is indeed the value of the corresponding restricted YB Hackenbush string.

We notice that since $0 < y < \frac{1}{2}$, $v(y)$ is positive. From $v(y)$, Right has a unique dominant incentive, $\uparrow[k]*$. Left has two dominant incentives, $*$ and $\downarrow^k$. However, Left's move of incentive $*$ reverses to another position whose incentive is dominated by $\downarrow^k$, and so we have the equation

$$v(y) = \{v(y) + \downarrow^k \mid v(y) - \uparrow[k] * \}. \tag{0-1}$$

If $n = m + 1$, this is canonical form. However, if $n \ge m + 2$, then Left's move is reversible. It continues to reverse to 0, and the canonical form is

$$v(y) = \{0 \mid v(y) - \uparrow[k]*\}, \quad \text{if } n \ge m + 2. \tag{0-2}$$

The canonical positions of the number y include:

$$y^L = y - 2^{-n},$$
$$y^R = y + 2^{-n},$$
$$y^{RL} = y + 2^{-n} - 2^{-m},$$
$$y^L = y^{RL} + (2^{-m} - 2^{1-n}).$$

The asserted YB values satisfy

$$v(y) = v(y^L) + \uparrow[k]*, \tag{0-3}$$
$$v(y^R) = v(y^{RL}) + \uparrow[k-1]*,$$
$$v(y) = v(y^{RL}) = (n-m).(\uparrow[k]*),$$
$$v(y^L) = v(y^{RL}) + (n-m-1).(\uparrow[k]*),$$

and

$$v(y) > v(y^R). \tag{0-4}$$

If $n = m + 1$, then in view of (0-3), (0-1) becomes

$$v(y) = \{v(y^R) \mid v(y^L)\} \tag{0-5}$$

We next show that (0-5) remains valid if $n \geq m + 2$.

From (0-2), we have

$$v(y) = \{0 \mid v(y^L)\}.$$

Relation (0-4) implies that $v(y^R) \not\geq v(y)$, and so the Gift Horse principle ensures that

$$v(y) = \{0 \mid v(y^L)\} = \{0, v(y^R) \mid v(y^L)\}$$
$$- \{v(y^R) \mid v(y^L)\}, \quad \text{because } v(y^R) \geq 0.$$

Thus, (0-5) is valid for any $n > m$. Since $y = \frac{1}{2} - x$, this recursion implies that if there is any value of y for which the asserted YB string value is incorrect, it must be degenerate. But we have verified the degenerate cases, and so the asserted values are correct for all restricted YB strings. □

Historical note

This game arose in the early 1990s when I was studying a variety of overheating operators and trying (in vain) to get more understanding of the conditions under which they are linear and/or monotonic. In my graduate seminars on game theory, we considered numbers overheated from 1 to infinity. We called them "vaporized numbers" and observed their close relationship to numbers overheated from 0 to 0. We then invented this game. Several graduate students participated in those discussions. In spring 1994, I wrote a preliminary version of this paper and distributed it to the class. Shortly thereafter Kuo-Yuen Kao studied some similar games, and included them in his unpublished doctoral thesis at UNC Charlotte.

The invention of Clobber in 2002 rekindled widespread interest in properties of infinitesimals and led me to begin a slow and sporadic search which culminated in finding this paper buried deep in my files.

References

"Sums of Hot and Tepid Combinatorial Games", by Kuo-Yuan Kao, Ph.D. Thesis, University of North Carolina at Charlotte, 1997.

Winning Ways for Your Mathematical Plays, 2nd edition (4 volumes), by Elwyn Berlekamp, John Conway and Richard Guy, A K Peters, Wellesley, MA, 2001–2004.

ELWYN BERLEKAMP
DEPARTMENT OF MATHEMATICS
UNIVERSITY OF CALIFORNIA
BERKELEY, CA 94720
UNITED STATES
berlek@math.berkeley.edu

Games of No Chance 3
MSRI Publications
Volume **56**, 2009

Ordinal partizan End Nim

ADAM DUFFY, GARRETT KOLPIN, AND DAVID WOLFE

Introduction

Partizan End Nim is a game played by two players called Left and Right. Initially there are n stacks of boxes in a row, each stack containing at least one box. Players take turns reducing the number of boxes from the stack on their respective side (Left removes from the leftmost stack, while Right removes from the rightmost stack). For example, the position 3 | 5 | 2 (or, denoted more tersely, 352) has three boxes in its leftmost pile, five boxes in the middle pile, and two boxes in the rightmost pile. When the game starts, Left can only remove boxes from the pile of size three, and Right can only remove boxes from the pile of size two. The first player that cannot move loses. This particular position should be a win for the first player (whether that be Left or Right). The first player should remove a whole pile, for the stack of size 5 dominates the remaining stack. Notice that if one player has a legal move from a position, the other player can also legally move, making the game *all small*, as defined in [3, p. 101] and [2, vol. 1, p. 221].

In our version of Partizan End Nim, piles are not limited to finite size. A move in Partizan End Nim requires the player to change the size of the closest pile to a smaller *ordinal* number (possibly 0). For instance, a move from a pile of size ω, the smallest nonfinite ordinal, consists of changing the size of that pile to some finite ordinal height. A player can move a pile of size $\omega + 1$ to any natural number or to a pile of size ω.

For an example of a position with ordinal pile sizes, the reader can confirm that Left can win from the three pile position $(\omega + \omega + 1) \mid (\omega + \omega) \mid (\omega + 1)$ whether she moves first or second. Moving first, Left can remove the leftmost pile since the middle pile is larger than the rightmost pile. The strategy that Left

This research was supported by a Research, Scholarship, and Creativity Grant from Gustavus Adolphus College.

uses when playing second depends on Right's move. If Right takes the whole rightmost pile, then Left's winning move is to remove one box from the leftmost pile. On the other hand, if Right changes the size of the rightmost pile to either ω or some natural number, then the winning move for Left is to remove the entire leftmost pile.

The results and proofs presented here parallel and tighten those found in [1] by allowing for ordinal pile sizes. We give an efficient recursive method to compute the outcome of (and winning moves from) any position. The reader who is not fond of ordinal numbers can safely skip the following section and assume all pile sizes are finite.

Ordinal numbers

Ordinal numbers generalize the natural numbers, allowing us to define trans-finite numbers. In this paper, we will represent ordinal numbers as sets by giving a standard recursive definition.

For sets X, Y, define $X < Y$ if $X \subset Y$, the natural partial ordering of sets.

DEFINITION 1. The segment of X determined by α, written X_α, is defined by $X_\alpha = \{x \in X \mid x < \alpha\}$.

DEFINITION 2. An ordinal is a well-ordered set X such that $\alpha = X_\alpha$ for all $\alpha \in X$.

From this definition, we are able to reach the following conclusions:

- If X is an ordinal, then for all $\alpha \in X$, α is also an ordinal.
- If X is an ordinal, then for all $\alpha \in X$, $\alpha \subset X$.
- The set of all ordinals is well-ordered.
- For ordinals α and β, $\alpha < \beta \iff \alpha \subset \beta \iff \alpha \in \beta$.

Now, we define the first ordinal, ϕ, to be 0. From this, we can define $1 = \{0\} = \{\phi\}$, $2 = \{0, 1\} = \{\phi, \{\phi\}\}$, and so on. The least transfinite ordinal is defined as $\omega = \{0, 1, 2, 3, 4, \ldots\}$. Since ordinals are well-ordered, the principle of mathematical induction applies to them. (See, for example, [4, 1.7; 3.1])

Next, we will describe some properties of ordinal numbers.

DEFINITION 3. If we have two ordinals, α and β, then the ordinal sum of α and β is defined by

$$\alpha + \beta = \alpha \cup \{\alpha + \beta'\}_{\beta' \in \beta}$$

and the ordinal difference of α and β, with $\alpha \geq \beta$, is defined by

$$\alpha - \beta = \{\alpha' - \beta\}_{\alpha' \geq \beta, \alpha' \in \alpha}.$$

Ordinal addition is not commutative, but it is associative. For example, $1 + \omega = \omega$, while $\omega + 1 = \{1, 2, 3, \ldots, \omega\} > \omega$. With these definitions, we are able to make the following observations:

OBSERVATION 4. *If $\beta' \in \beta$, then $(\alpha + \beta') \in (\alpha + \beta)$.*

OBSERVATION 5. *If $\alpha' \in \alpha$, then $(\alpha' - \beta) \in (\alpha - \beta)$.*

Now, we will show that our definitions for ordinal addition and subtraction observe the following identities:

LEMMA 6. *Let α and β be ordinals. Then*

$$(\beta + \alpha) - \beta = \alpha \quad and \quad \beta + (\alpha - \beta) = \alpha \quad (if \, \alpha \geq \beta).$$

PROOF.

$$
\begin{aligned}
(\beta + \alpha) - \beta &= (\beta \cup \{\beta + \alpha'\}) - \beta \\
&= \{(\beta + \alpha') - \beta\} \\
&= \{\alpha'\} \qquad\qquad \text{(by induction)} \\
&= \alpha;
\end{aligned}
$$

$$
\begin{aligned}
\beta + (\alpha - \beta) &= \beta + \{\alpha' - \beta\}_{\alpha' \geq \beta} \\
&= \beta \cup \{\beta + (\alpha' - \beta)\}_{\alpha' \geq \beta} \\
&= \beta \cup \{\alpha'\}_{\alpha' \geq \beta} \qquad \text{(by induction)} \\
&= \{\alpha'\} \\
&= \alpha. \qquad\qquad\qquad\qquad\qquad \square
\end{aligned}
$$

Note that $(\alpha + \beta) - \beta$ and $(\alpha - \beta) + \beta$ need not equal α. For example, $(1 + \omega) - \omega = 0$ and $(\omega - 1) + 1 = \omega + 1$.

Partizan End Nim

In this section, we will define a recursive algorithm that determines the *outcome class* of a game of Partizan End Nim.

DEFINITION 7. An outcome class describes which player has a winning strategy. The four possible outcome classes are $\mathcal{N}$, $\mathcal{P}$, $\mathcal{L}$, and $\mathcal{R}$. A game is in:

- $\mathcal{N}$ if the first player always has a winning strategy.
- $\mathcal{P}$ if the second player always has a winning strategy.
- $\mathcal{L}$ if the Left player always has a winning strategy.
- $\mathcal{R}$ if the Right player always has a winning strategy.

For the remainder of this paper, we'll encode a position by a string of ordinals, $\mathbf{x}$. We can append and/or prepend additional piles to the string by concatenation, as in $\alpha\mathbf{x}\beta$. For instance, if $\mathbf{x} = 3526$ is a position with 4 piles, we might construct a 6 pile game $\alpha\mathbf{x}\beta$, where $\alpha = \omega$ and $\beta = \omega + 3$.

DEFINITION 8. Let $R(\mathbf{x})$ be defined as the minimum ordinal β such that $\mathbf{x}\beta$ is a win for Right moving second, where $\beta \geq 0$. Similarly, define $L(\mathbf{x})$ as the minimum α such that $\alpha\mathbf{x}$ is a win for Left moving second.

OBSERVATION 9. Using these definitions, we are able to determine the outcome class of $\alpha\mathbf{x}\beta$ for ordinals $\alpha, \beta > 0$ as follows:

$$\alpha\mathbf{x}\beta \in \begin{cases} \mathcal{L} & \text{if } \alpha > L(\mathbf{x}\beta) \text{ and } \beta \leq R(\alpha\mathbf{x}), \\ \mathcal{R} & \text{if } \alpha \leq L(\mathbf{x}\beta) \text{ and } \beta > R(\alpha\mathbf{x}), \\ \mathcal{N} & \text{if } \alpha > L(\mathbf{x}\beta) \text{ and } \beta > R(\alpha\mathbf{x}), \\ \mathcal{P} & \text{if } \alpha \leq L(\mathbf{x}\beta) \text{ and } \beta \leq R(\alpha\mathbf{x}). \end{cases}$$

DEFINITION 10. The triple point of $\mathbf{x}$ is $(L(\mathbf{x}), R(\mathbf{x}))$.

We show in the next proposition that the triple point of $\mathbf{x}$ determines the outcome class of any game of the form $\alpha\mathbf{x}\beta$.

PROPOSITION 11. *We can determine the outcome class of any game $\alpha\mathbf{x}\beta$ for ordinals $\alpha, \beta > 0$ using just $L(\mathbf{x})$ and $R(\mathbf{x})$:*

$$\alpha\mathbf{x}\beta \in \begin{cases} \mathcal{N} & \text{if } \alpha \leq L(\mathbf{x}) \text{ and } \beta \leq R(\mathbf{x}), \\ \mathcal{P} & \text{if } \alpha = L(\mathbf{x}) + \gamma \text{ and } \beta = R(\mathbf{x}) + \gamma \text{ for some } \gamma > 0, \\ \mathcal{L} & \text{if } \alpha = L(\mathbf{x}) + \gamma \text{ and } \beta < R(\mathbf{x}) + \gamma \text{ for some } \gamma > 0, \\ \mathcal{R} & \text{if } \alpha < L(\mathbf{x}) + \gamma \text{ and } \beta = R(\mathbf{x}) + \gamma \text{ for some } \gamma > 0. \end{cases}$$

PROOF. First, assume that $\alpha \leq L(\mathbf{x})$ and $\beta \leq R(\mathbf{x})$. If Left removes all of α, Right cannot win since any move is to $\mathbf{x}\beta'$ where $\beta' < \beta$ and β was the least value such that Right wins moving second on $\mathbf{x}\beta$. Symmetrically, Right can also win moving first by removing all of β. Thus, $\alpha\mathbf{x}\beta \in \mathcal{N}$.

Next, assume that $\alpha = L(\mathbf{x}) + \gamma$ and $\beta = R(\mathbf{x}) + \gamma$ for some $\gamma > 0$. Also, assume that Left moves first. If Left changes the size of α to $\alpha = L(\mathbf{x}) + \gamma'$ where $0 < \gamma' < \gamma$, Right simply responds by moving on β to $\beta = R(\mathbf{x}) + \gamma'$. By induction, this position is in $\mathcal{P}$. On the other hand, if Left changes the size of α to α' where $\alpha' \leq L(\mathbf{x})$, Right can win by removing β as shown in the previous case. Thus, Left loses moving first. Symmetrically, Right also loses moving first. So, $\alpha\mathbf{x}\beta \in \mathcal{P}$.

Finally, assume $\alpha = L(\mathbf{x}) + \gamma$ and $\beta < R(\mathbf{x}) + \gamma$ for some $\gamma > 0$. If $\beta \leq R(\mathbf{x})$, Left can win moving first by removing all of α as shown in the first case. If $\beta > R(\mathbf{x})$, Left wins moving first by changing the size of α to $L(\mathbf{x}) + \gamma'$, where γ' is defined by $\beta = R(\mathbf{x}) + \gamma'$, which is in $\mathcal{P}$ by induction. Left can win moving

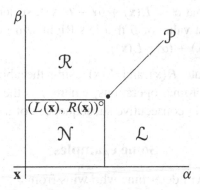

Figure 1. Outcome classes of the game $\alpha x \beta$ for all ordinals α and β. The unfilled circle represents the triple point, $(L(x), R(x))$, which has an outcome class of $\mathcal{N}$. The filled circle is the point $(L(x) + 1, R(x) + 1)$ and the games on the line originating from the filled circle, which are of the form $(L(x) + \gamma, R(x) + \gamma)$ where $\gamma > 0$, have an outcome class of $\mathcal{P}$.

second from $\alpha x \beta$ by making the same responses as in the previous case. Thus, $\alpha x \beta \in \mathcal{L}$. $\qquad\square$

OBSERVATION 12. Suppose we have a string of piles called x. If we were to list the outcome classes of $x \beta$ as β increases from 1 through the ordinals, we would get one of the following results:

- a string (possibly empty) of $\mathcal{N}$'s followed by $\mathcal{R}$'s, or,
- a string of $\mathcal{L}$'s (again possibly empty) followed by a single $\mathcal{P}$, and then $\mathcal{R}$'s.

OBSERVATION 13. If $\alpha x \beta \in \mathcal{P}$, then $(\alpha + \gamma) x (\beta + \gamma) \in \mathcal{P}$ for all ordinals $\gamma > 0$.

Notice that by using Proposition 11, the triple point of x determines the outcome classes of all games of the form $\alpha x \beta$. Figure 1 represents the outcome classes of the game $\alpha x \beta$.

Finally, we will give a recursive algorithm to compute $R(x)$ and $L(x)$, which allows us to efficiently analyze any Partizan End Nim position.

PROPOSITION 14. *The functions $R(x)$ and $L(x)$ can be computed recursively using*:

$$R(\alpha x) = \begin{cases} 0 & \text{if } \alpha \leq L(x), \\ R(x) + (\alpha - L(x)) & \text{if } \alpha > L(x), \end{cases}$$

$$L(x \beta) = \begin{cases} 0 & \text{if } \beta \leq R(x), \\ L(x) + (\beta - R(x)) & \text{if } \beta > R(x). \end{cases}$$

For the base case, $R(x) = L(x) = 0$ when x is empty.

PROOF. As already shown in Proposition 11, Left loses moving first on αx where $\alpha \leq L(x)$, so $R(\alpha x) = 0$. On the other hand, assume that $\alpha > L(x)$. We know from Lemma 6 that $\alpha = L(\mathbf{x}) + (\alpha - L(\mathbf{x}))$, so let $\gamma = (\alpha - L(\mathbf{x}))$. By Proposition 11, the least value of β that lets Right win moving second on $\alpha\mathbf{x}\beta$ is $\beta = R(\mathbf{x}) + \gamma = R(\mathbf{x}) + (\alpha - L(\mathbf{x}))$.   □

An algorithm to compute $R(\mathbf{x})$ and $L(\mathbf{x})$ using the above recurrence can be written to take $\Theta(n2)$ ordinal operations, where n is the number of piles in $\mathbf{x}$, since there are $\frac{1}{2}n(n+1)$ consecutive subsequences of a row of n piles.

Some examples

As an example, we will determine who wins from

$$3 \mid 5 \mid 2 \mid 3 \mid 3 \mid 1 \mid 9$$

when Left moves first and when Right moves first. Fix $\mathbf{x} = 52331$. We wish to compute $L(\mathbf{x})$ and $R(\mathbf{x})$ using Proposition 14. For single-pile positions, $L(\alpha) = R(\alpha) = \alpha$, because $R(\alpha) = R() + (\alpha - L()) = 0 + (\alpha - 0) = \alpha$. For 2-pile positions, we have

$$R(\alpha\beta) = \begin{cases} 0 & \text{if } \alpha \le \beta, \\ \alpha & \text{if } \alpha > \beta, \end{cases} \qquad L(\alpha\beta) = \begin{cases} 0 & \text{if } \alpha \ge \beta, \\ \beta & \text{if } \alpha < \beta. \end{cases}$$

We can compute $L(52331)$ by first calculating $L(\mathbf{w})$ and $R(\mathbf{w})$ for each shorter substring of piles:

w	L(w)	R(w)
523	0	2
233	6	2
331	1	6
5233	1	0
2331	0	7
52331	2	12

For instance, $R(523) = R(23) + (5 - L(23)) = 0 + (5 - 3) = 2$.

For the original position $\alpha\mathbf{x}\beta = 3\mathbf{x}9$, we have $3 > L(\mathbf{x}) = 2$ and $9 \le R(\mathbf{x}) = 12$, and hence, by Observation 9, 3523319 is an $\mathcal{L}$-position.

As a last example, we will tabulate $L(\mathbf{x})$ for three 3-pile positions, the first of which is from the introduction:

x			L(x)
left pile	middle pile	right pile	
$\omega+\omega+1$	$\omega+\omega$	$\omega+1$	0
$\omega+\omega+1$	$\omega+\omega+1$	$\omega+1$	$\omega+1$
$\omega+\omega+1$	$\omega+\omega+2$	$\omega+1$	$\omega+\omega+\omega+1$

A small change to an individual pile size (in this case, the middle pile) can have a large effect on the triple-point.

Open question

While the values of arbitrary Partizan End Nim positions appear quite complicated, we conjecture that the atomic weights are all integers.

References

[1] Michael Albert and Richard Nowakowski. "The game of End-Nim", *Electronic Journal of Combinatorics*, **8**(2):R1, 2001.

[2] Elwyn R. Berlekamp, John H. Conway, and Richard K. Guy, *Winning ways for your mathematical plays*. 2nd edition, A K Peters, Ltd., Natick, Massachusetts, 2001–2004. First edition published in 1982 by Academic Press.

[3] John H. Conway, *On numbers and games*, 2nd edition, A K Peters, Ltd., Natick, MA, 2001. First edition published in 1976 by Academic Press.

[4] Keith Devlin, *The joy of sets: fundamentals of contemporary set theory*, Springer, New York, 1992.

ADAM DUFFY, GARRETT KOLPIN, AND DAVID WOLFE
MATH/COMPUTER SCIENCE DEPARTMENT
GUSTAVUS ADOLPHUS COLLEGE
800 WEST COLLEGE AVENUE
ST. PETER, MN 56082-1498
UNITED STATES
    wolfe@gustavus.edu

Games of No Chance 3
MSRI Publications
Volume **56**, 2009

Reductions of partizan games

J.P. GROSSMAN AND AARON N. SIEGEL

ABSTRACT. The *reduced canonical form* of a game G, denoted by $\overline{G}$, is the simplest game infinitesimally close to G. Reduced canonical forms were introduced by Calistrate [2], who gave a simple construction for computing $\overline{G}$. We provide a new correctness proof of Calistrate's algorithm, and show that his techniques generalize to produce a family of reduction operators. In addition, we introduce a completely new construction of $\overline{G}$, motivated by Conway's original canonical-form construction.

1. Introduction

Although canonical forms sometimes reveal substantial information about the structure of combinatorial games, they are often too complicated to be of any great use. Many of the most interesting games — including Clobber, Amazons, and Hare and Hounds — give rise to some massively complex canonical forms even on relatively small boards. In such cases, a method for extracting more specific information is highly desirable. The familiar temperature theory, and the theory of atomic weights for all-small games, can be viewed as efforts to address this problem.

In 1996, Dan Calistrate [2] introduced another type of reduction. Calistrate observed that in certain situations, infinitesimal differences are of secondary importance. He proposed associating to each game G a *reduced canonical form*, $\overline{G}$, such that $\overline{G} = \overline{H}$ whenever $G - H$ is infinitesimal.

Calistrate's original construction defined $\overline{G}$ to be the *simplest* game equivalent to G modulo an infinitesimal, where simplicity is measured in terms of the number of edges in the complete game tree. He gave a method for calculating $\overline{G}$ and claimed that the map $G \mapsto \overline{G}$ was a homomorphism. However, his proof of this assertion contained a flaw.

Section 2 reviews Calistrate's construction, introduces some important definitions and notation, and establishes some basic results regarding infinitesimals. Section 3 introduces the group of *even-tempered* games. In Section 4, we give a natural definition of *reduced canonical form*, and show that it matches Calistrate's construction. In Section 5, we show that many of our results generalize to a broad family of homomorphisms of the group of games. Section 6 establishes the relationship between these homomorphisms and reduced canonical forms. Finally, Section 7 poses some interesting open problems.

2. Preliminaries

Throughout this paper we use the equivalent terms *form* and *representation* to denote a particular formal representation of a game G. Given a game $G = \{G^L \mid G^R\}$, we will use the terms *followers* to mean all subpositions of G, including G itself; *proper followers* to mean all subpositions of G, excluding G; and *options* to mean the immediate subpositions G^L, G^R.

A game ε is an *infinitesimal* if, for every positive number x, we have $-x < \varepsilon < x$. Let Inf denote the set of infinitesimals; clearly Inf is a subgroup of $\mathcal{G}$, the group of games. When $G - H$ is infinitesimal, we say that G and H are *infinitesimally close*, and write $G \equiv_{\text{Inf}} H$. We will sometimes say that H is G-ish (G infinitesimally *shifted*).

DEFINITION 2.1. If G is a game (in any form) and $\varepsilon \rhd 0$ is an infinitesimal, then G *reduced by* ε, denoted by G_ε, is defined by

$$G_\varepsilon = \begin{cases} G & \text{if } G \text{ is a number;} \\ \{G_\varepsilon^L - \varepsilon \mid G_\varepsilon^R + \varepsilon\} & \text{otherwise.} \end{cases}$$

When no restriction is placed on the game ε, this operation is commonly known as *unheating*. It is not immediately evident that $G_\varepsilon = H_\varepsilon$ whenever $G = H$, but this will emerge in Section 5. In fact, we will prove the stronger statement that $G_\varepsilon = H_\varepsilon$ if and only if G and H are infinitesimally close. The special case $\varepsilon = *$ was first considered by Calistrate, who defined an additional operator to effect a further reduction:

DEFINITION 2.2 (CALISTRATE). If G is a game (in any form), then the $*$-*projection* of G, denoted by $p(G)$, is defined by

$$p(G) = \begin{cases} x & \text{if } G = x \text{ or } x + * \text{ for some number } x, \\ \{p(G^L) \mid p(G^R)\} & \text{otherwise.} \end{cases}$$

We can then define $\overline{G} = p(G_*)$. Calistrate claimed that $\overline{G}$ is the *simplest* game infinitesimally close to G. While this statement is correct, Calistrate's proof relied on the assertion that $G \mapsto \overline{G}$ is a homomorphism, which is false. For example, let $G = \{1|0\}$ and $H = \{2\|1|0\}$. Then $G = \overline{G}$ and $H = \overline{H}$, so

$$\overline{G} + \overline{H} = G + H = \{2, \{3|2\}\|1\}, \text{ but } \overline{G + H} = \{3|2\|1\}.$$

We will give an alternate proof that $\overline{G}$ is the simplest game infinitesimally close to G. We shall have occasion to consider other mappings that select representatives of each Inf-equivalence class. Proceeding in maximum generality:

DEFINITION 2.3. A mapping $\rho : \mathcal{G} \to \mathcal{G}$ is a *reduction* (modulo Inf) if

(i) for all G, $\rho(G) \equiv_{\text{Inf}} G$;
(ii) if x is a number, then $\rho(x) = x$; and
(iii) for all G, H with $G \equiv_{\text{Inf}} H$, we have $\rho(G) = \rho(H)$.

If ρ is a reduction, then we say that $\rho(G)$ is the *reduced form of G (under ρ)*.

Note that we do *not* require our reductions to be homomorphisms. The definition on its own is not terribly restrictive, but serves as a useful checklist for verifying candidate mappings with other desirable properties. We will show that $G \mapsto \overline{G}$ is a reduction, and also that $G \mapsto G_\varepsilon$ is both a reduction and a homomorphism for any infinitesimal $\varepsilon \mathrel{\rhd} 0$.

Infinitesimals and stops. We will make extensive use of an equivalent definition of infinitesimal that is given in terms of the *stops* of a game. Recall, from Winning Ways, that the Left (Right) stop of G is equal to the first number reached when G is played optimally in isolation, with Left (Right) moving first. Formally:

DEFINITION 2.4. The *Left* and *Right stops* of G, denoted by $L_0(G)$ and $R_0(G)$, are defined recursively by

$$L_0(G) = \begin{cases} G & \text{if } G \text{ is a number,} \\ \max R_0(G^L) & \text{otherwise;} \end{cases}$$

$$R_0(G) = \begin{cases} G & \text{if } G \text{ is a number,} \\ \min L_0(G^R) & \text{otherwise.} \end{cases}$$

The following facts about stops, and their relationship to infinitesimals, will be used throughout the rest of this paper. Some proofs can be found in [3]; the rest are simple exercises left to the reader.

PROPOSITION 2.5. *Let G, H be any games.*

(a) *G is an infinitesimal if and only if $L_0(G) = R_0(G) = 0$.*
(b) *$R_0(G) + R_0(H) \le R_0(G + H) \le R_0(G) + L_0(H)$;*
    $L_0(G) + L_0(H) \ge L_0(G + H) \ge R_0(G) + L_0(H)$.

(c) If $G \geq H$, then $L_0(G) \geq L_0(H)$ and $R_0(G) \geq R_0(H)$.
(d) $L_0(G) \geq R_0(G^L)$ and $R_0(G) \leq L_0(G^R)$ for all G^L, G^R, even when G is a number.

Infinitesimal comparisons. We can define infinitesimal comparisons in a manner similar to our definition of infinitesimally close:

DEFINITION 2.6. $G \geq_{\text{Inf}} H$ if and only if $G \geq H + \varepsilon$ for some infinitesimal ε; $G \leq_{\text{Inf}} H$ is defined similarly.

Note that $G \equiv_{\text{Inf}} H$ if and only if $G \leq_{\text{Inf}} H$ and $G \geq_{\text{Inf}} H$. We will see shortly that in these definitions it suffices to take ε to be some multiple of $\uparrow$ or $\downarrow$. We begin with:

LEMMA 2.7. If $R_0(G) \geq 0$, then $G \geq n \cdot \downarrow$ for some n.

PROOF. Choose $n > \text{birthday}(G) + 1$. Then Left, playing second, can win $G + n \cdot \uparrow$ as follows: He makes all of his moves in G, playing optimally, until that component reaches a number x. Then, since $L_0(H) \geq R_0(H)$ for all followers H of G, we must have $x \geq 0$. By the assumptions on n, Left still has a move to 0 available in the $n \cdot \uparrow$ component (even if all of Right's moves were in that component), so he wins on his next move.   □

By symmetry, if $L_0(G) \leq 0$, then $G \leq n \cdot \uparrow$ for some n.

COROLLARY 2.8. Let G and H be games.

(a) $G \geq_{\text{Inf}} H$ if and only if $G - H \geq n \cdot \downarrow$ for some n.
(b) G is infinitesimal if and only if $n \cdot \downarrow \leq G \leq n \cdot \uparrow$ for some n.

PROOF. Follows immediately from Lemma 2.7 and Proposition 2.5(a).   □

Lemma 2.7 also allows us to restate a well-known incentive theorem in terms of infinitesimal comparisons:

THEOREM 2.9. If G is not a number, then G has at least one Left incentive and at least one Right incentive that are $\geq_{\text{Inf}} 0$.

PROOF. Let G^L be any Left option with $R_0(G^L) = L_0(G)$. Then

$$R_0(G^L - G) \geq R_0(G^L) + R_0(-G) = L_0(G) - L_0(G) = 0.$$

Hence $G^L - G \geq_{\text{Inf}} 0$ by Lemma 2.7. The proof for Right incentives is identical.
                    □

We conclude with a theorem that is intuitive yet difficult to prove without the preceding machinery; the theorem states that if the options of a game are infinitesimally perturbed, the resulting game is infinitesimally close to the original.

THEOREM 2.10. *If* $G = \{G^L \mid G^R\}$ *is not a number and* $G' = \{G^{L'} \mid G^{R'}\}$ *is a game with* $G^{L'} \equiv_{\text{Inf}} G^L$ *and* $G^{R'} \equiv_{\text{Inf}} G^R$, *then* $G' \equiv_{\text{Inf}} G$.

PROOF. By symmetry it suffices to show that $G - G' \geq_{\text{Inf}} 0$, or equivalently (Corollary 2.8) that $G - G' + n \cdot \uparrow \geq 0$ for sufficiently large n. If Right moves in G or $-G'$, Left answers with the corresponding move in $-G'$ or G and wins with n large enough by Corollary 2.8. If Right moves to $G - G' + (n-1) \cdot \uparrow *$, then by Theorem 2.9 there is some $G^R \leq_{\text{Inf}} G$, so Left moves in $-G'$ to $-G^{R'} \equiv_{\text{Inf}} -G^R$. Again, Left wins with n large enough by Corollary 2.8 since $G - G^{R'} \geq_{\text{Inf}} 0$. $\square$

Note that Theorem 2.10 fails if G is a number. For example, $0 \equiv_{\text{Inf}} *$, but $\{0 \mid 1\} = \frac{1}{2}$ and $\{* \mid 1\} = 0$.

3. Temper

By Proposition 2.5, infinitesimal differences do not change the final score of a game; they affect only who has the move when that score is reached. This observation motivates one of the most natural reduced-form constructions. Loosely speaking, call a game G *even-tempered* if, no matter how G is played, the first player will have the move when G reaches a number. If G and H are infinitesimally close and even-tempered then we should expect that $G - H$, since we have effectively discarded the particularities of who has the move and when. This is indeed the case, and in fact we can prove a stronger statement: For any game G, G_* is the unique even-tempered game infinitesimally close to G. We begin with a formal definition of temper.

DEFINITION 3.1. Let G be a fixed representation of a game.

(a) G is *even-tempered* if G a number, or every option of G is odd-tempered;
(b) G is *odd-tempered* if G is not a number and every option of G is even-tempered;
(c) G is *well-tempered* if G is even-tempered or odd-tempered.

We will call a game even- (odd-, well-) tempered if it has some even- (odd-, well-) tempered representation. Although temper is a property of the form of a game and can be destroyed by adding new dominated options, the following provides justification for treating it as a property of games:

THEOREM 3.2. *Let* G *be a game in any form. If* G *is even- (odd-) tempered, then so is its canonical form.*

PROOF. If G is a number then both G and its canonical form are necessarily even-tempered, so assume that it is not. By induction we may assume that all proper followers of G are canonical. It then suffices to show that temper is preserved when dominated options are eliminated or reversible moves bypassed. For dominated options this is trivial, so suppose some $G^{LR} \leq G$.

Consider the case where G^{LR} is a number. Then $G \neq G^{LR}$, so by the Number Avoidance Theorem, Left has a winning move from $G - G^{LR}$ to some $G^{L'} - G^{LR}$. Thus $G^{L'} \geq G^{LR}$. But since G^{LR} is a number, we have $G^{LR} > G^{LRL}$, so any G^{LRL} is dominated by $G^{L'}$ and hence does not contribute to the canonical form of G.

If G^{LR} is not a number then neither is G^L (since it was assumed canonical), and since G^L is odd- (even-) tempered, so is every G^{LRL}. $\qquad\square$

A simple corollary is that a game cannot be both odd- and even- tempered since its canonical form cannot be both. The main theorem of this section is the following:

THEOREM 3.3. *Let G be a game in any form. Then G_* is the unique G-ish even-tempered game.*

Theorem 3.3 implies that $G_* = H_*$ whenever $G = H$, which is not immediately clear from the definition of G_*. For now, we must specify a representation for G in order to compute G_*. Several lemmas are critical to the proof of Theorem 3.3.

LEMMA 3.4. *Let G, H be any games.*

(a) *If G and H are both even- (odd-) tempered, then $G + H$ is even-tempered.*
(b) *If G is even-tempered and H is odd-tempered, then $G + H$ is odd-tempered.*

PROOF. If G and H are both numbers then so is $G + H$, and the conclusion follows. If G is a number and H is not, then by the Number Avoidance Theorem

$$G + H = \{G + H^L \mid G + H^R\}.$$

By induction $G + H^L$ and $G + H^R$ have the same temper as H^L and H^R. Finally, if neither G nor H is a number, then

$$G + H = \{G^L + H, G + H^L \mid G^R + H, G + H^R\}.$$

By induction and assumption on G, H all options have the same temper. Furthermore, they are odd-tempered if G and H have the same temper, and even-tempered otherwise. $\qquad\square$

LEMMA 3.5. *Let G be a game in any form. Then G_* is even-tempered and infinitesimally close to G.*

PROOF. The conclusion is trivial if G is a number. If G is not a number then $G_* = \{G_*^L + * \mid G_*^R + *\}$, and the result follows from induction, Lemma 3.4, Theorem 2.10, and the fact that $*$ is odd-tempered. $\qquad\square$

LEMMA 3.6. *If G is even-tempered and $R_0(G) \geq 0$, then $G \geq 0$.*

PROOF. We may assume that G is in canonical form. Since $R_0(G) \geq 0$, Left, playing second, can assure that when the play reaches a number, the result is ≥ 0. By Theorem 3.2, this necessarily happens after an even number of moves, so Left has made the last move. Thus, Left can win G as second player.   □

PROOF OF THEOREM 3.3. Lemma 3.5 shows that G_* is even-tempered. For uniqueness, suppose G and H are infinitesimally close even-tempered games. Then $R_0(G - H) = 0$. But $G - H$ is even-tempered, so by Lemma 3.6, we have $G - H \geq 0$; by symmetry $G - H = 0$.   □

As a simple corollary, $G_* + *$ is the unique *odd*-tempered game infinitesimally close to G. A more substantial corollary is the following theorem.

THEOREM 3.7. *The map $G \mapsto G_*$ is a well-defined homomorphism of the group of games.*

PROOF. First, Theorem 3.3 shows that G_* does not depend on the form of G. Now fix games G, H. Lemma 3.5 implies that

$$(G + H)_* \equiv_{\text{Inf}} G + H \equiv_{\text{Inf}} G_* + H_*.$$

Also, from Lemmas 3.4 and 3.5, we know that $(G + H)_*$ and $G_* + H_*$ are both even-tempered. It follows from Theorem 3.3 that they are equal.   □

As a final note, Lemma 3.4 shows that the well-tempered games $\mathcal{W}$ and the even-tempered games $\mathcal{E}$ are subgroups of the group of games. Moreover, the mapping $G \mapsto G + *$ induces a perfect pairing of even- and odd- tempered games, so that

$$\mathcal{W} = \mathcal{E} \cup \{G + * : G \in \mathcal{E}\}.$$

Thus, the index of $\mathcal{E}$ in $\mathcal{W}$ is 2.

4. Reduced canonical forms

In this section we will show that every game G has a *reduced canonical form* $\overline{G}$. $\overline{G}$ is infinitesimally close to G, and it is the simplest such game, in a sense that we will define shortly.

DEFINITION 4.1. Let G be any game.

(a) A Left option G^L is Inf-*dominated* if $G^L \leq_{\text{Inf}} G^{L'}$ for some other Left option $G^{L'}$.

(b) A Left option G^L is Inf-*reversible* if $G^{LR} \leq_{\text{Inf}} G$ for some G^{LR}.

The definitions for Right options are similar.

EXAMPLE (i). Let $G = \{1, \{2|1\}\|0\}$. Then 1 is an Inf-dominated Left option of G, since $\{2|1\} - 1 + \uparrow \geq 0$.

EXAMPLE (ii). Let $H = \{1, \{2|0\}\|0\}$. Then $\{2|0\}$ is Inf-reversible through 0, since $H - 0 + \uparrow \geq 0$.

DEFINITION 4.2. A game G is said to be in *reduced canonical form* if, for every follower H of G, either

(i) H is a number in simplest form, or
(ii) H is not numberish, and contains no Inf-dominated or Inf-reversible options.

This definition of reduced canonical form appears radically different from Calistrate's, but we will soon see that his construction meets our criteria. The main theorems exactly parallel the corresponding results for canonical forms [3]:

THEOREM 4.3. *For any game G, there is a game $\overline{G}$ in reduced canonical form with $\overline{G} \equiv_{\text{Inf}} G$.*

THEOREM 4.4. *Suppose that G and H are in reduced canonical form. If $G \equiv_{\text{Inf}} H$, then $G = H$.*

Theorem 4.4 guarantees that the $\overline{G}$ found in Theorem 4.3 is unique. The following lemma is instrumental to the proof of Theorem 4.3.

LEMMA 4.5. *Let G be a well-tempered game.*

(a) *If $G \equiv_{\text{Inf}} x$ for some number x, then $G = x$ or $x + *$.*
(b) *If G is in canonical form, then G has no Inf-dominated or Inf-reversible options.*
(c) *$p(G) \equiv_{\text{Inf}} G$.*

PROOF. (a) x is the *unique* even-tempered x-ish game, so if G is even-tempered then $G = x$. Likewise, $x + *$ is the unique odd-tempered x-ish game, so if G is odd-tempered then $G = x + *$.

(b) First suppose (for contradiction) that $G^L \leq_{\text{Inf}} G^{L'}$ for distinct Left options $G^L, G^{L'}$. Since G is even- (odd-) tempered, both G^L and $G^{L'}$ are odd- (even-) tempered. Therefore $G^{L'} - G^L$ is even-tempered. But since $G^L \leq_{\text{Inf}} G^{L'}$, we know that $R_0(G^{L'} - G^L) \geq 0$. By Lemma 3.6, this implies that $G^{L'} - G^L \geq 0$, contradicting the assumption that G is in canonical form.

Next suppose (for contradiction) that $G^{LR} \leq_{\text{Inf}} G$. Consider the case where G^L is a number. Then G^{LR} is also a number and $G^L < G^{LR}$. Furthermore, by Theorem 2.9, there is some $G^{L'} \geq_{\text{Inf}} G$. Hence

$$G^L < G^{LR} \leq_{\text{Inf}} G \leq_{\text{Inf}} G^{L'}.$$

Since G^{LR} and G^L necessarily differ by more than an infinitesimal, this gives us $G^L < G^{L'}$, contradicting the assumption that G is in canonical form.

Finally, if G^L is not a number, then G^{LR} is well-tempered and has the same temper as G. So $G - G^{LR}$ is even-tempered. But $R_0(G - G^{LR}) \geq 0$, so by Lemma 3.6 we have $G - G^{LR} \geq 0$, again contradicting the assumption that G is in canonical form.

(c) follows immediately from induction and Theorem 2.10.   □

PROOF OF THEOREM 4.3. We will show that $p(G_*)$ has the desired properties, where we use the canonical form of G_* to compute $p(G_*)$. Let H be a follower of $p(G_*)$; then $H = p(H')$ for some follower H' of G_*. Since G_* is well-tempered and in canonical form, the same is true of H'.

If H is numberish, then by Lemma 4.5(c), so is H'. By Lemma 4.5(a), we have $H' = x$ or $x + *$ for some number x; then by definition, $H = x$. This verifies condition (i) in the definition of reduced canonical form.

If H is not numberish, then by Lemma 4.5(c), neither is H'. By Lemma 4.5(b), H' has no Inf-dominated or Inf-reversible options. Since all followers of H are infinitesimally close to followers of H', the same must be true of H. This completes the proof.   □

We are now ready to prove uniqueness (Theorem 4.4).

PROOF OF THEOREM 4.4. Suppose G and H are in reduced canonical form and $G \equiv_{\mathrm{Inf}} H$. If either of G, H is numberish then both must be, so by the definition of reduced canonical form, both are numbers; hence $G = H$.

Now suppose that neither G nor H is numberish, and consider $H - G$. Since $R_0(H - G) \geq 0$, by Corollary 2.8 we have

$$H - G + n \cdot \uparrow \geq 0$$

for suitably large n. Consider the game after Right moves to $-G^L$:

$$H - G^L + n \cdot \uparrow$$

Left must have a winning response. It cannot be to any $H - G^{LR} + n \cdot \uparrow$, since this would imply

$$G \equiv_{\mathrm{Inf}} H \geq_{\mathrm{Inf}} G^{LR},$$

contradicting the assumption that G has no Inf-reversible options. Furthermore, by Theorem 2.9, H has a Left incentive that exceeds $n \cdot \downarrow$ (assuming n is sufficiently large). Since $n \cdot \downarrow$ is the unique Left incentive of $n \cdot \uparrow$, Left must have a winning move in H, to

$$H^L - G^L + n \cdot \uparrow.$$

Therefore $H^L \geq_{\text{Inf}} G^L$. By an identical argument, there is some $G^{L'}$ with $G^{L'} \geq_{\text{Inf}} H^L$. But G has no Inf-dominated options, so in fact

$$G^{L'} \equiv_{\text{Inf}} H^L \equiv_{\text{Inf}} G^L.$$

By induction, $G^L = H^L$, so every Left option of G is a Left option of H. Symmetrical arguments show that G and H have exactly the same Left and Right options.   □

If G is any game, the value of G is unchanged when dominated options are eliminated or reversible ones bypassed. We can similarly eliminate Inf-dominated options and bypass Inf-reversible ones, preserving the value of G up to an infinitesimal.

LEMMA 4.6. *If G is not a number and G' is obtained from G by eliminating an Inf-dominated option, then $G' \equiv_{\text{Inf}} G$.*

PROOF. Suppose that $G^{L'} \leq_{\text{Inf}} G^L$ for Left options G^L, $G^{L'}$ of G. Then we can find $H \equiv_{\text{Inf}} G^L$ such that $G^{L'} \geq H$, so, by Theorem 2.10,

$$G = \{G^L, G^{L'}, \ldots \mid G^R\} \equiv_{\text{Inf}} \{G^{L'}, H, \ldots \mid G^R\} = \{G^{L'}, \ldots \mid G^R\}. □$$

LEMMA 4.7. *If G is not numberish and G' is obtained from G by bypassing an Inf-reversible option, then $G' \equiv_{\text{Inf}} G$.*

Note that the assumption of Lemma 4.7 (G not numberish) is stronger than the assumption of Lemma 4.6 (G not a number). If G is numberish, Inf-reversible moves *cannot* in general be bypassed. For example, let $G = -2 = \{2 \mid 0 \| 0\}$. Then $G^{LR} = 0 \leq_{\text{Inf}} G$, but if we replace $G^L = \{2 \mid 0\}$ with the Left options of 0, then the resulting game is $G' = \{\mid 0\} = -1 \not\equiv_{\text{Inf}} G$.

PROOF OF LEMMA 4.7. Suppose that G is not numberish, and G' is obtained from G by bypassing some Inf-reversible option G^{L_0} through $G^{L_0 R_0}$. We must show that $G - G' + n \cdot \uparrow \geq 0$ and $G' - G + n \cdot \uparrow \geq 0$ for sufficiently large n.

First consider $G - G' + n \cdot \uparrow \geq 0$. Right has three possible opening moves; we show that, in each case, Left has a winning response.

(a) If Right moves to $G - G' + (n-1) \cdot \uparrow *$, then by Theorem 2.9 Left can move to $G - G^R + (n-1) \cdot \uparrow *$ with $G^R \leq_{\text{Inf}} G$ which wins for large enough n.

(b) Suppose Right moves to $G - G^{L_0 R_0 L} + n \cdot \uparrow$. Since $G^{L_0 R_0} \leq_{\text{Inf}} G$, we have $G - G^{L_0 R_0} + n \cdot \uparrow \geq 0$ for large enough n, so $G - G^{L_0 R_0 L} + n \cdot \uparrow \rhd 0$ and therefore Left has a winning move.

(c) If Right makes any other move in G or $-G'$, Left makes the corresponding move in the other component leaving $n \cdot \uparrow$.

Now consider $G - G' + n \cdot \uparrow \geq 0$. Once again, there are three cases.

(d) Suppose that Right moves to $G' - G + (n-1) \cdot {\uparrow}*$. Since G is not a number, by Theorem 2.9 we can choose some $G^L \geq_{\text{Inf}} G$. If we can choose $G^L \neq G^{L_0}$, Left moves to $G^L - G + (n-1) \cdot {\uparrow}*$ and wins for large n. If G^{L_0} is the only choice, the proof of Theorem 2.9 shows that $R_0(G^{L_0}) = L_0(G)$. We also have from Proposition 2.5 that $R_0(G^{L_0 R_0}) \leq R_0(G)$ since $G^{L_0 R_0} \leq_{\text{Inf}} G$. Finally, since G is not numberish, we have $R_0(G) < L_0(G)$. Putting these inequalities together gives us

$$R_0(G^{L_0 R_0}) \leq R_0(G) < L_0(G) = R_0(G^{L_0}) \leq L_0(G^{L_0 R_0})$$

and therefore $G^{L_0 R_0}$ is not a number. It follows that $L_0(G^{L_0 R_0}) \leq L_0(G')$ since every Left option of $G^{L_0 R_0}$ is also a Left option of G' (by Proposition 2.5(d) this is true even if G' is a number). Similarly, $R_0(G') \leq R_0(G)$, since they have the same Right options and G is not a number. Hence

$$R_0(G') \leq R_0(G) < L_0(G) \leq L_0(G^{L_0 R_0}) \leq L_0(G')$$

so in fact G' is also not a number. It follows from Theorem 2.9 that Left can win by moving to $G' - G^R + (n-1) \cdot {\uparrow}*$, where $G^R \leq_{\text{Inf}} G'$.

(e) If Right moves to $G' - G^{L_0} + n \cdot {\uparrow}$ then Left moves to $G' - G^{L_0 R_0} + n \cdot {\uparrow}$, which we now show is a loss for Right.

If Right moves to either $G^R - G^{L_0 R_0} + n \cdot {\uparrow}$ or $G' - G^{L_0 R_0} + (n-1) \cdot {\uparrow}*$, Left can win with n large enough since $-G^{L_0 R_0} \geq_{\text{Inf}} -G$ and we have already shown that $G^R - G + n \cdot {\uparrow} \,\triangleright\, 0$ and $G' - G + (n-1) \cdot {\uparrow}* \,\triangleright\, 0$. If Right moves to $G' - G^{L_0 R_0 L} + n \cdot {\uparrow}$, Left makes the corresponding move in G' leaving $n \cdot {\uparrow}$ and wins.

(f) Finally, if Right makes any other move in G' or $-G$, Left makes the corresponding move in the other component leaving $n \cdot {\uparrow}$. $\qquad\square$

Lemmas 4.6 and 4.7 suggest an algorithm for computing $\overline{G}$. If G is numberish, we take $\overline{G} = L_0(G)$. Otherwise, by Theorem 2.10, we may assume that every option of G is in reduced canonical form. $\overline{G}$ can then be obtained by iteratively eliminating Inf-dominated options and bypassing Inf-reversible ones until none remain.

This algorithm bears a pleasing similarity to the classical procedure for computing the canonical form of G, but it is somewhat less efficient than simply calculating $p(G_*)$. Nonetheless, Theorem 2.10 and Lemmas 4.6, 4.7 can be useful in practice; see, for example, Mesdal [4, Section 7] in this volume.

THEOREM 4.8. *If G is not numberish, then $\overline{G}$ can be computed by repeating the following steps in any order*:

(i) *Replacing options with simpler options infinitesimally close to the original*;
(ii) *Eliminating Inf-dominated options*;

(iii) *Bypassing Inf-reversible options.*

PROOF. Since each step simplifies the game, the steps must at some point come to an end. Theorem 2.10 shows that step (i) does not change the game by more than an infinitesimal, and Lemmas 4.6 and 4.7 show that neither do steps (ii) or (iii). Thus, the final game is infinitesimally close to the original, and if no more steps are possible then all numberish followers are numbers and there are no Inf-dominated or Inf-reversible moves. It follows by Theorem 4.4 that this game is $\overline{G}$. ☐

5. Reduction by ε

Having observed one reduction that is also a homomorphism, it is natural to look for others. In this section we show that for any infinitesimal $\varepsilon \, \triangleright \, 0$, the mapping $G \mapsto G_\varepsilon$ is both a reduction and a homomorphism. In the process we will show that G_ε is well-defined, but for now we must still agree upon a specific representation of G in order to compute G_ε.

LEMMA 5.1. $L_0(G_\varepsilon) = L_0(G)$ and $R_0(G_\varepsilon) = R_0(G)$.

PROOF. This is immediate if G is a number. Otherwise, by induction and Proposition 2.5, we have

$$L_0(G_\varepsilon) = \max\{R_0(G_\varepsilon^L - \varepsilon)\} = \max\{R_0(G_\varepsilon^L)\} = \max\{R_0(G^L)\} = L_0(G),$$

and similarly for R_0. ☐

LEMMA 5.2. *If* $R_0(G) \geq 0$, *then* $G_\varepsilon \geq 0$; *if* $L_0(G) \leq 0$, *then* $G_\varepsilon \leq 0$.

PROOF. Suppose $R_0(G) \geq 0$. If G_ε is a number, then $G_\varepsilon = R_0(G_\varepsilon) = R_0(G) \geq 0$. Otherwise, suppose that Right moves in G_ε to $G_\varepsilon^R + \varepsilon$. If G_ε^R is a number, then

$$G_\varepsilon^R = L_0(G_\varepsilon^R) = L_0(G^R) \geq R_0(G) \geq 0,$$

so by choice of ε, $G_\varepsilon^R + \varepsilon \, \triangleright \, 0$ and Left has a winning move. Finally, if G_ε^R is not a number, Left moves to G_ε^{RL}, choosing G^{RL} so that $R_0(G^{RL}) = L_0(G^R)$. Since $L_0(G^R) \geq R_0(G) \geq 0$, by induction $G_\varepsilon^{RL} \geq 0$, so Left wins. The proof for $L_0(G)$ is identical. ☐

THEOREM 5.3. G *is infinitesimal if and only if* $G_\varepsilon = 0$.

PROOF. If G is infinitesimal then $R_0(G) = L_0(G) = 0$, so $0 \leq G_\varepsilon \leq 0$ by Lemma 5.2. Conversely, suppose $G_\varepsilon = 0$. Then by Lemma 5.1,

$$R_0(G) = R_0(G_\varepsilon) = 0 = L_0(G_\varepsilon) = L_0(G),$$

so G is infinitesimal. ☐

THEOREM 5.4. *Using the representation* $-G = \{-G^R \mid -G^L\}$, *we have*

$$(-G)_\varepsilon = -(G_\varepsilon).$$

PROOF. Trivial if G is a number; otherwise, by induction:

$$
\begin{aligned}
(-G)_\varepsilon &= \{(-G^R)_\varepsilon - \varepsilon \mid (-G^L)_\varepsilon + \varepsilon\} \\
&= \{-(G^R_\varepsilon) - \varepsilon \mid -(G^L_\varepsilon) + \varepsilon\} \\
&= -\{G^L_\varepsilon - \varepsilon \mid G^R_\varepsilon + \varepsilon\} \\
&= -(G_\varepsilon). \quad\quad\quad\quad\quad\quad\quad\quad\Box
\end{aligned}
$$

LEMMA 5.5. *Let G, H be any games.*

(a) *If $G + H \geq 0$, then $G_\varepsilon + H_\varepsilon \geq 0$.*

(b) *If $G + H \vartriangleright 0$, then $G_\varepsilon + H_\varepsilon + \varepsilon \vartriangleright 0$.*

PROOF. If G and H are numbers, (a) and (b) are immediate. Otherwise, we may assume that G is not a number, and we proceed by simultaneous induction on (a) and (b).

(a) Suppose that $G + H \geq 0$ and that Right moves in $G_\varepsilon + H_\varepsilon$ to $G^R_\varepsilon + H_\varepsilon + \varepsilon$ (we can assume that Right's move is in G_ε by symmetry and the Number Avoidance Theorem). Since $G^R + H \vartriangleright 0$, by induction $G^R_\varepsilon + H_\varepsilon + \varepsilon \vartriangleright 0$, so Left wins; hence $G_\varepsilon + H_\varepsilon \geq 0$.

(b) If $G + H \vartriangleright 0$ then Left has a winning move, say to $G^L + H \geq 0$ (again, by symmetry and the Number Avoidance Theorem we can assume that Left's winning move is in G). Then $G^L_\varepsilon + H_\varepsilon \geq 0$ by induction, so Left also has a winning move from $G_\varepsilon + H_\varepsilon + \varepsilon$. $\quad\quad\Box$

COROLLARY 5.6. *If $G \geq H$, then $G_\varepsilon \geq H_\varepsilon$.*

PROOF. Using Theorem 5.4 and Lemma 5.5,

$$G + (-H) \geq 0 \;\Rightarrow\; G_\varepsilon + (-H)_\varepsilon \geq 0 \;\Rightarrow\; G_\varepsilon \geq H_\varepsilon. \quad\quad\Box$$

THEOREM 5.7. *If $G = H$, then $G_\varepsilon = H_\varepsilon$.*

PROOF. Follows immediately from Corollary 5.6 since $G \geq H$ and $H \geq G$. $\quad\Box$

By Theorem 5.7, G_ε does not depend on the formal representation of G. Therefore, the mapping $G \mapsto G_\varepsilon$ is well-defined. We next show that it is a homomorphism.

LEMMA 5.8. *If x is a number, then $(G + x)_\varepsilon = G_\varepsilon + x$.*

PROOF. The conclusion is trivial if G is a number. Otherwise, $G + x$ is not a number, so by induction and the Number Translation Theorem

$$(G+x)_\varepsilon = \{G^L + x \mid G^R + x\}_\varepsilon$$
$$= \{(G^L + x)_\varepsilon - \varepsilon \mid (G^R + x)_\varepsilon + \varepsilon\}$$
$$= \{G^L_\varepsilon - \varepsilon + x \mid G^R_\varepsilon + \varepsilon + x\}$$

If G_ε is not a number, then this is equal to $G_\varepsilon + x$ by the Number Translation Theorem. Otherwise, there is some number z such that

$$G^L_\varepsilon - \varepsilon \vartriangleleft z \vartriangleleft G^R_\varepsilon + \varepsilon.$$

Translating by x, we have that

$$G^L_\varepsilon - \varepsilon + x \vartriangleleft z + x \vartriangleleft G^R_\varepsilon + \varepsilon + x,$$

and since $(G + x)_\varepsilon = \{G^L_\varepsilon - \varepsilon + x \mid G^R_\varepsilon + \varepsilon + x\}$, it follows that $(G+x)_\varepsilon$ is a number. Then by Lemma 5.1, we conclude that

$$(G+x)_\varepsilon = L_0((G+x)_\varepsilon) = L_0(G+x) = L_0(G) + x = L_0(G_\varepsilon) + x = G_\varepsilon + x.$$

$\square$

The following theorem establishes that reduction by ε is a *projection*: an idempotent homomorphism.

THEOREM 5.9. *Let G, H be any games. Then:*

(a) $(G + H)_\varepsilon = G_\varepsilon + H_\varepsilon$.
(b) $G_\varepsilon \equiv_{\text{Inf}} G$.
(c) $(G_\varepsilon)_\varepsilon = G_\varepsilon$.

PROOF. (a) This reduces to Lemma 5.8 if either G or H is a number. Otherwise, there are two cases. First, if $x = G + H$ is a number, then by Theorem 5.4 and Lemma 5.8 we have

$$(G + H)_\varepsilon - G_\varepsilon = x - G_\varepsilon = (x - G)_\varepsilon = H_\varepsilon.$$

Finally, if none of G, H, $G + H$ are numbers, then by induction

$$(G+H)_\varepsilon = \{(G^L + H)_\varepsilon - \varepsilon, (G + H^L)_\varepsilon - \varepsilon \mid (G^R + H)_\varepsilon + \varepsilon, (G + H^R)_\varepsilon + \varepsilon\}$$
$$= \{(G^L_\varepsilon - \varepsilon) + H_\varepsilon, G_\varepsilon + (H^L_\varepsilon - \varepsilon) \mid (G^R_\varepsilon + \varepsilon) + H_\varepsilon, G_\varepsilon + (H^R + \varepsilon)\}$$
$$= G_\varepsilon + H_\varepsilon.$$

(b) follows by induction and Theorem 2.10.

(c) Using (a), (b), and Theorem 5.3, we obtain

$$0 = (G_\varepsilon - G)_\varepsilon = (G_\varepsilon)_\varepsilon - G_\varepsilon.$$

$\square$

THEOREM 5.10. *The mapping $G \mapsto G_\varepsilon$ is a reduction.*

PROOF. $G \equiv_{\mathrm{Inf}} G_\varepsilon$ by Theorem 5.9, and $x_\varepsilon = x$ for numbers x by definition. Finally, if $G \equiv_{\mathrm{Inf}} H$, then $G = H + \delta$ for some infinitesimal δ, so by Theorems 5.3 and 5.9, $G_\varepsilon = H_\varepsilon + \delta_\varepsilon = H_\varepsilon$. $\square$

We conclude with a theorem on incentives for games reduced by ε whose elegant proof makes it deserving of presentation.

THEOREM 5.11. *If G_ε is not a number, then G_ε has at least one Left incentive and at least one Right incentive that are $\geq -\varepsilon$.*

PROOF. By Theorem 2.9, there is some Left option $G_\varepsilon^L - \varepsilon$ of G_ε and some infinitesimal δ such that

$$(G_\varepsilon^L - \varepsilon) - G_\varepsilon \geq \delta.$$

Reducing by ε gives $(G_\varepsilon^L)_\varepsilon - \varepsilon_\varepsilon - (G_\varepsilon)_\varepsilon \geq \delta_\varepsilon$ by Theorem 5.9(a) and Corollary 5.6, which implies $G_\varepsilon^L - G_\varepsilon \geq 0$ by Theorems 5.3 and 5.9(c), and finally

$$(G_\varepsilon^L - \varepsilon) - G_\varepsilon \geq -\varepsilon.$$

The proof for Right incentives is identical. $\square$

6. All-small reductions

Our final task is to relate the two reductions presented thus far. Recall that a game is *all-small* if all of its followers are infinitesimal. We will see that when ε is all-small, G_ε can be computed by adding appropriate multiples of ε to the stops of $\overline{G}$. We begin with a simple observation:

LEMMA 6.1. *If G_ε is not a number, then there is some G_ε^L with $R_0(G_\varepsilon^L) > R_0(G_\varepsilon)$.*

PROOF. Since G_ε is not a number, G is not numberish. Hence $L_0(G) > R_0(G)$, so we can choose any G_ε^L where $R_0(G^L) = L_0(G)$. $\square$

The following theorem strengthens Theorem 2.9 when ε is all-small.

THEOREM 6.2 (ALL-SMALL AVOIDANCE THEOREM). *If $\varepsilon \rhd 0$ is all-small and G_ε is not a number, then G_ε has at least one Left incentive and at least one Right incentive that exceed every incentive (Left and Right) of ε.*

The hypothesis that ε is all-small is essential. For example, if we take $G = \pm 1$ and $\varepsilon = 2$, then $G_\varepsilon = \pm\{3|1\|1\}$. We can verify that the Left incentive of G_ε is less than the Right incentive of 2.

PROOF OF THEOREM 6.2. We will show that G_ε has a Left incentive that exceeds every incentive of ε; the proof for Right incentives is identical. First, for any game H, we define the Left and Right *parities* $p^L(H)$, $p^R(H)$ as follows. If H is numberish, then $p^L(H) = p^R(H) = 0$. Otherwise,

$$p^L(H) = \begin{cases} 1 & \text{if } p^R(H^L) = 0 \text{ for every } H^L \text{ with } R_0(H^L) = L_0(H), \\ 0 & \text{if } p^R(H^L) = 1 \text{ for any such } H^L. \end{cases}$$

$p^R(H)$ is defined similarly, with L and R interchanged.

Since G_ε is not a number, neither is G, so there is some G^L with $R_0(G^L) = L_0(G)$. If $p^L(G) = 0$ then we can choose such G^L with $p^R(G^L) = 1$, so we can guarantee that at least one of $p^R(G^L)$, $p^L(G)$ is 1. We claim that for each (Left or Right) incentive Δ of ε,

$$(G_\varepsilon^L - \varepsilon) - G_\varepsilon \geq \Delta.$$

Observe that $p^R(-H) = p^L(H)$ and $p^L(-H) = p^R(H)$ (proof by simple induction). It therefore suffices to prove the following: if Right moves from any position of the form

$$A_\varepsilon + B_\varepsilon - \varepsilon - \Delta, \quad \text{with } R_0(A) + R_0(B) = 0, \ p^R(A) + p^R(B) \geq 1, \quad (6\text{-}1)$$

then Left can either win outright, or respond to another position of the same form. The following proof makes heavy use of the inequality $R_0(A + B) \geq R_0(A) + R_0(B)$ from Proposition 2.5(b).

Since $p^R(A) + p^R(B) \geq 1$, A_ε and B_ε cannot both be numbers. It follows that Right can never move in $(-\varepsilon - \Delta)$, since by Lemma 6.1 this would allow Left to move in one of A_ε or B_ε leaving a position H with $R_0(H) > 0$ (here we use the fact that ε is all-small which implies that none of its incentives or followers contribute to the stops of the entire game).

Next suppose that Right moves in A_ε from (6-1). By the Number Avoidance Theorem, we may assume that A_ε is not a number. There are three cases.

Case 1: A_ε^R is not a number. Then Left moves to A_ε^{RL} with $R_0(A^{RL}) = L_0(A^R) \geq R_0(A)$. In this case the position is

$$H = A_\varepsilon^{RL} + B_\varepsilon - \varepsilon - \Delta, \qquad \text{with } R_0(A^{RL}) + R_0(B) \geq 0.$$

If the inequality is strict, then since $-\varepsilon - \Delta$ is infinitesimal, $R_0(H) > 0$ and Left wins outright. Otherwise, either $p^R(B) = 1$ or $p^R(A) = 1$; in the latter case $p^L(A^R) = 0$, so Left can choose A^{RL} with $p^R(A^{RL}) = p^R(A) = 1$. In both cases, Left leaves a smaller position of the form (6-1).

Case 2: A_ε^R is a number, but B_ε is not. Then $R_0(A^R) = L_0(A^R) \geq R_0(A)$, so by Lemma 6.1 Left can move in B_ε leaving a position H with $R_0(H) > 0$, and Left wins outright.

Case 3: Both A_ε^R and B_ε are numbers. Then $A_\varepsilon^R = L_0(A^R) \geq R_0(A) = -R_0(B) = -B_\varepsilon$. If the inequality is strict, then Left wins outright. Otherwise, the overall position is

$$(A_\varepsilon^R + \varepsilon) + B_\varepsilon - \varepsilon - \Delta = -\Delta.$$

But Δ is an incentive, so necessarily $\Delta \lhd 0$, whence $-\Delta \rhd 0$.

This exhausts Right's moves in A_ε from (6-1). The situation is identical if Right moves in B_ε, so the proof is complete. $\qquad\square$

COROLLARY 6.3 (ALL-SMALL TRANSLATION THEOREM). *If G is not a number, n is an integer and $\varepsilon \rhd 0$ is all-small, then $G_\varepsilon + n \cdot \varepsilon = \{G_\varepsilon^L + (n-1) \cdot \varepsilon \mid G_\varepsilon^R + (n+1) \cdot \varepsilon\}$.*

PROOF. This is trivial if $n = 0$. Otherwise the game is either $G_\varepsilon + \varepsilon + \varepsilon + \cdots$ or $G_\varepsilon - \varepsilon - \varepsilon - \cdots$; in either case the Left and Right incentives of $\pm\varepsilon$ are dominated by incentives of G_ε by Theorem 6.2. $\qquad\square$

This translation theorem allows us to quickly compute G_ε from its definition by absorbing the $\pm\varepsilon$ terms in the followers of G_ε until we reach the stops. It is straightforward (although the notation is cumbersome) to determine the multiple of ε which must be added to each stop:

DEFINITION 6.4. Let $G^{XYZ\cdots}$ be a follower of G where each of $X, Y, Z, \ldots$ denotes a Left or Right option. The *weight* $w_G(G^{XYZ\cdots})$ of $G^{XYZ\cdots}$ is the number of Left options in $X, Y, Z, \ldots$ minus the number of Right options in $X, Y, Z, \ldots$

Note that we have already encountered the concept of weight disguised as temper in Section 3; a representation of a game is even- (odd-) tempered if the stops all have even (odd) weight. As an example, in the game $\{2\|1|0\}$, the stop 2 has weight 1 since it is reached by a single Left move, the stop 1 has weight 0 since it is reached by a Right move followed by a Left move, and the stop 0 has weight -2 since it is reached by two Right moves.

THEOREM 6.5. *If $\varepsilon \rhd 0$ is all-small, then G_ε is the game obtained from $\overline{G}$ by replacing each stop H with $H - w_{\overline{G}}(H) \cdot \varepsilon$.*

PROOF. Since $\overline{G} \equiv_{\mathrm{Inf}} G$, we have $G_\varepsilon = (\overline{G})_\varepsilon$. With this observation, the theorem follows immediately from Corollary 6.3 and the definition of reduction by ε applied to $\overline{G}$. $\qquad\square$

For example, if $G = \{2\|1|0\}$ then $G_\uparrow = \{2\downarrow\|1|\uparrow\Uparrow\}$. Note that for $\varepsilon = *$, Theorem 6.5 agrees with our results from Sections 3 and 4 as it states that G_* is obtained from $\overline{G}$ by adding $*$ to the stops having odd weight.

7. Conclusion and open problems

The reduced canonical form is a valuable tool in the study of combinatorial games; see [4, Section 7] in this volume for an example of its successful application. However, there are several potentially useful directions in which these ideas can be extended.

Section 5 does not completely characterize the reductions that are also homomorphisms. For example, the reader might wish to verify that the mapping ρ given by

$$\rho(G) = \begin{cases} G & \text{if } G \text{ is a number,} \\ \{\rho(G^L)-*, \rho(G^L)-*2 \mid \rho(G^R)+*, \rho(G^R)+*2\} & \text{otherwise} \end{cases}$$

is both a reduction and a homomorphism. In fact, we could replace $*$ and $*2$ by any finite set of infinitesimals $\rhd\ 0$: the results of Section 5 all apply with virtually unchanged proofs. It would be interesting to investigate other reduction-homomorphisms (if indeed they exist).

OPEN PROBLEM. Give a complete characterization of all reduction-homomorphisms $\rho : \mathcal{G} \to \mathcal{G}$.

Another important question is: to what extent can these constructions be generalized to groups other than Inf? In particular, if $\mathcal{K}$ is any subgroup of $\mathcal{G}$, then we can define a *reduction modulo $\mathcal{K}$* as a map that isolates a unique element of each $\mathcal{K}$-equivalence class.

As a typical example, consider the group of infinitesimals of order n:

$$\text{Inf}^n = \{G : k \cdot \downarrow_n < G < k \cdot \uparrow^n \text{ for some } k\}.$$

OPEN PROBLEM. Give an effective construction for reduction modulo Inf^n (or some other useful class of games).

There are many interesting games in which *all* positions are infinitesimals, and reduction modulo Inf is obviously unhelpful in studying such games. The theory of atomic weights is sometimes useful, but quite often one encounters large classes of positions with atomic weight zero. In such cases, reduction modulo Inf^2 could be a productive tool.

Generalizing beyond short games, reduced canonical forms can be suitably defined for a certain class of well-behaved loopy games known as *stoppers* (see [1] or [6] for a definition). That construction is beyond the scope of this paper, but see [5, Section 5.4] for a complete discussion.

References

[1] E. R. Berlekamp, J. H. Conway, and R. K. Guy. *Winning Ways for Your Mathematical Plays*. A. K. Peters, Ltd., Natick, MA, second edition, 2001.

[2] D. Calistrate. The reduced canonical form of a game. In R. J. Nowakowski, editor, *Games of No Chance*, number 29 in MSRI Publications, pages 409–416. Cambridge University Press, Cambridge, 1996.

[3] J. H. Conway. *On Numbers and Games*. A. K. Peters, Ltd., Natick, MA, second edition, 2001.

[4] G. A. Mesdal. Partizan Splittles. In this volume. 2006.

[5] A. N. Siegel. *Loopy Games and Computation*. PhD thesis, University of California at Berkeley, 2005.

[6] A. N. Siegel. Coping with cycles. In this volume. 2006.

J.P. GROSSMAN
D. E. SHAW RESEARCH
jpg@alum.mit.edu

AARON N. SIEGEL
aaron.n.siegel@gmail.com

Games of No Chance 3
MSRI Publications
Volume **56**, 2009

Partizan Splittles

G. A. MESDAL III*

ABSTRACT. Splittles is a nim variant in which a move consists of removing tokens from one heap, and (optionally) splitting the remaining heap into two. The possible numbers of tokens that can legally be removed are fixed, but the two players might have different *subtraction sets*. The nature of the game, and the analysis techniques employed, vary dramatically depending on the subtraction sets.

1. Introduction

Partizan Splittles is a game played by two players, conventionally called Left and Right. A position in the game consists of a number of heaps of tokens and a move requires a player to choose a heap, remove some positive number, s, of tokens from the heap and optionally to split the remaining heap (if there are two or more tokens remaining) into two heaps. Two sets of positive integers S_L and S_R are fixed in advance, and there is an additional restriction that when Left moves she must choose $s \in S_L$, while Right must choose $s \in S_R$ at his turn. The sets S_L and S_R are called the *subtraction sets* of Left and Right respectively.

It is sometimes convenient to represent a position pictorially by one-dimensional blocks of boxes rather than heaps of tokens. A move is to remove a contiguous block of boxes; moves in the middle of a block are tantamount to splitting a heap.

In this paper, we address several possible restrictions on S_L and S_R, each of which yields a game whose analysis requires different techniques from combinatorial game theory. For some choices of S_L and S_R, canonical forms are

This work was conducted at the third meeting of *Games at Dalhousie* in Halifax, Nova Scotia, in June of 2004. The authors are Michael Albert, Elwyn Berlekamp, William Fraser, J. P. Grossman, Richard Guy, Matt Herron, Lionel Levine, Richard Nowakowski, Paul Ottaway, Aaron Siegel, Angela Siegel, and David Wolfe.

readily available, while for others, canonical forms are complex and uninformative, while temperature theory and the relatively new techniques of reduced canonical forms reveal a great deal of information.

While impartial *octal games* [BCG01] are well-studied, there has been surprisingly little work on partizan variants. In a partizan octal game, the two players have different octal codes indicating their legal moves. Each code is a sequence of octal digits, $d_0.d_1d_2d_3\ldots$, where each d_i includes a **1**, **2**, and/or **4**, indicating whether it is legal to remove i tokens and leave 0, 1, and/or 2 heaps, respectively. In Partizan Splittles, the octal codes consist entirely of **0**s and **7**s. For instance, if $S_L = \{1, 4\}$ and $S_R = \{1, 5\}$, then the game is **0.7007** versus **0.70007**.

Thane Plambeck [Pla95], as well as Calistrate and Wolfe, have unpublished results in the game where players cannot split into two heaps; the octal codes for these games consist entirely of **0**s and **3**s.

2. $\{1, \textbf{odds}\}$ versus $\{1, \textbf{odds}\}$

Our first example is simple.

THEOREM 1. *If 1 is an element of both subtraction sets and all the elements of both subtraction sets are odd numbers, then*

$$G_n = \begin{cases} 0 & \text{if } n \text{ is even,} \\ * & \text{if } n \text{ is odd.} \end{cases}$$

PROOF. Each move changes the parity of the total number of tokens in all the heaps, and in the final position, which has zero tokens, this total is even. Thus, the game is she-loves-me she-loves-me-not. □

3. $\{1\}$ versus $\{k\}$

THEOREM 2. *If $S_L = \{1\}$ and $S_R = \{k\}$, then G_n is arithmetic-periodic with period k and saltus $\{k-1 \mid 0\}$. In particular,*

$$G_n = \begin{cases} n & \text{if } n < k, \\ \{k-1 \mid 0\} + G_{n-k} & \text{if } n \geq k. \end{cases}$$

We can write G_n more naturally with period $2k$ and saltus $k-1$ as

$$G_n = \begin{cases} n & \text{if } n < k, \\ \{n-1 \mid n-k\} & \text{if } k \leq n < 2k, \\ k-1+G_{n-2k} & \text{if } n \geq 2k. \end{cases}$$

PROOF. The proof follows in part from the fact that the conjectured saltus is exactly G_k. So the theorem asserts that one can treat a single heap as a collection of heaps of size k and (possibly) a single remaining heap of size less than k without changing its value. For instance:

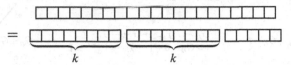

Clearly, $G_a = a$ for $0 \le a < k$, since only Left can move from such a position. Likewise, $G_k = \{k - 1 \mid 0\}$.

In a general position, it suffices to show that any move that straddles a period boundary is dominated by one that does not, for then the game reduces to its "decomposed" form. Left's moves never straddle a boundary. As for Right's moves, assume inductively that shorter positions achieve their conjectured values and decompose at period boundaries. Observe that if $a + b = k + c$ for $0 \le a, b, c < k$, then $G_a + G_b = a + b = k + c$ exceeds $G_k + G_c = \{k - 1 \mid 0\} + c$. Hence, Right prefers the latter move, avoiding a boundary. □

4. $\{1, 2k\}$ versus $\{1, 2k + 1\}$

In this case, too, we can find exact values for all G_n. The sequence is arithmetic-periodic with period $4k$ and saltus $\uparrow^{\rightarrow 2}$. (Note that $\uparrow^{\rightarrow 2} = \uparrow + \uparrow^2 = \{\uparrow \mid *\}$, and that $\uparrow^2 = \{0 \mid \downarrow *\}$ is positive and infinitesimal with respect to $\uparrow$.)

THEOREM 3. *If $S_L = \{1, 2k\}$ and $S_R = \{1, 2k + 1\}$ then*

$$G_{4jk+i} = \begin{cases} 0 + j.\uparrow^{\rightarrow 2} & \text{if } i \text{ is even and } 0 \le i < 2k, \\ * + j.\uparrow^{\rightarrow 2} & \text{if } i \text{ is odd and } 0 \le i < 2k, \\ \uparrow + j.\uparrow^{\rightarrow 2} & \text{if } i \text{ is even and } 2k \le i < 4k, \\ \uparrow* + j.\uparrow^{\rightarrow 2} & \text{if } i \text{ is odd and } 2k \le i < 4k. \end{cases}$$

That is, the values are given by

2k						
0	*	0	*	...	0	*
↑	↑*	↑	↑*	...	↑	↑*
Period $4k$, saltus $\uparrow^{\rightarrow 2}$						

PROOF. When $n < 4k$, it is easy to confirm that the proposed values of G_n are correct. It thus suffices to show that $G_{n+4k} - G_n = \uparrow^{\rightarrow 2}$. Assume, inductively, that the conjectured values are correct for heap sizes less than $n + 4k$. First, moves by either player that split $-G_n$ into $-G_a - G_b$ can be countered by splitting G_{n+4k} into $G_{a+4k} + G_b$, leaving zero by induction.

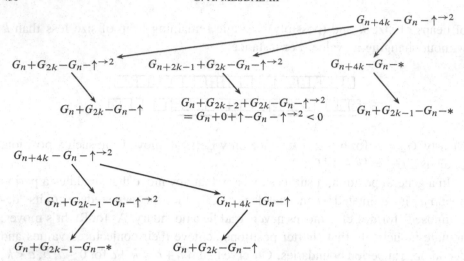

Figure 1. Diagrams showing $G_{n+4k} = G_n + \uparrow^{\to 2}$. The first tree shows Right's winning responses to Left's moves, while second shows how Left wins when Right moves first. Except in one case, the response leaves a game equal to 0.

Next, observe that Left's moves from G_{n+4k} that leave a heap of size $2k$ are at least as good as her other moves. Similarly, Right's dominant splitting moves from G_{n+4k} remove $2k + 1$ tokens, leaving a heap of size $2k - 1$. (When convincing yourself of these last assertions, it helps to keep in mind that all $G_a + G_b$ for fixed $a + b$ have the same $*$-parity, and alternate rows add $\uparrow$ and $\uparrow^2$.) Figure 1 summarizes how the second player wins in response to Left's (respectively, Right's) options not yet dispensed with. $\qquad \square$

COROLLARY 4. *The values in the last theorem remain unchanged when*

- *Left has additional odd options, and/or*
- *Left has additional options exceeding $2k$, and/or*
- *Right has additional odd options between 1 and $2k + 1$.*

PROOF. In all cases these new options are dominated. $\qquad \square$

5. $\{1, \text{others}\}$ versus $\{1, 3, 5, \ldots, 2k + 1\}$

While the actual values might be quite complex and depend on the specific choice for S_L, we can describe a few properties of the sequence G_n.

THEOREM 5. *Suppose $1 \in S_L$, and either $S_R = \{1, 3, 5, \ldots, 2k + 1\}$ for some k or $S_R = \{1, 3, 5, \ldots\}$. The following relations hold:*

$$G_n \leq G_{n+2} \tag{5-1}$$

$$G_{2n+1} = G_{2n} + * \tag{5-2}$$

$$G_n \leq G_i + G_j \quad \text{(for } S_R \text{ finite and } n - i - j \geq 2k \text{ even)} \tag{5-3}$$

PROOF. For (5-1), in most sequences of play, Left wins $G_{n+2} - G_n$ by matching options naturally, removing the same number of tokens as Right did from the opposite heap. Left can then leave a position of the form $G_{a+2} + G_b - G_a - G_b$, for $a, b \geq 0$, which is ≥ 0 by induction. The only exception is if Right removes $n + 1$ from the first heap. In this case, Left removes $n - 1$ from the second, leaving $G_1 - G_1 = 0$.

We show (5-1) and (5-1) in tandem by induction. In particular, we assume that (5-1) holds for $n' \leq n$ when proving (5-1), but that (5-1) holds for $n' < n$ when proving (5-1).

For (5-1), we wish to show the second player wins on the difference game

$$G_{2n+1} - G_{2n} - * = G_{2n+1} - G_{2n} - G_1.$$

We can depict this game as

Left's moves   **Right's moves**

$\{1, \text{others}\}$ ⬜⬜⬜⬜⬜⬜⬜⬜⬜⬜⬜⬜ $\{1, 3, 5, \ldots, 2k + 1\}$

$\{1, 3, 5, \ldots, 2k + 1\}$ ⬜⬜⬜⬜⬜⬜⬜⬜⬜⬜⬜⬜ □ $\{1, \text{others}\}$

The roles of the players are reversed in the second row, it being the negative of the game $G_{2n} + G_1$. So in the second row, Left removes elements from S_R while Right removes elements from S_L.

If either player removes r boxes from the top row, leaving a block of length i odd (and, perhaps a second block of either parity), the other player can counter symmetrically by removing r boxes from the bottom, leaving a block of length $i - 1$. The resulting position is 0 by induction. The reverse is also true; the second player can respond to moves on the bottom row that leave an even-length block. Pictorially, moves A on top leaving one end odd match up with moves A' on bottom leaving the same end even and one shorter.

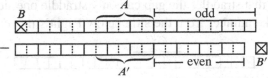

Also shown are moves B which take a single box from one end of the top row, which match up with the move B' taking the lone box on the bottom row.

So we are left with cases that split the top row into two even-length heaps or that split the bottom row into two odd-length heaps. Only Right can do the latter, for it requires removing an even number. Left's responses parallel Right's moves as below. If Right removes C^R on the top, Left's reply of C^{RL} wins by application of (5-1) then multiple applications of (5-1). Similarly Left wins after Right's D^R and Left's D^{RL}.

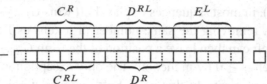

We are now left with the single case when Left removes an odd number from the top row, splitting it into two even-sized heaps, as in E^L above. Right responds by removing as large an (odd) number as possible from one of these two even-sized heaps. If one of the heaps is of size $\leq 2k+2$ (or S_R is infinite), Right leaves that heap a singleton, canceling the single box on the second row, and wins by (5-1). Otherwise, he has taken away $2k+1$, and wins by (5-1) and (5-1).

Lastly, to prove (5-1), we show that Left wins moving second on $G_n - G_i - G_j$:

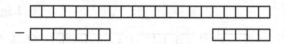

The gap in the bottom row is of even length and at least $2k$. Left can respond to moves that fail to straddle the gap as below:

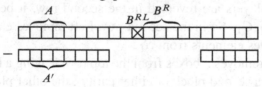

In particular moves outside the gap match up with moves in the other row, winning by induction. Left responds to moves inside the gap by responding on the *odd side:* Since the gap was of even length, and Right can remove only odd numbers, the gap is split into an even length and an odd length. Left then wins by application of (5-1) to both sides.

A Right move that straddles the gap can only straddle one side. Left responds by removing a like number from the side below Right's move:

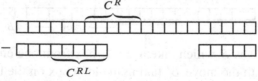

Since the parity of the number of boxes in each row is preserved, each segment can be shortened to an even length by an even number of applications of (5-1), which, since $*+* = 0$ and $* = -*$, leave the game value unchanged. Left then proceeds to win by (5-1) applied to both sides. □

6. Odd versus Even

In this section we consider the partizan splitting game where S_L is the set of all (positive) odd integers and S_R is the set of all (positive) even integers. A salient feature is that the endgame is overwhelmingly favorable to Left: $G_1 = 1$, but there are *no* positions with negative right stop, since Left has a move from every nonterminal position. One would expect, therefore, that Left should prefer to split each position into as many components as possible, preferably odd in size, while Right should aim to annihilate each component as quickly as possible. Since Right will naturally give preference to destroying the largest heaps, one might also expect that Left would prefer to split each heap as evenly as possible.

As is so often the case, the canonical forms of *Odd versus Even* are a mess, but the *orthodox* moves — as defined by Berlekamp [Ber96], Definition 10 — realize these intuitions precisely. Left's orthodox strategy is to split as evenly as possible; Right's is to consume the largest available heap. Furthermore, we will see that from positions of the form G_{2^k-1} — where it is most crucial that Left split evenly — these are the *unique* orthodox options (Theorem 11).

The game also exhibits a fascinating logarithmic behavior: if Left and Right play orthodox strategies, with Left splitting evenly and Right consuming what he can, then the game will last for $O(\log n)$ moves. Furthermore, from positions of the form G_{2^k-1} — where it is most crucial that Left split evenly — Left's *only* orthodox move is the even split. By contrast, we note that, G_{31} has seven canonical Left options.

The main result is the following theorem, which gives the mean, $m(G_n)$, and temperature, $t(G_n)$, of every single-heap Odd versus Even position.

THEOREM 6. *Fix $n \geq 1$ and let k be such that $2^k \leq n < 2^{k+1}$. Then $m(G_n) = \mu_n$ and $t(G_n) = \tau_n$, where*

$$\tau_n = \frac{\lfloor n/2 \rfloor + 1}{2^k} + \frac{k}{2} - 1; \qquad \mu_n = \begin{cases} \tau_n & \textit{if } n \textit{ is even,} \\ \tau_n + 1 & \textit{if } n \textit{ is odd.} \end{cases}$$

To prove Theorem 6, we will define H_n to be the auxiliary game where $S_L = \{1\}$ and S_R is the set of even integers. We will first show that $m(H_n) = \mu_n$ and $t(H_n) = \tau_n$, and then argue that the means and temperatures do not change when Left's subtraction set includes other odd integers.

We will need several lemmas. The first shows that if n is odd, then $H_n = H_{n-1} + 1$ (canonically). This reduces Theorem 6 to the case where n is even.

LEMMA 7. *Let n be odd. Then*

$$H_{n+1} - H_n < 1, \tag{6-1}$$

$$H_n - H_{n-1} = 1. \tag{6-2}$$

PROOF. We proceed by induction on n.

To prove (6-1), consider

$$1 + H_n - H_{n+1}.$$

Left can win by moving immediately to $1 + H_n$; this value is positive since $R_0(H_n) \geq 0$. If Right moves to $1 + H_a + H_b - H_{n+1}$, Left counters to $1 + H_a + H_b - H_{a+1} - H_b$. This is a winning move by induction on (6-1) or (6-1), depending on whether a is odd or even, respectively. If Right moves to $1 + H_n - H_a - H_b$, then since $n + 1$ is even, $a + b$ is odd and hence one of a, b (say a) must be odd. Left counters to $1 + H_{a-1} + H_b - H_a - H_b$, which is 0 by induction.

To prove (6-1) we show that

$$H_n - H_{n-1} - 1$$

is a second-player win. If Right moves to $H_a + H_b - H_{n-1} - 1$, then since n is odd, $a+b$ is also odd and hence one of a, b (say a) must be odd. Left counters to $H_a + H_b - H_{a-1} - H_b - 1$, which by induction is equal to 0. Likewise, if Right moves to $H_n - H_a - H_b - 1$, then since $n - 1$ is even, $a + b$ is odd and hence one of a, b (say a) must be even. Left counters to $H_{a+1} + H_b - H_a - H_b - 1$. Finally, if Right moves to $H_n - H_{n-1}$, Left simply responds with $H_{n-1} - H_{n-1}$.

Conversely, if Left moves to $H_a + H_b - H_{n-1} - 1$, then since $n > 1$ we can assume without loss of generality that $a > 0$. Right counters to $H_a + H_b - H_{a-1} - H_b - 1$. By induction, this is 0 if a is odd, and negative if a is even. If instead Left moves to $H_n - H_a - H_b - 1$, then since $n - 1$ is even, $a + b$ is odd and hence one of a, b (say a) must be odd. Right counters to $H_{a+1} + H_b - H_a - H_b - 1$, which is negative by induction on (6-1). □

The rather dry arithmetic of the τ_n and μ_n is described in the next two lemmas.

LEMMA 8. *Fix $n > 2$ and let k be such that $2^k \leq n < 2^{k+1}$. Then*

$$\tau_n - \tau_{n-2} = \frac{1}{2^k}.$$

PROOF. We may assume that n is even, for if $n' = n + 1$ is odd, we have $2^k \leq n, n' < 2^{k+1} \leq$ and

$$\tau_{n'} - \tau_{n'-2} = \tau_n - \tau_{n-2} = \frac{1}{2^k}.$$

We separate cases into $n = 2^k$ and $n \neq 2^k$.

If $n = 2^k$,

$$\tau_{2^k} - \tau_{2^k-2} = \left(\frac{2^{k-1}+1}{2^k} + \frac{k}{2} - 1\right) - \left(\frac{(2^{k-1}-1)+1}{2^{k-1}} + \frac{k-1}{2} - 1\right)$$

$$= \left(\frac{1}{2} + \frac{1}{2^k} + \frac{k}{2} - 1\right) - \left(1 + \frac{k}{2} - \frac{1}{2} - 1\right) = \frac{1}{2^k}.$$

If $n \neq 2^k$,

$$\tau_n - \tau_{n-2} = \left(\frac{n/2+1}{2^k} + \frac{k}{2} - 1\right) - \left(\frac{(n-2)/2+1}{2^k} + \frac{k}{2} - 1\right)$$

$$= \left(\frac{n}{2^{k+1}} + \frac{1}{2^k} + \frac{k}{2} - 1\right) - \left(\frac{n}{2^{k+1}} + \frac{k}{2} - 1\right) = \frac{1}{2^k}. \qquad \square$$

Lemma 8 shows that the τ_n are (nonstrictly) increasing; and therefore, up to parity, so are the μ_n. Furthermore, up to parity, the rate of increase is decreasing. This fact will be critical in the proof of Theorem 6, since it quantifies the intuition that Left prefers to split as evenly as possible.

LEMMA 9. *Fix $n > 2$ and let k be such that $2^k \leq n < 2^{k+1}$. Then*

$$\mu_n + \mu_{n-1} = \frac{n+1}{2^k} + k - 1.$$

PROOF. Again, we separate into the same two cases. If $n = 2^k$,

$$\mu_{2^k} + \mu_{2^k-1} = \left(\frac{2^{k-1}+1}{2^k} + \frac{k}{2} - 1\right) + \left(\frac{(2^{k-1}-1)+1}{2^{k-1}} + \frac{k-1}{2}\right)$$

$$= \left(\frac{1}{2} + \frac{1}{2^k} + \frac{k}{2} - 1\right) + \left(1 + \frac{k}{2} - \frac{1}{2}\right) = \frac{1}{2^k} + k.$$

Since $n/2^k = 1$, this yields the desired equality.

When $n \neq 2^k$, notice that exactly one of n, $n-1$ is odd, and in either case $\lfloor n/2 \rfloor + \lfloor (n-1)/2 \rfloor = n - 1$. So,

$$\tau_n + \tau_{n-1} = \left(\frac{\lfloor n/2 \rfloor + 1}{2^k} + \frac{k}{2} - 1\right) + \left(\frac{\lfloor (n-1)/2 \rfloor + 1}{2^k} + \frac{k}{2} - 1\right)$$

$$= \frac{(n-1)+2}{2^k} + k - 2 = \frac{n+1}{2^k} + k - 2.$$

Since exactly one of n, $n-1$ is odd, we have $\mu_n + \mu_{n-1} = \tau_n + \tau_{n-1} + 1$, as needed. $\qquad \square$

PROOF. (of Theorem 6) As noted in the exposition, we first show that $m(H_n) = \mu_n$ and $t(H_n) = \tau_n$, and then generalize to the G_n. The proof is by induction on n. The base cases $H_1 = 1$ and $H_2 = \{1 \mid 0\}$ are easily verified. At odd stages of the induction, the result is an immediate corollary of Lemma 7, so fix an even $n > 2$.

Left has a move to $H_n^L = H_{n/2} + H_{n/2-1}$. By induction, we know that

$$m(H_n^L) = m(H_{n/2}) + m(H_{n/2-1}) = \mu_{n/2} + \mu_{n/2-1}$$

and since $2^{k-1} \leq n/2 < 2^k$, Lemma 9 implies that

$$\mu_{n/2} + \mu_{n/2-1} = \frac{n/2+1}{2^{k-1}} + (k-1) - 1 = \frac{\lfloor n/2 \rfloor + 1}{2^{k-1}} + k - 2.$$

Furthermore,

$$t(H_n^L) \leq \max\{\tau_{n/2}, \tau_{n/2-1}\} = \tau_{n/2}.$$

Now Right can remove the entire heap, moving to $H_n^R = 0$. Therefore

$$\frac{m(H_n^L) - m(H_n^R)}{2} = \frac{m(H_n^L)}{2} = \frac{\lfloor n/2 \rfloor + 1}{2^k} + \frac{k}{2} - 1 = \tau_n.$$

Now certainly $\tau_n > \tau_{n/2}$. Therefore

$$\frac{m(H_n^L) - m(H_n^R)}{2} > \max\{t(H_n^L), t(H_n^R)\}.$$

If H_n^L and H_n^R were the *only* options of H_n, then by an elementary thermographic argument, we would have

$$t(H_n) = \tau_n \quad \text{and} \quad m(H_n) = \frac{m(H_n^L) + m(H_n^R)}{2} = \frac{m(H_n^L)}{2} = \tau_n = \mu_n.$$

Certainly both players have other options available, so we conclude the proof by showing that H_n^L and H_n^R are thermally optimal at all temperatures $t \geq \tau_{n/2}$. Since $\tau_{n/2}$ is an upper bound for $t(H_n^L)$, it suffices to show that, for any other options $H_n^{L'}$, $H_n^{R'}$, we have $m(H_n^{L'}) \leq m(H_n^L)$ and $m(H_n^{R'}) \geq m(H_n^R)$.

This is trivial in the case of Right options, since no *Odd versus Even* position can have negative mean.

Therefore, consider some arbitrary Left option $H_a + H_b$, with $a > b$ and $a + b = n - 1$. We necessarily have $a \geq n/2$ and $n/2 - 1 \geq b$, with $a - n/2 = (n/2-1) - b$. Now since exactly one of $n/2, n/2-1$ is odd, repeated applications of Lemma 8 imply that

$$\tau_a - \tau_{n/2} \leq \tau_{n/2-1} - \tau_b.$$

It follows that

$$\tau_a + \tau_b < \tau_{n/2} + \tau_{n/2-1}$$

and hence, since $a + b$ and $n/2 + (n/2 - 1)$ are both odd,

$$\mu_a + \mu_b < \mu_{n/2} + \mu_{n/2-1}.$$

This completes the proof for the H_n. To conclude, we show (again by induction on n) that Left's additional options in G_n convey no thermographic advantage. For suppose Left has a move from G_n to $G_a + G_b$, where $a+b = n - 2c - 1$ for some $c \geq 0$. By induction we may assume that $m(G_i) = \mu_i$ and $t(G_i) = \tau_i$ for all $i < n$. But just as before, we have

$$\mu_a + \mu_b \leq \mu_{a+2c} + \mu_b \leq \mu_{n/2} + \mu_{n/2-1}$$

so that $G_a + G_b$ is thermally dominated at temperatures $t \geq \tau_{n/2}$. $\qquad\square$

We conclude with a neat little theorem on orthodox moves.

DEFINITION 10. Let G be a game and fix $t \geq 0$. A Left option G^L is said to be *orthodox at temperature t* if $R_t(G^L) = L_t(G^L) + t$. Likewise, a Right option G^R is orthodox at temperature t if $L_t(G^R) = R_t(G) - t$. We say that an option is *orthodox* if it is orthodox at temperature $t(G)$.

That is, an orthodox move is one that achieves the best possible score at temperature $t(G)$.

THEOREM 11. *Let $k > 2$ and $n = 2^k - 1$. Left's only orthodox move from G_n is to $G_n^L = G_{2^{k-1}-1} + G_{2^{k-1}-1}$.*

PROOF. Since n is odd, Left must split G_n into two heaps that are either both even or both odd. It is easily seen that those options with both heaps even are badly dominated, so it suffices to show that G_n^L is strictly optimal among those options with both heaps odd.

By Lemma 8,

$$\mu_{2^{k-1}-1} - \mu_{2^{k-1}-3} = \frac{1}{2^{k-2}}, \quad \text{but} \quad \mu_{2^{k-1}+1} - \mu_{2^{k-1}-1} = \frac{1}{2^{k-1}}.$$

Therefore,

$$\mu_{2^{k-1}+1} + \mu_{2^{k-1}-3} < \mu_{2^{k-1}-1} + \mu_{2^{k-1}-1}.$$

Repeated application of Lemma 8 also shows that

$$\mu_a + \mu_b \leq \mu_{2^{k-1}+1} + \mu_{2^{k-1}-3}$$

for every other choice of a, b both odd with $a + b < n$. Therefore G_n^L has the *strictly* highest mean among the Left options of G_n. But we also know that

$$t(G_{2^{k-1}-1} + G_{2^{k-1}-1}) \leq \tau_{2^{k-1}-1} < \tau_n$$

so G_n^L is the unique optimal move at temperature τ_n. $\qquad\square$

7. $\{1, \textbf{odds}\}$ versus $\{2, 4\}$

Suppose S_L is any set of odd numbers containing 1, and $S_R = \{2, 4\}$. The values G_n for these games can be quite complex. For example, when $S_L = \{1, 3\}$ the canonical form of G_{14} contains 611 stops! Furthermore, the exact value of G_n depends strongly on the specific set S_L (if $S_L = \{1\}$ then the canonical form of G_{14} has only 6 stops). Although it is not practical to solve for G_n exactly, we can find a very good approximation for G_n. In particular, let $f(n)$ be the arithmetic-periodic sequence with period 4 and saltus 3/4 defined by

$$f(n) = \begin{cases} 0 & \text{if } n = 0, \\ 1 & \text{if } n = 1, \\ 1/2 & \text{if } n = 2, \\ 3/2 & \text{if } n = 3, \\ f(n-4) + 3/4 & \text{if } n \geq 4. \end{cases}$$

The main theorem of this section is that G_n is infinitesimally close to $f(n) \cdot 1*$ for any choice of S_L that contains 1 and zero or more other odd numbers. We begin by briefly reviewing some definitions and results that will be required.

Infinitesimals. Write $L_0(G)$ and $R_0(G)$ for the Left and Right stops of G, respectively. A game G is *infinitesimal* if $L_0(G) = R_0(G) = 0$. Write $G \equiv_{\text{Inf}} H$ when G and H differ by an infinitesimal; we also say that G is H-ish. If $G \equiv_{\text{Inf}} H$, then $L_0(G) = L_0(H)$ and $R_0(G) = R_0(H)$. The converse is in general not true, but if x is a number and $L_0(G) = R_0(G) = x$, then it is true that $G \equiv_{\text{Inf}} x$, in which case we say that G is *numberish*.

Write $G \geq_{\text{Inf}} H$ if there is some infinitesimal ε such that $G \geq H + \varepsilon$, and similarly for $G \leq_{\text{Inf}} H$. A Left option G^L of G is *Inf-dominated* if $G^{L'} \geq_{\text{Inf}} G^L$ for some other Left option $G^{L'}$, and similarly for Right options.

In [GS07] it is shown that if $G \geq_{\text{Inf}} H$, then $L_0(G) \geq L_0(H)$ and $R_0(G) \geq R_0(H)$. More importantly, they show:

PROPOSITION 12. *If G is not a number and G' is obtained from G by repeatedly*

(i) *eliminating Inf-dominated options, and*

(ii) *replacing any option H with $H' \equiv_{\text{Inf}} H$, then $G' \equiv_{\text{Inf}} G$.*

Norton multiplication. Fix a game $U > 0$. The Norton product $G \cdot U$ is defined by

$$G \cdot U = \begin{cases} 0 \text{ or } \overbrace{U + U + \cdots + U}^{G \text{ times}} \text{ or } \overbrace{-U - U - \cdots - U}^{-G \text{ times}} & \text{if } G \text{ is an integer,} \\ \{G^L \cdot U + (U + I) \mid G^R \cdot U - (U + I)\} & \text{otherwise.} \end{cases}$$

where I ranges over all Left and Right incentives of G. We will use the following properties of Norton multiplication, which are proved in [BCG01].

PROPOSITION 13. Let U be any positive game. Then:

(i) If $G = H$, then $G \cdot U = H \cdot U$ (independence of form).
(ii) $G \geq H$ if and only if $G \cdot U \geq H \cdot U$ (monotonicity).
(iii) $(G + H) \cdot U = G \cdot U + H \cdot U$ (distributivity).

For our purposes, we take $U = 1*$. Since the only Left or Right incentive of $1*$ is $*$, we have

$$G \cdot 1* = \{G^L \cdot 1* + 1 \mid G^R \cdot 1* - 1\}$$

when G is not an integer. We note that $G \cdot 1*$ is equal to G *overheated* from $1*$ to 1, an operation defined in [BCG01]. It is easy to verify by induction that if x is a number then $L_0(x \cdot 1*) = \lceil x \rceil$ and $R_0(x \cdot 1*) = \lfloor x \rfloor$.

LEMMA 14. *If $x = a/4$ for some integer a, then*

$$\{(x - 1/4) \cdot 1* + 1 \mid (x + 1/4) \cdot 1* - 1\} = x \cdot 1*$$

PROOF. If x is not an integer, then $x = \{x - 1/4 \mid x + 1/4\}$, so the result follows from the definition of Norton multiplication and Proposition 13(i). Otherwise, by symmetry, it suffices to show that Right has no winning move from

$$\{(x - 1/4) \cdot 1* + 1 \mid (x + 1/4) \cdot 1* - 1\} - x \cdot 1*.$$

If Right moves in the first component, then the resulting game is

$$(x + 1/4) \cdot 1* - 1 - x \cdot 1* = (1/4) \cdot 1* - 1$$

which we can verify is $\rhd 0$, so Left wins. If Right moves in the second component, which has the effect of subtracting $*$, then Left responds in the first, and the resulting game is

$$(x - 1/4) \cdot 1* + 1 - x \cdot 1* - * = (-1/4) \cdot 1* + 1* = (3/4) \cdot 1*$$

which we can verify is ≥ 0, so again Left wins. $\square$

PROOF OF MAIN RESULT. We will now show that $G_n \equiv_{\text{Inf}} f(n) \cdot 1*$. Our proof is by induction. Suppose the result holds for all $m < n$. It is convenient to assume that $n \geq 4$; for $n < 4$ we can easily validate the result by hand. We begin by showing that in the game G_n, there are only one Left and one Right option that need to be considered.

LEMMA 15. *The Left options of G_n are Inf-dominated by*

$$G_{n-4} + G_3 \equiv_{\text{Inf}} (f(n) + 3/4) \cdot 1*.$$

The Right options are Inf-dominated by

$$G_{n-4} \equiv_{\text{Inf}} (f(n) - 3/4) \cdot 1*.$$

PROOF. The Left options of G_n are $G_{n-k-a} + G_k$ with $a \in S_L$. Since $f(m) < f(m+2)$ for all m and $G_m \equiv_{\text{Inf}} f(m) \cdot 1*$ for $m < n$, $G_{n-k-a} + G_k \leq_{\text{Inf}} G_{n-k-1} + G_k$ so we may assume that $a = 1$. Next, since f is arithmetic-periodic with period 4, $G_{n-k-1} + G_k \equiv_{\text{Inf}} G_{n-k+3} + G_{k-4}$ for $k \geq 4$, so we may assume that $k < 4$. This leaves us with four options to consider, which are infinitesimally close to:

$$f(n-1) \cdot 1*, \quad (f(n-2) + 1) \cdot 1*, \quad (f(n-3) + 1/2) \cdot 1*, \quad (f(n-4) + 3/2) \cdot 1*$$

It is easy to verify that for all m we have

$$f(m) + 3/2 \geq f(m+3), \quad f(m) + 1/2 \geq f(m+2), \quad f(m) + 1 \geq f(m+1),$$

from which it follows that $(f(n-4)+3/2) \cdot 1* = (f(n)+3/4) \cdot 1*$ Inf-dominates the others.

The proof for the Right options is similar. Since $f(m) < f(m+2)$ and f is arithmetic-periodic with period 4, we need only consider the four options $G_{n-k-4} + G_k$ with $k < 4$, which are infinitesimally close to

$$f(n-4) \cdot 1*, \quad (f(n-5) + 1) \cdot 1*, \quad (f(n-6) + 1/2) \cdot 1*, \quad (f(n-7) + 3/2) \cdot 1*$$

The same three inequalities as before show that $f(n-4) \cdot 1* = (f(n)-3/4) \cdot 1*$ Inf-dominates the others (note that for $n = 4, 5, 6$, not all the other options exist, but this does not affect the result). □

Next we show that G_n has the same Left and Right stops as $f(n) \cdot 1*$.

LEMMA 16. $L_0(G_n) = \lceil f(n) \rceil$ and $R_0(G_n) = \lfloor f(n) \rfloor$.

PROOF. First we compute $\max\{R_0(G_n^L)\}$ and $\min\{L_0(G_n^R)\}$. By Lemma 15,

$$\max\{R_0(G_n^L)\} = R_0((f(n) + 3/4) \cdot 1*) = \lfloor f(n) + 3/4 \rfloor = \lceil f(n) \rceil.$$

The last equality follows from the fact that $f(n)$ is of the form $a/4$ for some integer a. Similarly,

$$\min\{L_0(G_n^R)\} = L_0((f(n) - 3/4) \cdot 1*) = \lceil f(n) - 3/4 \rceil = \lfloor f(n) \rfloor.$$

If G_n is not a number then we are done, as then $L_0(G_n) = \max\{R_0(G_n^L)\}$ and $R_0(G_n) = \min\{L_0(G_n^R)\}$. If G_n is a number then $G_n = L_0(G_n) = R_0(G_n)$, but

$$L_0(G_n) \geq \max\{R_0(G_n^L)\} = \lceil f(n) \rceil \geq \lfloor f(n) \rfloor = \min\{L_0(G_n^R)\} \geq R_0(G_n)$$

so in fact we must have equality throughout, which means that $f(n)$ is also an integer, and again we are done. □

From Lemma 16 it follows immediately that when $f(n)$ is an integer, $G_n \equiv_{\text{Inf}}$ $f(n) \equiv_{\text{Inf}} f(n) \cdot 1*$. Finally, if $f(n)$ is not an integer, then by Lemma 16, G_n is not numberish. So by Lemma 15, Proposition 12 and Lemma 14,

$$G_n \equiv_{\text{Inf}} \{(f(n) + 3/4) \cdot 1* \mid (f(n) - 3/4) \cdot 1*\}$$
$$\equiv_{\text{Inf}} \{(f(n) - 1/4) \cdot 1* + 1 \mid (f(n) + 1/4) \cdot 1* - 1\} = f(n) \cdot 1*. \qquad \square$$

References

[BCG01] Elwyn R. Berlekamp, John H. Conway, and Richard K. Guy. *Winning Ways for Your Mathematical Plays*. A K Peters, Ltd., Natick, Massachusetts, 2nd edition, 2001. First edition published in 1982 by Academic Press.

[Ber96] Elwyn Berlekamp. "The economist's view of combinatorial games". In Richard Nowakowski, editor, *Games of No Chance: Combinatorial Games at MSRI, 1994*, pages 101–120. Cambridge University Press, Mathematical Sciences Research Institute Publications 29, 1996.

[GS07] J. P. Grossman and A. N. Siegel. "Reductions of partisan games". In this volume.

[Pla95] Thane Plambeck. "Partisan subtraction games". Working notes, February 1995. http://www.plambeck.org/oldhtml/mathematics/games/subtraction/

G. A. MESDAL III
    malbert@cs.otago.ac.nz
    berlek@math.berkeley.edu
    bfraser@alumni.ucsd.edu
    jpg@ai.mit.edu
    rkg@cpsc.ucalgary.ca
    rjn@mscs.dal.ca
    ottaway@mathstat.dal.ca
    aaron.n.siegel@gmail.com
    siegel@mathstat.dal.ca
    wolfe@gustavus.edu

Columns

Games of No Chance 3
MSRI Publications
Volume **56**, 2009

Unsolved problems in Combinatorial Games

RICHARD K. GUY AND RICHARD J. NOWAKOWSKI

We have sorted the problems into sections:

- A. Taking and Breaking
- B. Pushing and Placing Pieces
- C. Playing with Pencil and Paper
- D. Disturbing and Destroying
- E. Theory of Games

They have been given new numbers. The numbers in parentheses are the old numbers used in each of the lists of unsolved problems given on pp. 183–189 of AMS *Proc. Sympos. Appl. Math.* **43** (1991), called PSAM **43** below; on pp. 475–491 of *Games of No Chance*, hereafter referred to as GONC; and on pp. 457–473 of *More Games of No Chance* (MGONC). Missing numbers are of problems which have been solved, or for which we have nothing new to add. References [year] may be found in Fraenkel's Bibliography at the end of this volume. References [#] are at the end of this article. A useful reference for the rules and an introduction to many of the specific games mentioned below is M. Albert, R. J. Nowakowski and D. Wolfe, *Lessons in Play: An Introduction to the Combinatorial Theory of Games*, A K Peters, 2007 (LIP).

A. Taking and breaking games

A1 (1). Subtraction games with finite subtraction sets are known to have periodic nim-sequences. Investigate the relationship between the subtraction set and the length and structure of the period. The same question can be asked about **partizan** subtraction games, in which each player is assigned an individual subtraction set. See Fraenkel and Kotzig [1987].

[A move in the game $S(s_1, s_2, s_3, \dots)$ is to take a number of beans from a heap, provided that number is a member of the **subtraction-set**, $\{s_1, s_2, s_3, \dots\}$. Analysis of such a game and of many other heap games is conveniently recorded by a **nim-sequence**,

$$n_0 n_1 n_2 n_3 \ldots,$$

meaning that the nim-value of a heap of h beans is n_h; i.e., that the value of a heap of h beans in this particular game is the **nimber** $*n_h$.]

For examples see Table 2 in §4 on p. 67 of the Impartial Games paper in GONC.

It would now seem feasible to give the complete analysis for games whose subtraction sets have just three members, though this has so far eluded us. Several people, including Mark Paulhus and Alex Fink, have given a complete analysis for all sets $\{1, b, c\}$ and for sets $\{a, b, c\}$ with $a < b < c < 32$.

In general, period lengths can be surprisingly long, and it has been suggested that they could be superpolynomial in terms of the size of the subtraction set. However, Guy conjectures that they are bounded by polynomials of degree at most $\binom{n}{2}$ in s_n, the largest member of a subtraction set of cardinality n. It would also be of interest to characterize the subtraction sets which yield a purely periodic nim-sequence, i.e., for which there is no preperiod.

Angela Siegel [18] considered infinite subtraction sets which are the complement of finite ones and showed that the nim-sequences are always arithmetic periodic. That is, the nim-values belong to a finite set of arithmetic progressions with the same common difference. The number of progressions is the period and their common difference is called the **saltus**. For instance, the game $S\{\hat{4}, \hat{9}, \widehat{26}, \widehat{30}\}$ (in which a player may take any number of beans except 4, 9, 26 or 30) has a preperiod of length 243, period-length 13014 and saltus 4702.

For infinite subtraction games in general there are corresponding questions about the length and purity of the period.

We note that Question A2 on the 2006-12-02 Putnam exam is the subtraction game with subtraction set $\{p - 1 : p \text{ prime}\}$. Show that there are infinitely many heap sizes which are $\mathcal{P}$-positions.

A2 (2). Are all finite **octal games** ultimately periodic?

[If the binary expansion of the k-th code digit in the game with code

$$\mathbf{d}_0 \cdot \mathbf{d}_1 \mathbf{d}_2 \mathbf{d}_3 \ldots$$

is

$$\mathbf{d}_k = 2^{a_k} + 2^{b_k} + 2^{c_k} + \ldots,$$

where $0 \le a_k < b_k < c_k < \ldots$, then it is legal to remove k beans from a heap, provided that the rest of the heap is left in exactly a_k or b_k or c_k or $\ldots$ nonempty heaps. See WW, 81–115. Some specimen games are exhibited in Table 3 of §5 of the Impartial Games paper in GONC.]

Resolve any number of outstanding particular cases, e.g., ·6 (Officers), ·04, ·06, ·14, ·36, ·37, ·64, ·74, ·76, ·004, ·005, ·006, ·007, ·014, ·015, ·016, ·024, ·026, ·034, ·064, ·114, ·125, ·126, ·135, ·136, ·142, ·143, ·146, ·162, ·163, ·164, ·166, ·167, ·172, ·174, ·204, ·205, ·206, ·207, ·224, ·244, ·245, ·264, ·324, ·334, ·336, ·342, ·344, ·346, ·362, ·364, ·366, ·371, ·374, ·404, ·414, ·416, ·444, ·564, ·604, ·606, ·744, ·764, ·774, ·776 and **Grundy's Game** (split a heap into two unequal heaps; WW, pp. 96–97, 111–112; LIP, p. 142), which has been analyzed, first by Dan Hoey and later by Achim Flammenkamp, as far as heaps of 2^{35} beans.

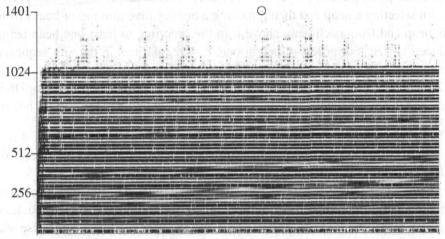

Figure 1. Plot of 11000000 nim-values of the octal game ·**007**.

Perhaps the most notorious and deserving of attention is the game ·**007**, one-dimensional Tic-Tac-Toe, or Treblecross, which Flammenkamp has pushed to 2^{25}. Figure 1 shows the first 11 million nim-values, a small proportion of which are ≥ 1024; the largest, $\mathcal{G}(6193903) = 1401$ is shown circled. Will 2048 ever be reached?

Achim Flammenkamp has settled ·**106**: it has the remarkable period and preperiod lengths of 328226140474 and 465384263797. For information on the current status of each of these games, see Flammenkamp's web page at http:// www.uni-bielefeld.de/~achim/octal.html.

A game similar to Grundy's, and which is also unsolved, is John Conway's **Couples-Are-Forever** (LIP, p. 142) where a move is to split any heap except a heap of two. The first 50 million nim-values haven't displayed any periodicity. See Caines et al. [1999]. More generally, Bill Pulleyblank suggests looking at splitting games in which you may only split heaps of size $> h$, so that $h = 1$ is She-Loves-Me-She-Loves-Me-Not and $h = 2$ is Couples-Are-Forever. David Singmaster suggested a similar generalization: you may split a heap provided that the resulting two heaps each contain at least k beans: $k = 1$ is the same as $h = 1$, while $k = 2$ is the third cousin of Dawson's Chess.

Explain the structure of the periods of games known to be periodic.

In *Discrete Math.*, 44(1983) 331–334, Problem 38, Fraenkel raised questions concerning the computational complexity (see **E1** below) of octal games. In Problem 39, he and Kotzig define **partizan octal games** in which distinct octals are assigned to the two players. The article by Mesdal, in this volume, shows that in many cases, if the game is "all-small" (WW, pp. 229–262, LIP, pp. 183–207), then the atomic weights are arithmetic periodic. In Problem 40, Fraenkel introduces **poset games**, played on a partially ordered set of heaps, each player in turn selecting a heap and then removing a nonnegative number of beans from this heap and from each heap above it in the ordering, at least one heap being reduced in size. For posets of height one, new regularities in the nim-sequence can occur; see Horrocks and Nowakowski [2003].

Note that this includes, as particular cases, Subset Takeaway, Chomp or Divisors, and Green Hackenbush forests. Compare Problems **A3**, **D1** and **D2** below.

A3 (3). Hexadecimal games have code digits d_k in the interval from 0 to f

(= 15), so that there are options splitting a heap into three heaps. See WW, 116–117.

Such games may be arithmetically periodic. Nowakowski has calculated the first 100000 nim-values for each of the 1-, 2- and 3-digit games. Richard Austin's theorem 6.8 in his thesis [1976] and the generalization by Howse and Nowakowski [2004] suffice to confirm the arithmetic periodicity of several of these games.

Some interesting specimens are $\cdot\mathbf{28} = \cdot\mathbf{29}$, which have period 53 and saltus 16, the only exceptional value being $\mathcal{G}(0) = 0$; $\cdot\mathbf{9c}$, which has period 36, preperiod 28 and saltus 16; and $\cdot\mathbf{f6}$ with period 43 and saltus 32, but its apparent preperiod of 604 and failure to satisfy one of the conditions of the theorem prevent us from verifying the ultimate periodicity. The game $\cdot\mathbf{205200c}$ is arithmetic periodic with preperiod length of 4, period length of 40, saltus 16 except that $40k + 19$ has nim-value 6 and $40k + 39$ has nim-value 14. This regularity, (which also seems to be exhibited by $\cdot\mathbf{660060008}$ with a period length of approximately 300,000), was first reported in Horrocks and Nowakowski [2003] (see Problem **A2**.) Grossman and Nowakowski [7] have shown that the nim-sequences for $\cdot\mathbf{200\ldots0048}$, with an odd number of zero code digits, exhibit "ruler function" patterns. The game $\cdot\mathbf{9}$ has not so far yielded its complete analysis, but, as far as analyzed (to heaps of size 100000), exhibits a remarkable fractal-like set of nim-values. See Howse and Nowakowski [2004]. Also of special interest are $\cdot\mathbf{e}$; $\cdot\mathbf{7f}$ (which has a strong tendency to period 8, saltus 4, but, for $n \leq 100,000$, has 14 exceptional values, the largest being $\mathcal{G}(94156) = 26614$); $\cdot\mathbf{b6}$ (which "looks octal"); ·**b33b** (where a heap of size n has nim-value n except for 27 heap sizes which appear to be random); and ·**817264517**

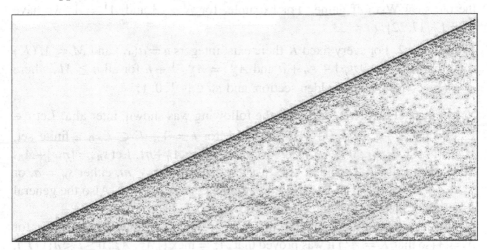

Figure 2. Plot of 200000 nim-values for the hexadecimal game ·**817264517**

[why 817264517 ?] whose nim-values appear to form a lattice of ruler functions with slopes slightly less than $\frac{1}{2}$ and $-\frac{1}{2}$ (see Figure 2). The largest value in the range calculated is $\mathcal{G}(206265) = 101458$.

Other unsolved hexadecimal games are ·**1x**, where $\mathbf{x} \in \{8, 9, c, d, e, f\}$; ·**2x**, $a \le \mathbf{x} \le f$; ·**3x**, $8 \le \mathbf{x} \le e$; ·**4x**, $\mathbf{x} \in \{9, b, d, f\}$; ·**5x**, $8 \le \mathbf{x} \le f$; ·**6x**, $8 \le \mathbf{x} \le f$; ·**7x**, $8 \le \mathbf{x} \le f$; ·**9x**, $1 \le \mathbf{x} \le a$; ·**9d**; ·**bx**, $\mathbf{x} \in \{6, 9, d\}$; ·**dx**, $1 \le \mathbf{x} \le f$; ·**fx** with $\mathbf{x} \in \{4, 6, 7\}$.

A4 (53). N-heap Wythoff Game. Given $N \ge 2$ heaps of finitely many tokens, whose sizes are $p_1, \ldots, p_N$ with $p_1 \le \cdots \le p_N$. Players take turns removing any positive number of tokens from a *single* heap or removing $(a_1, \ldots, a_N)$ from *all* the heaps — a_i from the i-th heap — subject to the conditions (i) $0 \le a_i \le p_i$ for each i, (ii) $\sum_{i=1}^{N} a_i > 0$, (iii) $a_1 \oplus \cdots \oplus a_N = 0$, where $\oplus$ is nim addition. The player making the last move wins and the opponent loses. Note that the classical Wythoff game is the case $N = 2$.

For $N \ge 3$, Fraenkel makes the following conjectures.

Conjecture 1. For every fixed set $K := (A^1, \ldots, A^{N-2})$ there exists an integer $m = m(K)$ (i.e, m depends only on K), such that

$$(A^1, \ldots, A^{N-2}, A_n^{N-1}, A_n^N), \quad A_n^{N-2} \le A_n^{N-1} \le A_n^N$$

with $A_n^{N-1} < A_{n+1}^{N-1}$ for all $n \ge 1$, is the n-th $\mathcal{P}$-position, and

$$A_n^{N-1} = \text{mex}\left(\{A_i^{N-1}, A_i^N : 0 \le i < n\} \cup T\right), \quad A_n^N = A_n^{N-1} + n$$

for all $n \geq m$, where $T = T(K)$ is a (small) set of integers.

That is, if you fix $N - 2$ of the heaps, the $\mathcal{P}$-positions resemble those for the classical Wythoff game. For example, for $N = 3$ and $A^1 = 1$, we have $T = \{2, 17, 22\}$, $m = 23$.

Conjecture 2. For every fixed K there exist integers $a = a(K)$ and $M = M(K)$ such that $A_n^{N-1} = \lfloor n\phi \rfloor + \varepsilon_n + a$ and $A_n^N = A_n^{N-1} + n$ for all $n \geq M$, where $\phi = (1 + \sqrt{5})/2$ is the golden section, and $\varepsilon_n \in \{-1, 0, 1\}$.

In Fraenkel and Krieger [2004] the following was shown, inter alia: Let $t \in \mathbb{Z}_{\geq 1}$, $\alpha = (2 - t + \sqrt{t^2 + 4})/2$ ($\alpha = \phi$ for $t = 1$), $T \subset \mathbb{Z}_{\geq 0}$ a finite set, $A_n = (\text{mex}\,\{A_i, B_i : 0 \leq i < n\} \cup T)$, where $B_n = A_n + nt$. Let $s_n := \lfloor n\alpha \rfloor - A_n$. Then there exist $a \in \mathbb{Z}$ and $m \in \mathbb{Z}_{\geq 1}$, such that for all $n \geq m$, either $s_n = a$, or $s_n = a + \varepsilon_n$, $\varepsilon_n \in \{-1, 0, 1\}$. If $\varepsilon_n \neq 0$, then $\varepsilon_{n-1} = \varepsilon_{n+1} = 0$. Also the general structure of the ε_n was characterized succinctly.

This result was then applied to the N-heap Wythoff game. In particular, for $N = 3$ (so that $K = A^1$) it was proved that $A_n^2 = \text{mex}\,(\{A_i^2, A_i^3 : 0 \leq i < n\} \cup T)$, where $T =$

$$\{x \geq K : \exists\, 0 \leq k < K \text{ s.t. } (k, K, x) \text{ is a } P - \text{position}\} \cup \{0, \ldots, K - 1\}$$

The following upper bound for A_n^3 was established: $A_n^3 \leq (K + 3)A_n^2 + 2K + 2$. It was also proved that Conjecture 1 implies Conjecture 2.

In Sun and Zeilberger [2004], a sufficient condition for the conjectures to hold was given. It was then proved that the conjectures are true for the case $N = 3$, where the first heap has up to 10 tokens. For those 10 cases, the parameter values m, M, a, T were listed in a table.

Sun [2005] obtained results similar to those in Fraenkel and Krieger [2004], but the proofs are different. It was also proved that Conjecture 1 implies Conjecture 2. A method was given to compute a in terms of certain indexes of the A_i and B_j.

A5 (23). Burning-the-Candle-at-Both-Ends.
Conway and Fraenkel ask us to analyze Nim played with a row of heaps. A move may only be made in the leftmost or in the rightmost heap. When a heap becomes empty, then its neighbor becomes the end heap.

Albert and Nowakowski [2001] have determined the outcome classes in impartial and partizan versions (called **End-Nim**, LIP, pp. 210, 263) with finite heaps, and Duffy, Kolpin and Wolfe, in this volume, extend the partizan case to infinite ordinal heaps. Wolfe asks for the actual values.

Nowakowski suggested to analyze impartial and partizan **End-Wythoff**: take from either end-pile, or the *same* number from both ends. The impartial game is solved by Fraenkel and Reisner, in this volume, Fraenkel [1982] asks a similar question about a generalized Wythoff game: take from either end-pile or take $k > 0$ from one end-pile and $\ell > 0$ from the other, subject to $|k - \ell| < a$, where a is a fixed integer parameter ($a = 1$ is End-Wythoff).

There is also **Hub-and-Spoke Nim**, proposed by Fraenkel. One heap is the hub and the others are arranged in rows forming spokes radiating from the hub. Albert notes that this game can be generalized to playing on a forest, i.e., a graph each of whose components is a tree. The most natural variant is that beans may only be taken from a leaf (valence 1) or isolated vertex (valence 0).

The partizan game of **Red-Blue Cherries** is played on an arbitrary graph. A player picks an appropriately colored cherry from a vertex of minimum degree, which disappears at the same time. Albert et al.[1] show that if the graph has a leaf, then the value is an integer. See also McCurdy [10].

A6 (17). Extend the analysis of **Kotzig's Nim** (WW, 515–517). Is the game eventually periodic in terms of the length of the circle for every finite move set? Analyze the misère version of Kotzig's Nim.

A7 (18). Obtain asymptotic estimates for the proportions of $\mathcal{N}$-, $\mathcal{O}$- and $\mathcal{P}$-positions in Epstein's **Put-or-Take-a-Square** game (WW, 518–520).

A8. Gale's Nim. This is Nim played with four heaps, but the game ends when three of the heaps have vanished, so that there is a single heap left. Brouwer and Guy have independently given a partial analysis, but the situation where the four heaps have distinct sizes greater than 2 is open. An obvious generalization is to play with h heaps and play finishes when k of them have vanished.

A9. Euclid's Nim is played with a pair of positive integers, a move being to diminish the larger by any multiple of the smaller. The winner is the player who reduces a number to zero. Analyses have been given by Cole and Davie [1969], Spitznagel [1973], Lengyel [2003], Collins [2005], Fraenkel [2005] and Nivasch [2006]. Gurvich [8] shows that the nim-value, $g^+(a, b)$ for the pair (a, b) in normal play is the same as the misère nim-value, $g^-(a, b)$ except for $(a, b) = (kF_i, kF_{i+1})$ where $k > 0$ and F_i is the i-th Fibonacci number. In this case, $g^+(kF_i, kF_{i+1}) = 0$ and $g^-(kF_i, kF_{i+1}) = 1$ if i is even and the values are reversed if i is odd.

We are not aware of an analysis of the game played with three or more integers.

A10 (20). Some advance in the analysis of **D.U.D.E.N.E.Y** (WW, 521–523) has been made by Marc Wallace, Alex Fink and Kevin Saff.

[The game is Nim, but with an upper bound, Y, on the number of beans that may be taken, and with the restriction that a player may not repeat his opponent's last move. If Y is even, the analysis is easy.]

We can, for example, extend the table of strings of pearls given in WW, p. 523, with the following values of Y which have the pure periods shown, where D=$Y+2$, E=$Y+1$. The first entry corrects an error of $128r + 31$ in WW.

$256r + 31$	DEE	$512r + 153$ DEE	$1024r + 415$ DEE
$512r + 97$	DDEDDDE	$512r + 159$ DEE	$512r + 425$ DE
$1024r + 103$	DE	$512r + 225$ DDE	$512r + 487$ DEE
$128r + 119$	DEE	$512r + 255$ E	$1024r + 521$ DDDE
$1024r + 127$	DEEE	$512r + 257$ DDDDE	$1024r + 607$ DDE
$512r + 151$	DDDEE	$512r + 297$ DDEDEDE	$1024r + 735$ DEEE

It seems likely that the string for $Y = 2^{2k+1} + 2^{2k} - 1$ has the simple period E for all values of k. But the following evidence of the fraction, among 2^k cases, that remain undetermined:

$k =$	3	5	6	7	8	9	10	11	12	13	14	15	16	17
fraction	$\frac{1}{2}$	$\frac{5}{16}$	$\frac{9}{32}$	$\frac{11}{64}$	$\frac{21}{128}$	$\frac{33}{256}$	$\frac{60}{512}$	$\frac{97}{1024}$	$\frac{177}{2048}$	$\frac{304}{4096}$	$\frac{556}{8192}$	$\frac{974}{16384}$	$\frac{1576}{32768}$	$\frac{2763}{65536}$

suggests that an analysis will never be complete.

Moreover, the periods of the pearl-strings appear to become arbitrarily long.

A11 (21). Schuhstrings is the same as D.U.D.E.N.E.Y, except that a deduction of zero is also allowed, but cannot be immediately repeated (WW, 523–524). In Winning Ways it was stated that it was not known whether there is any Schuh-string game in which three or more strings terminate simultaneously. Kevin Saff has found three such strings (when the maximum deduction is $Y = 3430$, the three strings of multiples of 2793, 3059, 3381 terminate simultaneously) and he conjectures that there can be arbitrarily many such simultaneous terminations.

A12 (22). Analyze **Dude**, i.e., unbounded D.U.D.E.N.E.Y, or Nim in which you are not allowed to repeat your opponent's last move.

Let $[h_1, h_2, \ldots, h_k; m]$, $h_i \leq h_{i+1}$, be the game with heaps of size h_1 through h_k, where m is the move just made and $m = 0$ denotes a starting position. Then [4], for $k = 1$ the $\mathcal{P}$-positions are $[(2s + 1)2^{2j}; (2s + 1)2^{2j}]$; for $k = 2$ they are $[(2s + 1)2^{2j}, (2s + 1)2^{2j}; 1]$; and for $k \geq 3$ the heap sizes are arbitrary, the only condition being that the previous move was 1. The nim-values do not seem to show an easily described pattern.

A13. Nim with pass. David Gale would like to see an analysis of Nim played with the option of a single pass by either of the players, which may be made at any time up to the penultimate move. It may not be made at the end of the game. Once a player has passed, the game is as in ordinary Nim. The game ends when all heaps have vanished.

A14. Games with a Muller twist. In such games, each player specifies a condition on the set of options available to her opponent on his next move.

In **Odd-or-Even Nim**, for example, each player specifies the parity of the opponent's next move. This game was analyzed by Smith and Stănică [2002], who propose several other such games which are still open (see also Gavel and Strimling [2004]).

The game of **Blocking Nim** proceeds in exactly the same way as ordinary Nim with N heaps, except that before a given player takes his turn, his opponent is allowed to announce a **block**, $(a_1, \ldots, a_N)$; i.e., for each pile of counters, he has the option of specifying a positive number of counters which may not be removed from that pile. Flammenkamp, Holshouser and Reiter [2003, 2004] give the $\mathcal{P}$-positions for three-heap Blocking Nim with an incomplete block containing only one number, and ask for an analysis of this game with a block on just two heaps, or on all three. There are corresponding questions for games with more than three heaps.

A15 (13). Misère analysis has been revolutionized by Thane Plambeck and Aaron Siegel with their concept of the **misère quotient** of a game [13], though the number of unsolved problems continues to increase.

Let $\mathcal{A}$ be some set of games played under misère rules. Typically, $\mathcal{A}$ is the set of positions that arise in a particular game, such as Dawson's Chess. Games $H, K \in \mathcal{A}$ are said to be equivalent, denoted by $H \equiv K$, if $H + X$ and $K + X$ have the same outcome for all games $X \in \mathcal{A}$. The relation $\equiv$ is an equivalence relation, and a set of representatives, one from each equivalence class, forms the **misère quotient**, $\mathcal{Q} = \mathcal{A}/\equiv$. A **quotient map** $\Phi : \mathcal{A} \to \mathcal{Q}$ is defined, for $G \in \mathcal{A}$, by $\Phi : \mathcal{G} = [G]_\equiv$.

Plambeck and Siegel ask the specific questions:

(1) The misère quotient of $\cdot 07$ (Dawson's Kayles) has order 638 at heap size 33. Is it infinite at heap size 34? Even if the misère quotient is infinite at heap 34 then, by Redei's theorem [6, p. 142], [14], it must be isomorphic to a finitely-presented commutative monoid. Call this monoid D_{34}. Exhibit a monoid presentation of D_{34}, and having done that, exhibit D_{35}, D_{36}, etc, and explain what is going on in general. Given a set of games $\mathcal{A}$, describe an algorithm to determine whether the misère quotient of $\mathcal{A}$ is infinite. Much harder: if the quotient is infinite, give an algorithm to compute a presentation for it.

(2) A quotient map $\Phi : \mathcal{A} \to \mathcal{Q}$ is said to be **faithful** if, whenever $\Phi(G) = \Phi(H)$, then G and H have the same normal-play Grundy value. Is every quotient map faithful?

(3) Let $(\mathcal{Q}, \mathcal{P})$ be a quotient and $\mathcal{S}$ a maximal subgroup of $\mathcal{Q}$. Must $\mathcal{S} \cap \mathcal{P}$ be nonempty? (Note: it's easy to get a "yes" answer in the special case when $\mathcal{S}$ is the kernel)

(4) Give complete misère analyses for any of the (normal-play periodic) octal games that show "algebraic-periodicity" in misère play. Some examples are ·54, ·261, ·355, ·357, ·516 and ·724. Give a precise definition of algebraic periodicity and describe an algorithm for detecting and generalizing it. This is a huge question: if such an algorithm exists, it would likely instantly give solutions to at least a half-dozen unsolved 2- and 3-digit octals.

(5) Extend the classification of misère quotients. We have preliminary results on the number of quotients of order $n \leq 18$ but believe that this can be pushed far higher.

(6) Exhibit a misère quotient with a period-5 element. Same question for period 8, etc. We've detected quotients with elements of periods 1, 2, 3, 4, 6, and infinity, and we conjecture that there is no restriction on the periods of quotient elements.

(7) In the flavor of both (5) and (6): What is the smallest quotient containing a period 4 (or 3 or 6) element?

Plambeck also offers prizes of US$500.00 for a complete analysis of Dawson's Chess, ·137 (alias Dawson's Kayles, ·07); US$200.00 for the "wild quaternary game", ·3102; and US$25.00 each for ·3122, ·3123 and ·3312.

The website http://www.miseregames.org contains thousands of misère quotients for octal games.

Siegel notes that Dawson first proposed his problem in 1935, making it perhaps the oldest open problem in combinatorial game theory. [Michael Albert offers the alternative "Is chess a first player win?"] It may be of historical interest to note that Dawson showed the problem to one of the present authors around 1947. Fortunately, he forgot that Dawson proposed it as a losing game, was able to analyze the normal play version, rediscover the Sprague–Grundy theory, and get Conway interested in games.

B. Pushing and placing pieces

B1 (5). The game of **Go** is of particular interest, partly because of the loopiness induced by the "ko" rule, and many problems involve general theory: see **E4** and **E5**.

Elwyn writes:

I attach one region that has been studied intermittently over the past several years. The region occurs in the southeast corner of the board (Figure 3). At move 85 Black takes the ko at L6. What then is the temperature at N4 ? This position is copied from the game Jiang and Rui played at MSRI in July 2000. In 2001, Bill Spight and I worked out a purported solution by hand, assuming either Black komaster or White komaster. I've

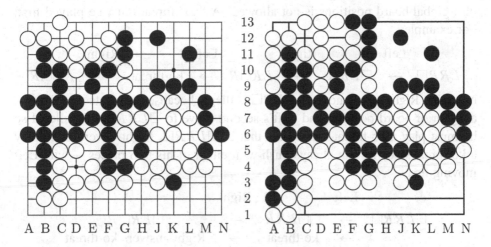

Figure 3. Jiang v. Rui, MSRI, July, 2000.

recently been trying to get that rather complicated solution confirmed by GoExplorer, which would then presumably also be able to calculate the dogmatic solution. I've been actively pursuing this off and on for the past couple weeks, and haven't gotten there yet.

Elwyn also writes:

Nakamura has shown [this volume] how capturing races in Go can be analyzed by treating liberties as combinatorial games. Like atomic weights, when the values are integers, each player's best move reduces his opponent's resources by one. The similarities between atomic weights and Nakamura's liberties are striking.

Theoretical problem: Either find a common formulation which includes much or all of atomic weight theory and Nakamura's theory of liberties, OR find some significant differences.

Important practical applied problem: Extend Nakamura's theory to include other complications which often arise in Go, such as simple kos, either internal and/or external.

B2. A simpler game involving kos is **Woodpush** (see LIP, pp. 214, 275). This is played on a finite strip of squares. Each square is empty or occupied by a black or white piece. A piece of the current player's color retreats: Left retreats to the left and Right to the right — to the next empty square, or off the board if there is no empty square; except, if there is a contiguous string containing an opponent's piece then it can move in the opposite direction *pushing* the string ahead of it. Pieces can be moved off the end of the strip. Immediate repetition of a global board positions is not allowed. A "ko" threat must be played first. For example

Left	Right	Left	Right
$LRR\square$ →	$\square LRR$ →	$LR\square R$ →	ko-threat →	$R\square\square R$

Note that Right's first move to $LRR\square$ is illegal because it repeats the immediately prior board position and Left's second move to $\square LRR$ is also illegal so he must play a ko-threat. Also note that in $\square LRR\square$, Right never has to play a ko-threat since he can always push with either of his two pieces — with Left moving first,

	Left		Right
$\square LRR\square$	→	$\square\square LRR$	→	$\square LR\square R$
	→	ko-threat	→	Right answers ko-threat
	→	$\square\square LRR$	→	$\square LRR\square$

Berlekamp, Plambeck, Ottaway, Aaron Siegel and Spight (work in progress) use top-down thermography to analyze the three piece positions. What about more pieces?

B3 (40). Chess. Noam Elkies [2002] has examined Dawson's Chess, but played under usual Chess rules, so that capture is not obligatory.

He would still welcome progress with his conjecture that the value $*k$ occurs for all k in (ordinary Chess) pawn endings on sufficiently large chessboards.

Thea van Roode has suggested **Impartial Chess**, in which the players may make moves of either color. Checks need not be responded to and Kings may be captured. The winner could be the first to promote a pawn.

B4 (30). Low and Stamp [2006] have given a strategy in which White wins the King and Rook vs. King problem within an 11×9 region.

B5. Nonattacking Queens. Noon and Van Brummelen [2006] alternately place queens on an $n \times n$ chessboard so that no queen attacks another. The winner is the last queen placer. They give nim-values for boards of sizes $1 \le n \le 10$ as 1121312310 and ask for the values of larger boards.

B6 (55). Amazons. Martin Müller [11] has shown that the 5×5 game is a first player win and asks about the 6×6 game.

B7. Conway's Philosopher's Football, or **Phutball**, is usually played on a Go board with positions (i, j), $-9 \le i, j \le 9$ and the ball starting at (0,0). For the rules, see WW, pp.752–755. The game is loopy (see **E5** below), and Nowakowski, Ottaway and Siegel (see [17]) discovered positions that contained tame cycles, i.e., cycles with only two strings, one each of Left and Right moves. Aaron Siegel asks if there are positions in such combinatorial games which are stoppers but contain a **wild cycle**, i.e., one which contains more than one alternation between Left and Right moves. Demaine, Demaine and Eppstein [2002] show that it is NP-complete to decide if a player can win on the next move.

Phlag Phutball is a variant played on an $n \times n$ board with the initial position of the ball at $(0, 0)$ except that now only the ball may occupy the positions $(2i, 2j)$ with both coordinates even. This eliminates "tackling", and is an extension of 1-dimensional **Oddish Phutball**, analyzed in Grossman and Nowakowski [2002]. The $(3, 2n+1)$ board (i.e. (i, j), $i = 0, 1, 2$ and $-n \le j \le n$) is already interesting and requires a different strategy from that appropriate to Oddish Phutball.

B8. Hex. (LIP, pp. 264–265) Nash's strategy stealing argument shows that Hex is a first player win but few quantitative results are known.

Garikai Campbell [2004] asks:

(1) For each n, what is the shortest path on an $n \times n$ board with which the first player can guarantee a win?

(2) What is the least number of moves in which the first player can guarantee a win?

B9 (54). Fox and Geese. Berlekamp and Siegel [17, Chapter 2] and WW pp.669–710, "analysed the game fairly completely, relying in part on results obtained using *CGSuite*." On p. 710 of WW the following open problems are given.

1. Define a position's **span** as the maximum occupied row-rank minus its minimum occupied row-rank. Then quantify and prove an assertion such as the following: If the backfield is sufficiently large, and the span is sufficiently large, and if the separation is sufficiently small, and if the Fox is neither already trapped in a daggered position along the side of the board, nor immediately about to be so trapped, then the Fox can escape and the value is **off**.

2. Show that any formation of three Geese near the centre of a very tall board has a "critical rank" with the following property: If the northern Goose is far above, and the Fox is far below, then the value of the position is either positive, HOT, or **off**, according as the northern Goose is closer, equidistant, or further from the critical rank than the Fox.

3. Welton asks what happens if the Fox is empowered to retreat like a Bishop, going back several squares at a time in a straight line? More generally, suppose his straight-line retreating moves are confined to some specific set of sizes. Does $\{1,3\}$, which maintains parity, give him more or less advantage than $\{1,2\}$?

4. What happens if the number of Geese and board widths are changed?

In Aaron Siegel's thesis there are several other questions:

5. In the critical position, with Geese at [we use the algebraic Chess notation of a, b, c, d, ... for the files and 1, 2, 3, ..., n for the ranks] (b,n), (d,n), (e,$n-1$),

(g,$n-1$), and Fox at (c,$n-1$), which has value $1 + 2^{-(n-8)}$ on an $n \times 8$ board with $n \geq 8$ in the usual game, is the value $-2n+11$ for all $n \geq 6$ when played with "Ceylonese rules"? (Fox allowed two moves at each turn.)

6. On an $n \times 4$ board with $n \geq 5$ and Geese at (b,n) and (c,$n-1$) do all Fox positions have value **over**? With the Geese on (b,n) and (d,n) are only other values 0 at (c,$n-1$) and {**over**|0} at (b,$n-2$) and (d,$n-2$)?

7. On an $n \times 6$ board with $n \geq 8$ and Geese at (b,n), (d,n) and (e,$n-1$) do the positions (a,$n-2k+1$), (c,$n-2k+1$), (e,$n-2k+1$), all have value 0, and those at (b,$n-2k$), (d,$n-2k$), (f,$n-2k$) all have value Star? And if the Geese are at (b,n), (d,n) and (f,n) are the zeroes and Stars interchanged?

B10. Hare and Hounds. Aaron Siegel asks if the positions of increasing board length shown in Figure 4, on the left, are increasingly hot, and, on the right, have arbitrarily large negative atomic weight. He also conjectures that the starting position on a $6n+5 \times 3$ board, for $n > 0$, has value

$$-(n-1) + \left\{ b, c \,|0\|0\|0\|0 \ldots \,\Big\| 0\right\}$$

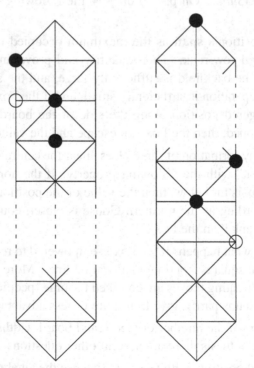

Figure 4. Sequences of Hare and Hounds positions.

where there are $2n + 4$ zeroes and slashes and

$$b = \{0, a \| 0, \{0 | \mathbf{off}\}\}, \quad c = \{0 \| \downarrow_{\to 2} * | 0 \| 0\}, \quad a = \{0, \downarrow_{\to 2} * \mid 0, \downarrow_{\to 2} *\}.$$

B11 (4). Extend the analysis of **Domineering** (WW, pp. 119–122, 138–142; LIP pp. 1–7, 260).

[Left and Right take turns to place dominoes on a checker-board. Left orients her dominoes North-South and Right orients his East-West. Each domino exactly covers two squares of the board and no two dominoes overlap. A player unable to play loses.]

See Berlekamp [1988] and the second edition of WW, 138–142, where some new values are given. For example David Wolfe and Dan Calistrate have found the values (to within '-ish', i.e., infinitesimally shifted) of 4×8, 5×6 and 6×6 boards. The value for a 5×7 board is

$$\left\{ \tfrac{3}{2} \middle| \{\tfrac{5}{4} | -\tfrac{1}{2}\}, \{\tfrac{3}{2} | -\tfrac{1}{2}, \{\tfrac{3}{2} | -1\} \middle\| -1 | -3\right\} \middle\| -1, \{\tfrac{3}{2} | -\tfrac{1}{2} \| -1\} \middle| -3\right\}.$$

Lachmann, Moore and Rapaport [2002] discovered who wins on rectangular, toroidal and cylindrical boards of widths 2, 3, 5 and 7, but do not find their values. Bullock [3, p. 84] showed that 19×4, 21×4, 14×6 and 10×8 are wins for Left and that 10×10 is a first player win.

Berlekamp notes that the value of a $2 \times n$ board, for n even, is only known to within"ish", and that there are problems on $3 \times n$ and $4 \times n$ boards that are still open.

Berlekamp asks, as a hard problem, to characterize all hot Domineering positions to within "ish". As a possibly easier problem he asks for a Domineering position with a new temperature, i.e., one not occurring in Table 1 on GONC, p. 477. Gabriel Drummond-Cole (2002) found values with temperatures between 1.5 and 2. Figure 5 shows a position of value $\pm 2*$ and temperature 2.

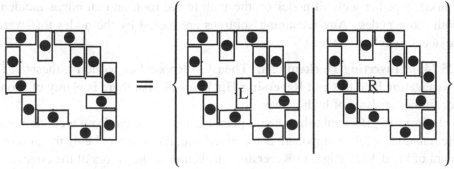

Figure 5. A Domineering position of value $\pm 2*$.

Shankar and Sridharan [2005] have found many Domineering positions with temperatures other than those shown in Table 1 on p. 477 of GONC. Blanco and Fraenkel [2] have obtained partial results for the game of Tromineering, played with trominoes in place of (or, alternatively, in addition to) dominoes.

C. Playing with pencil and paper

C1 (51). Elwyn Berlekamp asks for a complete theory of "Icelandic" $1 \times n$ **Dots-and-Boxes**, i.e., with starting position as in Figure 7.

Figure 6. Starting position for "Icelandic" $1 \times n$ Dots-and-Boxes.

See Berlekamp's book [2000] for more problems about this popular children's (and adults') game and see also WW, pp. 541–584; LIP, pp. 21–28, 260.

C2 (25). Extend the analysis of the Conway–Paterson game of **Sprouts** in either the normal or misère form. (WW, pp. 564–568).

[A move joins two spots, or a spot to itself by a curve which doesn't meet any other spot or previously drawn curve. When a curve is drawn, a new spot must be placed on it. The valence of any spot must not exceed three.]

C3 (26). Extend the analysis of **Sylver Coinage** (WW, 575–597).

[Players alternately name different positive integers, but may not name a number which is the sum of previously named ones, with repetitions allowed. Whoever names 1 loses.] Sicherman [2002] contains recent information.

C4 (28). Extend Úlehla's or Berlekamp's analysis of **von Neumann's game** from directed forests to directed acyclic graphs (WW, 570–572; Úlehla [1980]).

[Von Neumann's game, or Hackendot, is played on one or more rooted trees. The roots induce a direction, towards the root, on each edge. A move is to delete a node, together with all nodes on the path to the root, and all edges incident with those nodes. Any remaining subtrees are rooted by the nodes that were adjacent to deleted nodes.]

C5 (43). Inverting Hackenbush. Thea van Roode has written a thesis [15] investigating both this and **Reversing Hackenbush**, but there is plenty of room for further analysis of both games.

In Inverting Hackenbush, when a player deletes an edge from a component, the remainder of the component is replanted with the new root being the pruning point of the deleted edge. In Reversing Hackenbush, the colors of the edges are all changed after each deletion. Both games are hot, in contrast to Blue-Red

Hackenbush (WW, pp. 1–7; LIP, pp. 82, 88, 111–112, 212, 266) which is cold, and Green Hackenbush (WW, pp. 189–196), which is tepid.

C6 (42). Beanstalk and **Beans-Don't-Talk** are games invented respectively by John Isbell and John Conway. See Guy [1986]. Beanstalk is played between Jack and the Giant. The Giant chooses a positive integer, n_0. Then J. and G. play alternately n_1, n_2, n_3, ... according to the rule $n_{i+1} = n_i/2$ if n_i is even, $= 3n_i \pm 1$ if n_i is odd; i.e. if n_i is even, there's only one option, while if n_i is odd there are just two. The winner is the person moving to 1.

We still don't know if there are any $\mathcal{O}$-**positions** (positions of infinite remoteness).

C7 (63). The **Erdős–Szekeres game** [5] (and see Schensted [16]) was introduced by Harary, Sagan and West [1985]. From a deck of cards labelled from 1 through n, Alexander and Bridget alternately choose a card and append it to a sequence of cards. The game ends when there is an ascending subsequence of a cards or a descending subsequence of d cards.

The game appears to have a strong bias towards the first player. Albert et al., in this volume, show that for $d = 2$ and $a \le n$ the outcome is $\mathcal{N}$ or $\mathcal{P}$ according as n is odd or even, and is $\mathcal{O}$ (drawn) if $n < a$. They conjecture that for $a \ge d \ge 3$ and all sufficiently large n, it is $\mathcal{N}$ with both normal and misère play, and also with normal play when played with the rationals in place of the first n integers.

They also suggest investigating the form of the game in which players take turns naming pairs (i, π_i) subject to the constraint that the chosen values form part of the graph of some permutation of $\{1, 2, \ldots, n\}$.

D. Disturbing and destroying

D1 (27). Extend the analysis of **Chomp** (WW, 598–599, LIP 19, 46, 216).

David Gale offers \$300.00 for the solution of the infinite 3-D version where the board is the set of all triples (x, y, z) of non-negative integers, that is, the lattice points in the positive octant of $\mathbb{R}^3$. The problem is to decide whether it is a win for the first or second player.

Chomp (Gale [1974]) is equivalent to **Divisors** (Schuh [1952]). Chomp is easily solved for $2 \times n$ arrays, Sun [2002], and indeed a recent result by Steven Byrnes [2003] shows that any poset game eventually displays periodic behavior if it has two rows, and a fixed finite number of other elements. See also the Fraenkel poset games mentioned near the end of **A2**.

Thus, most of the work in recent years has been on three-rowed Chomp. The situation becomes quite complicated when a third row is added, see Zeilberger [2001] and Brouwer et al. [2005]. A novel approach (renormalization) is taken by Friedman and Landsberg in their article in this volume (see also [12]). They demonstrate that three-rowed Chomp exhibits certain scaling and self-similarity patterns similar to chaotic systems. Is there a deterministic proof that there is a unique winning move from a $3 \times n$ rectangle? The renormalization approach is based on statistical methods and has caused some controversy and so the technique seems worthy of further investigation.

Transfinite Chomp has been investigated by Huddleston and Shurman [2002]. An open question is to calculate the nim value of the position $\omega \times 4$; they conjecture it to be $\omega \cdot 2$, but it could be as low as 46, or even uncomputable! Perhaps the most fascinating open question in Transfinite Chomp is their *Stratification Conjecture*, which states that if the number of elements taken in a move is $< \omega^i$, then the change in the nim-value is also $< \omega^i$.

D2 (33). Subset Take-away. Given a finite set, players alternately choose proper subsets subject to the rule that once a subset has been chosen no proper subset can be removed. Last player to move wins.

Many people play the dual, i.e. a nonempty subset must be chosen and no proper superset of this can be chosen. We discuss this version of the game which now can be considered a poset game with the sets ordered by inclusion.

The $(n; k)$ Subset Take-away game is played using all subsets of sizes 1 through k of a n-element set. In the $(n; n)$ game one has the whole set (i.e. the set of size n) as an option, so a strategy-stealing argument shows this must be a first player win.

1. Gale and Neyman [1982], in their original paper on the game, conjectured that the winning move in the $(n; n)$ game is to remove just the whole set. This is equivalent to the statement that the $(n; n-1)$ game is a second-player win, which has been verified only for $n \leq 5$.

2. A stronger conjecture states that $(n; k)$ is a second player win if and only if $k + 1$ divides n. This was proved in the original paper only for $k = 1$ or 2.

See also Fraenkel and Scheinerman [1991].

D3 (39). Sowing or **Mancala** games. There appears to have been no advance on the papers mentioned in MGONC, although we feel that this should be a fruitful field of investigation at several different levels.

D4. Annihilation games. k-**Annihilation.** Initially place tokens on some of the vertices of a finite digraph. Denote by $\rho_{\text{out}}(u)$ the outvalence of a vertex u. A move consists of removing a token from some vertex u, and "complementing" $t := \min(k, \rho_{\text{out}}(u))$ (immediate) followers of u, say $v_1, \ldots, v_t$: if there is a token on v_j, remove it; if there is no token there, put one on it. The player making the last move wins. If there is no last move, the outcome is a draw. For $k = 1$, there is an $O(n^6)$ algorithm for deciding whether any given position is in $\mathcal{P}$, $\mathcal{N}$, or $\mathcal{O}$; and for computing an optimal next move in the last 2 cases (Fraenkel and Yesha [1982]). Fraenkel asks: Is there a polynomial algorithm for $k > 1$? For an application of k-annihilation games to lexicodes, see Fraenkel and Rahat [2003].

D5. Toppling dominoes (LIP, pp. 110–112, 274) is played with a row of vertical dominoes each of which is either blue or red. A player topples one of his/her dominoes to the left or to the right.

David Wolfe asks if all dyadic rationals occur as a unique single row of dominoes and if that row is always palindromic (symmetrical).

D6. Hanoi Stick-up is played with the disks of the Towers of Hanoi puzzle, starting with each disk in a separate stack. A move is to place one stack on top of another such that the size of the bottom of the first stack is less than the size of the top of the second; the two stacks then fuse (and) into one. The only relevant information about a stack are its top and bottom sizes, and it's often possible to collapse the labelling of positions: so for instance, starting with 8 disks and fusing 1and7 and 2and5

> we have stacks   0   1 and 7   2 and 5   3   4   6
> which can be relabelled 0   1 and 3   1 and 2   1   2   3

in which the legal moves are still the same. John Conway, Alex Fink and others have found that the $\mathcal{P}$-positions of height ≤ 3 in normal Hanoi Stickup are exactly those which, after collapsing, are of the form $0^a \, 01^b \, 1^c \, 12^d \, 2^e$ with $\min(a + b + c, c + d + e, a + e)$ even, except that if $a + e \leq a + b + c$ and $a + e \leq c + d + e$ then both a and e must be even (02 can't be involved in a legal move so can be dropped).

They also found the normal and misère outcomes of all positions with up to six stacks, but there is more to be discovered.

D7 (56). Are there any draws in **Beggar-my-Neighbor** ? Marc Paulhus showed that there are no cycles when using a half-deck of two suits, but the problem for the whole deck (one of Conway's "anti-Hilbert" problems) is still open.

E. Theory of games

E1 (49). Fraenkel updates Berlekamp's earlier questions on computational complexity as follows:

Demaine, Demaine and Eppstein [2002] proved that deciding whether a player can win in a *single* move in Phutball (WW, pp. 752–755; LIP, p. 212) is NP-complete. Grossman and Nowakowski [2002] gave constructive partial strategies for 1-dimensional Phutball. Thus, these papers do not show that Phutball has the required properties.

Perhaps Nimania (Fraenkel and Nešetřil [1985]) and Multivision (Fraenkel [1998]) satisfy the requirements. Nimania begins with a single positive integer, but after a while there is a multiset of positive integers on the table. At move k, a copy of an existing integer m is selected, and 1 is subtracted from it. If $m = 1$, the copy is deleted. Otherwise, k copies of $m - 1$ are adjoined to the copy $m - 1$. The player first unable to move loses and the opponent wins. It was proved: (i) The game terminates. (ii) Player I can win. In Fraenkel, Loebl and Nešetřil [1988], it was shown that the max number of moves in Nimania is an Ackermann function, and the min number satisfies $2^{2^{n-2}} \leq \text{Min}(n) \leq 2^{2^{n-1}}$.

The game is thus intractable simply because of the length of its play. This is a *provable* intractability, much stronger than NP-hardness, which is normally only a *conditional* intractability. One of the requirements for the tractability of a game is that a winner can consummate a win in at most $O(c^n)$ moves, where $c > 1$ is a constant, and n a sufficiently succinct encoding of the input (this much is needed for nim on 2 equal heaps of size n).

To consummate a win in Nimania, player I can play randomly most of the time, but near the end of play, a winning strategy is needed, given explicitly. Whether or not this is an intricate solution depends on the beholder. But it seems that it's of even greater interest to construct a game with a very *simple* strategy which still has high complexity!

Also every play of Multivision terminates, the winner can be determined in linear time, and the winning moves can be computed linearly. But the length of play can be arbitrarily long. So let's ask the following: Is there a game which has

1. simple, playable rules,
2. a simple explicit strategy,
3. length of play at most exponential; and
4. is NP-hard or harder.

Theorem [Tung 1987]. Given a polynomial $P(x, y) \in \mathbb{Z}[x, y]$, the problem of deciding whether $\forall x \exists y [P(x, y) = 0]$ holds over $\mathbb{Z}_{\geq 0}$, is co-NP-complete.

Define the following game of length 2: player I picks $x \in \mathbb{Z}_{\geq 0}$, player II picks $y \in \mathbb{Z}_{\geq 0}$. Player I wins if $P(x, y) \neq 0$, otherwise player II wins. For winning, player II has only to compute y such that $P(x, y) = 0$, given x, and there are many algorithms for doing so.

Also Jones and Fraenkel [1995] produced games, with small length of play, which satisfy these conditions.

So we are led to the following reformulation of Berlekamp's question: Is there a game which has

1. simple, playable rules,

2. a finite set of options at every move,
3. a simple explicit strategy,
4. length of play at most exponential;
5. and is NP-hard or harder.

E2. Complexity closure. Aviezri Fraenkel asks: Are there partizan games G_1, G_2, G_3 such that: (i) G_1, G_2, G_3, $G_1 + G_2$, $G_2 + G_3$ and all their options have polynomial-time strategies, (ii) $G_1 + G_3$ is NP-hard?

E3. Sums of switch games. David Wolfe considers a sum of games G, each of the form $a\|b|c$ or $a|b\|c$ where a, b, and c are integers specified in unary. Is there a polynomial time algorithm to determine who wins in G, or is the problem NP-hard?

E4 (52). How does one play **sums** of games with varied overheating operators?

Sentestrat and Top-down thermography (LIP, p. 214):

David Wolfe would like to see a formal proof that sentestrat works, an algorithm for top-down thermography, and conditions for which top-down thermography is computationally efficient.

Aaron Siegel asks the following generalized thermography questions.

(1) Show that the Left scaffold of a dogmatic (neutral ko-threat environment; LIP, p. 215) thermograph is decreasing as function of t. (Note, this is NOT true for komaster thermographs.) [Dogmatic thermography was invented by Berlekamp and Spight. See [19] for a good introduction.]

(2) Develop the machinery for computing dogmatic thermographs of double kos (multiple alternating 2-cycles joined at a single node).

In the same vein as (2):

(3) Develop a temperature theory that applies to all loopy games.

Siegel thinks that (3) is among the most important open problems in combinatorial game theory. The temperature theory of Go appears radically different from the classical combinatorial theory of loopy games (where infinite plays are draws). It would be a huge step forward if these could be reconciled into a "grand unified temperature theory". Problem (2) seems to be the obvious next step toward (3).

Conway asks for a natural set of conditions under which the mapping $G \mapsto \int^* G$ is the *unique* homomorphism that annihilates all infinitesimals.

E5. Loopy games (WW, pp. 334–377; LIP, pp. 213–214) are partizan games that do not satisfy the ending condition. A **stopper** is a game that, when played on its own, has no ultimately alternating, Left and Right, infinite sequence of legal moves. Aaron Siegel reminds us of WW, 2nd ed., p.369, where the authors tried hard to prove that every loopy game had stoppers, until Clive Bach found the Carousel counterexample. Is there an alternative notion of simplest form that works for *all* finite loopy games, and, in particular, for the Carousel? The simplest form theorem for stoppers is at WW, p.351.

Siegel conjectures that, if Q is an arbitrary cycle of Left and Right moves that contains at least two moves for each player, and is not strictly alternating, then there is a stopper consisting of a single cycle that matches Q, together with various exits to enders, i.e., games which end in a finite, though possibly unbounded, number of moves. [Note that games normally have Left and Right playing alternately, but if the game is a sum, then play in one component can have arbitrary sequences of Left and Right moves, not just alternating ones.]

A long cycle is *tame* if it alternates just once between Left and Right, otherwise it is *wild*. Aaron Siegel writes:

> I can produce wild cycles "in the laboratory," by specifying their game graphs explicitly. So the question is to detect one "in nature", i.e., in an actual game with (reasonably) playable rules such as Phutball [Problem **B7**].

Siegel also asks under what conditions does a given infinitesimal have a well-defined atomic weight, and asks to specify an algorithm to calculate the atomic weight of an infinitesimal stopper g. The algorithm should succeed whenever the atomic weight is well-defined, i.e., whenever g can be sandwiched between loopfree all-smalls of equal atomic weight.

E6 (45). Elwyn Berlekamp asks for the **habitat** of $*2$, where $*2 = \{0, *|0, *\}$. Gabriel Drummond-Cole [2005] has found Domineering positions with this value. See, for example, Figure 7, which also shows a Go position, found by Nakamura and Berlekamp [2003], whose chilled value is $*2$. The Black and White groups are both connected to life via unshown connections emanating upwards from the second row. Either player can move to $*$ by placing a stone at E2, or to 0 by going to E1.

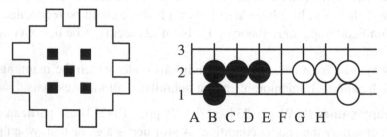

Figure 7. A Domineering position and a Chilled Go position of value $*2$.

E7. Partial ordering of games. David Wolfe lets $g(n)$ be the number of games born by day n, notes that an upper bound is given by $g(n+1) \le g(n) + 2^{g(n)} + 2$, and a lower bound for each $\alpha < 0$ is given by $g(n+1) \geq 2^{g(n)^{\alpha}}$, for n sufficiently large, and asks us to tighten these bounds.

He also asks what group is generated by the all-small games (or — much harder — of all games) born by day 3. Describe the partial order of games born by day 3, identifying all the largest "hypercubes" (Boolean sublattices) and how they are interconnected. These questions have been answered for day 2; see Wolfe's article "On day n" in this volume.

Berlekamp suggests other possible definitions for games born by day n, $\mathcal{G}_n$, depending on how one defines $\mathcal{G}_0$. Our definition is 0-based, as $\mathcal{G}_0 = \{0\}$. Other natural definitions are integer-based (where $\mathcal{G}_0$ are integers) or number-based. These two alternatives do not form a lattice, for if G_1 and G_2 are born by day k, then the games

$$H_n := \left\{ G_1, G_2 \;\middle\|\; G_1, \{G_2|-n\} \right\}$$

form a decreasing sequence of games born by day $k+2$ exceeding any game $G \geq G_1, G_2$, and the day $k+2$ join of G_1 and G_2 cannot exist. What is the structure of the partial order given by one of these alternative definitions of birthday?

The set of all short games does not form a lattice, but Calistrate, Paulhus and Wolfe [2002] have shown that the games born by day n form a distributive lattice $\mathcal{L}_n$ under the usual partial order. They ask for a description of the exact structure of $\mathcal{L}_3$. Siegel describes $\mathcal{L}_4$ as "truly gigantic and exceedingly difficult to penetrate" but suggests that it may be possible to find its dimension and the maximum **longitude**, $\mathrm{long}_4(G)$, of a game in $\mathcal{L}_4$, which he defines as

$$\mathrm{long}_n(G) = \mathrm{rank}_n(G \vee G^{\bullet}) - \mathrm{rank}_n(G)$$

where $\mathrm{rank}_n(G)$ is the rank of G in $\mathcal{L}_n$ and $G^{\bullet}$ is the **companion** of G,

$$G^{\bullet} = \begin{cases} * & \text{if } G = 0 \\ \{0, (G^L)^{\bullet} \mid (G^R)^{\bullet}\} & \text{if } G > 0 \\ \{(G^L)^{\bullet} \mid 0, (G^R)^{\bullet}\} & \text{if } G < 0 \\ \{(G^L)^{\bullet} \mid (G^R)^{\bullet}\} & \text{if } G \parallel 0 \end{cases}$$

The set of all-small games does not form a lattice, but Siegel forms a lattice $\mathcal{L}_n^0$ by adjoining least and greatest elements $\triangle$ and $\triangledown$ and asks: do the elements of $\mathcal{L}_n^0$ have an intrinsic "handedness" that distinguishes, say, $(n-1)\uparrow$ from $(n-1)\uparrow + *$?

E8. Aaron Siegel asks, given a group or monoid, $\mathcal{K}$, of games, to specify a technique for calculating the simplest game in each $\mathcal{K}$-equivalence class. He notes that some restriction on $\mathcal{K}$ might be needed; for example, $\mathcal{K}$ might be the monoid of games absorbed by a given idempotent.

E9. Siegel also would like to investigate how search methods might be integrated with a canonical-form engine.

E10 (9). Develop a **misère theory** for unions of partizan games (WW, p. 312).

E11. Four-outcome-games. Guy has given a brute force analysis of a parity subtraction game [9] which didn't allow the use of Sprague–Grundy theory because it wasn't impartial, nor the Conway theory, because it was not last-player-winning. Is there a class of games in which there are four outcomes, $\mathcal{N}$ext, $\mathcal{P}$revious, $\mathcal{L}$eft and $\mathcal{R}$ight, and for which a general theory can be given?

Acknowledgement

We have had help in compiling this collection from all those mentioned, and from others. We would especially like to mention Elwyn Berlekamp, Aviezri Fraenkel, Thane Plambeck, Aaron Siegel and David Wolfe. All mistakes are deliberate and designed to keep the reader alert.

References

[those not listed here may be found in Fraenkel's Bibliography]

[1] M. H. Albert, J. P. Grossman, S. McCurdy, R. J. Nowakowski and D. Wolfe, Cherries, preprint, 2005. [Problem **A5**]

[2] Saúl A. Blanco and Aviezri S. Fraenkel, Tromineering, Tridomineering and L-Tridomineering, August 2006 preprint. [Problem **B11**]

[3] N. Bullock, Domineering: Solving large combinatorial search spaces, *ICGA J.*, **25**(2002) 67–85; also MSc thesis, Univ. of Alberta, 2002. [Problem **B11**]

[4] N. Comeau, J. Cullis, R. J. Nowakowski and J. Paek, personal communication (class project). [Problem **A12**]

[5] Paul Erdős and George Szekeres, A combinatorial problem in geometry, *Compositio Math.*, **2**(1935) 464–470; *Zbl* **12** 270–271. [Problem **C7**]

[6] P. A. Grillet, *Commutative Semigroups*, *Advances in Mathematics*, **2**, Springer 2001. [Problem **A15**]

[7] J. P. Grossman and R. J. Nowakowski, A ruler regularity in hexadecimal games, preprint 2005. [Problem **A3**]

[8] Vladimir Gurvich, On the misère version of game Euclid and miserable games, *Discrete Math.*, (to appear). [Problem **A9**]

[9] Richard Guy, A parity subtraction game, *Crux Math.*, **33**(2007) (to appear) [Problem **E11**]

[10] Sarah McCurdy, Two Combinatorial Games, MSc thesis, Dalhousie Univ., 2004. [Problem **A5**]

[11] M. Müller. Solving 5 × 5 Amazons. In *The 6th Game Programming Workshop 2001*, **14** in IPSJ Symposium Series Vol.2001, pp. 64–71, Hakone, Japan, 2001. [Problem **B6**]

[12] Ivars Peterson, Chaotic Chomp, the mathematics of crystal growth sheds light on a tantalizing game, *Science News*, **170** (2006-07-22) 58–60. [Problem **D1**]

[13] Thane E. Plambeck and Aaron N. Siegel, Misère quotients of impartial games, *J. Combin. Theory, Ser. A* (submitted). [Problem **A15**]

[14] L. Rédei, *The Theory of Finitely Generated Commutative Semigroups*, Pergamon, 1965. [Problem **A15**]

[15] Thea van Roode, *Partizan Forms of Hackenbush*, MSc. thesis, The University of Calgary, 2002. [Problem **C5**]

[16] C. Schensted, Longest increasing and decreasing subsequences, *Canad. J. Math.*, **13**(1961) 179–191; *MR* **22** #12047. [Problem **C7**]

[17] Aaron Nathan Siegel, *Loopy Games and Computation*, PhD dissertation, Univ. of California, Berkeley, Spring 2005. [passim]

[18] Angela Siegel, Finite excluded subtraction sets and Infinite Geography, MSc thesis, Dalhousie Univ., 2005. [Problem **A1**]

[19] W. L. Spight, Evaluating kos in a neutral threat environment: Preliminary results. In J. Schaeffer, M. Müller and Y. Björnsson, editors, *Computers and Games: Third Internat. Conf., CG'02*, Lect Notes Comput. Sci., **2883** Springer, Berlin, 2003, pp.413–428. [Problem **E4**]

RICHARD K. GUY
DEPARTMENT OF MATHEMATICS AND STATISTICS
THE UNIVERSITY OF CALGARY
CALGARY, ALBERTA
CANADA T2N 1N4
rkg@cpsc.ucalgary.ca

RICHARD J. NOWAKOWSKI
DEPARTMENT OF MATHEMATICS AND STATISTICS
DALHOUSIE UNIVERSITY
HALIFAX, NS
CANADA B3H 3J5
rjn@mathstat.dal.ca

Games of No Chance 3
MSRI Publications
Volume 56, 2009

Combinatorial Games: selected bibliography with a succinct gourmet introduction

AVIEZRI S. FRAENKEL

1. What are combinatorial games?

Roughly speaking, the family of *combinatorial games* consists of two-player games with perfect information (no hidden information as in some card games), no chance moves (no dice) and outcome restricted to (lose, win), (tie, tie) and (draw, draw) for the two players who move alternately. Tie is an end position such as in tic-tac-toe, where no player wins, whereas draw is a dynamic tie: any position from which a player has a nonlosing move, but cannot force a win. Both the easy game of Nim and the seemingly difficult chess are examples of combinatorial games. And so is go. The shorter terminology *game*, *games* is used below to designate combinatorial games.

2. Why are games intriguing and tempting?

Amusing oneself with games may sound like a frivolous occupation. But the fact is that the bulk of interesting and natural mathematical problems that are hardest in complexity classes beyond NP, such as Pspace, Exptime and Expspace, are two-player games; occasionally even one-player games (puzzles) or even zero-player games (Conway's "Life"). Some of the reasons for the high complexity of two-player games are outlined in the next section. Before that we note that in addition to a natural appeal of the subject, there are applications or connections to various areas, including complexity, logic, graph and matroid theory, networks, error-correcting codes, surreal numbers, on-line algorithms, biology — and analysis and design of mathematical and commercial games!

But when the chips are down, it is this "natural appeal" that lures both amateurs and professionals to become addicted to the subject. What is the essence of this appeal? Perhaps the urge to play games is rooted in our primal beastly instincts; the desire to corner, torture, or at least dominate our peers. A common expression of these vile cravings is found in the passions roused by local, national and international tournaments. An intellectually refined version of these dark desires, well hidden beneath the façade of scientific research, is the consuming drive "to beat them all", to be more clever than the most clever, in short — to create the tools to *Math*ter them all in hot *comb*inatorial *comb*at! Reaching this goal is particularly satisfying and sweet in the context of combinatorial games, in view of their inherent high complexity.

With a slant towards artificial intelligence, Pearl wrote that games "offer a perfect laboratory for studying complex problem-solving methodologies. With a few parsimonious rules, one can create complex situations that require no less insight, creativity, and expertise than problems actually encountered in areas such as business, government, scientific, legal, and others. Moreover, unlike these applied areas, games offer an arena in which computerized decisions can be evaluated by absolute standards of performance and in which proven human experts are both available and willing to work towards the goal of seeing their expertise emulated by a machine. Last, but not least, games possess addictive entertaining qualities of a very general appeal. That helps maintain a steady influx of research talents into the field and renders games a convenient media for communicating powerful ideas about general methods of strategic planning."

To further explore the nature of games, we consider, informally, two subclasses.

(i) Games People Play (*playgames*): games that are challenging to the point that people will purchase them and play them.
(ii) Games Mathematicians Play (*mathgames*): games that are challenging to mathematicians or other scientists to play with and ponder about, but not necessarily to "the man in the street".

Examples of playgames are chess, go, hex, reversi; of mathgames: Nim-type games, Wythoff games, annihilation games, octal games.

Some "rule of thumb" properties, which seem to hold for the majority of playgames and mathgames are listed below.

I. Complexity. Both playgames and mathgames tend to be computationally intractable. There are a few tractable mathgames, such as Nim, but most games still live in *Wonderland*: we are wondering about their as yet unknown complexity. Roughly speaking, however, NP-hardness is a necessary but not a sufficient condition for being a playgame! (Some games on Boolean formulas are Exptime-complete, yet none of them seems to have the potential of commercial marketability.)

II. Boardfeel. None of us may know an exact strategy from a midgame position of chess, but even a novice, merely by looking at the board, gets some feel who of the two players is in a stronger position – even what a strong or weak next move is. This is what we loosely call *boardfeel*. Our informal definition of playgames and mathgames suggests that the former do have a boardfeel, whereas the latter don't. For many mathgames, such as Nim, a player without prior knowledge of the strategy has no inkling whether any given position is "strong" or "weak" for a player. Even when defeat is imminent, only one or two moves away, the player sustaining it may be in the dark about the outcome, which will stump him. The player has no boardfeel. (Even many mathgames, including Nim-type games, can be played, equivalently, on a board.)

Thus, in the boardfeel sense, simple games are complex and complex games are simple! This paradoxical property also doesn't seem to have an analog in the realm of decision problems. The boardfeel is the main ingredient which makes PlayGames interesting to play.

III. Math Appeal. Playgames, in addition to being interesting to play, also have considerable mathematical appeal. This has been exposed recently by the theory of partizan games established by Conway and applied to endgames of go by Berlekamp, students and associates. On the other hand, mathgames have their own special combinatorial appeal, of a somewhat different flavor. They appeal to and are created by mathematicians of various disciplines, who find special intellectual challenges in analyzing them. As Peter Winkler called a subset of them: "games people don't play". We might also call them, in a more positive vein, "games mathematicians play". Both classes of games have applications to areas outside game theory. Examples: surreal numbers (playgames), error correcting codes (mathgames). Both provide enlightenment through bewilderment, as David Wolfe and Tom Rodgers put it.

IV. Existence. There are relatively few successful playgames around. It seems to be hard to invent a playgame that catches the masses. In contrast, mathgames abound. They appeal to a large subclass of mathematicians and other scientists, who cherish producing them and pondering about them. The large proportion of mathgames-papers in the games bibliography below reflects this phenomenon.

We conclude, inter alia, that for playgames, high complexity is desirable. Whereas in all respectable walks of life we strive towards solutions or at least approximate solutions which are polynomial, there are two less respectable human activities in which high complexity is appreciated. These are cryptography (covert warfare) and games (overt warfare). The desirability of high complexity in cryptography — at least for the encryptor! — is clear. We claim that it is also desirable for playgames.

It's no accident that games and cryptography team up: in both there are adversaries, who pit their wits against each other! But games are, in general, considerably harder than cryptography. For the latter, the problem whether the designer of a cryptosystem has a safe system can be expressed with two quantifiers only: $\exists$ a cryptosystem such that $\forall$ attacks on it, the cryptosystem remains unbroken? In contrast, the decision problem whether White can win if White moves first in a chess game, has the form: "$\exists\forall\exists\forall\cdots$ move: White wins?", expressing the question whether White has an opening winning move — with an unbounded number of alternating quantifiers. This makes games the more challenging and fascinating of the two, besides being fun! See also the next section.

Thus, it's no surprise that the skill of playing games, such as checkers, chess, or go has long been regarded as a distinctive mark of human intelligence.

3. Why are combinatorial games hard?

Existential decision problems, such as graph hamiltonicity and Traveling Salesperson (Is there a round tour through specified cities of cost $\leq C$?), involve a single existential quantifier ("Is there...?"). In mathematical terms an existential problem boils down to finding a path — sometimes even just verifying its existence — in a large "decision-tree" of all possibilities, that satisfies specified properties. The above two problems, as well as thousands of other interesting and important combinatorial-type problems are NP-*complete*. This means that they are *conditionally intractable*, i.e., the best way to solve them seems to require traversal of most if not all of the decision tree, whose size is exponential in the input size of the problem. No essentially better method is known to date at any rate, and, roughly speaking, if an efficient solution will ever be found for any NP-complete problem, then all NP-complete problems will be solvable efficiently.

The decision problem whether White can win if White moves first in a chess game, on the other hand, has the form: Is there a move of White such that for *every* move of Black there is a move of White such that for *every* move of Black there is a move of White ... such that White can win? Here we have a large number of alternating existential and universal quantifiers rather than a single existential one. We are looking for an entire subtree rather than just a path in the decision tree. Because of this, most nonpolynomial games are at least *P*space-hard. The problem for generalized chess on an $n \times n$ board, and even for a number of seemingly simpler mathgames, is, in fact, Exptime-complete, which is a *provable intractability*.

Put in simple language, in analyzing an instance of Traveling Salesperson, the problem itself is passive: it does not resist your attempt to attack it, yet it is difficult. In a game, in contrast, there is your opponent, who, at every step, attempts to foil your effort to win. It's similar to the difference between an autopsy and surgery. Einstein, contemplating the nature of physics said, "Der Allmächtige ist nicht boshaft; Er ist raffiniert" (The Almighty is not mean; He is sophisticated). NP-complete existential problems are perhaps sophisticated. But your opponent in a game can be very mean!

Another manifestation of the high complexity of games is associated with a most basic tool of a game : its *game-graph*. It is a directed graph G whose vertices are the positions of the game, and (u, v) is an edge if and only if there is a move from position u to position v. Since every combination of tokens in the given game is a *single* vertex in G, the latter has normally exponential size. This holds, in particular, for both Nim and chess. Analyzing a game means reasoning about its game-graph. We are thus faced with a problem that is *a priori* exponential, quite unlike many present-day interesting existential problems.

A fundamental notion is the *sum* (disjunctive compound) of games. A sum is a finite collection of disjoint games; often very basic, simple games. Each of the two players, at every turn, selects one of the games and makes a move in it. If the outcome is not a draw, the sum-game ends when there is no move left in any of the component games. If the outcome is not a tie either, then in *normal* play, the player first unable to move loses and the opponent wins. The outcome is reversed in *misère* play.

If a game decomposes into a *disjoint* sum of its components, either from the beginning (Nim) or after a while (domineering), the potential for its tractability increases despite the exponential size of the game graph. As Elwyn Berlekamp remarked, the situation is similar to that in other scientific endeavors, where we often attempt to decompose a given system into its functional components. This approach may yield improved insights into hardware, software or biological systems, human organizations, and abstract mathematical objects such as groups.

If a game doesn't decompose into a sum of disjoint components, it is more likely to be intractable (Geography or Poset Games). Intermediate cases happen when the components are not quite fixed (which explains why misère play of sums of games is much harder than normal play) or not quite disjoint (Welter). Thane Plambeck has recently made progress with misère play, and we will be hearing more about this shortly.

The hardness of games is eased somewhat by the efficient freeware package "Combinatorial Game Suite", courtesy of Aaron Siegel.

4. Breaking the rules

As the experts know, some of the most exciting games are obtained by breaking some of the rules for combinatorial games, such as permitting a player to pass a bounded or unbounded number of times, i.e., relaxing the requirement that players play alternately; or permitting a number of players other than two.

But permitting a payoff function other than $(0,1)$ for the outcome (lose, win) and a payoff of $(\frac{1}{2}, \frac{1}{2})$ for either (tie, tie) or (draw, draw) usually, but not always, leads to games that are not considered to be combinatorial games; or to borderline cases.

5. Why is the bibliography vast?

In the realm of existential problems, such as sorting or Traveling Salesperson, most present-day interesting decision problems can be classified into tractable, conditionally intractable, and provably intractable ones. There are exceptions, to be sure, such as graph isomorphism, whose complexity is still unknown. But the exceptions are few. In contrast, most games are still in Wonderland, as pointed out in § 2(I) above. Only a few games have been classified into the complexity classes they belong to. Despite recent impressive progress, the tools for reducing Wonderland are still few and inadequate.

To give an example, many interesting games have a very succinct input size, so a polynomial strategy is often more difficult to come by (Richard Guy and Cedric Smith's octal games; Grundy's game). Succinctness and non-disjointness of games in a sum may be present simultaneously (Poset games). In general, the alternating quantifiers, and, to a smaller measure, "breaking the rules", add to the volume of Wonderland. We suspect that the large size of Wonderland, a fact of independent interest, is the main contributing factor to the bulk of the bibliography on games.

6. Why isn't it larger?

The bibliography below is a *partial* list of books and articles on combinatorial games and related material. It is partial not only because I constantly learn of additional relevant material I did not know about previously, but also because of certain self-imposed restrictions. The most important of these is that only papers with some original and nontrivial mathematical content are considered. This excludes most historical reviews of games and most, but not all, of the work on heuristic or artificial intelligence approaches to games, especially the large literature concerning computer chess. I have, however, included the compendium Levy [1988], which, with its 50 articles and extensive bibliography, can serve as a first guide to this world. Also some papers on chess-endgames and clever exhaustive computer searches of some games have been included.

On the other hand, papers on games that break some of the rules of combinatorial games are included liberally, as long as they are interesting and retain a combinatorial flavor. These are vague and hard to define criteria, yet combinatorialists usually recognize a combinatorial game when they see it. Besides, it is interesting to break also this rule sometimes! We have included some references to one-player games, e.g., towers of Hanoi, n-queen problems, 15-puzzle and peg-solitaire, but only few zero-player games (such as Life and games on "sand piles"). We have also included papers on various applications of games, especially when the connection to games is substantial or the application is interesting or important.

High-class meetings on combinatorial games, as in Columbus, OH (1990), at MSRI (1994, 2000, 2005), at BIRS (2005) resulted in books, or a special issue of a journal — for the Dagstuhl seminar (2002). During 1990–2001, *Theoretical Computer Science* ran a special Mathematical Games Section whose main purpose was to publish papers on combinatorial games. TCS still solicits papers on games. In 2002, *Integers — Electronic J. of Combinatorial Number Theory* began publishing a Combinatorial Games Section. The combinatorial games community is growing in quantity and quality!

7. The dynamics of the literature

The game bibliography below is very dynamic in nature. Previous versions have been circulated to colleagues, intermittently, since the early 1980's. Prior to every mailing updates were prepared, and usually also afterwards, as a result of the comments received from several correspondents. The listing can never be "complete". Thus also the present form of the bibliography is by no means complete.

Because of its dynamic nature, it is natural that the bibliography became a "dynamic survey" in the Dynamic Surveys (DS) section of the *Electronic Journal of Combinatorics* (ElJC) and *The World Combinatorics Exchange* (WCE). The ElJC and WCE are on the World Wide Web (WWW), and the DS can be accessed at http://www.combinatorics.org (click on "Surveys"). The ElJC has mirrors at various locations. Furthermore, the European Mathematical Information Service (EMIS) mirrors this Journal, as do all of its mirror sites (currently over forty of them). See http://www.emis.de/tech/mirrors.html.

8. An appeal

I ask readers to continue sending to me corrections and comments; and inform me of significant omissions, remembering, however, that it is a *selected* bibliography. I prefer to get reprints, preprints or URLs, rather than only titles — whenever possible.

Material on games is mushrooming on the Web. The URLs can be located using a standard search engine, such as Google.

9. Idiosyncrasies

Most of the bibliographic entries refer to items written in English, though there is a sprinkling of Danish, Dutch, French, German, Japanese, Slovakian and Russian, as well as some English translations from Russian. The predominance of English may be due to certain prejudices, but it also reflects the fact that nowadays the *lingua franca* of science is English. In any case, I'm soliciting also papers in languages other than English, especially if accompanied by an abstract in English.

On the administrative side, technical reports, submitted papers and unpublished theses have normally been excluded; but some exceptions have been made. Abbreviations of book series and journal names usually follow the *Math Reviews* conventions. Another convention is that de Bruijn appears under D, not B; von Neumann under V, not N, McIntyre under M not I, etc.

Earlier versions of this bibliography have appeared, under the title "Selected bibliography on combinatorial games and some related material", as the master bibliography for the book *Combinatorial Games*, AMS Short Course Lecture Notes, Summer 1990, Ohio State University, Columbus, OH, *Proc. Symp. Appl. Math.* **43** (R. K. Guy, ed.), AMS 1991, pp. 191–226 with 400 items, and in the *Dynamic Surveys* section of the *Electronic J. of Combinatorics* in November 1994 with 542 items (updated there at odd times). It also appeared as the master bibliography in *Games of No Chance*, Proc. MSRI Workshop on Combinatorial Games, July, 1994, Berkeley, CA (R. J. Nowakowski, ed.), MSRI Publ. Vol. 29, Cambridge University Press, Cambridge, 1996, pp. 493–537, under the present title, containing 666 items. The version published in the palindromic year 2002 contained the palindromic number 919 of references. It constituted a growth of 38%. It appeared in ElJC and as the master bibliography in *More Games of No Chance*, Proc. MSRI Workshop on Combinatorial Games, July, 2000, Berkeley, CA (R. J. Nowakowski, ed.), MSRI Publ. Vol. 42, Cambridge University Press, Cambridge, pp. 475-535. The current update (mid-2003), in ElJC, contains 1001 items, another palindrome.

10. Acknowledgments

Many people have suggested additions to the bibliography, or contributed to it in other ways. Ilan Vardi distilled my *Math-master* (§ 2) into *Math*ter. Among those that contributed more than two or three items are: Akeo Adachi, Ingo Althöfer, Thomas Andreae, Eli Bachmupsky, Adriano Barlotti, József Beck, the late Claude Berge, Gerald E. Bergum, H. S. MacDonald Coxeter, Thomas S. Ferguson, James A. Flanigan, Fred Galvin, Martin Gardner, Alan J. Goldman, Solomon W. Golomb, Richard K. Guy, Shigeki Iwata, David S. Johnson, Victor Klee, Donald E. Knuth, Anton Kotzig, Jeff C. Lagarias, Michel Las Vergnas, Hendrik W. Lenstra, Hermann Loimer, F. Lockwood Morris, Richard J. Nowakowski, Judea Pearl, J. Michael Robson, David Singmaster, Wolfgang Slany, Cedric A. B. Smith, Rastislaw Telgársky, Mark D. Ward, Yōhei Yamasaki and others. Thanks to all and keep up the game! Special thanks to Mark Ward who went through the entire file with a fine comb in late 2005, when it contained 1,151 items, correcting errors and typos. Many thanks also to various anonymous helpers who assisted with the initial TEX file, to Silvio Levy, who has edited and transformed it into LATEX2e in 1996, and to Wolfgang Slany, who has transformed it into a BIBTeX file at the end of the previous millenium, and solved a "new millenium" problem encountered when the bibliography grew beyond 999 items. Keen users of the bibliography will notice that there is a beginning of MR references, due to Richard Guy's suggestion, facilitated by his student Alex Fink.

11. The bibliography

1. S. Abbasi and N. Sheikh [2007], Some hardness results for question/answer games, *Integers, Electr. J of Combinat. Number Theory* **7**, #G08, 29 pp., MR2342186. http://www.integers-ejcnt.org/vol7.html

2. B. Abramson and M. Yung [1989], Divide and conquer under global constraints: a solution to the n-queens problem, *J. Parallel Distrib. Comput.* **6**, 649–662.

3. A. Adachi, S. Iwata and T. Kasai [1981], Low level complexity for combinatorial games, *Proc. 13th Ann. ACM Symp. Theory of Computing (Milwaukee, WI, 1981)*, Assoc. Comput. Mach., New York, NY, pp. 228–237.

4. A. Adachi, S. Iwata and T. Kasai [1984], Some combinatorial game problems require $\Omega(n^k)$ time, *J. Assoc. Comput. Mach.* **31**, 361–376.

5. H. Adachi, H. Kamekawa and S. Iwata [1987], Shogi on $n \times n$ board is complete in exponential time, *Trans. IEICE* **J70-D**, 1843–1852 (in Japanese).

6. E. W. Adams and D. C. Benson [1956], Nim-Type Games, *Technical Report No. 31*, Department of Mathematics, Pittsburgh, PA.

7. W. Ahrens [1910], *Mathematische Unterhaltungen und Spiele*, Vol. I, Teubner, Leipzig, Zweite vermehrte und verbesserte Auflage. (There are further editions and related game-books of Ahrens).

8. O. Aichholzer, D. Bremmer, E. D. Demaine, F. Hurtado, E. Kranakis, H. Krasser, S. Ramaswami, S. Sethia and J. Urrutia [2005], Games on triangulations, *Theoret. Comput. Sci.* **259**, 42–71, special issue: Game Theory Meets Theoretical Computer Science, MR2168844 (2006d:91037).

9. S. Aida, M. Crasmaru, K. Regan and O. Watanabe [2004], Games with uniqueness properties, *Theory Comput. Syst.* **37**, 29–47, Symposium on Theoretical Aspects of Computer Science (Antibes-Juan les Pins, 2002), MR2038401 (2004m:68055).

10. M. Aigner [1995], Ulams Millionenspiel, *Math. Semesterber.* **42**, 71–80.

11. M. Aigner [1996], Searching with lies, *J. Combin. Theory* (Ser. A) **74**, 43–56.

12. M. Aigner and M. Fromme [1984], A game of cops and robbers, *Discrete Appl. Math.* **8**, 1–11.

13. M. Ajtai, L. Csirmaz and Z. Nagy [1979], On a generalization of the game Go-Moku I, *Studia Sci. Math. Hungar.* **14**, 209–226.

14. E. Akin and M. Davis [1985], Bulgarian solitaire, *Amer. Math. Monthly* **92**, 237–250, MR786523 (86m:05014).

15. M. H. Albert, R. E. L. Aldred, M. D. Atkinson, C. C. Handley, D. A. Holton, D. J. McCaughan and B. E. Sagan [2009], Monotonic sequence games, in: *Games of No Chance III*, Proc. BIRS Workshop on Combinatorial Games, July, 2005, Banff, Alberta, Canada, MSRI Publ. (M. H. Albert and R. J. Nowakowski, eds.), Vol. 56, Cambridge University Press, Cambridge, pp. 309–327.

16. M. H. Albert, J. P. Grossman, R. J. Nowakowski and D. Wolfe [2005], An introduction to Clobber, *Integers, Electr. J of Combinat. Number Theory* **5(2)**, #A01, 12 pp., MR2192079. http://www.integers-ejcnt.org/vol5(2).html

17. M. H. Albert and R. J. Nowakowski [2001], The game of End-Nim, *Electr. J. Combin.* **8(2)**, #R1, 12 pp., Volume in honor of Aviezri S. Fraenkel, MR1853252 (2002g:91044).

18. M. H. Albert and R. J. Nowakowski [2004], Nim restrictions, *Integers, Electr. J of Combinat. Number Theory* **4**, #G1, 10 pp., Comb. Games Sect., MR2056015. http://www.integers-ejcnt.org/vol4.html

19. M. Albert, R. J. Nowakowski and D. Wolfe [2007], *Lessons in Play: An Introduction to Combinatorial Game Theory*, A K Peters.

20. R. E. Allardice and A. Y. Fraser [1884], La tour d'Hanoï, *Proc. Edinburgh Math. Soc.* **2**, 50–53.

21. D. T. Allemang [1984], Machine computation with finite games, M.Sc. Thesis, Cambridge University.

22. D. T. Allemang [2001], Generalized genus sequences for misère octal games, *Intern. J. Game Theory*, **30**, 539–556, MR1907264 (2003h:91003).

23. J. D. Allen [1989], A note on the computer solution of Connect-Four, *Heuristic Programming in Artificial Intelligence* 1: *The First Computer Olympiad* (D. N. L. Levy and D. F. Beal, eds.), Ellis Horwood, Chichester, England, pp. 134–135.

24. M. R. Allen [2007], On the periodicity of genus sequences of quaternary games, *Integers, Electr. J. of Combinat. Number Theory* 7, #G04, 11 pp., MR2299810 (2007k:91050). http://www.integers-ejcnt.org/vol7.html

25. N. L. Alling [1985], Conway's field of surreal numbers, *Trans. Amer. Math. Soc.* 287, 365–386.

26. N. L. Alling [1987], *Foundations of Analysis Over Surreal Number Fields*, North-Holland, Amsterdam.

27. N. L. Alling [1989], Fundamentals of analysis over surreal number fields, *Rocky Mountain J. Math.* 19, 565–573.

28. N. L. Alling and P. Ehrlich [1986], An alternative construction of Conway's surreal numbers, *C. R. Math. Rep. Acad. Sci.* 8, 241–246.

29. N. L. Alling and P. Ehrlich [1986], An abstract characterization of a full class of surreal numbers, *C. R. Math. Rep. Acad. Sci.* 8, 303–308.

30. L. V. Allis [1994], Searching for solutions in games and artificial intelligence, Ph.D. Thesis, University of Limburg. ftp://ftp.cs.vu.nl/pub/victor/PhDthesis/thesis.ps.Z

31. L. V. Allis and P. N. A. Schoo [1992], Qubic solved again, *Heuristic Programming in Artificial Intelligence* 3: *The Third Computer Olympiad* (H. J. van den Herik and L. V. Allis, eds.), Ellis Horwood, New York, pp. 192–204.

32. L. V. Allis, H. J. van den Herik and M. P. H. Huntjens [1993], Go-Moku solved by new search techniques, *Proc. 1993 AAAI Fall Symp. on Games: Planning and Learning*, AAAI Press Tech. Report FS93–02, Menlo Park, CA, pp. 1–9.

33. J.-P. Allouche, D. Astoorian, J. Randall and J. Shallit [1994], Morphisms, square-free strings, and the tower of Hanoi puzzle, *Amer. Math. Monthly* 101, 651–658, MR1289274 (95g:68090).

34. J.-P. Allouche and A. Sapir [2005], Restricted towers of Hanoi and morphisms, in: *Developments in Language Theory*, Vol. 3572 of *Lecture Notes in Comput. Sci.*, Springer, Berlin, pp. 1–10, MR2187246 (2006g:68266).

35. N. Alon, J. Balogh, B. Bollobás and T. Szabó [2002], Game domination number, *Discrete Math.* 256, 23–33, MR1927054 (2003f:05086).

36. N. Alon, M. Krivelevich, J. Spencer and T. Szabó [2005], Discrepancy games, *Electr. J. Combin.* 12(1), #R51, 9 pp., MR2176527.

37. N. Alon and Z. Tuza [1995], The acyclic orientation game on random graphs, *Random Structures Algorithms* 6, 261–268.

38. S. Alpern and A. Beck [1991], Hex games and twist maps on the annulus, *Amer. Math. Monthly* 98, 803–811.

39. I. Althöfer [1988], Nim games with arbitrary periodic moving orders, *Internat. J. Game Theory* 17, 165–175.

40. I. Althöfer [1988], On the complexity of searching game trees and other recursion trees, *J. Algorithms* 9, 538–567.

41. I. Althöfer [1989], Generalized minimax algorithms are no better error correctors than minimax is itself, in: *Advances in Computer Chess* (D. F. Beal, ed.), Vol. 5, Elsevier, Amsterdam, pp. 265–282.

42. I. Althöfer and J. Bültermann [1995], Superlinear period lengths in some subtraction games, *Theoret. Comput. Sci.* (Math Games) **148**, 111–119.

43. G. Ambrus and J. Barát [2006], A contribution to queens graphs: a substitution method, *Discrete Math.* **306**(12), 1105–1114, MR2245636 (2007b:05174).

44. M. Anderson and T. Feil [1998], Turning lights out with linear algebra, *Math. Mag.* **71**, 300–303.

45. M. Anderson and F. Harary [1987], Achievement and avoidance games for generating abelian groups, *Internat. J. Game Theory* **16**, 321–325.

46. R. Anderson, L. Lovász, P. Shor, J. Spencer, E. Tardós and S. Winograd [1989], Disks, balls and walls: analysis of a combinatorial game, *Amer. Math. Monthly* **96**, 481–493.

47. T. Andreae [1984], Note on a pursuit game played on graphs, *Discrete Appl. Math.* **9**, 111–115.

48. T. Andreae [1986], On a pursuit game played on graphs for which a minor is excluded, *J. Combin. Theory* (Ser. B) **41**, 37–47.

49. T. Andreae, F. Hartenstein and A. Wolter [1999], A two-person game on graphs where each player tries to encircle his opponent's men, *Theoret. Comput. Sci.* (Math Games) **215**, 305–323.

50. S. D. Andres [2003], The game chromatic index of forests of maximum degree 5, in: *2nd Cologne-Twente Workshop on Graphs and Combinatorial Optimization*, Vol. 13 of *Electron. Notes Discrete Math.*, Elsevier, Amsterdam, p. 4 pp. (electronic), MR2153344.

51. S. D. Andres [2006], Game-perfect graphs with clique number 2, in: *CTW2006 — Cologne-Twente Workshop on Graphs and Combinatorial Optimization*, Vol. 25 of *Electron. Notes Discrete Math.*, Elsevier, Amsterdam, pp. 13–16 (electronic), MR2301125 (2008a:05080).

52. S. D. Andres [2006], The game chromatic index of forests of maximum degree $\Delta \geq 5$, *Discrete Appl. Math.* **154**, 1317–1323, MR2221551.

53. V. V. Anshelevich [2000], The Game of Hex: an automatic theorem proving approach to game programming, *Proc. 17-th National Conference on Artificial Intelligence (AAAI-2000)*, AAAI Press, Menlo Park, CA, pp. 189–194, MR1973011 (2004b:91039).

54. V. V. Anshelevich [2002], The game of Hex: the hierarchical approach, in: *More Games of No Chance*, Proc. MSRI Workshop on Combinatorial Games, July, 2000, Berkeley, CA, MSRI Publ. (R. J. Nowakowski, ed.), Vol. 42, Cambridge University Press, Cambridge, pp. 151–165.

55. R. P. Anstee and M. Farber [1988], On bridged graphs and cop-win graphs, *J. Combin. Theory* (Ser. B) **44**, 22–28.

56. A. Apartsin, E. Ferapontova and V. Gurvich [1998], A circular graph — counterexample to the Duchet kernel conjecture, *Discrete Math.* **178**, 229–231, MR1483752 (98f:05122).

57. D. Applegate, G. Jacobson and D. Sleator [1999], Computer analysis of Sprouts, in: *The Mathemagician and Pied Puzzler*, honoring Martin Gardner; E. Berlekamp and T. Rodgers, eds., A K Peters, Natick, MA, pp. 199-201.

58. A. A. Arakelyan [1982], *D*-products and compositions of Nim games, *Akad. Nauk Armyan. SSR Dokl.* **74**, 3–6, (Russian).

59. A. F. Archer [1999], A modern treatment of the 15 puzzle, *Amer. Math. Monthly* **106**, 793–799.

60. P. Arnold, ed. [1993], *The Book of Games*, Hamlyn, Chancellor Press.

61. A. A. Arratia-Quesada and I. A. Stewart [1997], Generalized Hex and logical characterizations of polynomial space, *Inform. Process. Lett.* **63**, 147–152.

62. A. A. Arratia and I. A. Stewart [2003], A note on first-order projections and games, *Theoret. Comput. Sci.* **290**, 2085–2093, MR1937766 (2003i:68034).

63. M. Ascher [1987], Mu Torere: An analysis of a Maori game, *Math. Mag.* **60**, 90–100.

64. I. M. Asel'derova [1974], On a certain discrete pursuit game on graphs, *Cybernetics* **10**, 859–864, trans. of *Kibernetika* **10** (1974) 102–105.

65. J. A. Aslam and A. Dhagat [1993], On-line algorithms for 2-coloring hypergraphs via chip games, *Theoret. Comput. Sci.* (Math Games) **112**, 355–369.

66. M. D. Atkinson [1981], The cyclic towers of Hanoi, *Inform. Process. Lett.* **13**, 118–119.

67. J. M. Auger [1991], An infiltration game on *k* arcs, *Naval Res. Logistics* **38**, 511–529.

68. V. Auletta, A. Negro and G. Parlati [1992], Some results on searching with lies, *Proc. 4th Italian Conf. on Theoretical Computer Science*, L'Aquila, Italy, pp. 24–37.

69. J. Auslander, A. T. Benjamin and D. S. Wilkerson [1993], Optimal leapfrogging, *Math. Mag.* **66**, 14–19.

70. R. Austin [1976], Impartial and partisan games, M.Sc. Thesis, Univ. of Calgary.

71. J. O. A. Ayeni and H. O. D. Longe [1985], Game people play: Ayo, *Internat. J. Game Theory* **14**, 207–218.

72. D. Azriel and D. Berend [2006], On a question of Leiss regarding the Hanoi Tower problem, *Theoret. Comput. Sci.* **369**(1-3), 377–383, MR2277583 (2007h:68226).

73. L. Babai [1985], Trading group theory for randomness, *Proc. 17th Ann. ACM Symp. Theory of Computing,* Assoc. Comput. Mach., New York, NY, pp. 421 – 429.

74. L. Babai and S. Moran [1988], Arthur–Merlin games: a randomized proof system, and a hierarchy of complexity classes, *J. Comput. System Sci.* **36**, 254–276.

75. R. J. R. Back and J. von Wright [1995], Games and winning strategies, *Inform. Process. Lett.* **53**, 165–172.

76. R. Backhouse and D. Michaelis [2004], Fixed-point characterisation of winning strategies in impartial games in: *Relational and Kleene-Algebraic Methods in Computer Science*, Vol. 3051/2004, Lecture Notes in Computer Scienc, Springer Berlin, Heidelberg, pp. 34–47.

77. C. K. Bailey and M. E. Kidwell [1985], A king's tour of the chessboard, *Math. Mag.* **58**, 285–286.

78. W. W. R. Ball and H. S. M. Coxeter [1987], *Mathematical Recreations and Essays*, Dover, New York, NY, 13th edn.

79. B. A. Balof and J. J. Watkins [1996], Knight's tours and magic squares, *Congr. Numer.* **120**, 23–32, Proc. 27th Southeastern Internat. Conf. on Combinatorics, Graph Theory and Computing (Baton Rouge, LA, 1996), MR1431952.

80. B. Banaschewski and A. Pultr [1990/91], Tarski's fixpoint lemma and combinatorial games, *Order* **7**, 375–386.

81. R. B. Banerji [1971], Similarities in games and their use in strategy construction, *Proc. Symp. Computers and Automata* (J. Fox, ed.), Polytechnic Press, Brooklyn, NY, pp. 337–357.

82. R. B. Banerji [1980], *Artificial Intelligence, A Theoretical Approach*, Elsevier, North-Holland, New York, NY.

83. R. B. Banerji and C. A. Dunning [1992], On misere games, *Cybernetics and Systems* **23**, 221–228.

84. R. B. Banerji and G. W. Ernst [1972], Strategy construction using homomorphisms between games, *Artificial Intelligence* **3**, 223–249.

85. R. Bar Yehuda, T. Etzion and S. Moran [1993], Rotating-table games and derivatives of words, *Theoret. Comput. Sci.* (Math Games) **108**, 311–329.

86. I. Bárány [1979], On a class of balancing games, *J. Combin. Theory* (Ser. A) **26**, 115–126.

87. J. G. Baron [1974], The game of nim — a heuristic approach, *Math. Mag.* **47**, 23–28.

88. T. Bartnicki, B. Brešar, J. Grytczuk, M. Kovše, Z. Miechowicz and I. Peterinz [2008], Game chromatic number of Cartesian product graphs, *Electr. J. Combin.* **15(1)**, #R72, 13 pp.

89. T. Bartnicki, J. A. Grytczuk and H. A. Kierstead [2008], The game of arboricity, *Discrete Math.* **308**, 1388–1393, MR2392056.

90. T. Bartnicki, J. Grytczuk, H. A. Kierstead and X. Zhu [2007], The map-coloring game, *Amer. Math. Monthly* **114**, 793–803.

91. R. Barua and S. Ramakrishnan [1996], σ-game, σ^+-game and two-dimensional additive cellular automata, *Theoret. Comput. Sci.* (Math Games) **154**, 349–366.

92. V. J. D. Baston and F. A. Bostock [1985], A game locating a needle in a cirular haystack, *J. Optimization Theory and Applications* **47**, 383–391.

93. V. J. D. Baston and F. A. Bostock [1986], A game locating a needle in a square haystack, *J. Optimization Theory and Applications* **51**, 405–419.

94. V. J. D. Baston and F. A. Bostock [1987], Discrete hamstrung squad car games, *Internat. J. Game Theory* **16**, 253–261.

95. V. J. D. Baston and F. A. Bostock [1988], A simple cover-up game, *Amer. Math. Monthly* **95**, 850–854.

96. V. J. D. Baston and F. A. Bostock [1989], A one-dimensional helicopter-submarine game, *Naval Res. Logistics* **36**, 479–490.

97. V. J. D. Baston and F. A. Bostock [1993], Infinite deterministic graphical games, *SIAM J. Control Optim.* **31**, 1623–1629.

98. J. Baumgartner, F. Galvin, R. Laver and R. McKenzie [1975], Game theoretic versions of partition relations, in: *Colloquia Mathematica Societatis János Bolyai* **10**, *Proc. Internat. Colloq. on Infinite and Finite Sets,* Vol. 1, Keszthely, Hungary, 1973 (A. Hajnal, R. Radu and V. T. Sós, eds.), North-Holland, pp. 131–135.

99. J. D. Beasley [1985], *The Ins & Outs of Peg Solitaire*, Oxford University Press, Oxford.

100. J. D. Beasley [1989], *The Mathematics of Games*, Oxford University Press, Oxford.

101. P. Beaver [1995], *Victorian Parlour Games*, Magna Books.

102. A. Beck [1969], Games, in: *Excursions into Mathematics* (A. Beck, M. N. Bleicher and D. W. Crowe, eds.), A K Peters, Natick, MA, millennium edn., Chap. 5, pp. 317–387, With a foreword by Martin Gardner; first appeared in 1969, Worth Publ., MR1744676 (2000k:00002).

103. J. Beck [1981], On positional games, *J. Combin. Theory* (Ser. A) **30**, 117–133.

104. J. Beck [1981], Van der Waerden and Ramsey type games, *Combinatorica* **1**, 103–116.

105. J. Beck [1982], On a generalization of Kaplansky's game, *Discrete Math.* **42**, 27–35.

106. J. Beck [1982], Remarks on positional games, I, *Acta Math. Acad. Sci. Hungar.* **40**(1–2), 65–71.

107. J. Beck [1983], Biased Ramsey type games, *Studia Sci. Math. Hung.* **18**, 287–292.

108. J. Beck [1985], Random graphs and positional games on the complete graph, *Ann. Discrete Math.* **28**, 7–13.

109. J. Beck [1993], Achievement games and the probabilistic method, in: *Combinatorics, Paul Erdős is Eighty*, Vol. 1, Bolyai Soc. Math. Stud., János Bolyai Math. Soc., Budapest, pp. 51–78.

110. J. Beck [1994], Deterministic graph games and a probabilistic intuition, *Combin. Probab. Comput.* **3**, 13–26.

111. J. Beck [1996], Foundations of positional games, *Random Structures Algorithms* **9**, 15–47, appeared first in: Proc. Seventh International Conference on Random Structures and Algorithms, Atlanta, GA, 1995.

112. J. Beck [1997], Games, randomness and algorithms, in: *The Mathematics of Paul Erdős* (R. L. Graham and J. Nešetřil, eds.), Vol. I, Springer, pp. 280–310.

113. J. Beck [1997], Graph games, *Proc. Int. Coll. Extremal Graph Theory*, Balatonlelle, Hungary.

114. J. Beck [2002], Positional games and the second moment method, *Combinatorica* **22**, 169–216, special issue: Paul Erdös and his mathematics, MR1909083 (2003i:91027).

115. J. Beck [2002], Ramsey games, *Discrete Math.* **249**, 3–30, MR1898254 (2003e:05084).

116. J. Beck [2002], The Erdös-Selfridge theorem in positional game theory, *Bolyai Soc. Math. Stud.* **11**, 33–77, Paul Erdös and his mathematics, II, János Bolyai Math. Soc., Budapest, MR1954724 (2004a:91028).

117. J. Beck [2002], Tic-Tac-Toe, *Bolyai Soc. Math. Stud.* **10**, 93–137, János Bolyai Math. Soc., Budapest, MR1919569 (2003e:05137).

118. J. Beck [2005], Positional games, *Combin. Probab. Comput.* **14**(5-6), 649–696, MR2174650 (2006f:91030).

119. J. Beck [2008], *Tic-Tac-Toe Theory*, Cambridge University Press, Cambridge, to appear.

120. J. Beck and L. Csirmaz [1982], Variations on a game, *J. Combin. Theory* (Ser. A) **33**, 297–315.

121. R. Beigel and W. I. Gasarch [1991], The mapmaker's dilemma, *Discrete Appl. Math.* **34**, 37–48.

122. L. W. Beineke, I. Broereand and M. A. Henning [1999], Queens graphs, *Discrete Math.* **206**, 63–75, Combinatorics and number theory (Tiruchirappalli, 1996), MR1665386 (2000f:05066).

123. A. Bekmetjev, G. Brightwell, A. Czygrinow and G. Hurlbert [2003], Thresholds for families of multisets, with an application to graph pebbling, *Discrete Math.* **269**, 21–34, MR1989450 (2004i:05152).

124. G. I. Bell [2007], A fresh look at peg solitaire, *Math. Mag.* **80**(1), 16–28, MR2286485.

125. G. I. Bell [2007], Diagonal peg solitaire, *Integers, Electr. J. of Combinat. Number Theory* **7**, #G01, 20 pp., MR2282183 (2007j:05018). http://www.integers-ejcnt.org/vol7.html

126. G. I. Bell, D. S. Hirschberg and P. Guerrero-García [2007], The minimum size required of a solitaire army, *Integers, Electr. J. of Combinat. Number Theory* **7**, #G07, 22 pp., MR2342185. http://www.integers-ejcnt.org/vol7.html

127. R. C. Bell [1960, 1969], *Board and Table Games from Many Civilisations*, Vol. I & II, Oxford University Press, revised in 1979, Dover.

128. R. Bell and M. Cornelius [1988], *Board Games Round the World: A Resource Book for Mathematical Investigations*, Cambridge University Press, Cambridge, reprinted 1990.

129. A. J. Benjamin, M. T. Fluet and M. L. Huber [2001], Optimal Token Allocations in Solitaire Knock 'm Down, *Electr. J. Combin.* **8**(2), #R2, 8 pp., Volume in honor of Aviezri S. Fraenkel, MR1853253 (2002g:91048).

130. S. J. Benkoski, M. G. Monticino and J. R. Weisinger [1991], A survey of the search theory literature, *Naval Res. Logistics* **38**, 469–494.

131. G. Bennett [1994], Double dipping: the case of the missing binomial coefficient identities, *Theoret. Comput. Sci.* (Math Games) **123**, 351–375.

132. H.-J. Bentz [1987], Proof of the Bulgarian Solitaire conjectures, *Ars Combin.* **23**, 151–170, MR886950 (88k:05018).

133. D. Berend and A. Sapir [2006], The cyclic multi-peg tower of Hanoi, *ACM Trans. Algorithms* **2**, 297–317, MR2253783.

134. D. Berend and A. Sapir [2006], The diameter of Hanoi graphs, *Inform. Process. Lett.* **98**, 79–85, MR2207581 (2006m:68187).

135. D. Berengut [1981], A random hopscotch problem or how to make Johnny read more, in: *The Mathematical Gardner* (D. A. Klarner, ed.), Wadsworth Internat., Belmont, CA, pp. 51–59.

136. B. Berezovskiy and A. Gnedin [1984], The best choice problem, *Akad. Nauk, USSR, Moscow* (in Russian) .

137. C. Berge [1956], La fonction de Grundy d'un graphe infini, *C. R. Acad. Sci. Paris* **242**, 1404–1407, MR0089115 (19,621d).

138. C. Berge [1957], *Théorie générale des jeux à n personnes*, Mémor. Sci. Math., no. 138, Gauthier-Villars, Paris, MR0099259 (20 #5700).

139. C. Berge [1976], Sur les jeux positionnels, *Cahiers du Centre Études Rech. Opér.* **18**, 91–107.

140. C. Berge [1977], Vers une théorie générale des jeux positionnels, in: *Mathematical Economics and Game Theory, Essays in Honor of Oskar Morgenstern*, Lecture Notes in Economics (R. Henn and O. Moeschlin, eds.), Vol. 141, Springer Verlag, Berlin, pp. 13–24.

141. C. Berge [1981], Some remarks about a Hex problem, in: *The Mathematical Gardner* (D. A. Klarner, ed.), Wadsworth Internat., Belmont, CA, pp. 25–27.

142. C. Berge [1982], Les jeux combinatoires, *Cahiers Centre Études Rech. Opér.* **24**, 89–105, MR687875 (84d:90112).

143. C. Berge [1985], *Graphs*, North-Holland, Amsterdam, Chap. 14.

144. C. Berge [1989], *Hypergraphs*, Elsevier (French: Gauthier Villars 1988), Chap. 4.

145. C. Berge [1992], Les jeux sur un graphe, *Cahiers Centre Études Rech. Opér.* **34**, 95–101, MR1226531 (94h:05089).

146. C. Berge [1996], Combinatorial games on a graph, *Discrete Math.* **151**, 59–65.

147. C. Berge and P. Duchet [1988], Perfect graphs and kernels, *Bull. Inst. Math. Acad. Sinica* **16**, 263–274.

148. C. Berge and P. Duchet [1990], Recent problems and results about kernels in directed graphs, *Discrete Math.* **86**, 27–31, appeared first under the same title in *Applications of Discrete Mathematics* (Clemson, SC, 1986), 200–204, SIAM, Philadelphia, PA, 1988.

149. C. Berge and M. Las Vergnas [1976], Un nouveau jeu positionnel, le "Match-It", ou une construction dialectique des couplages parfaits, *Cahiers du Centre Études Rech. Opér.* **18**, 83–89.

150. C. Berge and M. P. Schützenberger [1956], Jeux de Nim et solutions, *Acad. Sci. Paris* **242**, 1672–1674, (French).

151. E. R. Berlekamp [1972], Some recent results on the combinatorial game called Welter's Nim, *Proc. 6th Ann. Princeton Conf. Information Science and Systems*, pp. 203–204.

152. E. R. Berlekamp [1974], The Hackenbush number system for compression of numerical data, *Inform. and Control* **26**, 134–140.

153. E. R. Berlekamp [1988], Blockbusting and domineering, *J. Combin. Theory* (Ser. A) **49**, 67–116, an earlier version, entitled Introduction to blockbusting and domineering, appeared in: *The Lighter Side of Mathematics*, Proc. E. Strens Memorial Conf. on Recr. Math. and its History, Calgary, 1986, Spectrum Series (R. K. Guy and R. E. Woodrow, eds.), Math. Assoc. of America, Washington, DC, 1994, pp. 137–148.

154. E. Berlekamp [1990], Two-person, perfect-information games, in: *The Legacy of John von Neumann* (Hempstead NY, 1988), *Proc. Sympos. Pure Math.*, Vol. 50, Amer. Math. Soc., Providence, RI, pp. 275–287.

155. E. R. Berlekamp [1991], Introductory overview of mathematical Go endgames, in: *Combinatorial Games,* Proc. Symp. Appl. Math. (R. K. Guy, ed.), Vol. 43, Amer. Math. Soc., Providence, RI, pp. 73–100.

156. E. R. Berlekamp [1996], The economist's view of combinatorial games, in: *Games of No Chance,* Proc. MSRI Workshop on Combinatorial Games, July, 1994, Berkeley, CA, MSRI Publ. (R. J. Nowakowski, ed.), Vol. 29, Cambridge University Press, Cambridge, pp. 365–405.

157. E. R. Berlekamp [2000], Sums of $N \times 2$ Amazons, in: *Institute of Mathematical Statistics Lecture Notes–Monograph Series* (F.T. Bruss and L.M. Le Cam, eds.), Vol. 35, Beechwood, Ohio: Institute of Mathematical Statistics, pp. 1–34, Papers in honor of Thomas S. Ferguson, MR1833848 (2002e:91033).

158. E. R. Berlekamp [2000], *The Dots-and-Boxes Game: Sophisticated Child's Play*, A K Peters, Natick, MA, MR1780088 (2001i:00005).

159. E. R. Berlekamp [2002], Four games for Gardner, in: *Puzzler's Tribute: a Feast for the Mind*, pp. 383–386, honoring Martin Gardner (D. Wolfe and T. Rodgers, eds.), A K Peters, Natick, MA.

160. E. R. Berlekamp [2002], Idempotents among partisan games, in: *More Games of No Chance,* Proc. MSRI Workshop on Combinatorial Games, July, 2000, Berkeley, CA, MSRI Publ. (R. J. Nowakowski, ed.), Vol. 42, Cambridge University Press, Cambridge, pp. 3–23.

161. E. R. Berlekamp [2002], The 4G4G4G4G4 problems and solutions, in: *More Games of No Chance,* Proc. MSRI Workshop on Combinatorial Games, July, 2000, Berkeley, CA, MSRI Publ. (R. J. Nowakowski, ed.), Vol. 42, Cambridge University Press, Cambridge, pp. 231–241.

162. E. R. Berlekamp [2009], Yellow-Brown Hackenbush, in: *Games of No Chance III,* Proc. BIRS Workshop on Combinatorial Games, July, 2005, Banff, Alberta, Canada, MSRI Publ. (M. H. Albert and R. J. Nowakowski, eds.), Vol. 56, Cambridge University Press, Cambridge, pp. 413–418.

163. E. R. Berlekamp, J. H. Conway and R. K. Guy [2001-2004], *Winning Ways for your Mathematical Plays*, Vol. 1-4, A K Peters, Wellesley, MA, 2nd edition: vol. 1 (2001), vols. 2, 3 (2003), vol. 4 (2004); translation of 1st edition (1982) into German: *Gewinnen, Strategien für Mathematische Spiele* by G. Seiffert, Foreword by K. Jacobs, M. Reményi and Seiffert, Friedr. Vieweg & Sohn, Braunschweig (four volumes), 1985.

164. E. R. Berlekamp and Y. Kim [1996], Where is the "Thousand-Dollar Ko?", in: *Games of No Chance,* Proc. MSRI Workshop on Combinatorial Games, July,

1994, Berkeley, CA, MSRI Publ. (R. J. Nowakowski, ed.), Vol. 29, Cambridge University Press, Cambridge, pp. 203–226.

165. E. Berlekamp and T. Rodgers, eds. [1999], *The Mathemagician and Pied Puzzler*, A K Peters, Natick, MA, A collection in tribute to Martin Gardner. Papers from the Gathering for Gardner Meeting (G4G1) held in Atlanta, GA, January 1993.

166. E. Berlekamp and K. Scott [2002], Forcing your opponent to stay in control of a loony dots-and-boxes endgame, in: *More Games of No Chance*, Proc. MSRI Workshop on Combinatorial Games, July, 2000, Berkeley, CA, MSRI Publ. (R. J. Nowakowski, ed.), Vol. 42, Cambridge University Press, Cambridge, pp. 317–330, MR1973020.

167. E. Berlekamp and D. Wolfe [1994], *Mathematical Go: Chilling Gets the Last Point*, A K Peters, Natick, MA.

168. P. Berloquin [1976], *100 Jeux de Table*, Flammarion, Paris.

169. P. Berloquin [1995], *100 Games of Logic*, Barnes & Noble.

170. P. Berloquin and D. Dugas (Illustrator) [1999], *100 Perceptual Puzzles*, Barnes & Noble.

171. T. C. Biedl, E. D. Demaine, M. L. Demaine, R. Fleischer, L. Jacobsen and I. Munro [2002], The complexity of Clickomania, in: *More Games of No Chance*, Proc. MSRI Workshop on Combinatorial Games, July, 2000, Berkeley, CA, MSRI Publ. (R. J. Nowakowski, ed.), Vol. 42, Cambridge University Press, Cambridge, pp. 389–404, MR1973107 (2004h:91041).

172. N. L. Biggs [1999], Chip-firing and the critical group of a graph, *J. Algebr. Comb.* **9**, 25–45.

173. K. Binmore [1992], *Fun and Games: a Text on Game Theory*, D.C. Heath, Lexington.

174. J. Bitar and E. Goles [1992], Parallel chip firing games on graphs, *Theoret. Comput. Sci.* **92**, 291–300.

175. A. Björner and L. Lovász [1992], Chip-firing games on directed graphs, *J. Algebraic Combin.* **1**, 305–328.

176. A. Björner, L. Lovász and P. Chor [1991], Chip-firing games on graphs, *European J. Combin.* **12**, 283–291.

177. Y. Björnsson, R. Hayward, M. Johanson and J. van Rijswijck [2007], Dead cell analysis in Hex and the Shannon game, in: *Graph theory in Paris*, Trends Math., Birkhäuser, Basel, pp. 45–59, MR2279166 (2007j:91023).

178. N. M. Blachman and D. M. Kilgour [2001], Elusive optimality in the box problem, *Math. Mag.* **74**(3), 171–181, MR2104911.

179. D. Blackwell and M. A. Girshick [1954], *Theory of Games and Statistical Decisions*, Wiley, New York, NY.

180. L. Blanc, E. Duchêne and S. Gravier [2006], A deletion game on graphs: "le pic arête", *Integers, Electr. J. of Combinat. Number Theory* **6**, #G02, 10 pp., Comb. Games Sect., MR2215359. http://www.integers-ejcnt.org/vol5.html

181. U. Blass and A. S. Fraenkel [1990], The Sprague-Grundy function for Wythoff's game, *Theoret. Comput. Sci.* **75**(3), 311–333, MR1080539 (92a:90101).

182. U. Blass, A. S. Fraenkel and R. Guelman [1998], How far can Nim in disguise be stretched?, *J. Combin. Theory* (Ser. A) **84**, 145–156, MR1652900 (2000d:91029).

183. M. Blidia [1986], A parity digraph has a kernel, *Combinatorica* **6**, 23–27.

184. M. Blidia, P. Duchet, H. Jacob, F. Maffray and H. Meyniel [1999], Some operations preserving the existence of kernels, *Discrete Math.* **205**, 211–216.

185. M. Blidia, P. Duchet and F. Maffray [1993], On kernels in perfect graphs, *Combinatorica* **13**, 231–233.

186. J.-P. Bode and H. Harborth [1998], Achievement games on Platonic solids, *Bull. Inst. Combin. Appl.* **23**, 23–32, MR1621748 (99d:05020).

187. J.-P. Bode and H. Harborth [2000], Hexagonal polyomino achievement, *Discrete Math.* **212**, 5–18, MR1748669 (2000k:05082).

188. J.-P. Bode and H. Harborth [2000], Independent chess pieces on Euclidean boards, *J. Combin. Math. Combin. Comput.* **33**, 209–223, MR1772763 (2001c:05105).

189. J.-P. Bode and H. Harborth [2000], Triangular mosaic polyomino achievement, *Congr. Numer.* **144**, 143–152, Proc. 31st Southeastern Internat. Conf. on Combinatorics, Graph Theory and Computing (Boca Raton, FL, 2000), MR1817929 (2001m:05082).

190. J.-P. Bode and H. Harborth [2001], Triangle polyomino set achievement, *Congr. Numer.* **148**, 97–101, Proc.32nd Southeastern Internat. Conf. on Combinatorics, Graph Theory and Computing (Boca Raton, FL, 2002), MR1887377 (2002k:05057).

191. J.-P. Bode and H. Harborth [2002], Triangle and hexagon gameboard Ramsey numbers, *Congr. Numer.* **158**, 93–98, Proc.33rd Southeastern Internat. Conf. on Combinatorics, Graph Theory and Computing (Boca Raton, FL, 2002), MR1985149 (2004d:05129).

192. J.-P. Bode and H. Harborth [2003], Independence for knights on hexagon and triangle boards, *Discrete Math.* **272**, 27–35, MR2019197 (2004i:05115).

193. J.-P. Bode, H. Harborth and M. Harborth [2003], King independence on triangle boards, *Discrete Math.* **266**, 101–107, Presented at 18th British Combinatorial Conference (Brighton, 2001), MR1991709 (2004f:05129).

194. J.-P. Bode, H. Harborth and M. Harborth [2004], King graph Ramsey numbers, *J. Combin. Math. Combin. Comput.* **50**, 47–55, MR2075855.

195. J.-P. Bode, H. Harborth and H. Weiss [1999], Independent knights on hexagon boards, *Congr. Numer.* **141**, 31–35, Proc. 30th Southeastern Internat. Conf. on Combinatorics, Graph Theory, and Computing (Boca Raton, FL, 1999), MR1745222 (2000k:05201).

196. J.-P. Bode and A. M. Hinz [1999], Results and open problems on the Tower of Hanoi, *Congr. Numer.* **139**, 113–122, Proc.30th Southeastern Internat. Conf. on Combinatorics, Graph Theory, and Computing (Boca Raton, FL, 1999).

197. H. L. Bodlaender [1991], On the complexity of some coloring games, *Internat. J. Found. Comput. Sci.* **2**, 133–147, MR1143920 (92j:68042).

198. H. L. Bodlaender [1993], Complexity of path forming games, *Theoret. Comput. Sci.* (Math Games) **110**, 215–245.

199. H. L. Bodlaender [1993], Kayles on special classes of graphs: an application of Sprague-Grundy theory, in: *Graph-Theoretic Concepts in Computer Science* (Wiesbaden-Naurod, 1992), Lecture Notes in Comput. Sci., Vol. 657, Springer, Berlin, pp. 90–102, MR1244129 (94i:90189).

200. H. L. Bodlaender and D. Kratsch [1992], The complexity of coloring games on perfect graphs, *Theoret. Comput. Sci.* (Math Games) **106**, 309–326.

201. H. L. Bodlaender and D. Kratsch [2002], Kayles and nimbers, *J. Algorithms* **43**, 106–119, MR1900711 (2003d:05201).

202. T. Bohman, R. Holzman and D. Kleitman [2001], Six Lonely Runners, *Electr. J. Combin.* **8**(2), #R3, 49 pp., Volume in honor of Aviezri S. Fraenkel, MR1853254 (2002g:11095).

203. K. D. Boklan [1984], The *n*-number game, *Fibonacci Quart.* **22**, 152–155.

204. B. Bollobás and I. Leader [2005], The devil and the angel in three dimensions, *J. Combin. Theory* (Ser. A) to appear.

205. B. Bollobás and T. Szabó [1998], The oriented cycle game, *Discrete Math.* **186**, 55–67.

206. D. L. Book [1998, Sept. 9-th], What the Hex, *The Washington Post* p. H02.

207. E. Borel [1921], La théorie du jeu et les équations integrales à noyau symmetrique gauche, *C. R. Acad. Sci. Paris* **173**, 1304–1308.

208. E. Boros and V. Gurvich [1996], Perfect graphs are kernel solvable, *Discrete Math.* **159**, 35–55.

209. E. Boros and V. Gurvich [1998], A corrected version of the Duchet kernel conjecture, *Discrete Math.* **179**, 231–233.

210. E. Boros and V. Gurvich [2006], Perfect graphs, kernels, and cores of cooperative games, *Discrete Math.* **306**(19-20), 2336–2354, MR2261906 (2007g:05069).

211. M. Borowiecki, S. Jendrol', D. Král and J. Miškuf [2006], List coloring of Cartesian products of graphs, *Discrete Math.* **306**, 1955–1958, MR2251575 (2007b:05067).

212. M. Borowiecki and E. Sidorowicz [2007], Generalised game colouring of graphs, *Discrete Math.* **307**(11-12), 1225–1231, MR2311092.

213. M. Borowiecki, E. Sidorowicz and Z. Tuza [2007], Game list colouring of graphs, *Electron. J. Combin.* **14**(1), #R26, 11 pp., MR2302533.

214. E. Boudreau, B. Hartnell, K. Schmeisser and J. Whiteley [2004], A game based on vertex-magic total labelings, *Australas. J. Combin.* **29**, 67–73, MR2037334 (2004k:91059).

215. C. L. Bouton [1902], Nim, a game with a complete mathematical theory, *Ann. of Math.* **3**(2), 35–39.

216. J. Boyce [1981], A Kriegspiel endgame, in: *The Mathematical Gardner* (D. A. Klarner, ed.), Wadsworth Internat., Belmont, CA, pp. 28–36.

217. S. J. Brams and D. M. Kilgour [1995], The box problem: to switch or not to switch, *Math. Mag.* **68**(1), 27–34.

218. G. Brandreth [1981], *The Bumper Book of Indoor Games*, Victorama, Chancellor Press.

219. D. M. Breuker, J. W. H. M. Uiterwijk and H. J. van den Herik [2000], Solving 8×8 Domineering, *Theoret. Comput. Sci.* (Math Games) **230**, 195–206, MR1725637 (2001i:68148).

220. D. M. Broline and D. E. Loeb [1995], The combinatorics of Mancala-type games: Ayo, Tchoukaillon, and $1/\pi$, *UMAP J.* **16**(1), 21–36.

221. A. Brousseau [1976], Tower of Hanoi with more pegs, *J. Recr. Math.* **8**, 169–176.

222. A. E. Brouwer, G. Horváth, I. Molnár-Sáska and C. Szabó [2005], On three-rowed chomp, *Integers, Electr. J. of Combinat. Number Theory* **5**, #G07, 11 pp., Comb. Games Sect., MR2192255 (2006g:11051). http://www.integers-ejcnt.org/vol5.html

223. C. Browne [2000], *HEX Strategy: Making the Right Connections*, A K Peters, Natick, MA.

224. C. Browne [2005], *Connection Games: Variations on a Theme*, A K Peters, Natick, MA.

225. C. Browne [2006], Fractal board games., *Computers & Graphics* **30**(1), 126–133.

226. R. A. Brualdi and V. S. Pless [1993], Greedy codes, *J. Combin. Theory* (Ser. A) **64**, 10–30.

227. A. A. Bruen and R. Dixon [1975], The n-queen problem, *Discrete Math.* **12**, 393–395.

228. J. Bruno and L. Weinberg [1970], A constructive graph-theoretic solution of the Shannon switching game, *IEEE Trans. Circuit Theory* **CT-17**, 74–81.

229. G. P. Bucan and L. P. Varvak [1966], On the question of games on a graph, in: *Algebra and Math. Logic: Studies in Algebra (Russian)*, Izdat. Kiev. Univ., Kiev, pp. 122–138, MR0207584 (34 #7399).

230. P. Buneman and L. Levy [1980], The towers of Hanoi problem, *Inform. Process. Lett.* **10**, 243–244.

231. A. P. Burger, E. J. Cockayne and C. M. Mynhardt [1997], Domination and irredundance in the queen's graph, *Discrete Math.* **163**, 47–66, MR1428557 (97m:05130).

232. A. P. Burger and C. M. Mynhardt [1999], Queens on hexagonal boards, *J. Combin. Math. Combin. Comput.* **31**, 97–111, paper in honour of Stephen T. Hedetniemi, MR1726950 (2000h:05158).

233. A. P. Burger and C. M. Mynhardt [2000], Properties of dominating sets of the queens graph Q_{4k+3}, *Util. Math.* **57**, 237–253, MR1760187 (2000m:05166).

234. A. P. Burger and C. M. Mynhardt [2000], Small irredundance numbers for queens graphs, *J. Combin. Math. Combin. Comput.* **33**, 33–43, paper in honour of Ernest J. Cockayne, MR1772752 (2001c:05106).

235. A. P. Burger and C. M. Mynhardt [2000], Symmetry and domination in queens graphs, *Bull. Inst. Combin. Appl.* **29**, 11–24.

236. A. P. Burger and C. M. Mynhardt [2002], An upper bound for the minimum number of queens covering the $n \times n$ chessboard, *Discrete Appl. Math.* **121**, 51–60.

237. A. P. Burger and C. M. Mynhardt [2003], An improved upper bound for queens domination numbers, *Discrete Math.* **266**, 119–131.

238. A. P. Burger, C. M. Mynhardt and E. J. Cockayne [1994], Domination numbers for the queen's graph, *Bull. Inst. Combin. Appl.* **10**, 73–82.

239. A. P. Burger, C. M. Mynhardt and E. J. Cockayne [2001], Queens graphs for chessboards on the torus, *Australas. J. Combin.* **24**, 231–246.

240. M. Buro [2001], Simple Amazon endgames and their connection to Hamilton circuits in cubic subgrid graphs, *Proc. 2nd Intern. Conference on Computers and Games CG'2000* (T. Marsland and I. Frank, eds.), Vol. 2063, Hamamatsu, Japan, Oct. 2000, Lecture Notes in Computer Science, Springer, pp. 251–261, MR1909614.

241. D. W. Bushaw [1967], On the name and history of Nim, *Washington Math.* **11**, Oct. 1966. Reprinted in: *NY State Math. Teachers J.*, **17**, pp. 52–55.

242. P. J. Byrne and R. Hesse [1996], A Markov chain analysis of jai alai, *Math. Mag.* **69**, 279–283.

243. S. Byrnes [2003], Poset game periodicity, *Integers, Electr. J. of Combinat. Number Theory* **3**, #G3, 16 pp., Comb. Games Sect., MR2036487 (2005c:91031). http://www.integers-ejcnt.org/vol3.html

244. J.-Y. Cai, A. Condon and R. J. Lipton [1992], On games of incomplete information, *Theoret. Comput. Sci.* **103**, 25–38.

245. L. Cai and X. Zhu [2001], Game chromatic index of k-degenerate graphs, *J. Graph Theory* **36**, 144–155, MR1814531 (2002b.05058).

246. G. Cairns [2002], Pillow chess, *Math. Mag.* **75**, 173–186, MR2075210 (2005b:91011).

247. D. Calistrate [1996], The reduced canonical form of a game, in: *Games of No Chance,* Proc. MSRI Workshop on Combinatorial Games, July, 1994, Berkeley, CA, MSRI Publ. (R. J. Nowakowski, ed.), Vol. 29, Cambridge University Press, Cambridge, pp. 409–416, MR1427979 (97m:90122).

248. D. Calistrate, M. Paulhus and D. Wolfe [2002], On the lattice structure of finite games, in: *More Games of No Chance,* Proc. MSRI Workshop on Combinatorial Games, July, 2000, Berkeley, CA, MSRI Publ. (R. J. Nowakowski, ed.), Vol. 42, Cambridge University Press, Cambridge, pp. 25–30, MR1973001.

249. G. Campbell [2004], On optimal play in the game of Hex, *Integers, Electr. J. of Combinat. Number Theory* **4**, #G2, 23 pp., Comb. Games Sect., MR2056016 (2005c:91032). http://www.integers-ejcnt.org/vol4.html

250. C. Cannings and J. Haigh [1992], Montreal solitaire, *J. Combin. Theory* (Ser. A) **60**, 50–66.

251. J. Carlson and D. Stolarski [2004], The correct solution to Berlekamp's switching game, *Discrete Math.* **287**, 145–150, MR2094708 (2005d:05005).

252. A. Chan and A. Tsai [2002], $1 \times n$ Konane: a summary of results, in: *More Games of No Chance,* Proc. MSRI Workshop on Combinatorial Games, July, 2000, Berkeley, CA, MSRI Publ. (R. J. Nowakowski, ed.), Vol. 42, Cambridge University Press, Cambridge, pp. 331–339, MR1973021.

253. T.-H. Chan [1989], A statistical analysis of the towers of Hanoi problem, *Intern. J. Computer Math.* **28**, 57–65.

254. A. K. Chandra, D. C. Kozen and L. J. Stockmeyer [1981], Alternation, *J. Assoc. Comput. Mach.* **28**, 114–133.

255. A. K. Chandra and L. J. Stockmeyer [1976], Alternation, *Proc. 17th Ann. Symp. Foundations of Computer Science* (Houston, TX, Oct. 1976), IEEE Computer Soc., Long Beach, CA, pp. 98–108.

256. H. Chang and X. Zhu [2006], The d-relaxed game chromatic index of k-degenerated graphs, *Australas. J. Combin.* **36**, 73–82, MR2262608.

257. G. Chartrand, F. Harary, M. Schultz and D. W. VanderJagt [1995], Achievement and avoidance of a strong orientation of a graph, *Congr. Numer.* **108**, 193–203.

258. S. M. Chase [1972], An implemented graph algorithm for winning Shannon switching games, *Commun. Assoc. Comput. Mach.* **15**, 253–256.

259. M. Chastand, F. Laviolette and N. Polat [2000], On constructible graphs, infinite bridged graphs and weakly cop-win graphs, *Discrete Math.* **224**, 61–78, MR1781285 (2002g:05152).

260. G. Chen, R. H. Schelp and W. E. Shreve [1997], A new game chromatic number, *European J. Combin.* **18**, 1–9.

261. V. Chepoi [1997], Bridged graphs are cop-win graphs: an algorithmic proof, *J. Combin. Theory* (Ser. B) **69**, 97–100.

262. B. S. Chlebus [1986], Domino-tiling games, *J. Comput. System Sci.* **32**, 374–392.

263. C.-Y. Chou, W. Wang and X. Zhu [2003], Relaxed game chromatic number of graphs, *Discrete Math.* **262**, 89–98, MR1951379 (2003m:05062).

264. J. D. Christensen and M. Tilford [1997], Unsolved Problems: David Gale's subset take-away game, *Amer. Math. Monthly* **104**, 762–766, Unsolved Problems Section.

265. F. R. K. Chung [1989], Pebbling in hypercubes, *SIAM J. Disc. Math.* **2**, 467–472.

266. F. R. K. Chung, J. E. Cohen and R. L. Graham [1988], Pursuit-evasion games on graphs, *J. Graph Theory* **12**, 159–167.

267. F. Chung and R. B. Ellis [2002], A chip-firing game and Dirichlet eigenvalues, *Discrete Math.* **257**, 341–355, MR1935732 (2003i:05087).

268. F. Chung, R. Graham, J. Morrison and A. Odlyzko [1995], Pebbling a chessboard, *Amer. Math. Monthly* **102**, 113–123.

269. V. Chvátal [1973], On the computational complexity of finding a kernel, Report No. CRM-300, Centre de Recherches Mathématiques, Université de Montréal.

270. V. Chvátal [1981], Cheap, middling or dear, in: *The Mathematical Gardner* (D. A. Klarner, ed.), Wadsworth Internat., Belmont, CA, pp. 44–50.

271. V. Chvátal [1983], Mastermind, *Combinatorica* **3**, 325–329.

272. V. Chvátal and P. Erdős [1978], Biased positional games, *Ann. Discrete Math.* **2**, 221–229, Algorithmic Aspects of Combinatorics, (B. Alspach, P. Hell and D. J. Miller, eds.), Qualicum Beach, BC, Canada, 1976, North-Holland.

273. F. Cicalese and C. Deppe [2003], Quasi-perfect minimally adaptive q-ary search with unreliable tests, in: *Algorithms and Computation*, Vol. 2906 of

Lecture Notes in Comput. Sci., Springer, Berlin, pp. 527–536, MR2088232 (2005d:68101).

274. F. Cicalese, C. Deppe and D. Mundici [2004], Q-ary Ulam-Rényi game with weighted constrained lies, in: *Computing and Combinatorics*, Vol. 3106 of *Lecture Notes in Comput. Sci.*, Springer, Berlin, pp. 82–91, MR2162023 (2006c:68037).

275. F. Cicalese and D. Mundici [1999], Optimal binary search with two unreliable tests and minimum adaptiveness, in: *Algorithms — ESA '99 (Prague)*, Vol. 1643 of *Lecture Notes in Comput. Sci.*, Springer, Berlin, pp. 257–266, MR1729129 (2000i:68035).

276. F. Cicalese, D. Mundici and U. Vaccaro [2000], Least adaptive optimal search with unreliable tests, in: *Algorithm Theory — SWAT 2000 (Bergen)*, Vol. 1851 of *Lecture Notes in Comput. Sci.*, Springer, Berlin, pp. 549–562, MR1793099 (2001k:91005).

277. F. Cicalese, D. Mundici and U. Vaccaro [2002], Least adaptive optimal search with unreliable tests, *Theoret. Comput. Sci.* (Math Games) **270**, 877–893, MR1871101 (2003g:94052).

278. F. Cicalese and U. Vaccaro [2000], An improved heuristic for the "Ulam-Rényi game", *Inform. Process. Lett.* **73**, 119–124, MR1741817 (2000m:91026).

279. F. Cicalese and U. Vaccaro [2000], Optimal strategies against a liar, *Theoret. Comput. Sci.* **230**, 167–193, MR1725636 (2001g:91040).

280. F. Cicalese and U. Vaccaro [2003], Binary search with delayed and missing answers, *Inform. Process. Lett.* **85**, 239–247, MR1952901 (2003m:68031).

281. A. Cincotti [2005], Three-player partizan games, *Theoret. Comput. Sci.* **332**, 367–389, MR2122510 (2005j:91018).

282. C. Clark [1996], On achieving channels in a bipolar game, in: *African Americans in Mathematics* (Piscataway, NJ, 1996), DIMACS Ser. Discrete Math. Theoret. Comput. Sci., Vol. 34, Amer. Math. Soc., Providence, RI, pp. 23–27.

283. D. S. Clark [1986], Fibonacci numbers as expected values in a game of chance, *Fibonacci Quart.* **24**, 263–267.

284. N. E. Clarke and R. J. Nowakowski [2000], Cops, robber, and photo radar, *Ars Combin.* **56**, 97–103, MR1768605 (2001e:91040).

285. N. E. Clarke and R. J. Nowakowski [2001], Cops, robber and traps, *Util. Math.* **60**, 91–98, MR1863432 (2002i:91014).

286. N. E. Clarke and R. J. Nowakowski [2005], Tandem-win graphs, *Discrete Math.* **299**, 56–64, MR2168695.

287. T. A. Clarke, R. A. Hochberg and G. H. Hurlbert [1997], Pebbling in diameter two graphs and products of paths, *J. Graph Theory* **25**, 119–128.

288. N. Claus (=E. Lucas) [1884], La tour d'Hanoi, jeu de calcul, *Science et Nature* **1**, 127–128.

289. A. Clausing [2001], Das Trisentis Spiel (The Trisentis game), *Math. Semesterberichte* **48**, 49–66, MR1950212 (2003k:91045).

290. E. J. Cockayne [1990], Chessboard domination problems, *Discrete Math.* **86**, 13–20.

291. E. J. Cockayne and S. T. Hedetniemi [1986], On the diagonal queens domination problem, *J. Combin. Theory* (Ser. A) **42**, 137–139.

292. E. J. Cockayne and C. M. Mynhardt [2001], Properties of queens graphs and the irredundance number of Q_7, *Australas. J. Combin.* **23**, 285–299, MR1815019 (2002a:05189).

293. A. J. Cole and A. J. T. Davie [1969], A game based on the Euclidean algorithm and a winning strategy for it, *Math. Gaz.* **53**, 354–357.

294. D. B. Coleman [1978], Stretch: a geoboard game, *Math. Mag.* **51**, 49–54.

295. D. Collins [2005], Variations on a theme of Euclid, *Integers, Electr. J. of Combinat. Number Theory* **5**, #G3, 12 pp., Comb. Games Sect., MR2139166 (2005k:91080). http://www.integers-ejcnt.org/vol5.html

296. D. Collins and T. Lengyel [2008], The game of 3-Euclid, *Discrete Math.* **308**, 1130–1136, MR2382351.

297. A. Condon [1989], *Computational Models of Games*, ACM Distinguished Dissertation, MIT Press, Cambridge, MA.

298. A. Condon [1991], Space-bounded probabilistic game automata, *J. Assoc. Comput. Mach.* **38**, 472–494.

299. A. Condon [1992], The complexity of Stochastic games, *Information and Computation* **96**, 203–224.

300. A. Condon [1993], On algorithms for simple stochastic games, in: *Advances in Computational Complexity Theory* (New Brunswick, NJ, 1990), DIMACS Ser. Discrete Math. Theoret. Comput. Sci., Vol. 13, Amer. Math. Soc., Providence, RI, pp. 51–71.

301. A. Condon, J. Feigenbaum, C. Lund and P. Shor [1993], Probabilistically checkable debate systems and approximation algorithms for PSPACE-hard functions, *Proc. 25th Ann. ACM Symp. Theory of Computing,* Assoc. Comput. Mach., New York, NY, pp. 305–314.

302. A. Condon and R. E. Ladner [1988], Probabilistic game automata, *J. Comput. System Sci.* **36**, 452–489, preliminary version in: Proc. Structure in complexity theory (Berkeley, CA, 1986), Lecture Notes in Comput. Sci., Vol. 223, Springer, Berlin, pp. 144–162.

303. I. G. Connell [1959], A generalization of Wythoff's game, *Canad. Math. Bull.* **2**, 181–190.

304. J. H. Conway [1972], All numbers great and small, Res. Paper No. 149, Univ. of Calgary Math. Dept.

305. J. H. Conway [1977], All games bright and beautiful, *Amer. Math. Monthly* **84**, 417–434.

306. J. H. Conway [1978], A gamut of game theories, *Math. Mag.* **51**, 5–12.

307. J. H. Conway [1978], Loopy Games, *Ann. Discrete Math.* **3**, 55–74, Proc. Symp. Advances in Graph Theory, Cambridge Combinatorial Conf. (B. Bollobás, ed.), Cambridge, May 1977.

308. J. H. Conway [1990], Integral lexicographic codes, *Discrete Math.* **83**, 219–235.

309. J. H. Conway [1991], More ways of combining games, in: *Combinatorial Games,* Proc. Symp. Appl. Math. (R. K. Guy, ed.), Vol. 43, Amer. Math. Soc., Providence, RI, pp. 57–71.

310. J. H. Conway [1991], Numbers and games, in: *Combinatorial Games,* Proc. Symp. Appl. Math. (R. K. Guy, ed.), Vol. 43, Amer. Math. Soc., Providence, RI, pp. 23–34.

311. J. H. Conway [1994], The surreals and the reals. Real numbers, generalizations of the reals, and theories of continua, in: *Synthese Lib.,* Vol. 242, Kluwer Acad. Publ., Dordrecht, pp. 93–103.

312. J. H. Conway [1996], The angel problem, in: *Games of No Chance,* Proc. MSRI Workshop on Combinatorial Games, July, 1994, Berkeley, CA, MSRI Publ. (R. J. Nowakowski, ed.), Vol. 29, Cambridge University Press, Cambridge, pp. 3–12.

313. J. H. Conway [1997], M_{13}, in: *Surveys in combinatorics,* London Math. Soc., Lecture Note Ser. 241, Cambridge Univ. Press, Cambridge, pp. 1–11.

314. J. H. Conway [2001], *On Numbers and Games,* A K Peters, Natick, MA, 2nd edition; translation of 1st edition (1976) into German: *Über Zahlen und Spiele* by Brigitte Kunisch, Friedr. Vieweg & Sohn, Braunschweig, 1983.

315. J. H. Conway [2002], More infinite games, in: *More Games of No Chance,* Proc. MSRI Workshop on Combinatorial Games, July, 2000, Berkeley, CA, MSRI Publ. (R. J. Nowakowski, ed.), Vol. 42, Cambridge University Press, Cambridge, pp. 31–36, MR1973002.

316. J. H. Conway [2003], Integral lexicographic codes, in: *MASS selecta* (S. Katok and S. Tabachnikov, eds.), Amer. Math. Soc., pp. 185–189, MR2027176.

317. J. H. Conway and H. S. M. Coxeter [1973], Triangulated polygons and frieze patterns, *Math. Gaz.* **57**, 87–94; 175–183.

318. J. H. Conway and N. J. A. Sloane [1986], Lexicographic codes: error-correcting codes from game theory, *IEEE Trans. Inform. Theory* **IT-32**, 337–348.

319. M. L. Cook and L. E. Shader [1979], A strategy for the Ramsey game "TRI-TIP", *Congr. Numer.* **23**, 315–324, Utilitas Math., Proc. 10th Southeastern Conf. on Combinatorics, Graph Theory and Computing (Florida Atlantic Univ., Boca Raton, Fla., 1979), MR561058 (83e:90167).

320. M. Copper [1993], Graph theory and the game of sprouts, *Amer. Math. Monthly* **100**, 478–482.

321. M. Cornelius and A. Parr [1991], *What's Your Game?,* Cambridge University Press, Cambridge.

322. T. Cover [1987], Pick the largest number, *Open Problems in Communication and Computation* (T. M. Cover and B. Gopinath, eds.), Springer-Verlag, New York, p. 152.

323. H. S. M. Coxeter [1953], The golden section, phyllotaxis and Wythoff's game, *Scripta Math.* **19**, 135–143.

324. M. Crâşmaru [2001], On the complexity of Tsume-Go, *Proc. Intern. Conference on Computers and Games CG'1998, Tsukuba, Japan, Nov. 1998* (H. J. van den Herik, ed.), Vol. LNCS 1558, Lecture Notes in Computer Science, Springer, pp. 222–231.

325. M. Crâşmaru and J. Tromp [2001], Ladders are PSPACE-complete, *Proc. 2nd Intern. Conference on Computers and Games CG'2000* (T. Marsland and I. Frank, eds.), Vol. 2063, Hamamatsu, Japan, Oct. 2000, Lecture Notes in Computer Science, Springer, pp. 241–249, MR1909613.

326. J. W. Creely [1987], The length of a two-number game, *Fibonacci Quart.* **25**, 174–179.

327. J. W. Creely [1988], The length of a three-number game, *Fibonacci Quart.* **26**, 141–143.

328. H. T. Croft [1964], 'Lion and man': a postscript, *J. London Math. Soc.* **39**, 385–390.

329. D. W. Crowe [1956], The *n*-dimensional cube and the tower of Hanoi, *Amer. Math. Monthly* **63**, 29–30.

330. B. Crull, T. Cundiff, P. Feltman, G. H. Hurlbert, L. Pudwell, Z. Szaniszlo and Z. Tuza [2005], The cover pebbling number of graphs, *Discrete Math.* **296**, 15–23, MR2148478.

331. L. Csirmaz [1980], On a combinatorial game with an application to Go-Moku, *Discrete Math.* **29**, 19–23.

332. L. Csirmaz [2002], Connected graph game, *Studia Sci. Math. Hungar.* **39**, 129–136, MR1909151 (2003d:91028).

333. L. Csirmaz and Z. Nagy [1979], On a generalization of the game Go-Moku II, *Studia Sci. Math. Hung.* **14**, 461–469.

334. P. Csorba [2005], On the biased *n*-in-a-row game, *Discrete Math.* **503**, 100–111.

335. J. Culberson [1999], Sokoban is PSPACE complete, in: *Fun With Algorithms*, Vol. 4 of *Proceedings in Informatics*, Carleton Scientific, University of Waterloo, Waterloo, Ont., pp. 65–76, Conference took place on the island of Elba, June 1998.

336. P. Cull and E. F. Ecklund, Jr. [1982], On the towers of Hanoi and generalized towers of Hanoi problems, *Congr. Numer.* **35**, 229–238.

337. P. Cull and E. F. Ecklund, Jr. [1985], Towers of Hanoi and analysis of algorithms, *Amer. Math. Monthly* **92**, 407–420.

338. P. Cull and C. Gerety [1985], Is towers of Hanoi really hard?, *Congr. Numer.* **47**, 237–242.

339. P. Cull and I. Nelson [1999], Error-correcting codes on the towers of Hanoi graphs, *Discrete Math.* **208/209**, 157–175.

340. P. Cull and I. Nelson [1999], Perfect codes, NP-completeness, and towers of Hanoi graphs, *Bull. Inst. Combin. Appl.* **26**, 13–38.

341. A. Czygrinow, N. Eaton, G. Hurlbert and P. M. Kayll [2002], On pebbling threshold functions for graph sequences, *Discrete Math.* **247**, 93–105, MR1877652 (2002m:05058).

342. A. Czygrinow and G.Hurlbert [2003], Pebbling in dense graphs, *Australas. J. Combin.* **28**, 201–208.

343. A. Czygrinow, G. Hurlbert, H. A. Kierstead and W. T. Trotter [2002], A note on graph pebbling, *Graphs Combin.* **18**, 219–225, MR1913664 (2004d:05170).

344. J. Czyzowicz, K. B. Lakshmanan and A. Pelc [1991], Searching with a forbidden lie pattern in responses, *Inform. Process. Lett.* **37**, 127–132, MR1095694 (92e:68027).

345. J. Czyzowicz, K. B. Lakshmanan and A. Pelc [1994], Searching with local constraints on error patterns, *European J. Combin.* **15**, 217–222, MR1273940 (95a:05005).

346. J. Czyzowicz, D. Mundici and A. Pelc [1989], Ulam's searching game with lies, *J. Combin. Theory Ser. A* **52**, 62–76, MR1008160 (90k:94026).

347. J. Czyzowicz, A. Pelc and D. Mundici [1988], Solution of Ulam's problem on binary search with two lies, *J. Combin. Theory Ser. A* **49**, 384–388, MR964397 (90k:94025).

348. G. Danaraj and V. Klee [1977], The connectedness game and the c-complexity of certain graphs, *SIAM J. Appl. Math.* **32**, 431–442.

349. C. Darby and R. Laver [1998], Countable length Ramsey games, *Set Theory: Techniques and Applications.* Proc. of the conferences, Curacao, Netherlands Antilles, June 26–30, 1995 and Barcelona, Spain, June 10–14, 1996 (C. A. Di Prisco et al., eds.), Kluwer, Dordrecht, pp. 41–46.

350. A. L. Davies [1970], Rotating the fifteen puzzle, *Math. Gaz.* **54**, 237–240.

351. M. Davis [1963], Infinite games of perfect information, *Ann. of Math. Stud., Princeton* **52**, 85–101.

352. R. W. Dawes [1992], Some pursuit-evasion problems on grids, *Inform. Process. Lett.* **43**, 241–247.

353. T. R. Dawson [1934], Problem 1603, *Fairy Chess Review* p. 94, Dec.

354. T. R. Dawson [1935], Caissa's Wild Roses, reprinted in: *Five Classics of Fairy Chess*, Dover, 1973.

355. N. G. de Bruijn [1972], A solitaire game and its relation to a finite field, *J. Recr. Math.* **5**, 133–137.

356. N. G. de Bruijn [1981], Pretzel Solitaire as a pastime for the lonely mathematician, in: *The Mathematical Gardner* (D. A. Klarner, ed.), Wadsworth Internat., Belmont, CA, pp. 16–24.

357. F. de Carteblanche [1970], The princess and the roses, *J. Recr. Math.* **3**, 238–239.

358. F. deCarte Blanche [1974], The roses and the princes, *J. Recr. Math.* **7**, 295–298.

359. A. P. DeLoach [1971], Some investigations into the game of SIM, *J. Recr. Math.* **4**, 36–41.

360. E. D. Demaine [2001], Playing games with algorithms: algorithmic combinatorial game theory, *Mathematical Foundations of Computer Science* (J. Sgall, A. Pultr and P. Kolman, eds.), Vol. 2136 of *Lecture Notes in Comput. Sci.*, Springer, Berlin, pp. 18–32, MR1906998 (2003d:68076).

361. E. D. Demaine, M. L. Demaine and D. Eppstein [2002], Phutball Endgames are Hard, in: *More Games of No Chance,* Proc. MSRI Workshop on Combinatorial Games, July, 2000, Berkeley, CA, MSRI Publ. (R. J. Nowakowski, ed.), Vol. 42, Cambridge University Press, Cambridge, pp. 351–360, MR1973023 (2004b:91042).

362. E. D. Demaine, M. L. Demaine and R. Fleischer [2004], Solitaire clobber, *Theoret. Comp. Sci.* **313**, 325–338, special issue of Dagstuhl Seminar "Algorithmic Combinatorial Game Theory", Feb. 2002, MR2056930.

363. E. D. Demaine, M. L. Demaine, R. Fleischer, R. A. Hearn and T. von Oertzen [2009], The complexity of the Dyson telescopes puzzle, in: *Games of No Chance III*, Proc. BIRS Workshop on Combinatorial Games, July, 2005, Banff, Alberta, Canada, MSRI Publ. (M. H. Albert and R. J. Nowakowski, eds.), Vol. 56, Cambridge University Press, Cambridge, pp. 271–285.

364. E. Demaine, M. L. Demaine and J. O'Rourke [2000], PushPush and Push-1 are NP-hard in 2D, *Proc. 12th Annual Canadian Conf. on Computational Geometry*, Fredericton, New Brunswick, Canada, pp. 17–20.

365. E. D. Demaine, M. L. Demaine and H. A. Verrill [2002], Coin-moving puzzles, in: *More Games of No Chance*, Proc. MSRI Workshop on Combinatorial Games, July, 2000, Berkeley, CA, MSRI Publ. (R. J. Nowakowski, ed.), Vol. 42, Cambridge University Press, Cambridge, pp. 405–431, MR1973108 (2004b:91043).

366. E. D. Demaine, R. Fleischer, A. S. Fraenkel and R. J. Nowakowski [2004], Open problems at the 2002 Dagstuhl Seminar on algorithmic combinatorial game theory, *Theoret. Comp. Sci.* **313**, 539–543, special issue of Dagstuhl Seminar "Algorithmic Combinatorial Game Theory" (Appendix B), Feb. 2002, MR2056945.

367. E. Demaine and R. A. Hearn [2008], Constraint Logic: A Uniform Framework for Modeling Computation as Games, *Proc. 23rd Annual IEEE Conf. on Computational Complexity*, Univ. of Maryland, College Park, to appear.

368. E. D. Demaine and R. A. Hearn [2009], Playing games with algorithms: algorithmic combinatorial game theory, in: *Games of No Chance III*, Proc. BIRS Workshop on Combinatorial Games, July, 2005, Banff, Alberta, Canada, MSRI Publ. (M. H. Albert and R. J. Nowakowski, eds.), Vol. 56, Cambridge University Press, Cambridge, pp. 3–56.

369. E. D. Demaine, R. A. Hearn and M. Hoffmann [2002], Push-2-f is PSPACE-Complete, *Proc. 14th Canad. Conf. Computational Geometry*, Lethbridge, Alberta, Canada, pp. 31–35.

370. J. DeMaio [2007], Which chessboards have a closed knight's tour within the cube?, *Electr. J. Combin.* **14**, #R32, 9 pp.

371. H. de Parville [1884], La tour d'Hanoï et la question du Tonkin, *La Nature* **12**, 285–286.

372. C. Deppe [2000], Solution of Ulam's searching game with three lies or an optimal adaptive strategy for binary three-error-correcting codes, *Discrete Math.* **224**, 79–98, MR1781286 (2001j:94054).

373. C. Deppe [2004], Strategies for the Renyi-Ulam game with fixed number of lies, *Theoret. Comput. Sci.* **314**, 45–55, MR2033744 (2004k:05012).

374. B. Descartes [1953], Why are series musical?, *Eureka* **16**, 18–20, reprinted *ibid.* **27** (1964) 29–31.

375. W. Deuber and S. Thomassé [1996], Grundy Sets of Partial Orders, *Technical Report No. 96-123*, Diskrete Strukturen in der Mathematik, Universität Bielefeld.

376. A. K. Dewdney [1984 – 1991], Computer Recreations, a column in Scientific American (May, 1984 – September 1991).

377. A. K. Dewdney [1988], *The Armchair Universe: An Exploration of Computer Worlds*, W. H. Freeman and Company, New York.

378. A. K. Dewdney [1989], *The Turing Omnibus: 61 Excursions in Computer Science*, Computer Science Press, Rockville, MD.

379. A. K. Dewdney [1993], *The (new) Turing Omnibus: 66 Excursions in Computer Science*, Computer Science Press, New York.

380. A. Dhagat, P. Gács and P. Winkler [1992], On playing "twenty questions" with a liar, *Proc. Third Annual ACM-SIAM Sympos. on Discrete Algorithms*, (Orlando, FL, 1992), ACM, New York, pp. 16–22.

381. C. S. Dibley and W. D. Wallis [1981], The effect of starting position in jai-alai, *Congr. Numer.* **32**, 253–259, Proc. 12-th Southeastern Conf. on Combinatorics, Graph Theory and Computing, Vol. I (Baton Rouge, LA, 1981).

382. C. G. Diderich [1995], Bibliography on minimax game theory, sequential and parallel algorithms. http://diwww.epfl.ch/~diderich/bibliographies.html

383. T. Dinski and X. Zhu [1999], A bound for the game chromatic number of graphs, *Discrete Math.* **196**(1-3), 109–115, MR1664506 (99k:05077).

384. R. Diestel and I. Leader [1994], Domination games on infinite graphs, *Theoret. Comput. Sci.* (Math Games) **132**, 337–345.

385. T. Dinski and X. Zhu [1999], A bound for the game chromatic number of graphs, *Discrete Math.* **196**(1-3), 109–115, MR1664506 (99k:05077).

386. Y. Dodis and P. Winkler [2001], Universal configurations in light-flipping games, *Proc. 12th Annual ACM-SIAM Sympos. on Discrete Algorithms*, (Washington, DC, 2001), ACM, New York, pp. 926–927.

387. B. Doerr [2001], Vector balancing games with aging, *J. Combin. Theory Ser. A* **95**, 219–233, MR2154483.

388. A. P. Domoryad [1964], *Mathematical Games and Pastimes*, Pergamon Press, Oxford, translated by H. Moss.

389. D. Dor and U. Zwick [1999], SOKOBAN and other motion planning problems, *Comput. Geom.* **13**, 215–228.

390. M. Dresher [1951], Games of strategy, *Math. Mag.* **25**, 93–99.

391. A. Dress, A. Flammenkamp and N. Pink [1999], Additive periodicity of the Sprague-Grundy function of certain Nim games, *Adv. in Appl. Math.* **22**, 249–270.

392. G. C. Drummond-Cole [2005], Positions of value *2 in generalized domineering and chess, *Integers, Electr. J. of Combinat. Number Theory* **5**, #G6, 13 pp., Comb. Games Sect., MR2192254. http://www.integers-ejcnt.org/vol5.html

393. E. Duchêne and S. Gravier [2008], Geometrical extensions of Wythoff's game, *Discrete Math.* to appear.

394. P. Duchet [1980], Graphes noyau-parfaits, *Ann. Discrete Math.* **9**, 93–101.

395. P. Duchet [1987], A sufficient condition for a digraph to be kernel-perfect, *J. Graph Theory* **11**, 81–85.

396. P. Duchet [1987], Parity graphs are kernel-M-solvable, *J. Combin. Theory* (Ser. B) **43**, 121–126.

397. P. Duchet and H. Meyniel [1981], A note on kernel-critical graphs, *Discrete Math.* **33**, 103–105.

398. P. Duchet and H. Meyniel [1983], Une généralisation du théorème de Richardson sur l'existence de noyaux dans le graphes orientés, *Discrete Math.* **43**, 21–27.

399. P. Duchet and H. Meyniel [1993], Kernels in directed graphs: a poison game, *Discrete Math.* **115**, 273–276.

400. H. E. Dudeney [1958], *The Canterbury Puzzles and Other Curious Problems*, Dover, Mineola, NY, 4th edn., 1st edn: W. Heinemann, 1907.

401. H. E. Dudeney [1970], *Amusements in Mathematics*, Dover, Mineola, NY, 1st edn: Dover, 1917, reprinted by Dover in 1959.

402. A. Duffy, G. Kolpin and D. Wolfe [2009], Ordinal partizan End Nim, in: *Games of No Chance III*, Proc. BIRS Workshop on Combinatorial Games, July, 2005, Banff, Alberta, Canada, MSRI Publ. (M. H. Albert and R. J. Nowakowski, eds.), Vol. 56, Cambridge University Press, Cambridge, pp. 419–425.

403. I. Dumitriu and J. Spencer [2004], A halfliar's game, *Theoret. Comp. Sci.* **313**, 353–369, special issue of Dagstuhl Seminar "Algorithmic Combinatorial Game Theory", Feb. 2002, MR2056932.

404. I. Dumitriu and J. Spencer [2005], The liar game over an arbitrary channel, *Combinatorica* **25**(5), 537–559, MR2176424 (2006m:91003).

405. I. Dumitriu and J. Spencer [2005], The two-batch liar game over an arbitrary channel, *SIAM J. Discrete Math.* **19**(4), 1056–1064 (electronic), MR2206379 (2006m:91037).

406. C. Dunn [2007], The relaxed game chromatic index of k-degenerate graphs, *Discrete Math.* **307**(14), 1767–1775, MR2316815 (2007m:05085).

407. C. Dunn and H. A. Kierstead [2004], A simple competitive graph coloring algorithm. II, *J. Combin. Theory Ser. B* **90**(1), 93–106, Dedicated to Adrian Bondy and U. S. R. Murty, MR2041319 (2005h:05072).

408. C. Dunn and H. A. Kierstead [2004], A simple competitive graph coloring algorithm. III, *J. Combin. Theory Ser. B* **92**(1), 137–150, MR2078498 (2007b:05069).

409. C. Dunn and H. A. Kierstead [2004], The relaxed game chromatic number of outerplanar graphs, *J. Graph Theory* **46**(1), 69–78, MR2051470 (2004m:05097).

410. N. Duvdevani and A. S. Fraenkel [1989], Properties of k-Welter's game, *Discrete Math.* **76**, 197–221.

411. J. Edmonds [1965], Lehman's switching game and a theorem of Tutte and Nash–Williams, *J. Res. Nat. Bur. Standards* **69B**, 73–77.

412. R. Ehrenborg and E. Steingrímsson [1996], Playing Nim on a simplicial complex, *Electr. J. Combin.* **3**(1), #R9, 33 pp.

413. A. Ehrenfeucht and J. Mycielski [1979], Positional strategies for mean payoff games, *Internat. J. Game Theory* **8**, 109–113.

414. N. D. Elkies [1996], On numbers and endgames: combinatorial game theory in chess endgames, in: *Games of No Chance,* Proc. MSRI Workshop on Combina-

torial Games, July, 1994, Berkeley, CA, MSRI Publ. (R. J. Nowakowski, ed.), Vol. 29, Cambridge University Press, Cambridge, pp. 135–150, MR1427963.

415. N. D. Elkies [2002], Higher nimbers in pawn endgames on large chessboards, in: *More Games of No Chance*, Proc. MSRI Workshop on Combinatorial Games, July, 2000, Berkeley, CA, MSRI Publ. (R. J. Nowakowski, ed.), Vol. 42, Cambridge University Press, Cambridge, pp. 61–78, MR1973005 (2004c:91029).

416. R. B. Ellis, V. Ponomarenko and C. H. Yan [2005], The Rényi-Ulam pathological liar game with a fixed number of lies, *J. Combin. Theory Ser. A* **112**, 328–336, MR2177490 (2006g:91040).

417. R. B. Ellis and C. H. Yan [2004], Ulam's pathological liar game with one half-lie, *Int. J. Math. Math. Sci.* pp. 1523–1532, MR2085073 (2005c:91033).

418. D. Engel [1972], DIM: three-dimensional Sim, *J. Recr. Math.* **5**, 274–275.

419. B. Engels and T. Kamphans [2006], Randolph's robot game is NP-hard!, in: *CTW2006 — Cologne-Twente Workshop on Graphs and Combinatorial Optimization*, Vol. 25 of *Electron. Notes Discrete Math.*, Elsevier, Amsterdam, pp. 49–53 (electronic), MR2301136.

420. R. J. Epp and T. S. Ferguson [1980], A note on take-away games, *Fibonacci Quart.* **18**, 300–303.

421. D. Eppstein [2002], Searching for spaceships, in: *More Games of No Chance*, Proc. MSRI Workshop on Combinatorial Games, July, 2000, Berkeley, CA, MSRI Publ. (R. J. Nowakowski, ed.), Vol. 42, Cambridge University Press, Cambridge, pp. 433–453.

422. R. A. Epstein [1977], *Theory of Gambling and Statistial Logic*, Academic Press, New York, NY.

423. M. C. Er [1982], A representation approach to the tower of Hanoi problem, *Comput. J.* **25**, 442–447.

424. M. C. Er [1983], An analysis of the generalized towers of Hanoi problem, *BIT* **23**, 429–435.

425. M. C. Er [1983], An iterative solution to the generalized towers of Hanoi problem, *BIT* **23**, 295–302.

426. M. C. Er [1984], A generalization of the cyclic towers of Hanoi, *Intern. J. Comput. Math.* **15**, 129–140.

427. M. C. Er [1984], The colour towers of Hanoi: a generalization, *Comput. J.* **27**, 80–82.

428. M. C. Er [1984], The cyclic towers of Hanoi: a representation approach, *Comput. J.* **27**, 171–175.

429. M. C. Er [1984], The generalized colour towers of Hanoi: an iterative algorithm, *Comput. J.* **27**, 278–282.

430. M. C. Er [1984], The generalized towers of Hanoi problem, *J. Inform. Optim. Sci.* **5**, 89–94.

431. M. C. Er [1985], The complexity of the generalized cyclic towers of Hanoi problem, *J. Algorithms* **6**, 351–358.

432. M. C. Er [1987], A general algorithm for finding a shortest path between two *n*-configurations, *Information Sciences* **42**, 137–141.

433. M. C. Er [1987], A time and space efficient algorithm for the cyclic Towers of Hanoi problem, *J. Inform. Process.* **9**, 163–165.

434. M. C. Er [1988], A minimal space algorithm for solving the towers of Hanoi problem, *J. Inform. Optim. Sci.* **9**, 183–191.

435. M. C. Er [1989], A linear space algorithm for solving the Towers of Hanoi problem by using a virtual disc., *Inform. Sci.* **47**, 47–52.

436. C. Erbas, S. Sarkeshik and M. M. Tanik [1992], Different perspectives of the N-queens problem, *Proc. ACM Computer Science Conf.*, Kansas City, MO, pp. 99–108.

437. C. Erbas and M. M. Tanik [1994], Parallel memory allocation and data alignment in SIMD machines, *Parallel Algorithms and Applications* **4**, 139–151, preliminary version appeared under the title: Storage schemes for parallel memory systems and the N-queens problem, in: Proc. 15th Ann. Energy Tech. Conf., Houston, TX, Amer. Soc. Mech. Eng., Vol. 43, 1992, pp. 115–120.

438. C. Erbas, M. M. Tanik and Z. Aliyazicioglu [1992], Linear conguence equations for the solutions of the N-queens problem, *Inform. Process. Lett.* **41**, 301–306.

439. P. L. Erdös, U. Faigle, W. Hochstättler and W. Kern [2004], Note on the game chromatic index of trees, *Theoret. Comp. Sci.* **313**, 371–376, special issue of Dagstuhl Seminar "Algorithmic Combinatorial Game Theory", Feb. 2002, MR2056933.

440. P. Erdős, W. R. Hare, S. T. Hedetniemi and R. C. Laskar [1987], On the equality of the Grundy and ochromatic numbers of a graph, *J. Graph Theory* **11**, 157–159.

441. P. Erdős, S. T. Hedetniemi, R. C. Laskar and G. C. E. Prins [2003], On the equality of the partial Grundy and upper ochromatic numbers of graphs, *Discrete Math.* **272**, 53–64, MR2019200 (2004i:05048).

442. P. Erdős and J. L. Selfridge [1973], On a combinatorial game, *J. Combin. Theory* (Ser. A) **14**, 298–301.

443. J. Erickson [1996], New toads and frogs results, in: *Games of No Chance,* Proc. MSRI Workshop on Combinatorial Games, July, 1994, Berkeley, CA, MSRI Publ. (R. J. Nowakowski, ed.), Vol. 29, Cambridge University Press, Cambridge, pp. 299–310.

444. J. Erickson [1996], Sowing games, in: *Games of No Chance,* Proc. MSRI Workshop on Combinatorial Games, July, 1994, Berkeley, CA, MSRI Publ. (R. J. Nowakowski, ed.), Vol. 29, Cambridge University Press, Cambridge, pp. 287–297.

445. M. Erickson and F. Harary [1983], Picasso animal achievement games, *Bull. Malaysian Math. Soc.* **6**, 37–44.

446. N. Eriksen, H. Eriksson and K. Eriksson [2000], Diagonal checker-jumping and Eulerian numbers for color-signed permutations, *Electr. J. Combin.* **7**, #R3, 11 pp.

447. H. Eriksson [1995], Pebblings, *Electr. J. Combin.* **2**, #R7, 18 pp.

448. H. Eriksson, K. Eriksson, J. Karlander, L. Svensson and J. Wästlund [2001], Sorting a bridge hand, *Discrete Math.* **241**, 289–300, Selected papers in honor of Helge Tverberg.

449. H. Eriksson, K. Eriksson and J. Sjöstrand [2001], Note on the lamp lighting problem, *Adv. in Appl. Math.* **27**, 357–366, Special issue in honor of Dominique Foata's 65th birthday (Philadelphia, PA, 2000), MR1868970 (2002i:68082).

450. H. Eriksson and B. Lindström [1995], Twin jumping checkers in $\mathbb{Z}^d$, *European J. Combin.* **16**, 153–157.

451. K. Eriksson [1991], No polynomial bound for the chip firing game on directed graphs, *Proc. Amer. Math. Soc.* **112**, 1203–1205.

452. K. Eriksson [1992], Convergence of Mozes' game of numbers, *Linear Algebra Appl.* **166**, 151–165.

453. K. Eriksson [1994], Node firing games on graphs, *Contemp. Math.* **178**, 117–127.

454. K. Eriksson [1994], Reachability is decidable in the numbers game, *Theoret. Comput. Sci.* (Math Games) **131**, 431–439.

455. K. Eriksson [1995], The numbers game and Coxeter groups, *Discrete Math.* **139**, 155–166.

456. K. Eriksson [1996], Chip-firing games on mutating graphs, *SIAM J. Discrete Math.* **9**, 118–128.

457. K. Eriksson [1996], Strong convergence and a game of numbers, *European J. Combin.* **17**, 379–390.

458. K. Eriksson [1996], Strong convergence and the polygon property of 1-player games, *Discrete Math.* **153**, 105–122, Proc. 5th Conf. on Formal Power Series and Algebraic Combinatorics (Florence 1993).

459. G. Etienne [1991], Tableaux de Young et solitaire bulgare, *J. Combin. Theory* (Ser. A) **58**, 181–197, MR1129115 (93a:05134).

460. J. M. Ettinger [2000], A metric for positional games, *Theoret. Comput. Sci.* (Math Games) **230**, 207–219, MR1725638 (2001g:91041).

461. M. Euwe [1929], Mengentheoretische Betrachtungen über das Schachspiel, *Proc. Konin. Akad. Wetenschappen* **32**, 633–642.

462. R. J. Evans [1974], A winning opening in reverse Hex, *J. Recr. Math.* **7**, 189–192.

463. R. J. Evans [1975–76], Some variants of Hex, *J. Recr. Math.* **8**, 120–122.

464. R. J. Evans [1979], Silverman's game on intervals, *Amer. Math. Monthly* **86**, 277–281.

465. R. J. Evans and G. A. Heuer [1992], Silverman's game on discrete sets, *Linear Algebra Appl.* **166**, 217–235.

466. S. Even and R. E. Tarjan [1976], A combinatorial problem which is complete in polynomial space, *J. Assoc. Comput. Mach.* **23**, 710–719, also appeared in Proc. 7th Ann. ACM Symp. Theory of Computing (Albuquerque, NM, 1975), Assoc. Comput. Mach., New York, NY, 1975, pp. 66–71.

467. G. Exoo [1980-81], A new way to play Ramsey games, *J. Recr. Math.* **13**(2), 111–113.

468. U. Faigle, W. Kern, H. Kierstead and W. T. Trotter [1993], On the game chromatic number of some classes of graphs, *Ars Combin.* **35**, 143–150.

469. U. Faigle, W. Kern and J. Kuipers [1998], Computing the nucleolus of min-cost spanning tree games is NP-hard, *Internat. J. Game Theory* **27**, 443–450.

470. E. Falkener [1961], *Games Ancient and Oriental and How to Play Them*, Dover, New York, NY. (Published previously by Longmans Green, 1892.).

471. B.-J. Falkowski and L. Schmitz [1986], A note on the queens' problem, *Inform. Process. Lett.* **23**, 39–46.

472. G. E. Farr [2003], The Go polynomials of a graph, *Theoret. Comp. Sci.* **306**, 1–18, MR2000162 (2004e:05074).

473. J. Farrell, M. Gardner and T. Rodgers [2005], Configuration games, in: *Tribute to a Mathemagician*, honoring Martin Gardner (B. Cipra, E. D. Demaine, M. L. Demaine and T. Rodgers, eds.), A K Peters, Wellesley, MA, pp. 93-99.

474. T. Feder [1990], Toetjes, *Amer. Math. Monthly* **97**, 785–794.

475. T. Feder and C. Subi [2005], Disks on a tree: analysis of a combinatorial game, *SIAM J. Discrete Math.* **19**, 543–552 (electronic), MR2191279 (2006i:05022).

476. S. P. Fekete, R. Fleischer, A. S. Fraenkel and M. Schmitt [2004], Traveling salesmen in the presence of competition, *Theoret. Comp. Sci.* **313**, 377–392, special issue of Dagstuhl Seminar "Algorithmic Combinatorial Game Theory", Feb. 2002, MR2056934 (2005a:90168).

477. T. S. Ferguson [1974], On sums of graph games with last player losing, *Internat. J. Game Theory* **3**, 159–167.

478. T. S. Ferguson [1984], Misère annihilation games, *J. Combin. Theory* (Ser. A) **37**, 205–230.

479. T. S. Ferguson [1989], Who solved the secretary problem?, *Statistical Science* **4**, 282–296.

480. T. S. Ferguson [1992], Mate with bishop and knight in kriegspiel, *Theoret. Comput. Sci.* (Math Games) **96**, 389–403.

481. T. S. Ferguson [1998], Some chip transfer games, *Theoret. Comp. Sci.* (Math Games) **191**, 157–171.

482. T. S. Ferguson [2001], Another form of matrix Nim, *Electr. J. Combin.* **8(2)**, #R9, 9 pp., Volume in honor of Aviezri S. Fraenkel, MR1853260 (2002g:91046).

483. A. S. Finbow and B. L. Hartnell [1983], A game related to covering by stars, *Ars Combinatoria* **16-A**, 189–198.

484. A. Fink and R. Guy [2007], The Number-pad Game, *Coll. Math. J.* **38**, 260–264, MR2340919.

485. M. J. Fischer and R. N. Wright [1993], An application of game-theoretic techniques to cryptography, *Advances in Computational Complexity Theory* (New Brunswick, NJ, 1990), DIMACS Ser. Discrete Math. Theoret. Comput. Sci., Vol. 13, pp. 99–118.

486. P. C. Fishburn and N. J. A. Sloane [1989], The solution to Berlekamp's switching game, *Discrete Math.* **74**, 263–290.

487. D. C. Fisher and J. Ryan [1992], Optimal strategies for a generalized "scissors, paper, and stone" game, *Amer. Math. Monthly* **99**, 935–942.

488. D. C. Fisher and J. Ryan [1995], Probabilities within optimal strategies for tournament games, *Discrete Appl. Math.* **56**, 87–91.

489. D. C. Fisher and J. Ryan [1995], Tournament games and positive tournaments, *J. Graph Theory* **19**, 217–236.

490. S. L. Fitzpatrick and R. J. Nowakowski [2001], Copnumber of graphs with strong isometric dimension two, *Ars Combin.* **59**, 65–73, MR1832198 (2002b:05053).

491. G. W. Flake and E. B. Baum [2002], *Rush Hour* is PSPACE-complete, or "Why you should generously tip parking lot attendants", *Theoret. Comput. Sci.* (Math Games) **270**, 895–911, MR1871102 (2002h:68068).

492. A. Flammenkamp [1996], Lange Perioden in Subtraktions-Spielen, Ph.D. Thesis, University of Bielefeld.

493. A. Flammenkamp, A. Holshouser and H. Reiter [2003], Dynamic one-pile blocking Nim, *Electr. J. Combinatorics* **10**, #N4, 6 pp., MR1975777 (2004b:05027).

494. J. A. Flanigan [1978], Generalized two-pile Fibonacci nim, *Fibonacci Quart.* **16**, 459–469.

495. J. A. Flanigan [1981], On the distribution of winning moves in random game trees, *Bull. Austr. Math. Soc.* **24**, 227–237.

496. J. A. Flanigan [1981], Selective sums of loopy partizan graph games, *Internat. J. Game Theory* **10**, 1–10.

497. J. A. Flanigan [1982], A complete analysis of black-white Hackendot, *Internat. J. Game Theory* **11**, 21–25.

498. J. A. Flanigan [1982], One-pile time and size dependent take-away games, *Fibonacci Quart.* **20**, 51–59.

499. J. A. Flanigan [1983], Slow joins of loopy games, *J. Combin. Theory* (Ser. A) **34**, 46–59.

500. R. Fleischer and G. Trippen [2006], Kayles on the way to the stars, *Proc. 4th Intern. Conference on Computers and Games CG'2004* (H. J. van den Herik, Y. Björnsson and N. S. Netanyahu, eds.), Bar-Ilan University, Ramat-Gan, Israel, July 2004, Lecture Notes in Computer Science Vol. 3846, Springer, pp. 232–245.

501. J. O. Flynn [1973], Lion and man: the boundary constraint, *SIAM J. Control* **11**, 397–411.

502. J. O. Flynn [1974], Lion and man: the general case, *SIAM J. Control* **12**, 581–597.

503. J. O. Flynn [1974], Some results on max-min pursuit, *SIAM J. Control* **12**, 53–69.

504. F. V. Fomin [1998], Helicopter search problems, bandwidth and pathwidth, *Discrete Appl. Math.* **85**, 59–70.

505. F. V. Fomin [1999], Note on a helicopter search problem on graphs, *Discrete Appl. Math.* **95**, 241–249, Proc. Conf. on Optimal Discrete Structures and Algorithms — ODSA '97 (Rostock).

506. F. V. Fomin and N. N. Petrov [1996], Pursuit-evasion and search problems on graphs, *Congr. Numer.* **122**, 47–58, Proc. 27-th Southeastern Intern. Conf. on Combinatorics, Graph Theory and Computing (Baton Rouge, LA, 1996).

507. L. R. Foulds and D. G. Johnson [1984], An application of graph theory and integer programming: chessboard non-attacking puzzles, *Math. Mag.* **57**, 95–104.

508. A. S. Fraenkel [1974], Combinatorial games with an annihilation rule, in: *The Influence of Computing on Mathematical Research and Education,* Missoula MT, August 1973, Proc. Symp. Appl. Math., (J. P. LaSalle, ed.), Vol. 20, Amer. Math. Soc., Providence, RI, pp. 87–91.

509. A. S. Fraenkel [1977], The particles and antiparticles game, *Comput. Math. Appl.* **3**, 327–328.

510. A. S. Fraenkel [1980], From Nim to Go, *Ann. Discrete Math.* **6**, 137–156, Proc. Symp. on Combinatorial Mathematics, Combinatorial Designs and Their Applications (J. Srivastava, ed.), Colorado State Univ., Fort Collins, CO, June 1978.

511. A. S. Fraenkel [1981], Planar kernel and Grundy with $d \leq 3$, $d_{out} \leq 2$, $d_{in} \leq 2$ are NP-complete, *Discrete Appl. Math.* **3**, 257–262.

512. A. S. Fraenkel [1982], How to beat your Wythoff games' opponent on three fronts, *Amer. Math. Monthly* **89**, 353–361.

513. A. S. Fraenkel [1983], 15 Research problems on games, *Discrete Math.* in "Research Problems" section, Vols. **43-46**.

514. A. S. Fraenkel [1984], Wythoff games, continued fractions, cedar trees and Fibonacci searches, *Theoret. Comput. Sci.* **29**, 49–73, an earlier version appeared in Proc. 10th Internat. Colloq. on Automata, Languages and Programming (J. Diaz, ed.), Vol. 154, Barcelona, July 1983, Lecture Notes in Computer Science, Springer Verlag, Berlin, 1983, pp. 203–225.

515. A. S. Fraenkel [1988], The complexity of chess, Letter to the Editor, *J. Recr. Math.* **20**, 13–14.

516. A. S. Fraenkel [1991], Complexity of games, in: *Combinatorial Games,* Proc. Symp. Appl. Math. (R. K. Guy, ed.), Vol. 43, Amer. Math. Soc., Providence, RI, pp. 111–153.

517. A. S. Fraenkel [1994], Even kernels, *Electr. J. Combinatorics* **1**, #R5, 13 pp.

518. A. S. Fraenkel [1994], Recreation and depth in combinatorial games, in: *The Lighter Side of Mathematics,* Proc. E. Strens Memorial Conf. on Recr. Math. and its History, Calgary, 1986, Spectrum Series (R. K. Guy and R. E. Woodrow, eds.), Math. Assoc. of America, Washington, DC, pp. 176–194.

519. A. S. Fraenkel [1996], Error-correcting codes derived from combinatorial games, in: *Games of No Chance,* Proc. MSRI Workshop on Combinatorial Games, July, 1994, Berkeley, CA, MSRI Publ. (R. J. Nowakowski, ed.), Vol. 29, Cambridge University Press, Cambridge, pp. 417–431.

520. A. S. Fraenkel [1996], Scenic trails ascending from sea-level Nim to alpine chess, in: *Games of No Chance,* Proc. MSRI Workshop on Combinatorial Games, July, 1994, Berkeley, CA, MSRI Publ. (R. J. Nowakowski, ed.), Vol. 29, Cambridge University Press, Cambridge, pp. 13–42.

521. A. S. Fraenkel [1997], Combinatorial game theory foundations applied to digraph kernels, *Electr. J. Combinatorics* **4**(2), #R10, 17 pp., Volume in honor of Herbert Wilf.

522. A. S. Fraenkel [1998], Heap games, numeration systems and sequences, *Ann. Comb.* **2**, 197–210, an earlier version appeared in: *Fun With Algorithms*, Vol. 4 of *Proceedings in Informatics* (E. Lodi, L. Pagli and N. Santoro, eds.), Carleton Scientific, University of Waterloo, Waterloo, Ont., pp. 99–113, 1999. Conference took place on the island of Elba, June 1998., MR1681514 (2000b:91001).

523. A. S. Fraenkel [1998], Multivision: an intractable impartial game with a linear winning strategy, *Amer. Math. Monthly* **105**, 923–928.

524. A. S. Fraenkel [2000], Recent results and questions in combinatorial game complexities, *Theoret. Comput. Sci.* **249**, 265–288, Conference version in: Proc. AWOCA98 — Ninth Australasian Workshop on Combinatorial Algorithms, C.S. Iliopoulos, ed., Perth, Western Australia, 27–30 July, 1998, special AWOCA98 issue, pp. 124-146, MR1798313 (2001j:91033).

525. A. S. Fraenkel [2001], Virus versus mankind, *Proc. 2nd Intern. Conference on Computers and Games CG'2000* (T. Marsland and I. Frank, eds.), Vol. 2063, Hamamatsu, Japan, Oct. 2000, Lecture Notes in Computer Science, Springer, pp. 204–213.

526. A. S. Fraenkel [2002], Arrays, numeration systems and Frankenstein games, *Theoret. Comput. Sci.* **282**, 271–284, special "Fun with Algorithms" issue, MR1909052 (2003h:91036).

527. A. S. Fraenkel [2002], Mathematical chats between two physicists, in: *Puzzler's Tribute: a Feast for the Mind*, honoring Martin Gardner (D. Wolfe and T. Rodgers, eds.), A K Peters, Natick, MA, pp. 383-386.

528. A. S. Fraenkel [2002], Two-player games on cellular automata, in: *More Games of No Chance*, Proc. MSRI Workshop on Combinatorial Games, July, 2000, Berkeley, CA, MSRI Publ. (R. J. Nowakowski, ed.), Vol. 42, Cambridge University Press, Cambridge, pp. 279–306, MR1973018 (2004b:91004).

529. A. S. Fraenkel [2004], Complexity, appeal and challenges of combinatorial games, *Theoret. Comp. Sci.* **313**, 393–415, Expanded version of a keynote address at Dagstuhl Seminar "Algorithmic Combinatorial Game Theory", Feb. 2002, special issue on Algorithmic Combinatorial Game Theory, MR2056935.

530. A. S. Fraenkel [2004], New games related to old and new sequences, *Integers, Electr. J of Combinat. Number Theory* **4**, #G6, 18 pp., Comb. Games Sect., 1st version in Proc.10-th Advances in Computer Games (ACG-10 Conf.), H. J. van den Herik, H. Iida and E. A. Heinz eds., Graz, Austria, Nov. 2003, Kluwer, pp. 367-382, MR2042724. http://www.integers-ejcnt.org/vol4.html

531. A. S. Fraenkel [2005], Euclid and Wythoff games, *Discrete Math.* **304**, 65–68, MR2184445 (2006f:91006).

532. A. S. Fraenkel [2006], Nim is easy, chess is hard — but why??, *J. Internat. Computer Games Assoc.* **29**(4), 203–206, earlier version appeared in Plus Mag. (electronic), pluschat section, http://plus.maths.org/issue40/editorial/index.html.

533. A. S. Fraenkel [2007], *The Raleigh game,* in: Combinatorial Number Theory, de Gruyter, pp. 199–208, Proc. Integers Conference, Carrollton, Georgia, October 27-30,2005, in celebration of the 70th birthday of Ronald Graham. B. Landman, M. Nathanson, J. Nešetřil, R. Nowakowski, C. Pomerance eds., appeared also in *Integers, Electr. J. of Combinat. Number Theory* **7**(2), special volume in honor of Ron Graham, #A13, 11 pp., MR2337047. http://www.integers-ejcnt.org/vol7(2).html

534. A. S. Fraenkel [2007], Why are games exciting and stimulating?, *Math Horizons* pp. 5–7; 32–33, special issue: "Games, Gambling, and Magic" February. German translation by Niek Neuwahl, poster-displayed at traveling exhibition "Games & Science, Science & Games", opened in Göttingen July 17 – Aug 21, 2005.

535. A. S. Fraenkel [2008], Games played by Boole and Galois, *Discrete Appl. Math.* **156**, 420–427.

536. A. S. Fraenkel [2009], *The cyclic Butler University game,* in: Mathematical Wizardry for a Gardner, Volume honoring Martin Gardner (E. Pegg Jr., A. H. Schoen, and T. Rodgers, eds.), A K Peters, Natick, MA, to appear.

537. A. S. Fraenkel and I. Borosh [1973], A generalization of Wythoff's game, *J. Combin. Theory* (Ser. A) **15**, 175–191.

538. A. S. Fraenkel, M. R. Garey, D. S. Johnson, T. Schaefer and Y. Yesha [1978], The complexity of checkers on an $n \times n$ board—preliminary report, *Proc. 19th Ann. Symp. Foundations of Computer Science* (Ann Arbor, MI, Oct. 1978), IEEE Computer Soc., Long Beach, CA, pp. 55–64.

539. A. S. Fraenkel and E. Goldschmidt [1987], Pspace-hardness of some combinatorial games, *J. Combin. Theory* (Ser. A) **46**, 21–38.

540. A. S. Fraenkel and F. Harary [1989], Geodetic contraction games on graphs, *Internat. J. Game Theory* **18**, 327–338.

541. A. S. Fraenkel and H. Herda [1980], Never rush to be first in playing Nimbi, *Math. Mag.* **53**, 21–26.

542. A. S. Fraenkel, A. Jaffray, A. Kotzig and G. Sabidussi [1995], Modular Nim, *Theoret. Comput. Sci.* (Math Games) **143**, 319–333.

543. A. S. Fraenkel and C. Kimberling [1994], Generalized Wythoff arrays, shuffles and interspersions, *Discrete Math.* **126**, 137–149, MR1264482 (95c:11028).

544. A. S. Fraenkel and A. Kontorovich [2007], *The Sierpiński sieve of Nim-varieties and binomial coefficients,* in: Combinatorial Number Theory, de Gruyter, pp. 209–227, Proc. Integers Conference, Carrollton, Georgia, October 27-30,2005, in celebration of the 70th birthday of Ronald Graham. B. Landman, M. Nathanson, J. Nešetřil, R. Nowakowski, C. Pomerance eds., appeared also in *Integers, Electr. J. of Combinat. Number Theory* **7(2)**, special volume in honor of Ron Graham, A14, 19 pp., MR2337048. http://www.integers-ejcnt.org/vol7(2).html

545. A. S. Fraenkel and A. Kotzig [1987], Partizan octal games: partizan subtraction games, *Internat. J. Game Theory* **16**, 145–154.

546. A. S. Fraenkel and D. Krieger [2004], The structure of complementary sets of integers: a 3-shift theorem, *Internat. J. Pure and Appl. Math.* **10**, 1–49, MR2020683 (2004h:05012).

547. A. S. Fraenkel and D. Lichtenstein [1981], Computing a perfect strategy for $n \times n$ chess requires time exponential in n, *J. Combin. Theory* (Ser. A) **31**, 199–214, preliminary version in Proc. 8th Internat. Colloq. Automata, Languages and Programming (S. Even and O. Kariv, eds.), Vol. 115, Acre, Israel, 1981, Lecture Notes in Computer Science, Springer Verlag, Berlin, pp. 278–293, MR629595 (83b:68044).

548. A. S. Fraenkel, M. Loebl and J. Nešetřil [1988], Epidemiography II. Games with a dozing yet winning player, *J. Combin. Theory* (Ser. A) **49**, 129–144.

549. A. S. Fraenkel and M. Lorberbom [1989], Epidemiography with various growth functions, *Discrete Appl. Math.* **25**, 53–71, special issue on Combinatorics and Complexity.

550. A. S. Fraenkel and M. Lorberbom [1991], Nimhoff games, *J. Combin. Theory* (Ser. A) **58**, 1–25.

551. A. S. Fraenkel and J. Nešetřil [1985], Epidemiography, *Pacific J. Math.* **118**, 369–381.

552. A. S. Fraenkel and M. Ozery [1998], Adjoining to Wythoff's game its *P*-positions as moves, *Theoret. Comput. Sci.* **205**, 283–296.

553. A. S. Fraenkel and Y. Perl [1975], Constructions in combinatorial games with cycles, *Coll. Math. Soc. János Bolyai* **10**, 667–699, Proc. Internat. Colloq. on Infinite and Finite Sets, Vol. 2 (A. Hajnal, R. Rado and V. T. Sós, eds.) Keszthely, Hungary, 1973, North-Holland.

554. A. S. Fraenkel and O. Rahat [2001], Infinite cyclic impartial games, *Theoret. Comput. Sci.* **252**, 13–22, special "Computers and Games" issue; first version appeared in Proc. 1st Intern. Conf. on Computer Games CG'98, Tsukuba, Japan, Nov. 1998, *Lecture Notes in Computer Science*, Vol. 1558, Springer, pp. 212–221, 1999., MR1715689 (2000m:91028).

555. A. S. Fraenkel and O. Rahat [2003], Complexity of error-correcting codes derived from combinatorial games, *Proc. Intern. Conference on Computers and Games CG'2002, Edmonton, Alberta, Canada, July 2002*, (Y. Björnsson, M. Müller and J. Schaeffer, eds.), Vol. LNCS 2883, Lecture Notes in Computer Science, Springer, pp. 201–21.

556. A. S. Fraenkel and E. Reisner [2009], The game of End-Wythoff, in: *Games of No Chance III*, Proc. BIRS Workshop on Combinatorial Games, July, 2005, Banff, Alberta, Canada, MSRI Publ. (M. H. Albert and R. J. Nowakowski, eds.), Vol. 56, Cambridge University Press, Cambridge, pp. 329–347.

557. A. S. Fraenkel and E. R. Scheinerman [1991], A deletion game on hypergraphs, *Discrete Appl. Math.* **30**, 155–162.

558. A. S. Fraenkel, E. R. Scheinerman and D. Ullman [1993], Undirected edge geography, *Theoret. Comput. Sci.* (Math Games) **112**, 371–381.

559. A. S. Fraenkel and S. Simonson [1993], Geography, *Theoret. Comput. Sci.* (Math Games) **110**, 197–214.

560. A. S. Fraenkel and U. Tassa [1975], Strategy for a class of games with dynamic ties, *Comput. Math. Appl.* **1**, 237–254.

561. A. S. Fraenkel and U. Tassa [1982], Strategies for compounds of partizan games, *Math. Proc. Camb. Phil. Soc.* **92**, 193–204.

562. A. S. Fraenkel, U. Tassa and Y. Yesha [1978], Three annihilation games, *Math. Mag.* **51**, 13–17, special issue on Recreational Math.

563. A. S. Fraenkel and Y. Yesha [1976], Theory of annihilation games, *Bull. Amer. Math. Soc.* **82**, 775–777.

564. A. S. Fraenkel and Y. Yesha [1979], Complexity of problems in games, graphs and algebraic equations, *Discrete Appl. Math.* **1**, 15–30.

565. A. S. Fraenkel and Y. Yesha [1982], Theory of annihilation games, I, *J. Combin. Theory* (Ser. B) **33**, 60–86.

566. A. S. Fraenkel and Y. Yesha [1986], The generalized Sprague–Grundy function and its invariance under certain mappings, *J. Combin. Theory* (Ser. A) **43**, 165–177.

567. A. S. Fraenkel and D. Zusman [2001], A new heap game, *Theoret. Comput. Sci.* **252**, 5–12, special "Computers and Games" issue; first version appeared in Proc. 1st Intern. Conf. on Computer Games CG'98, Tsukuba, Japan, Nov. 1998, *Lecture Notes in Computer Science*, Vol. 1558, Springer, pp. 205-211, 1999., MR1715688 (2000m:91027).

568. C. N. Frangakis [1981], A backtracking algorithm to generate all kernels of a directed graph, *Intern. J. Comput. Math.* **10**, 35–41.

569. P. Frankl [1987], Cops and robbers in graphs with large girth and Cayley graphs, *Discrete Appl. Math.* **17**, 301–305.

570. P. Frankl [1987], On a pursuit game on Cayley graphs, *Combinatorica* **7**, 67–70.

571. P. Frankl and N. Tokushige [2003], The game of n-times nim, *Discrete Math.* **260**, 205–209, MR1948387 (2003m:05017).

572. W. Fraser, S. Hirshberg and D. Wolfe [2005], The structure of the distributive lattice of games born by day n, *Integers, Electr. J of Combinat. Number Theory* **5(2)**, #A06, 11 pp., MR2192084 (2006g:91041). http://www.integers-ejcnt.org/vol5(2).html

573. D. Fremlin [1973], Well-founded games, *Eureka* **36**, 33–37.

574. G. H. Fricke, S. M. Hedetniemi, S. T. Hedetniemi, A. A. McRae, C. K. Wallis, M. S. Jacobson, H. W. Martin and W. D. Weakley [1995], Combinatorial problems on chessboards: a brief survey, in: *Graph Theory, Combinatorics, and Algorithms:* Proc. 7th Quadrennial Internat. Conf. on the Theory and Applications of Graphs (Y. Alavi and A. Schwenk, eds.), Vol. 1, Wiley, pp. 507–528.

575. E. J. Friedman and A. S. Landsberg [2007], Nonlinear dynamics in combinatorial games: renormalizing Chomp, *Chaos* **17**(2), 023117 1–14, MR2340612.

576. E. J. Friedman and A. S. Landsberg [2009], On the geometry of combinatorial games: a renormalization approach, in: *Games of No Chance III*, Proc. BIRS Workshop on Combinatorial Games, July, 2005, Banff, Alberta, Canada, MSRI Publ. (M. H. Albert and R. J. Nowakowski, eds.), Vol. 56, Cambridge University Press, Cambridge, pp. 349–376.

577. A. Frieze, M. Krivelevich, O. Pikhurko and T. Szabó [2005], The game of JumbleG, *Combin. Probab. Comput.* **14**, 783–793, MR2174656 (2006k:05207).

578. M. Fukuyama [2003], A Nim game played on graphs, *Theoret. Comput. Sci.* **304**, 387–399, MR1992342 (2004f:91041).

579. M. Fukuyama [2003], A Nim game played on graphs II, *Theoret. Comput. Sci.* **304**, 401–419, MR1992343 (2004g:91036).

580. W. W. Funkenbusch [1971], SIM as a game of chance, *J. Recr. Math.* **4**(4), 297–298.

581. Z. Füredi and Á. Seress [1994], Maximal triangle-free graphs with restrictions on the degrees, *J. Graph Theory* **18**, 11–24.

582. H. N. Gabow and H. H. Westermann [1992], Forests, frames, and games: algorithms for matroid sums and applications, *Algorithmica* **7**, 465–497.

583. D. Gale [1974], A curious Nim-type game, *Amer. Math. Monthly* **81**, 876–879.

584. D. Gale [1979], The game of Hex and the Brouwer fixed-point theorem, *Amer. Math. Monthly* **86**, 818–827.

585. D. Gale [1986], Problem 1237 (line-drawing game), *Math. Mag.* **59**, 111, solution by J. Hutchinson and S. Wagon, *ibid.* **60** (1987) 116.

586. D. Gale [1991 – 1996], Mathematical Entertainments, *Math. Intelligencer* **13 - 18**, column on mathematical games and gems, Winter 1991 – Fall 1996.

587. D. Gale [1998], *Tracking the Automatic Ant and Other Mathematical Explorations*, Springer-Verlag, New York, A collection of Mathematical Entertainments columns from The Mathematical Intelligencer.

588. D. Gale and A. Neyman [1982], Nim-type games, *Internat. J. Game Theory* **11**, 17–20.

589. D. Gale, J. Propp, S. Sutherland and S. Troubetzkoy [1995], Further travels with my ant, *Math. Intelligencer* **17** issue 3, 48–56.

590. D. Gale and F. M. Stewart [1953], Infinite games with perfect information, *Ann. of Math. Stud.* (Contributions to the Theory of Games), Princeton **2**(28), 245–266.

591. H. Galeana-Sánchez [1982], A counterexample to a conjecture of Meyniel on kernel-perfect graphs, *Discrete Math.* **41**, 105–107.

592. H. Galeana-Sánchez [1986], A theorem about a conjecture of Meyniel on kernel-perfect graphs, *Discrete Math.* **59**, 35–41.

593. H. Galeana-Sánchez [1988], A new method to extend kernel-perfect graphs to kernel-perfect critical graphs, *Discrete Math.* **69**, 207–209, MR937787 (89d:05086).

594. H. Galeana-Sánchez [1990], On the existence of (k, l)-kernels in digraphs, *Discrete Math.* **85**, 99–102, MR1078316 (91h:05056).

595. H. Galeana-Sánchez [1992], On the existence of kernels and h-kernels in directed graphs, *Discrete Math.* **110**, 251–255.

596. H. Galeana-Sánchez [1995], B_1 and B_2-orientable graphs in kernel theory, *Discrete Math.* **143**, 269–274.

597. H. Galeana-Sánchez [1996], On claw-free M-oriented critical kernel-imperfect digraphs, *J. Graph Theory* **21**, 33–39, MR1363686 (96h:05089).

598. H. Galeana-Sánchez [1998], Kernels in edge-colored digraphs, *Discrete Math.* **184**, 87–99, MR1609359 (99a:05055).

599. H. Galeana-Sánchez [2000], Semikernels modulo F and kernels in digraphs, *Discrete Math.* **218**, 61–71, MR1754327 (2001d:05068).

600. H. Galeana-Sánchez [2002], Kernels in digraphs with covering number at most 3, *Discrete Math.* **259**, 121–135, MR1948776 (2003m:05090).

601. H. Galeana-Sánchez [2004], Some sufficient conditions on odd directed cycles of bounded length for the existence of a kernel, *Discuss. Math. Graph Theory* **24**, 171–182, MR2120561 (2005m:05104).

602. H. Galeana-Sánchez [2006], Kernels and perfectness in arc-local tournament digraphs, *Discrete Math.* **306**(19-20), 2473–2480, special issue: Creation and Recreation: A Tribute to the Memory of Claude Berge, MR2261913 (2007g:05071).

603. H. Galeana-Sánchez and J. de J. García-Ruvalcaba [2000], Kernels in the closure of coloured digraphs, *Discuss. Math. Graph Theory* **20**, 243–254, MR1817495 (2001m:05123).

604. H. Galeana-Sánchez and M.-k. Guevara [2007], Kernel perfect and critical kernel imperfect digraphs structure, in: *6th Czech-Slovak International Symposium on Combinatorics, Graph Theory, Algorithms and Applications*, Vol. 28 of *Electron. Notes Discrete Math.*, Elsevier, Amsterdam, pp. 401–408, MR2324045.

605. H. Galeana-Sánchez and X. Li [1998], Kernels in a special class of digraphs, *Discrete Math.* **178**, 73–80, MR1483740 (98f:05073).

606. H. Galeana-Sánchez and X. Li [1998], Semikernels and (k, l)-kernels in digraphs, *SIAM J. Discrete Math.* **11**, 340–346 (electronic), MR1617163 (99c:05078).

607. H. Galeana-Sánchez and V. Neuman-Lara [1984], On kernels and semikernels of digraphs, *Discrete Math.* **48**, 67–76.

608. H. Galeana-Sánchez and V. Neumann Lara [1986], On kernel-perfect critical digraphs, *Discrete Math.* **59**, 257–265, MR842278 (88b:05069).

609. H. Galeana-Sánchez and V. Neuman-Lara [1991], Extending kernel perfect digraphs to kernel perfect critical digraphs, *Discrete Math.* **94**, 181–187.

610. H. Galeana-Sánchez and V. Neumann Lara [1991], Orientations of graphs in kernel theory, *Discrete Math.* **87**, 271–280, MR1095472 (92d:05134).

611. H. Galeana-Sánchez and V. Neuman-Lara [1994], New extensions of kernel perfect digraphs to kernel imperfect critical digraphs, *Graphs Combin.* **10**, 329–336.

612. H. Galeana-Sánchez and V. Neumann-Lara [1998], New classes of critical kernel-imperfect digraphs, *Discuss. Math. Graph Theory* **18**, 85–89, MR1646233 (99g:05087).

613. H. Galeana-Sánchez and V. Neumann-Lara [1998], On the dichromatic number in kernel theory, *Math. Slovaca* **48**, 213–219, MR1647678 (99k:05080).

614. H. Galeana-Sánchez and L. Pastrana Ramírez [1998], Kernels in edge coloured line digraph, *Discuss. Math. Graph Theory* **18**, 91–98, MR1646234 (99g:05088).

615. H. Galeana-Sánchez, L. Pastrana Ramírez and H. A. Rincón-Mejía [1991], Semikernels, quasi kernels, and Grundy functions in the line digraph, *SIAM J. Discrete Math.* **4**, 80–83, MR1090291 (92c:05068).

616. H. Galeana-Sánchez, L. P. Ramírez and H. A. Rincón Mejía [2005], Kernels in monochromatic path digraphs, *Discuss. Math. Graph Theory* **25**, 407–417, MR2233005.

617. H. Galeana-Sánchez and H. A. Rincón-Mejía [1998], A sufficient condition for the existence of k-kernels in digraphs, *Discuss. Math. Graph Theory* **18**, 197–204, MR1687843 (2000c:05074).

618. H. Galeana-Sánchez and R. Rojas-Monroy [2004], Kernels in pretransitive digraphs, *Discrete Math.* **275**, 129–136, MR2026280 (2004j:05059).

619. H. Galeana-Sánchez and R. Rojas-Monroy [2006], Kernels in quasi-transitive digraphs, *Discrete Math.* **306**(16), 1969–1974, MR2251578 (2007b:05092).

620. H. Galeana-Sánchez and R. Rojas-Monroy [2007], Extensions and kernels in digraphs and in edge-coloured digraphs, *WSEAS Trans. Math.* **6**(2), 334–341, MR2273872.

621. H. Galeana-Sánchez and R. Rojas-Monroy [2007], Kernels in orientations of pre-transitive orientable graphs, in: *Graph theory in Paris*, Trends Math., Birkhäuser, Basel, pp. 197–208, MR2279176.

622. R. P. Gallant, G. Gunther, B. L. Hartnell and D. F. Rall [2006], A game of edge removal on graphs, *J. Combin. Math. Combin. Comput.* **57**, 75–82, MR2226684 (2006m:05245).

623. F. Galvin [1978], Indeterminacy of point-open games, *Bull. de l'Academie Polonaise des Sciences* (Math. astr. et phys.) **26**, 445–449.

624. F. Galvin [1985], Stationary strategies in topological games, *Proc. Conf. on Infinitistic Mathematics* (Lyon, 1984), Publ. Dép. Math. Nouvelle Sér. B, 85–2, Univ. Claude-Bernard, Lyon, pp. 41–43.

625. F. Galvin [1990], Hypergraph games and the chromatic number, in: *A Tribute to Paul Erdős*, Cambridge Univ Press, Cambridge, pp. 201–206.

626. F. Galvin, T. Jech and M. Magidor [1978], An ideal game, *J. Symbolic Logic* **43**, 284–292.

627. F. Galvin and M. Scheepers [1992], A Ramseyan theorem and an infinite game, *J. Combin. Theory* (Ser. A) **59**, 125–129.

628. F. Galvin and R. Telgársky [1986], Stationary strategies in topological games, *Topology Appl.* **22**, 51–69.

629. B. B. Gan and Y. N. Yeh [1995], A nim-like game and dynamic recurrence relations, *Stud. Appl. Math.* **95**, 213–228.

630. A. Gangolli and T. Plambeck [1989], A note on periodicity in some octal games, *Internat. J. Game Theory* **18**, 311–320.

631. T. E. Gantner [1988], The game of Quatrainment, *Math. Mag.* **61**, 29–34.

632. M. Gardner [1956], *Mathematics, Magic and Mystery*, Dover, New York, NY.

633. M. Gardner [Jan. 1957–Dec. 1981], Mathematical Games, a column in Scientific American; all 15 volumes of the column are on 1 CD-ROM, available from MAA.

634. M. Gardner [1959], *Fads and Fallacies in the Name of Science*, Dover, NY.

635. M. Gardner [1959], *Logic Machines and Diagrams*, McGraw-Hill, NY.

636. M. Gardner [1959], *Mathematical Puzzles of Sam Loyd*, Dover, New York, NY.

637. M. Gardner [1960], *More Mathematical Puzzles of Sam Loyd*, Dover, New York, NY.

638. M. Gardner [1966], *More Mathematical Puzzles and Diversions*, Harmondsworth, Middlesex, England (Penguin Books), translated into German: *Mathematische Rätsel und Probleme*, Vieweg, Braunschweig, 1964.

639. M. Gardner, ed. [1967], *536 Puzzles and Curious Problems*, Scribner's, NY, reissue of H. E. Dudeney's *Modern Puzzles* and *Puzzles and Curious Problems*.

640. M. Gardner [1968], *Logic Machines, Diagrams and Boolean Algebra*, Dover, NY.

641. M. Gardner [1970], *Further Mathematical Diversions*, Allen and Unwin, London.

642. M. Gardner [1977], *Mathematical Magic Show*, Knopf, NY.

643. M. Gardner [1978], *Aha! Insight*, Freeman, New York, NY.

644. M. Gardner [1979], *Mathematical Circus*, Knopf, NY.

645. M. Gardner [1981], *Entertaining Science Experiments with Everyday Objects*, Dover, NY.

646. M. Gardner [1981], *Science Fiction Puzzle Tales*, Potter.

647. M. Gardner [1982], *Aha! Gotcha!*, Freeman, New York, NY.

648. M. Gardner [1983], *New Mathematical Diversions from Scientific American*, University of Chicago Press, Chicago, before that appeared in 1971, Simon and Schuster, New York, NY. First appeared in 1966.

649. M. Gardner [1983], *Order and Surprise*, Prometheus Books, Buffalo, NY.

650. M. Gardner [1983], *Wheels, Life and Other Mathematical Amusements*, Freeman, New York, NY.

651. M. Gardner [1984], *Codes, Ciphers and Secret Writing*, Dover, NY.

652. M. Gardner [1984], *Puzzles from Other Worlds*, Random House.

653. M. Gardner [1984], *The Magic Numbers of Dr. Matrix*, Prometheus.

654. M. Gardner [1984], *The Sixth Book of Mathematical Games*, Univ. of Chicago Press. First appeared in 1971, Freeman, New York, NY.

655. M. Gardner [1986], *Knotted Doughnuts and Other Mathematical Entertainments*, Freeman, New York, NY.

656. M. Gardner [1987], *The Second Scientific American Book of Mathematical Puzzles and Diversions*, University of Chicago Press, Chicago. First appeared in 1961, Simon and Schuster, NY.

657. M. Gardner [1988], *Hexaflexagons and Other Mathematical Diversions*, University of Chicago Press, Chicago, 1988. A first version appeared under the title *The Scientific American Book of Mathematical Puzzles and Diversions*, Simon & Schuster, 1959, NY.

658. M. Gardner [1988], *Perplexing Puzzles and Tantalizing Teasers*, Dover, NY.

659. M. Gardner [1988], *Riddles of the Sphinx*, Math. Assoc. of America, Washington, DC.

660. M. Gardner [1988], *Time Travel and Other Mathematical Bewilderments*, Freeman, New York, NY.

661. M. Gardner [1989], *How Not to Test a Psychic*, Prometheus Books, Buffalo, NY.

662. M. Gardner [1989], *Mathematical Carnival*, 2nd edition, Math. Assoc. of America, Washington, DC. First appeared in 1975, Knopf, NY.

663. M. Gardner [1990], *The New Ambidextrous Universe*, Freeman, New York, NY. First appeared in 1964, Basic Books, then in 1969, New American Library.

664. M. Gardner [1991], *The Unexpected Hanging and Other Mathematical Diversions*, University of Chicago Press. First appeared in 1969, Simon and Schuster,

NY, adapted from "Further Mathematical Diversions"; translated into German: *Logik Unterm Galgen*, Vieweg, Braunschweig, 1980.

665. M. Gardner [1992], *Fractal Music, Hypercards and More*, Freeman, New York, NY.

666. M. Gardner [1992], *On the Wild Side*, Prometheus Books, Buffalo, NY.

667. M. Gardner [1993], *Book of Visual Illusions*, Dover, NY.

668. M. Gardner [1997], *Penrose Tiles to Trapdoor Ciphers*, The Math. Assoc. of America, Washington, DC. First appeared in 1989, Freeman, New York, NY. Freeman, New York, NY.

669. M. Gardner [1997], *The Last Recreations*, Copernicus, NY.

670. M. Gardner [2001], *A Gardner's Workout: Training the Mind and Entertaining the Spirit*, A K Peters, Natick, MA.

671. M. R. Garey and D. S. Johnson [1979], *Computers and Intractability: A Guide to the Theory of NP-Completeness*, Freeman, San Francisco, Appendix A8: Games and Puzzles, pp. 254-258.

672. R. Gasser [1996], Solving nine men's Morris, in: *Games of No Chance,* Proc. MSRI Workshop on Combinatorial Games, July, 1994, Berkeley, CA, MSRI Publ. (R. J. Nowakowski, ed.), Vol. 29, Cambridge University Press, Cambridge, pp. 101–114.

673. H. Gavel and P. Strimling [2004], Nim with a modular Muller twist, *Integers, Electr. J of Combinat. Number Theory* **4**, #G4, 9 pp., Comb. Games Sect., MR2116010. http://www.integers-ejcnt.org/vol4.html

674. B. Gerla [2000], Conditioning a state by Łukasiewicz event: a probabilistic approach to Ulam Games, *Theoret. Comput. Sci.* (Math Games) **230**, 149–166, MR1725635 (2001e:03117).

675. C. Germain and H. Kheddouci [2003], Grundy coloring for power graphs, in: *Electron. Notes Discrete Math.* (D. Ray-Chaudhuri, A. Rao and B. Roy, eds.), Vol. 15, Elsevier, Amsterdam, pp. 67–69.

676. P. B. Gibbons and J. A. Webb [1997], Some new results for the queens domination problem, *Australas. J. Combin.* **15**, 145–160.

677. J. R. Gilbert, T. Lengauer and R. E. Tarjan [1980], The pebbling problem is complete in polynomial space, *SIAM J. Comput.* **9**, 513–524, preliminary version in Proc. 11th Ann. ACM Symp. Theory of Computing (Atlanta, GA, 1979), Assoc. Comput. Mach., New York, NY, pp. 237–248.

678. M. Ginsberg [2002], Alpha-beta pruning under partial orders, in: *More Games of No Chance,* Proc. MSRI Workshop on Combinatorial Games, July, 2000, Berkeley, CA, MSRI Publ. (R. J. Nowakowski, ed.), Vol. 42, Cambridge University Press, Cambridge, pp. 37–48.

679. J. Ginsburg [1939], Gauss's arithmetization of the problem of 8 queens, *Scripta Math.* **5**, 63–66.

680. A. S. Goldstein and E. M. Reingold [1995], The complexity of pursuit on a graph, *Theoret. Comput. Sci.* (Math Games) **143**, 93–112.

681. J. Goldwasser and W. Klostermeyer [1997], Maximization versions of "lights out" games in grids and graphs, *Congr. Numer.* **126**, 99–111, Proc. 28th South-

eastern Internat. Conf. on Combinatorics, Graph Theory, and Computing (Boca Raton, FL, 1997), MR1604979 (98j:05119).

682. E. Goles [1991], Sand piles, combinatorial games and cellular automata, *Math. Appl.* **64**, 101–121.

683. E. Goles and M. A. Kiwi [1993], Games on line graphs and sand piles, *Theoret. Comput. Sci.* (Math Games) **115**, 321–349 (0–player game).

684. E. Goles, M. Latapy, C. Magnien, M. Morvan and H. D. Phan [2004], Sandpile models and lattices: a comprehensive survey, *Theoret. Comput. Sci.* **322**, 383–407, MR2080235 (2005h:05013).

685. E. Goles and M. Margenstern [1997], Universality of the chip-firing game, *Theoret. Comput. Sci.* (Math Games) **172**, 121–134.

686. E. Goles, M. Morvan and H. D. Phan [2002], The structure of a linear chip firing game and related models, *Theoret. Comput. Sci.* **270**, 827–841, MR1871097 (2002k:05013).

687. E. Goles and E. Prisner [2000], Source reversal and chip firing on graphs, *Theoret. Comput. Sci.* **233**, 287–295.

688. S. W. Golomb [1966], A mathematical investigation of games of "take-away", *J. Combin. Theory* **1**, 443–458.

689. S. W. Golomb [1994], *Polyominoes: Puzzles, Patterns, Problems, and Packings*, Princeton University Press. Original edition: *Polyominoes*, Scribner's, NY, 1965; Allen and Unwin, London, 1965.

690. S. W. Golomb and A. W. Hales [2002], Hypercube Tic-Tac-Toe, in: *More Games of No Chance*, Proc. MSRI Workshop on Combinatorial Games, July, 2000, Berkeley, CA, MSRI Publ. (R. J. Nowakowski, ed.), Vol. 42, Cambridge University Press, Cambridge, pp. 167–182, MR1973012 (2004c:91030).

691. H. Gonshor [1986], *An Introduction to the Theory of Surreal Numbers*, Cambridge University Press, Cambridge.

692. D. M. Gordon, R. W. Robinson and F. Harary [1994], Minimum degree games for graphs, *Discrete Math.* **128**, 151–163, MR1271861 (95b:05201).

693. E. Grädel [1990], Domino games and complexity, *SIAM J. Comput.* **19**, 787–804.

694. S. B. Grantham [1985], Galvin's "racing pawns" game and a well-ordering of trees, *Memoirs Amer. Math. Soc.* **53**(316), 63 pp.

695. S. Gravier, M. Mhalla and E. Tannier [2003], On a modular domination game, *Theoret. Comput. Sci.* **306**, 291–303, MR2000178 (2004i:91041).

696. R. Greenlaw, H. J. Hoover and W. L. Ruzzo [1995], *Limits to Parallel Computation: P-completeness Theory*, Oxford University Press, New York.

697. J. P. Grossman [2004], Periodicity in one-dimensional peg duotaire, *Theoret. Comp. Sci.* **313**, 417–425, special issue of Dagstuhl Seminar "Algorithmic Combinatorial Game Theory", Feb. 2002, MR2056936 (2005b:91055).

698. J. P. Grossman and R. J. Nowakowski [2002], One-dimensional Phutball, in: *More Games of No Chance,* Proc. MSRI Workshop on Combinatorial Games, July, 2000, Berkeley, CA, MSRI Publ. (R. J. Nowakowski, ed.), Vol. 42, Cambridge University Press, Cambridge, pp. 361–367, MR1973024.

699. J. P. Grossman and A. N. Siegel [2009], Reductions of partizan games, in: *Games of No Chance III*, Proc. BIRS Workshop on Combinatorial Games, July, 2005, Banff, Alberta, Canada, MSRI Publ. (M. H. Albert and R. J. Nowakowski, eds.), Vol. 56, Cambridge University Press, Cambridge, pp. 427–455.

700. J. W. Grossman [2000], A variant of Nim, *Math. Mag.* **73**, 323–324, Problem No. 1580; originally posed *ibid.* **72** (1999) 325.

701. P. M. Grundy [1964], Mathematics and Games, *Eureka* **27**, 9–11, originally published: *ibid.* **2** (1939) 6–8.

702. P. M. Grundy and C. A. B. Smith [1956], Disjunctive games with the last player losing, *Proc. Camb. Phil. Soc.* **52**, 527–533.

703. F. Grunfeld and R. C. Bell [1975], *Games of the World*, Holt, Rinehart and Winston.

704. C. D. Grupp [1976], *Brettspiele-Denkspiele*, Humboldt-Taschenbuchverlag, München.

705. D. J. Guan and X. Zhu [1999], Game chromatic number of outerplanar graphs, *J. Graph Theory* **30**, 67–70.

706. L. J. Guibas and A. M. Odlyzko [1981], String overlaps, pattern matching, and nontransitive games, *J. Combin. Theory* (Ser. A) **30**, 183–208.

707. S. Gunther [1874], Zur mathematischen Theorie des Schachbretts, *Arch. Math. Physik* **56**, 281–292.

708. V. A. Gurvich [2007], On the misere version of game Euclid and miserable games, *Discrete Math.* **307**(9–10), 1199–1204, MR2292548 (2007k:91052).

709. R. K. Guy [1976], Packing $[1, n]$ with solutions of $ax + by = cz$; the unity of combinatorics, *Atti Conv. Lincei #17, Acad. Naz. Lincei, Tomo II*, Rome, pp. 173–179.

710. R. K. Guy [1976], Twenty questions concerning Conway's sylver coinage, *Amer. Math. Monthly* **83**, 634–637.

711. R. K. Guy [1977], Games are graphs, indeed they are trees, *Proc. 2nd Carib. Conf. Combin. and Comput.*, Letchworth Press, Barbados, pp. 6–18.

712. R. K. Guy [1977], She loves me, she loves me not; relatives of two games of Lenstra, Een Pak met een Korte Broek (papers presented to H. W. Lenstra), Mathematisch Centrum, Amsterdam.

713. R. K. Guy [1978], Partizan and impartial combinatorial games, *Colloq. Math. Soc. János Bolyai* **18**, 437–461, Proc. 5th Hungar. Conf. Combin. Vol. I (A. Hajnal and V. T. Sós, eds.), Keszthely, Hungary, 1976, North-Holland.

714. R. K. Guy [1979], Partizan Games, *Colloques Internationaux C. N. R. No. 260 — Problèmes Combinatoires et Théorie des Graphes*, pp. 199–205.

715. R. K. Guy [1981], Anyone for twopins?, in: *The Mathematical Gardner* (D. A. Klarner, ed.), Wadsworth Internat., Belmont, CA, pp. 2–15.

716. R. K. Guy [1983], Graphs and games, in: *Selected Topics in Graph Theory* (L. W. Beineke and R. J. Wilson, eds.), Vol. 2, Academic Press, London, pp. 269–295.

717. R. K. Guy [1986], John Isbell's game of beanstalk and John Conway's game of beans-don't-talk, *Math. Mag.* **59**, 259–269.

718. R. K. Guy [1989], *Fair Game*, COMAP Math. Exploration Series, Arlington, MA.

719. R. K. Guy [1990], A guessing game of Bill Sands, and Bernardo Recamán's Barranca, *Amer. Math. Monthly* **97**, 314–315.

720. R. K. Guy [1991], Mathematics from fun & fun from mathematics; an informal autobiographical history of combinatorial games, in: *Paul Halmos: Celebrating 50 Years of Mathematics* (J. H. Ewing and F. W. Gehring, eds.), Springer Verlag, New York, NY, pp. 287–295.

721. R. K. Guy [1995], Combinatorial games, in: *Handbook of Combinatorics,* (R. L. Graham, M. Grötschel and L. Lovász, eds.), Vol. II, North-Holland, Amsterdam, pp. 2117–2162.

722. R. K. Guy [1996], Impartial Games, in: *Games of No Chance,* Proc. MSRI Workshop on Combinatorial Games, July, 1994, Berkeley, CA, MSRI Publ. (R. J. Nowakowski, ed.), Vol. 29, Cambridge University Press, Cambridge, pp. 61–78, earlier version in: *Combinatorial Games*, Proc. Symp. Appl. Math. (R. K. Guy, ed.), Vol. 43, Amer. Math. Soc., Providence, RI, 1991, pp. 35–55.

723. R. K. Guy [1996], Unsolved problems in combinatorial games, in: *Games of No Chance,* Proc. MSRI Workshop on Combinatorial Games, July, 1994, Berkeley, CA, MSRI Publ. (R. J. Nowakowski, ed.), Vol. 29, Cambridge University Press, Cambridge, pp. 475–491, update with 52 problems of earlier version with 37 problems, in: *Combinatorial Games*, Proc. Symp. Appl. Math. (R. K. Guy, ed.), Vol. 43, Amer. Math. Soc., Providence, RI, 1991, pp. 183–189.

724. R. K. Guy [1996], What is a game?, in: *Games of No Chance,* Proc. MSRI Workshop on Combinatorial Games, July, 1994, Berkeley, CA, MSRI Publ. (R. J. Nowakowski, ed.), Vol. 29, Cambridge University Press, Cambridge, pp. 43–60, earlier version in: *Combinatorial Games*, Proc. Symp. Appl. Math. (R. K. Guy, ed.), Vol. 43, Amer. Math. Soc., Providence, RI, 1991, pp. 1–21.

725. R. K. Guy [2007], A parity subtraction game, *Crux Math.* **33**, 37–39.

726. R. K. Guy and R. J. Nowakowski [1993], Mousetrap, in: *Combinatorics, Paul Erdös is eighty*, Vol. 1, Bolyai Soc. Math. Stud., János Bolyai Math. Soc., Budapest, pp. 193–206.

727. R. K. Guy and R. J. Nowakowski [2002], Unsolved problems in combinatorial games, in: *More Games of No Chance,* Proc. MSRI Workshop on Combinatorial Games, July, 2000, Berkeley, CA, MSRI Publ. (R. J. Nowakowski, ed.), Vol. 42, Cambridge University Press, Cambridge, pp. 457–473.

728. R. K. Guy, R. J. Nowakowski, I. Caines and C. Gates [1999], Unsolved Problems: periods in taking and splitting games, *Amer. Math. Monthly* **106**, 359–361, Unsolved Problems Section.

729. R. K. Guy and C. A. B. Smith [1956], The *G*-values of various games, *Proc. Camb. Phil. Soc.* **52**, 514–526.

730. R. K. Guy and R. E. Woodrow, eds. [1994], *The Lighter Side of Mathematics,* Proc. E. Strens Memorial Conf. on Recr. Math. and its History, Calgary, 1986, Spectrum Series, Math. Assoc. Amer., Washington, DC.

731. W. Guzicki [1990], Ulam's searching game with two lies, *J. Combin. Theory* (Ser. A) **54**, 1–19.

732. G. Hahn, F. Laviolette, N. Sauer and R. E. Woodrow [2002], On cop-win graphs, *Discrete Math.* **258**, 27–41, MR2002070 (2004f:05156).

733. G. Hahn and G. MacGillivray [2006], A note on k-cop, l-robber games on graphs, *Discrete Math.* **306**(19-20), 2492–2497, MR2261915 (2007d:05069).

734. A. Hajnal and Z. Nagy [1984], Ramsey games, *Trans. American Math. Soc.* **284**, 815–827.

735. D. R. Hale [1983], A variant of Nim and a function defined by Fibonacci representation, *Fibonacci Quart.* **21**, 139–142.

736. A. W. Hales and R. I. Jewett [1963], Regularity and positional games, *Trans. Amer. Math. Soc.* **106**, 222–229.

737. L. Halpenny and C. Smyth [1992], A classification of minimal standard-path 2×2 switching networks, *Theoret. Comput. Sci.* (Math Games) **102**, 329–354.

738. I. Halupczok and J.-C. Schlage-Puchta [2007], Achieving snaky, *Integers, Electr. J. of Combinat. Number Theory* **7**, #G02, 28 pp., MR2282185 (2007i:05045). http://www.integers-ejcnt.org/vol7.html

739. Y. O. Hamidoune [1987], On a pursuit game of Cayley digraphs, *Europ. J. Combin.* **8**, 289–295.

740. Y. O. Hamidoune and M. Las Vergnas [1985], The directed Shannon switching game and the one-way game, in: *Graph Theory with Applications to Algorithms and Computer Science* (Y. Alavi et al., eds.), Wiley, pp. 391–400.

741. Y. O. Hamidoune and M. Las Vergnas [1986], Directed switching games on graphs and matroids, *J. Combin. Theory* (Ser. B) **40**, 237–269.

742. Y. O. Hamidoune and M. Las Vergnas [1987], A solution to the box game, *Discrete Math.* **65**, 157–171.

743. Y. O. Hamidoune and M. Las Vergnas [1988], A solution to the misère Shannon switching game, *Discrete Math.* **72**, 163–166.

744. O. Hanner [1959], Mean play of sums of positional games, *Pacific J. Math.* **9**, 81–99.

745. F. Harary [1982], Achievement and avoidance games for graphs, *Ann. Discrete Math.* **13**, 111–119.

746. F. Harary [2002], Sum-free games, in: *Puzzler's Tribute: a Feast for the Mind*, pp. 395–398, honoring Martin Gardner (D. Wolfe and T. Rodgers, eds.), A K Peters, Natick, MA.

747. F. Harary, H. Harborth and M. Seemann [2000], Handicap achievement for polyominoes, *Congr. Numer.* **145**, 65–80, Proc. 31st Southeastern Internat. Conf. on Combinatorics, Graph Theory, and Computing (Boca Raton, FL, 1999), MR1817942 (2001m:05083).

748. F. Harary and K. Plochinski [1987], On degree achievement and avoidance games for graphs, *Math. Mag.* **60**, 316–321.

749. F. Harary, B. Sagan and D. West [1985], Computer-aided analysis of monotonic sequence games, *Atti Accad. Peloritana Pericolanti Cl. Sci. Fis. Mat. Natur.* **61**, 67–78.

750. F. Harary, W. Slany and O. Verbitsky [2002], A symmetric strategy in graph avoidance games, in: *More Games of No Chance,* Proc. MSRI Workshop on Combinatorial Games, July, 2000, Berkeley, CA, MSRI Publ. (R. J. Nowakowski, ed.), Vol. 42, Cambridge University Press, Cambridge, pp. 369–381, MR1973105 (2004b:91036).

751. F. Harary, W. Slany and O. Verbitsky [2004], On the length of symmetry breaking-preserving games on graphs, *Theoret. Comp. Sci.* **313**, 427–446, special issue of Dagstuhl Seminar "Algorithmic Combinatorial Game Theory", Feb. 2002, MR2056937 (2005a:91025).

752. H. Harborth and M. Seemann [1996], Snaky is an edge-to-edge loser, *Geombinatorics* **5**, 132–136, MR1380142.

753. H. Harborth and M. Seemann [1997], Snaky is a paving winner, *Bull. Inst. Combin. Appl.* **19**, 71–78, MR1427578 (97g:05052).

754. H. Harborth and M. Seemann [2003], Handicap achievement for squares, *J. Combin. Math. Combin. Comput.* **46**, 47–52, 15th MCCCC (Las Vegas, NV, 2001), MR2004838 (2004g:05044).

755. P. J. Hayes [1977], A note on the towers of Hanoi problem, *Computer J.* **20**, 282–285.

756. R. B. Hayward [2009], A puzzling Hex primer, in: *Games of No Chance III,* Proc. BIRS Workshop on Combinatorial Games, July, 2005, Banff, Alberta, Canada, MSRI Publ. (M. H. Albert and R. J. Nowakowski, eds.), Vol. 56, Cambridge University Press, Cambridge, pp. 151–161.

757. R. Hayward, Y. Björnsson, M. Johanson, M. Kan, N. Po and J. van Rijswijck [2005], Solving 7×7 Hex with domination, fill-in, and virtual connections, *Theoret. Comput. Sci.* **349**(2), 123–139, MR2184213 (2006f:68112).

758. R. B. Hayward and J. van Rijswijck [2006], Hex and combinatorics, *Discrete Math.* **306**(19-20), 2515–2528, special issue: Creation and Recreation: A Tribute to the Memory of Claude Berge, MR2261917 (2007e:91045).

759. W. He, X. Hou, K.-W. Lih, J. Shao, W. Wang and X. Zhu [2002], Edge-partitions of planar graphs and their game coloring numbers, *J. Graph Theory* **41**(4), 307–317, MR1936946 (2003h:05153).

760. W. He, J. Wu and X. Zhu [2004], Relaxed game chromatic number of trees and outerplanar graphs, *Discrete Math.* **281**, 209–219, MR2047768 (2005a:05088).

761. R. A. Hearn [2005], The complexity of sliding block puzzles and plank puzzles, in: *Tribute to a Mathemagician,* pp. 173–183, honoring Martin Gardner (B. Cipra, E. D. Demaine, M. L. Demaine and T. Rodgers, eds.), A K Peters, Wellesley, MA, pp. 93-99.

762. R. A. Hearn [2006], TipOver is NP-complete, *Math. Intelligencer* **28**, 10–14, Number 3, Summer 2006.

763. R. A. Hearn [2009], Amazons, Konane, and Cross Purposes are PSPACE-complete, in: *Games of No Chance III,* Proc. BIRS Workshop on Combinatorial Games, July, 2005, Banff, Alberta, Canada, MSRI Publ. (M. H. Albert and R. J. Nowakowski, eds.), Cambridge University Press, Cambridge. Vol. 56, Cambridge University Press, Cambridge, pp. 287–306.

764. R. A. Hearn and E. D. Demaine [2002], The nondeterministic constraint logic model of computation: reductions and applications, in: *Automata, Languages and Programming*, Vol. 2380 of *Lecture Notes in Comput. Sci.*, Springer, Berlin, pp. 401–413, MR2062475.

765. R. A. Hearn and E. D. Demaine [2005], PSPACE-completeness of sliding-block puzzles and other problems through the nondeterministic constraint logic model of computation, *Theoret. Comput. Sci.* **343**, 72–96, special issue: Game Theory Meets Theoretical Computer Science, MR2168845 (2006d:68070).

766. O. Heden [1992], On the modular *n*-queen problem, *Discrete Math.* **102**, 155–161.

767. O. Heden [1993], Maximal partial spreads and the modular *n*-queen problem, *Discrete Math.* **120**, 75–91.

768. O. Heden [1995], Maximal partial spreads and the modular *n*-queen problem II, *Discrete Math.* **142**, 97–106.

769. S. M. Hedetniemi, S. Hedetniemi and T. Beyer [1982], A linear algorithm for the Grundy (coloring) number of a tree, *Congr. Numer.* **36**, 351–363.

770. S. M. Hedetniemi, S. T. Hedetniemi and R. Reynolds [1998], Combinatorial problems on chessboards, II, in: *Domination in Graphs* (T. W. Haynes et al., ed.), M. Dekker, NY, pp. 133–162.

771. D. Hefetz, M. Krivelevich and T. Szabó [2007], Avoider-enforcer games, *J. Combin. Theory Ser. A* **114**(5), 840–853, MR2333136.

772. P. Hegarty and U. Larsson [2006], Permutations of the natural numbers with prescribed difference multisets, *Integers, Electr. J. of Combinat. Number Theory* **6**, #A03, 25 pp. http://www.integers-ejcnt.org/vol6.html

773. P. Hein [1942], Vil de lre Polygon?, *Politiken* (description of Hex in this Danish newspaper of Dec. 26) .

774. D. Hensley [1988], A winning strategy at Taxman, *Fibonacci Quart.* **26**, 262–270.

775. C. W. Henson [1970], Winning strategies for the ideal game, *Amer. Math. Monthly* **77**, 836–840.

776. J. C. Hernández-Castro, I. Blasco-Lopez, J. M. Estévez-Tapiador and A. Ribagorda-Garnacho [2006], Steganography in games: A general methodology and its application to the game of Go., *Computers & Security* **25**, 64–71.

777. R. I. Hess [1999], Puzzles from around the world, in: *The Mathemagician and Pied Puzzler*, honoring Martin Gardner; E. Berlekamp and T. Rodgers, eds., A K Peters, Natick, MA, pp. 53-84.

778. G. A. Heuer [1982], Odds versus evens in Silverman-type games, *Internat. J. Game Theory* **11**, 183–194.

779. G. A. Heuer [1989], Reduction of Silverman-like games to games on bounded sets, *Internat. J. Game Theory* **18**, 31–36.

780. G. A. Heuer [2001], Three-part partition games on rectangles, *Theoret. Comput. Sci.* (Math Games) **259**, 639–661.

781. G. A. Heuer and U. Leopold-Wildburger [1995], *Silverman's game, A special class of two-person zero-sum games,* with a foreword by Reinhard Selten, Vol.

424 of Lecture Notes in Economics and Mathematical Systems, Springer-Verlag, Berlin.

782. G. A. Heuer and W. D. Rieder [1988], Silverman games on disjoint discrete sets, *SIAM J. Disc. Math.* **1**, 485–525.

783. R. Hill [1995], Searching with lies, in: *Surveys in Combinatorics*, Vol. 218, Cambridge University Press, Cambridge, pp. 41–70.

784. R. Hill and J. P. Karim [1992], Searching with lies: the Ulam problem, *Discrete Math.* **106/107**, 273–283.

785. T. P. Hill and U. Krengel [1991], Minimax-optimal stop rules and distributions in secretary problems, *Ann. Probab.* **19**, 342–353.

786. T. P. Hill and U. Krengel [1992], On the game of Googol, *Internat. J. Game Theory* **21**, 151–160.

787. P. G. Hinman [1972], Finite termination games with tie, *Israel J. Math.* **12**, 17–22.

788. A. M. Hinz [1989], An iterative algorithm for the tower of Hanoi with four pegs, *Computing* **42**, 133–140.

789. A. M. Hinz [1989], The tower of Hanoi, *Enseign. Math.* **35**, 289–321.

790. A. M. Hinz [1992], Pascal's triangle and the tower of Hanoi, *Amer. Math. Monthly* **99**, 538–544.

791. A. M. Hinz [1992], Shortest paths between regular states of the tower of Hanoi, *Inform. Sci.* **63**, 173–181.

792. A. M. Hinz [1999], The tower of Hanoi, *Algebras and Combinatorics* (K. P. Shum, E. J. Taft and Z. X. Wan, eds.), Springer, Singapore, pp. 277–289.

793. A. M. Hinz and D. Parisse [2002], On the planarity of Hanoi graphs, *Expo. Math.* **20**, 263–268, MR1924112 (2003f:05034).

794. S. Hitotumatu [1968], *Sin Ishi tori gēmu no sūri* [*The Theory of Nim-Like Games*], Morikita Publishing Co., Tokyo, (Japanese).

795. S. Hitotumatu [1968], Some remarks on nim-like games, *Comment. Math. Univ. St. Paul* **17**, 85–98.

796. D. G. Hoffman and P. D. Johnson [1999], Greedy colorings and the Grundy chromatic number of the *n*-cube, *Bull. Inst. Combin. Appl.* **26**, 49–57.

797. E. J. Hoffman, J. C. Loessi and R. C. Moore [1969], Construction for the solution of the *n*-queens problem, *Math. Mag.* **42**, 66–72.

798. M. S. Hogan and D. G. Horrocks [2003], Geography played on an N-cycle times a 4-cycle, *Integers, Electr. J. of Combinat. Number Theory* **3**, #G2, 12 pp., Comb. Games Sect., MR1985672 (2004b:91037). http://www.integers-ejcnt.org/vol3.html

799. J. C. Holladay [1957], Cartesian products of termination games, *Ann. of Math. Stud.* (Contributions to the Theory of Games), Princeton **3**(39), 189–200.

800. J. C. Holladay [1958], Matrix nim, *Amer. Math. Monthly* **65**, 107–109.

801. J. C. Holladay [1966], A note on the game of dots, *Amer. Math. Monthly* **73**, 717–720.

802. A. Holshouser and H. Reiter [2001], A generalization of Beatty's Theorem, *Southwest J. Pure and Appl. Math.*, issue 2, 24–29.

803. A. Holshouser, H. Reiter and J. Rudzinski [2004], Pilesize dynamic one-pile Nim and Beatty's theorem, *Integers, Electr. J. of Combinat. Number Theory* **4**, #G3, 13 pp., Comb. Games Sect., MR2079848. http://www.integers-ejcnt.org/vol4.html

804. M. Holzer and W. Holzer [2004], $TANTRIX^{TM}$ rotation puzzles are intractable, *Discrete Appl. Math.* **144**, 345–358, MR2098189 (2005j:94043).

805. M. Holzer and S. Schwoon [2004], Assembling molecules in ATOMIX is hard, *Theoret. Comp. Sci.* **313**, 447–462, special issue of Dagstuhl Seminar "Algorithmic Combinatorial Game Theory", Feb. 2002, MR2056938.

806. J. E. Hopcroft, J. T. Schwartz and M. Sharir [1984], On the complexity of motion planning for multiple independent objects: PSPACE-hardness of the "Warehouseman's problem", *J. Robot. Res.* **3** issue 4, 76–88.

807. B. Hopkins and J. A. Sellers [2007], *Exact enumeration of Garden of Eden partitions*, de Gruyter, pp. 299–303, Proc. Integers Conference, Carrollton, Georgia, October 27-30,2005, in celebration of the 70th birthday of Ronald Graham. B. Landman, M. Nathanson, J. Nešetřil, R. Nowakowski, C. Pomerance eds., appeared also in *Integers, Electr. J. of Combinat. Number Theory* **7(20)**, special volume in honor of Ron Graham, #A19, 5 pp., MR2337053. http://www.integers-ejcnt.org/vol7(2).html

808. E. Hordern [1986], *Sliding Piece Puzzles*, Oxford University Press, Oxford.

809. D. G. Horrocks and R. J. Nowakowski [2003], Regularity in the $\mathcal{G}-$ sequences of octal games with a pass, *Integers, Electr. J. of Combinat. Number Theory* **3**, #G1, 10 pp., Comb. Games Sect., MR1985671 (2004d:91049). http://www.integers-ejcnt.org/vol3.html

810. S. Howse and R. J. Nowakoski [2004], Periodicity and arithmetic-periodicity in hexadecimal games, *Theoret. Comp. Sci.* **313**, 463–472, special issue of Dagstuhl Seminar "Algorithmic Combinatorial Game Theory", Feb. 2002, MR2056939 (2005f:91037).

811. S. Huddleston and J. Shurman [2002], Transfinite Chomp, in: *More Games of No Chance*, Proc. MSRI Workshop on Combinatorial Games, July, 2000, Berkeley, CA, MSRI Publ. (R. J. Nowakowski, ed.), Vol. 42, Cambridge University Press, Cambridge, pp. 183–212, MR1973013 (2004b:91046).

812. G. H. Hurlbert [1998], Two pebbling theorems, *Congr. Numer.* **135**, 55–63, Proc. 29-th Southeastern Internat. Conf. on Combinatorics, Graph Theory, and Computing (Boca Raton, FL, 1998).

813. G. H. Hurlbert [1999], A survey of graph pebbling, *Congr. Numer.* **139**, 41–64, Proc. 30-th Southeastern Internat. Conf. on Combinatorics, Graph Theory, and Computing (Boca Raton, FL, 1999).

814. G. Iba and J. Tanton [2003], Candy sharing, *Amer. Math. Monthly* **110**, 25–35, MR1952745.

815. K. Igusa [1985], Solution of the Bulgarian solitaire conjecture, *Math. Mag.* **58**(5), 259–271, MR810147 (87c:00003).

816. J. Isbell [1992], The Gordon game of a finite group, *Amer. Math. Monthly* **99**, 567–569.

817. H. Ito, G. Nakamura and S. Takata [2007], Winning ways of weighted poset games, *Graphs Combin.* **23**(suppl. 1), 291–306, MR2320636.

818. O. Itzinger [1977], The South American game, *Math. Mag.* **50**, 17–21.

819. S. Iwata and T. Kasai [1994], The Othello game on an $n \times n$ board is PSPACE-complete, *Theoret. Comput. Sci.* (Math Games) **123**, 329–340.

820. C. F. A. Jaenisch [1862], *Traité des Applications de l'Analyse mathématiques au Jeu des Echecs*, Petrograd, three volumes (Carl Friedrich Andreyevich Jaenisch from St Petersburg [changed its name to Petrograd, then to Leningrad, and then back to St Petersburg] is also known by his cyrillic transliteration "Yanich", or in the French form: de J, or German form: von J).

821. J. James and M. Schlatter [2008], Some observations and solutions to short and long global Nim, *Integers, Electr. J. of Combinat. Number Theory* **8**, #G01, 14 pp.. http://www.integers-ejcnt.org/vol8.html

822. J. Jeffs and S. Seager [1995], The chip firing game on n-cycles, *Graphs Combin.* **11**, 59–67.

823. T. A. Jenkyns and J. P. Mayberry [1980], The skeleton of an impartial game and the Nim-function of Moore's Nim_k, *Internat. J. Game Theory* **9**, 51–63.

824. T. R. Jensen and B. Toft [1995], *Graph Coloring Problems*, Wiley-Interscience Series in Discrete Mathematics and Optimization, Wiley, New York, NY.

825. L. Y. Jin and J. Nievergelt [2009], Tigers and Goats is a draw, in: *Games of No Chance III*, Proc. BIRS Workshop on Combinatorial Games, July, 2005, Banff, Alberta, Canada, MSRI Publ. (M. H. Albert and R. J. Nowakowski, eds.), Vol. 56, Cambridge University Press, Cambridge, pp. 163–176.

826. D. S. Johnson [1983], The NP-Completeness Column: An Ongoing Guide, *J. Algorithms* **4**, 397–411, 9th quarterly column (games); the column started in 1981; a further games update appeared in vol. **5**, 10th column, pp. 154-155.

827. W. W. Johnson [1879], Notes on the "15" puzzle, *Amer. J. Math.* **2**, 397–399.

828. J. P. Jones [1982], Some undecidable determined games, *Internat. J. Game Theory* **11**, 63–70.

829. J. P. Jones and A. S. Fraenkel [1995], Complexities of winning strategies in diophantine games, *J. Complexity* **11**, 435–455.

830. D. Joyner [2006], Mathematics of Ghaly's machine, *Integers, Electr. J. of Combinat. Number Theory* **6**, #G01, 27 pp., Comb. Games Sect. http://www.integers-ejcnt.org/vol6.html

831. M. Kac [1974], Hugo Steinhaus, a reminiscence and a tribute, *Amer. Math. Monthly* **81**, 572–581 (p. 577).

832. J. Kahane and A. S. Fraenkel [1987], k-Welter: a generalization of Welter's game, *J. Combin. Theory* (Ser. A) **46**, 1–20.

833. J. Kahane and A. J. Ryba [2001], The Hexad game, *Electr. J. Combin.* **8**(2), #R11, 9 pp., Volume in honor of Aviezri S. Fraenkel, MR1853262 (2002h:91031).

834. J. Kahn, J. C. Lagarias and H. S. Witsenhausen [1987], Single-suit two-person card play, *Internat. J. Game Theory* **16**, 291–320, MR0918819 (88m:90162).

835. J. Kahn, J. C. Lagarias and H. S. Witsenhausen [1988], Single-suit two-person card play, II. Dominance, *Order* **5**, 45–60, MR0953941 (89j:90279).

836. J. Kahn, J. C. Lagarias and H. S. Witsenhausen [1989], On Lasker's card game, in: *Differential Games and Applications*, Lecture Notes in Control and Information Sciences (T. S. Başar and P. Bernhard, eds.), Vol. 119, Springer Verlag, Berlin, pp. 1–8, MR1230186.

837. J. Kahn, J. C. Lagarias and H. S. Witsenhausen [1989], Single-suit two-person card play, III. The misère game, *SIAM J. Disc. Math.* **2**, 329–343, MR1002697 (90g:90197).

838. L. Kalmár [1928], Zur Theorie der abstrakten Spiele, *Acta Sci. Math. Univ. Szeged* **4**, 65–85.

839. B. Kalyanasundram [1991], On the power of white pebbles, *Combinatorica* **11**, 157–171.

840. B. Kalyanasundram and G. Schnitger [1988], On the power of white pebbles, *Proc. 20th Ann. ACM Symp. Theory of Computing* (Chicago, IL, 1988), Assoc. Comput. Mach., New York, NY, pp. 258–266.

841. J. H. Kan [1994], (0, 1) matrices, combinatorial games and Ramsey problems, *Math. Appl. (Wuhan)* **7**, 97–101, (Chinese; English summary).

842. M. Kano [1983], Cycle games and cycle cut games, *Combinatorica* **3**, 201–206.

843. M. Kano [1993], An edge-removing game on a graph (a generalization of Nim and Kayles), in: *Optimal Combinatorial Structures on Discrete Mathematical Models* (in Japanese, Kyoto, 1992), Sūrikaisekikenkyūsho Kōkyūroku, pp. 82–90, MR1259331 (94m:90132).

844. M. Kano [1996], Edge-removing games of star type, *Discrete Math.* **151**, 113–119.

845. M. Kano, T. Sasaki, H. Fujita and S. Hoshi [1993], Life games of Ibadai type, in: *Combinatorial Structure in Mathematical Models* (in Japanese, Kyoto, 1993), Sūrikaisekikenkyūsho Kōkyūroku, pp. 108–117.

846. S. Kanungo and R. M. Low [2007], Further analysis on the "king and rook vs. king on a quarter-infinite board" problem, *Integers* **7**, #G09, 7 pp. http://www.integers-ejcnt.org/vol7.html

847. K.-Y. Kao [2005], Sumbers – sums of ups and downs, *Integers, Electr. J. of Combinat. Number Theory* **5**, #G01, 13 pp., Comb. Games Sect., MR2139164 (2005k:91081). http://www.integers-ejcnt.org/vol5.html

848. R. M. Karp and Y. Zhang [1989], On parallel evaluation of game trees, *Proc. ACM Symp. Parallel Algorithms and Architectures*, pp. 409–420.

849. T. Kasai, A. Adachi and S. Iwata [1979], Classes of pebble games and complete problems, *SIAM J. Comput.* **8**, 574–586.

850. Y. Kawano [1996], Using similar positions to search game trees, in: *Games of No Chance, Proc. MSRI Workshop on Combinatorial Games, July, 1994, Berkeley, CA*, MSRI Publ. (R. J. Nowakowski, ed.), Vol. 29, Cambridge University Press, Cambridge, pp. 193–202.

851. R. Kaye [2000], Minesweeper is NP-complete, *Math. Intelligencer* **22** issue 2, 9–15, MR1764264.

852. M. D. Kearse and P. B. Gibbons [2001], Computational methods and new results for chessboard problems, *Australas. J. Combin.* **23**, 253–284, MR1815018 (2001m:05190).

853. M. D. Kearse and P. B. Gibbons [2002], A new lower bound on upper irredundance in the queens' graph, *Discrete Math.* **256**, 225–242, MR1927068 (2003g:05098).

854. J. C. Kenyon [1967], A Nim-like game with period 349, Res. Paper No. 13, Univ. of Calgary, Math. Dept.

855. J. C. Kenyon [1967], Nim-like games and the Sprague–Grundy theory, M.Sc. Thesis, Univ. of Calgary.

856. B. Kerst [1933], *Mathematische Spiele*, Reprinted by Dr. Martin Sändig oHG, Wiesbaden 1968.

857. H. A. Kierstead [2000], A simple competitive graph coloring algorithm, *J. Combin. Theory* (Ser. B) **78**, 57–68.

858. H. A. Kierstead [2005], Asymmetric graph coloring games, *J. Graph Theory* **48**(3), 169–185, MR2116838 (2006c:91033).

859. H. A. Kierstead [2006], Weak acyclic coloring and asymmetric coloring games, *Discrete Math.* **306**, 673–677.

860. H. A. Kierstead and W. T. Trotter [1994], Planar graph coloring with an uncooperative partner, *J. Graph Theory* **18**, 569–584, MR1292976 (95f:05043).

861. H. A. Kierstead and W. T. Trotter [2001], Competitive colorings of oriented graphs, *Electr. J. Combin.* **8**(2), #R12, 15 pp., Volume in honor of Aviezri S. Fraenkel, MR1853263 (2002g:05083).

862. H. A. Kierstead and Zs. Tuza [2003], Marking games and the oriented game chromatic number of partial k-trees, *Graphs Combin.* **19**, 121–129, MR1974374 (2003m:05069).

863. H. A. Kierstead and D. Yang [2005], Very asymmetric marking games, *Order* **22**(2), 93–107 (2006), MR2207192 (2007d:91045).

864. Y. Kim [1996], New values in domineering, *Theoret. Comput. Sci.* (Math Games) **156**, 263–280.

865. C. Kimberling [2007], Complementary equations, *J. Integer Seq.* **10**(1), Article 07.1.4, 14 pp. (electronic), MR2268455. http://www.cs.uwaterloo.ca/journals/JIS/VOL10/Kimberling/kimberling26.pdf

866. H. Kinder [1973], Gewinnstrategien des Anziehenden in einigen Spielen auf Graphen, *Arch. Math.* **24**, 332–336.

867. M. A. Kiwi, R. Ndoundam, M. Tchuente and E. Goles [1994], No polynomial bound for the period of the parallel chip firing game on graphs, *Theoret. Comput. Sci.* **136**, 527–532.

868. D. A. Klarner, ed. [1998], *Mathematical recreations. A collection in honor of Martin Gardner*, Dover, Mineola, NY, Corrected reprint of *The Mathematical Gardner*, Wadsworth Internat., Belmont, CA, 1981.

869. S. Klavžar and U. Milutinović [1997], Graphs $S(n, k)$ and a variant of the tower of Hanoi problem, *Czechoslovak Math. J.* **47**, 95–104, MR1435608 (97k:05061).

870. S. Klavžar and U. Milutinović [2002], Simple explicit formulas for the Frame-Stewart numbers, *Ann. Comb.* **6**, 157–167, MR1955516 (2003k:05004).

871. S. Klavžar, U. Milutinović and C. Petr [2001], Combinatorics of topmost discs of multi-peg tower of Hanoi problem, *Ars Combinatoria* **59**, 55–64, MR1832197 (2002b:05013).

872. S. Klavžar, U. Milutinović and C. Petr [2002], On the Frame-Stewart algorithm for the multi-peg tower of Hanoi problem, *Discrete Appl. Math.* **120**, 141–157, MR1966398.

873. M. M. Klawe [1985], A tight bound for black and white pebbles on the pyramids, *J. Assoc. Comput. Mach.* **32**, 218–228.

874. C. S. Klein and S. Minsker [1993], The super towers of Hanoi problem: large rings on small rings, *Discrete Math.* **114**, 283–295.

875. D. J. Kleitman and B. L. Rothschild [1972], A generalization of Kaplansky's game, *Discrete Math.* **2**, 173–178.

876. T. Kløve [1977], The modular n-queen problem, *Discrete Math.* **19**, 289–291.

877. T. Kløve [1981], The modular n-queen problem II, *Discrete Math.* **36**, 33–48.

878. A. Knopfmacher and H. Prodinger [2001], A Simple Card Guessing Game Revisited, *Electr. J. Combin.* **8**(2), #R13, 9 pp., Volume in honor of Aviezri S. Fraenkel, MR1853264 (2002g:05007).

879. M. Knor [1996], On Ramsey-type games for graphs, *Australas. J. Combin.* **14**, 199–206.

880. D. E. Knuth [1974], *Surreal Numbers*, Addison-Wesley, Reading, MA.

881. D. E. Knuth [1976], The computer as Master Mind, *J. Recr. Math.* **9**, 1–6.

882. D. E. Knuth [1977], Are toy problems useful?, *Popular Computing* **5** issue 1, 3-10; continued in **5** issue 2, 3-7.

883. D. E. Knuth [1993], *The Stanford Graph Base*: A Platform for Combinatorial Computing, ACM Press, New York.

884. P. Komjáth [1984], A simple strategy for the Ramsey-game, *Studia Sci. Math. Hung.* **19**, 231–232.

885. A. Kotzig [1946], O k-posunutiach (On k-translations; in Slovakian), *Časop. pro Pěst. Mat. a Fys.* **71**, 55–61, extended abstract in French, pp. 62–66.

886. G. Kowalewski [1930], *Alte und neue mathematische Spiele*, Reprinted by Dr. Martin Sändig oHG, Wiesbaden 1968.

887. K. Koyama and T. W. Lai [1993], An optimal Mastermind strategy, *J. Recr. Math.* **25**, 251–256.

888. M. Kraitchik [1953], *Mathematical Recreations*, Dover, New York, NY, 2nd edn.

889. D. Král, V. Majerech, J. Sgall, T. Tichý and G. Woeginger [2004], It is tough to be a plumber, *Theoret. Comp. Sci.* **313**, 473–484, special issue of Dagstuhl Seminar "Algorithmic Combinatorial Game Theory", Feb. 2002, MR2056940.

890. B. Kummer [1980], *Spiele auf Graphen*, Internat. Series of Numerical Mathematics, Birkhäuser Verlag, Basel.

891. D. Kunkle and G. Cooperman [2007], Twenty-Six Moves Suffice for Rubik's Cube, *Proc. Intern. Sympos. Symbolic and Algebraic Computation (ISSAC '07)*, ACM Press, to appear.

892. M. Kutz [2005], Conway's angel in three dimensions, *Theoret. Comp. Sci.* **349**, 443–451.

893. M. Lachmann, C. Moore and I. Rapaport [2002], Who wins Domineering on rectangular boards?, in: *More Games of No Chance,* Proc. MSRI Workshop on Combinatorial Games, July, 2000, Berkeley, CA, MSRI Publ. (R. J. Nowakowski, ed.), Vol. 42, Cambridge University Press, Cambridge, pp. 307–315, MR1973019 (2004b:91047).

894. R. E. Ladner and J. K. Norman [1985], Solitaire automata, *J. Comput. System Sci.* **30**, 116–129, MR 86i:68032.

895. J. C. Lagarias [1977], Discrete balancing games, *Bull. Inst. Math. Acad. Sinica* **5**, 363–373.

896. J. Lagarias and D. Sleator [1999], Who wins misère Hex?, in: *The Mathemagician and Pied Puzzler*, honoring Martin Gardner; E. Berlekamp and T. Rodgers, eds., A K Peters, Natick, MA, pp. 237-240.

897. S. P. Lalley [1988], A one-dimensional infiltration game, *Naval Res. Logistics* **35**, 441–446.

898. P. C. B. Lam, W. C. Shiu and B. Xu [1999], Edge game-coloring of graphs, *Graph Theory Notes N. Y.* **37**, 17–19, MR1724310.

899. T. K. Lam [1997], Connected sprouts, *Amer. Math. Monthly* **104**, 116–119.

900. H. A. Landman [1996], Eyespace values in Go, in: *Games of No Chance,* Proc. MSRI Workshop on Combinatorial Games, July, 1994, Berkeley, CA, MSRI Publ. (R. J. Nowakowski, ed.), Vol. 29, Cambridge University Press, Cambridge, pp. 227–257, MR1427967 (97j:90100).

901. H. Landman [2002], A simple FSM-based proof of the additive periodicity of the Sprague-Grundy function of Wythoff's game, in: *More Games of No Chance,* Proc. MSRI Workshop on Combinatorial Games, July, 2000, Berkeley, CA, MSRI Publ. (R. J. Nowakowski, ed.), Vol. 42, Cambridge University Press, Cambridge, pp. 383–386, MR1973106 (2004b:91048).

902. L. Larson [1977], A theorem about primes proved on a chessboard, *Math. Mag.* **50**, 69–74.

903. E. Lasker [1931], *Brettspiele der Völker, Rätsel und mathematische Spiele*, Berlin.

904. M. Latapy and C. Magnien [2002], Coding distributive lattices with edge firing games, *Inform. Process. Lett.* **83**, 125–128, MR1837204 (2002h:37022).

905. M. Latapy and H. D. Phan [2001], The lattice structure of chip firing games and related models, *Phys. D* **155**, 69–82, MR1837204 (2002h:37022).

906. I. Lavalée [1985], Note sur le problème des tours d'Hanoï, *Rev. Roumaine Math. Pures Appl.* **30**, 433–438.

907. E. L. Lawler and S. Sarkissian [1995], An algorithm for "Ulam's game" and its application to error correcting codes, *Inform. Process. Lett.* **56**, 89–93.

908. A. J. Lazarus, D. E. Loeb, J. G. Propp, W. R. Stromquist and D. Ullman [1999], Combinatorial games under auction play, *Games Econom. Behav.* **27**, 229–264, MR1685133 (2001f:91023).

909. A. J. Lazarus, D. E. Loeb, J. G. Propp and D. Ullman [1996], Richman Games, in: *Games of No Chance*, Proc. MSRI Workshop on Combinatorial Games, July, 1994, Berkeley, CA, MSRI Publ. (R. J. Nowakowski, ed.), Vol. 29, Cambridge University Press, Cambridge, pp. 439–449, MR1427981 (97j:90101).

910. D. B. Leep and G. Myerson [1999], Marriage, magic and solitaire, *Amer. Math. Monthly* **106**, 419–429.

911. A. Lehman [1964], A solution to the Shannon switching game, *SIAM J. Appl. Math.* **12**, 687–725.

912. E. L. Leiss [1983], On restricted Hanoi problems, *J. Combin. Inform. System Sci.* **8**, 277–285, MR783767 (86i:05072).

913. E. Leiss [1983], Solving the "Towers of Hanoi" on graphs, *J. Combin. Inform. System Sci.* **8**, 81–89, MR783741 (86i:05071).

914. E. L. Leiss [1984], Finite Hanoi problems: how many discs can be handled?, *Congr. Numer.* **44**, 221–229, Proc. 15th Southeastern Internat. Conf. on Combinatorics, Graph Theory and Computing (Baton Rouge, LA, 1984), MR777543 (87b:05062).

915. T. Lengauer and R. Tarjan [1980], The space complexity of pebble games on trees, *Inform. Process. Lett.* **10**, 184–188.

916. T. Lengauer and R. Tarjan [1982], Asymptotically tight bounds on time-space trade-offs in a pebble game, *J. Assoc. Comput. Mach.* **29**, 1087–1130.

917. T. Lengyel [2003], A Nim-type game and continued fractions, *Fibonacci Quart.* **41**, 310–320, MR2022411.

918. J. H. W. Lenstra [1977], On the algebraic closure of two, *Proc. Kon. Nederl. Akad. Wetensch.* (Ser. A) **80**, 389–396.

919. J. H. W. Lenstra [1977/1978], Nim multiplication, Séminaire de Théorie des Nombres No. 11, Université de Bordeaux, France.

920. L. Levine [2006], Fractal sequences and restricted Nim, *Ars Combin.* **80**, 113–127, MR2243082 (2007b:91029).

921. D. N. L. Levy, ed. [1988], *Computer Games* I, II, Springer-Verlag, New York, NY.

922. J. Lewin [1986], The lion and man problem revisited, *J. Optimization Theory and Applications* **49**, 411–430.

923. T. Lewis and S. Willard [1980], The rotating table, *Math. Mag.* **53**, 174–179.

924. C.-K. Li and I. Nelson [1998], Perfect codes on the towers of Hanoi graph, *Bull. Austral. Math. Soc.* **57**, 367–376.

925. S.-Y. R. Li [1974], Generalized impartial games, *Internat. J. Game Theory* **3**, 169–184.

926. S.-Y. R. Li [1976], Sums of Zuchswang games, *J. Combin. Theory* (Ser. A) **21**, 52–67.

927. S.-Y. R. Li [1977], N-person nim and N-person Moore's games, *Internat. J. Game Theory* **7**, 31–36.

928. D. Lichtenstein [1982], Planar formulae and their uses, *SIAM J. Comput.* **11**, 329–343.

929. D. Lichtenstein and M. Sipser [1980], Go is Polynomial-Space hard, *J. Assoc. Comput. Mach.* **27**, 393–401, earlier version appeared in Proc. 19th Ann. Symp. Foundations of Computer Science (Ann Arbor, MI, Oct. 1978), IEEE Computer Soc., Long Beach, CA, 1978, pp. 48–54.

930. H. Liebeck [1971], Some generalizations of the 14-15 puzzle, *Math. Mag.* **44**, 185–189.

931. C.-W. Lim [2005], Partial Nim, *Integers, Electr. J. of Combinat. Number Theory* **5**, #G02, 9 pp., Comb. Games Sect., MR2139165 (2005m:91046). http://www.integers-ejcnt.org/vol5.html

932. C. Löding and P. Rohde [2003], Solving the sabotage game is PSPACE-hard, in: *Mathematical Foundations of Computer Science 2003*, Vol. 2747 of *Lecture Notes in Comput. Sci.*, Springer, Berlin, pp. 531–540, MR2081603.

933. D. E. Loeb [1995], How to win at Nim, *UMAP J.* **16**, 367–388.

934. D. E. Loeb [1996], Stable winning coalitions, in: *Games of No Chance*, Proc. MSRI Workshop on Combinatorial Games, July, 1994, Berkeley, CA, MSRI Publ. (R. J. Nowakowski, ed.), Vol. 29, Cambridge University Press, Cambridge, pp. 451–471.

935. J. A. M. Lopez [1991], A prolog Mastermind program, *J. Recr. Math.* **23**, 81–93.

936. R. M. Low and M. Stamp [2006], King and rook vs. king on a quarter-infinite board, *Integers, Electr. J. of Combinat. Number Theory* **6**, #G3, 8 pp., MR2247807 (2007c:91038). http://www.integers-ejcnt.org/vol6.html

937. S. Loyd [1914], *Cyclopedia of Puzzles and Tricks*, Franklin Bigelow Corporation, Morningside Press, NY, reissued and edited by M. Gardner under the name *The Mathematical Puzzles of Sam Loyd* (two volumes), Dover, New York, NY, 1959.

938. X. Lu [1986], A Hamiltonian game, *J. Math. Res. Exposition* (*English Ed.*) **1**, 101–106.

939. X. Lu [1991], A matching game, *Discrete Math.* **94**, 199–207.

940. X. Lu [1992], A characterization on n-critical economical generalized tic-tac-toe games, *Discrete Math.* **110**, 197–203.

941. X. Lu [1992], Hamiltonian games, *J. Combin. Theory* (Ser. B) **55**, 18–32.

942. X. Lu [1995], A Hamiltonian game on $K_{n,n}$, *Discrete Math.* **142**, 185–191.

943. X. Lu [1995], A note on biased and non-biased games, *Discrete Appl. Math.* **60**, 285–291, Presented at ARIDAM VI and VII (New Brunswick, NJ, 1991/1992).

944. X.-M. Lu [1986], Towers of Hanoi graphs, *Intern. J. Comput. Math.* **19**, 23–38.

945. X.-M. Lu [1988], Towers of Hanoi problem with arbitrary $k \geq 3$ pegs, *Intern. J. Comput. Math.* **24**, 39–54.

946. X.-M. Lu [1989], An iterative solution for the 4-peg towers of Hanoi, *Comput. J.* **32**, 187–189.

947. X.-M. Lu and T. S. Dillon [1994], A note on parallelism for the towers of Hanoi, *Math. Comput. Modelling* **20** issue 3, 1–6.

948. X.-M. Lu and T. S. Dillon [1995], Parallelism for multipeg towers of Hanoi, *Math. Comput. Modelling* **21** issue 3, 3–17.

949. X.-M. Lu and T. S. Dillon [1996], Nonrecursive solution to parallel multipeg towers of Hanoi: a decomposition approach, *Math. Comput. Modelling* **24** issue 3, 29–35.

950. É. Lucas [1960], *Récréations Mathématiques*, Vol. I – IV, A. Blanchard, Paris. Previous editions: Gauthier-Villars, Paris, 1891–1894.

951. É. Lucas [1974], *Introduction aux Récréations Mathématiques: L'Arithmétique Amusante*, reprinted by A. Blanchard, Paris. Originally published by A. Blanchard, Paris, 1895.

952. A. L. Ludington [1988], Length of the 7-number game, *Fibonacci Quart.* **26**, 195–204.

953. A. Ludington-Young [1990], Length of the *n*-number game, *Fibonacci Quart.* **28**, 259–265.

954. W. F. Lunnon [1986], The Reve's puzzle, *Comput. J.* **29**, 478.

955. M. Maamoun and H. Meyniel [1987], On a game of policemen and robber, *Discrete Appl. Math.* **17**, 307–309.

956. P. A. MacMahon [1921], *New Mathematical Pastimes*, Cambridge University Press, Cambridge.

957. F. Maffray [1986], On kernels in *i*-triangulated graphs, *Discrete Math.* **61**, 247–251.

958. F. Maffray [1992], Kernels in perfect line-graphs, *J. Combin. Theory* (Ser. B) **55**, 1–8.

959. C. Magnien, H. D. Phan and L. Vuillon [2001], Characterization of lattices induced by (extended) chip firing games, *Discrete models: combinatorics, computation, and geometry (Paris, 2001), 229–244 (electronic), Discrete Math. and Theor. Comput. Sci. Proc.* pp. 229–244, MR1888776 (2003a:91029).

960. R. Mansfield [1996], Strategies for the Shannon switching game, *Amer. Math. Monthly* **103**, 250–252.

961. D. Marcu [1992], On finding a Grundy function, *Polytech. Inst. Bucharest Sci. Bull. Chem. Mater. Sci.* **54**, 43–47.

962. G. Martin [1991], *Polyominoes: Puzzles and Problems in Tiling*, Math. Assoc. America, Wasington, DC.

963. G. Martin [2002], Restoring fairness to Dukego, in: *More Games of No Chance*, Proc. MSRI Workshop on Combinatorial Games, July, 2000, Berkeley, CA, MSRI Publ. (R. J. Nowakowski, ed.), Vol. 42, Cambridge University Press, Cambridge, pp. 79–87, MR1973006 (2004c:91031).

964. O. Martín-Sánchez and C. Pareja-Flores [2001], Two reflected analyses of lights out, *Math. Mag.* **74**, 295–304.

965. F. Mäser [2002], Global threats in combinatorial games: a computational model with applications to chess endgames, in: *More Games of No Chance*, Proc. MSRI Workshop on Combinatorial Games, July, 2000, Berkeley, CA, MSRI Publ. (R. J. Nowakowski, ed.), Vol. 42, Cambridge University Press, Cambridge, pp. 137–149, MR1973010 (2004b:91049).

966. J. G. Mauldon [1978], Num, a variant of nim with no first player win, *Amer. Math. Monthly* **85**, 575–578.

967. R. McConville [1974], *The History of Board Games*, Creative Publications, Palo Alto, CA.

968. S. K. McCurdy and R. Nowakowski [2005], Cutthroat, an all-small game on graphs, *Integers, Electr. J of Combinat. Number Theory* **5(2)**, #A13, 13 pp., MR2192091 (2006g:05076). http://www.integers-ejcnt.org/vol5(2).html

969. D. P. McIntyre [1942], A new system for playing the game of nim, *Amer. Math. Monthly* **49**, 44–46.

970. R. McNaughton [1993], Infinite games played on finite graphs, *Ann. Pure Appl. Logic* **65**, 149–184.

971. J. W. A. McWorter [1981], Kriegspiel Hex, *Math. Mag.* **54**, 85–86, solution to Problem 1084 posed by the author in Math. Mag. **52** (1979) 317.

972. E. Mead, A. Rosa and C. Huang [1974], The game of SIM: A winning strategy for the second player, *Math. Mag.* **47**, 243–247.

973. N. Megiddo, S. L. Hakimi, M. R. Garey, D. S. Johnson and C. H. Papadimitriou [1988], The complexity of searching a graph, *J. Assoc. Comput. Mach.* **35**, 18–44.

974. K. Mehlhorn, S. Näher and M. Rauch [1990], On the complexity of a game related to the dictionary problem, *SIAM J. Comput.* **19**, 902–906, earlier draft appeared in Proc. 30th Ann. Symp. Foundations of Computer Science, pp. 546–548.

975. N. S. Mendelsohn [1946], A psychological game, *Amer. Math. Monthly* **53**, 86–88.

976. E. Mendelson [2004], *Introducing game theory and its applications*, Discrete Mathematics and its Applications (Boca Raton), Chapman & Hall/CRC, Boca Raton, FL, (Ch. 1 is on combinatorial games), MR2073442 (2005b:91003).

977. C. G. Méndez [1981], On the law of large numbers, infinite games and category, *Amer. Math. Monthly* **88**, 40–42.

978. C. Merino [2001], The chip firing game and matroid complexes, *Discrete models: combinatorics, computation, and geometry (Paris, 2001) (electronic), Discrete Math. and Theor. Comput. Sci. Proc.* pp. 245–255, MR1888777 (2002k:91048).

979. C. Merino [2005], The chip firing game, *Discrete Math.* **302**, 188–210, MR2179643.

980. C. Merino López [1997], Chip firing and the Tutte polynomial, *Ann. Comb.* **1**, 253–259, (C. Merino López = C.M. López = C. Merino is the same author), MR1630779 (99k:90232).

981. G. A. Mesdal III [2009], Partizan Splittles, in: *Games of No Chance III*, Proc. BIRS Workshop on Combinatorial Games, July, 2005, Banff, Alberta, Canada, MSRI Publ. (M. H. Albert and R. J. Nowakowski, eds.), Vol. 56, Cambridge University Press, Cambridge, pp. 447–461.

982. G. A. Mesdal and P. Ottaway [2007], Simplification of partizan games in misère play, *Integers, Electr. J of Combinat. Number Theory* **7**, #G06, 12 pp.), (G. A. Mesdal designates the following authors: M. Allen, J. P. Grossman, A. Hill, N. A. McKay, R. J. Nowakowski, T. Plambeck, A. A. Siegel, D. Wolfe, who par-

ticipated at the 2006 "Games-at-Dal 4" (Dalhousie) conference), MR2299812 (2007k:91054). http://www.integers-ejcnt.org/vol7.html

983. D. Mey [1994], Finite games for a predicate logic without contractions, *Theoret. Comput. Sci.* **123**, 341–349.

984. F. Meyer auf der Heide [1981], A comparison of two variations of a pebble game on graphs, *Theoret. Comput. Sci.* **13**, 315–322.

985. H. Meyniel and J.-P. Roudneff [1988], The vertex picking game and a variation of the game of dots and boxes, *Discrete Math.* **70**, 311–313, MR0955128 (89f:05111).

986. D. Michie and I. Bratko [1987], Ideas on knowledge synthesis stemming from the KBBKN endgame, *Internat. Comp. Chess Assoc. J.* **10**, 3–13.

987. J. Milnor [1953], Sums of positional games, *Ann. of Math. Stud.* (Contributions to the Theory of Games, H. W. Kuhn and A. W. Tucker, eds.), Princeton **2**(28), 291–301.

988. P. Min Lin [1982], Principal partition of graphs and connectivity games, *J. Franklin Inst.* **314**, 203–210.

989. S. Minsker [1989], The towers of Hanoi rainbow problem: coloring the rings, *J. Algorithms* **10**, 1–19.

990. S. Minsker [1991], The towers of Antwerpen problem, *Inform. Process. Lett.* **38**, 107–111.

991. J. Missigman and R. Weida [2001], An easy solution to mini lights out, *Math. Mag.* **74**, 57–59.

992. D. Moews [1991], Sums of games born on days 2 and 3, *Theoret. Comput. Sci.* (Math Games) **91**, 119–128, MR1142561 (93b:90130).

993. D. Moews [1992], Pebbling graphs, *J. Combin. Theory* (Ser. B) **55**, 244–252.

994. D. Moews [1996], Coin-sliding and Go, *Theoret. Comput. Sci.* (Math Games) **164**, 253–276.

995. D. Moews [1996], Infinitesimals and coin-sliding, in: *Games of No Chance*, Proc. MSRI Workshop on Combinatorial Games, July, 1994, Berkeley, CA, MSRI Publ. (R. J. Nowakowski, ed.), Vol. 29, Cambridge University Press, Cambridge, pp. 315–327.

996. D. Moews [1996], Loopy games and Go, in: *Games of No Chance*, Proc. MSRI Workshop on Combinatorial Games, July, 1994, Berkeley, CA, MSRI Publ. (R. J. Nowakowski, ed.), Vol. 29, Cambridge University Press, Cambridge, pp. 259–272, MR1427968 (98d:90152).

997. D. Moews [2002], The abstract structure of the group of games, in: *More Games of No Chance*, Proc. MSRI Workshop on Combinatorial Games, July, 2000, Berkeley, CA, MSRI Publ. (R. J. Nowakowski, ed.), Vol. 42, Cambridge University Press, Cambridge, pp. 49–57, MR1973004 (2004b:91034).

998. C. Moore and D. Eppstein [2002], 1-dimensional peg solitaire, and duotaire, in: *More Games of No Chance*, Proc. MSRI Workshop on Combinatorial Games, July, 2000, Berkeley, CA, MSRI Publ. (R. J. Nowakowski, ed.), Vol. 42, Cambridge University Press, Cambridge, pp. 341–350, MR1973022.

999. E. H. Moore [1909–1910], A generalization of the game called nim, *Ann. of Math.* (Ser. 2) **11**, 93–94.

1000. A. H. Moorehead and G. Mott-Smith [1963], *Hoyle's Rules of Games*, Signet, New American Library.

1001. F. L. Morris [1981], Playing disjunctive sums is polynomial space complete, *Internat. J. Game Theory* **10**, 195–205.

1002. M. Morse and G. A. Hedlund [1944], Unending chess, symbolic dynamics and a problem in semigroups, *Duke Math. J.* **11**, 1–7.

1003. M. Müller [2003], Conditional combinatorial games and their application to analyzing capturing races in Go, *Information Sciences* **154**, 189–202.

1004. M. Müller and R. Gasser [1996], Conditional combinatorial games and their application to analyzing capturing races in Go, in: *Games of No Chance,* Proc. MSRI Workshop on Combinatorial Games, July, 1994, Berkeley, CA, MSRI Publ. (R. J. Nowakowski, ed.), Vol. 29, Cambridge University Press, Cambridge, pp. 273–284.

1005. M. Müller and T. Tegos [2002], Experiments in computer Amazons, in: *More games of no chance (Berkeley, CA, 2000),* Vol. 42 of *Math. Sci. Res. Inst. Publ.,* Cambridge Univ. Press, Cambridge, pp. 243–260, MR1973016 (2004b:91050).

1006. D. Mundici [1991], Logic and algebra in Ulam's searching game with lies, *Jbuch. Kurt-Gödel-Ges.* pp. 109–114 (1993), MR1213606 (94j:03130).

1007. D. Mundici [1992], The logic of Ulam's game with lies, in: *Knowledge, Belief, and Strategic Interaction (Castiglioncello, 1989),* Cambridge Stud. Probab. Induc. Decis. Theory, Cambridge Univ. Press, Cambridge, pp. 275–284, MR1206717 (94c:90133).

1008. D. Mundici and T. Trombetta [1997], Optimal comparison strategies in Ulam's searching game with two errors, *Theoret. Comput. Sci.* (Math Games) **182**, 217–232.

1009. H. J. R. Murray [1952], *A History of Board Games Other Than Chess*, Oxford University Press.

1010. B. Nadel [1990], Representation selection for constraint satisfaction: a case study using n-queens, *IEEE Expert* **5** issue 3, 16–23.

1011. T. Nakamura [2009], Counting liberties in Go capturing races, in: *Games of No Chance III,* Proc. BIRS Workshop on Combinatorial Games, July, 2005, Banff, Alberta, Canada, MSRI Publ. (M. H. Albert and R. J. Nowakowski, eds.), Vol. 56, Cambridge University Press, Cambridge, pp. 177–196.

1012. T. Nakamura and E. Berlekamp [2003], Analysis of composite corridors, *Proc. Intern. Conference on Computers and Games CG'2002, Edmonton, Alberta, Canada, July 2002,* (Y. Björnsson, M. Müller and J. Schaeffer, eds.), Vol. LNCS 288, Lecture Notes in Computer Science, Springer, pp. 213–229.

1013. B. Nalebuff [1989], The other person's envelope is always greener, *J. Economic Perspectives* **3** issue 1, 171–181.

1014. A. Napier [1970], A new game in town, Empire Mag., Denver Post, May 2.

1015. A. Negro and M. Sereno [1992], Solution of Ulam's problem on binary search with three lies, *J. Combin. Theory Ser. A* **59**, 149–154, MR1141331 (92h:05013).

1016. A. Negro and M. Sereno [1992], Ulam's searching game with three lies, *Adv. in Appl. Math.* **13**, 404–428, MR1190120 (93i:90129).

1017. J. Nešetřil and F. Sopena [2001], On the oriented game chromatic number, *Electr. J. Combin.* **8(2)**, #R14, 13 pp., Volume in honor of Aviezri S. Fraenkel, MR1853265 (2002h:05069).

1018. J. Nešetřil and R. Thomas [1987], Well quasi ordering, long games and combinatorial study of undecidability, in: *Logic and Combinatorics, Contemp. Math.* (S. G. Simpson, ed.), Vol. 65, Amer. Math. Soc., Providence, RI, pp. 281–293.

1019. S. Neufeld and R. J. Nowakowski [1993], A vertex-to-vertex pursuit game played with disjoint sets of edges, in: *Finite and Infinite Combinatorics in Sets and Logic* (N. W. Sauer et al., eds.), Kluwer, Dordrecht, pp. 299–312.

1020. S. Neufeld and R. J. Nowakowski [1998], A game of cops and robbers played on products of graphs, *Discrete Math.* **186**, 253–268.

1021. J. Nievergelt, F. Maeser, B. Mann, K. Roeseler, M. Schulze and C. Wirth [1999], CRASH! Mathematik und kombinatorisches Chaos prallen aufeinander, *Informatik Spektrum* **22**, 45–48, (German).

1022. G. Nivasch [2006], The Sprague-Grundy function of the game Euclid, *Discrete Math.* **306**(21), 2798–2800, MR2264378 (2007i:91039).

1023. G. Nivasch [2009], More on the Sprague-Grundy function for Wythoff's game, in: *Games of No Chance III*, Proc. BIRS Workshop on Combinatorial Games, July, 2005, Banff, Alberta, Canada, MSRI Publ. (M. H. Albert and R. J. Nowakowski, eds.), Vol. 56, Cambridge University Press, Cambridge, pp. 377–410.

1024. I. Niven [1988], Coding theory applied to a problem of Ulam, *Math. Mag.* **61**, 275–281.

1025. H. Noon and G. V. Brummelen [2006], The non-attacking queens game, *College Math. J.* **37**, 223–227.

1026. R. J. Nowakowski [1991], . . ., Welter's game, Sylver coinage, dots-and-boxes, . . ., in: *Combinatorial Games,* Proc. Symp. Appl. Math. (R. K. Guy, ed.), Vol. 43, Amer. Math. Soc., Providence, RI, pp. 155–182.

1027. R. J. Nowakowski and P. Ottaway [2005], Vertex Deletion Games with Parity Rules, *Integers, Electr. J of Combinat. Number Theory* **5(2)**, #A15, 16 pp., MR2192093. http://www.integers-ejcnt.org/vol5(2).html

1028. R. J. Nowakowski and D. G. Poole [1996], Geography played on products of directed cycles, in: *Games of No Chance,* Proc. MSRI Workshop on Combinatorial Games, July, 1994, Berkeley, CA, MSRI Publ. (R. J. Nowakowski, ed.), Vol. 29, Cambridge University Press, Cambridge, pp. 329–337.

1029. R. J. Nowakowski and A. A. Siegel [2007], *Partizan geography on $K_n \times K_2$*, de Gruyter, pp. 389–401, Proc. Integers Conference, Carrollton, Georgia, October 27-30,2005, in celebration of the 70th birthday of Ronald Graham. B. Landman, M. Nathanson, J. Nešetřil, R. Nowakowski, C. Pomerance eds., appeared also in *Integers, Electr. J. of Combinat. Number Theory* **7(20)**, special volume in honor of Ron Graham, #A29, 14 pp., MR2337063. http://www.integers-ejcnt.org/vol7(2).html

1030. R. Nowakowski and P. Winkler [1983], Vertex-to-vertex pursuit in a graph, *Discrete Math.* **43**, 235–239.

1031. S. P. Nudelman [1995], The modular n-queens problem in higher dimensions, *Discrete Math.* **146**, 159–167.

1032. T. H. O'Beirne [1984], *Puzzles and Paradoxes*, Dover, New York, NY. (Appeared previously by Oxford University Press, London, 1965.).

1033. G. L. O'Brian [1978-79], The graph of positions in the game of SIM, *J. Recr. Math.* **11**, 3–9.

1034. J. Olson [1985], A balancing strategy, *J. Combin. Theory Ser. A* **40**, 175–178.

1035. T. Ooya and J. Akiyama [2003], Impact of binary and Fibonacci expansions of numbers on winning strategies for Nim-like games, *Internat. J. Math. Ed. Sci. Tech.* **34**, 121–128, MR1958736.

1036. H. K. Orman [1996], Pentominoes: a first player win, in: *Games of No Chance*, Proc. MSRI Workshop on Combinatorial Games, July, 1994, Berkeley, CA, MSRI Publ. (R. J. Nowakowski, ed.), Vol. 29, Cambridge University Press, Cambridge, pp. 339–344.

1037. P. R. J. Östergård and W. D. Weakley [2001], Values of domination numbers of the queen's graph, *Electr. J. Combin.* **8(1)**, #R29, 19 pp.

1038. J. Pach [1992], On the game of Misery, *Studia Sci. Math. Hungar.* **27**, 355–360.

1039. E. W. Packel [1987], The algorithm designer versus nature: a game-theoretic approach to information-based complexity, *J. Complexity* **3**, 244–257.

1040. C. H. Papadimitriou [1985], Games against nature, *J. Comput. System Sci.* **31**, 288–301.

1041. C. H. Papadimitriou [1994], *Computational Complexity*, Addison-Wesley, Chapter 19: Polynomial Space.

1042. C. H. Papadimitriou, P. Raghavan, M. Sudan and H. Tamaki [1994], Motion planning on a graph (extended abstract), *Proc. 35-th Annual IEEE Symp. on Foundations of Computer Science*, Santa Fe, NM, pp. 511–520.

1043. A. Papaioannou [1982], A Hamiltonian game, *Ann. Discrete Math.* **13**, 171–178.

1044. T. Pappas [1994], *The Magic of Mathematics*, Wide World, San Carlos.

1045. D. Parlett [1999], *The Oxford History of Board Games*, Oxford University Press.

1046. T. D. Parsons [1978], Pursuit-evasion in a graph, in: *Theory and Applications of Graphs* (Y. Alavi and D. R. Lick, eds.), Springer-Verlag, pp. 426–441.

1047. T. D. Parsons [1978], The search number of a connected graph, *Proc. 9th South-Eastern Conf. on Combinatorics, Graph Theory, and Computing,*, pp. 549–554.

1048. O. Patashnik [1980], Qubic: $4 \times 4 \times 4$ Tic-Tac-Toe, *Math. Mag.* **53**, 202–216.

1049. J. L. Paul [1978], Tic-Tac-Toe in n dimensions, *Math. Mag.* **51**, 45–49.

1050. W. J. Paul, E. J. Prauss and R. Reischuk [1980], On alternation, *Acta Informatica* **14**, 243–255.

1051. W. J. Paul and R. Reischuk [1980], On alternation, II, *Acta Informatica* **14**, 391–403.

1052. W. J. Paul, R. E. Tarjan and J. R. Celoni [1976/77], Space bounds for a game on graphs, *Math. Systems Theory* **10**, 239–251, correction ibid. **11** (1977/78),

85. First appeared in Eighth Annual ACM Symposium on Theory of Computing (Hershey, Pa., 1976), Assoc. Comput. Mach., New York, NY, 1976, pp 149–160.

1053. M. M. Paulhus [1999], Beggar my neighbour, *Amer. Math. Monthly* **106**, 162–165.

1054. J. Pearl [1980], Asymptotic properties of minimax trees and game-searching procedures, *Artificial Intelligence* **14**, 113–138.

1055. J. Pearl [1984], *Heuristics: Intelligent Search Strategies for Computer Problem Solving*, Addison-Wesley, Reading, MA.

1056. A. Pedrotti [2002], Playing by searching: two strategies against a linearly bounded liar, *Theoret. Comput. Sci.* **282**, 285–302, special "Fun With Algorithms" issue, MR1909053 (2003m:91010).

1057. A. Pekeč [1996], A winning strategy for the Ramsey graph game, *Combin. Probab. Comput.* **5**, 267–276.

1058. A. Pelc [1986], Lie patterns in search procedures, *Theoret. Comput. Sci.* **47**, 61–69, MR871464 (88b:68020).

1059. A. Pelc [1987], Solution of Ulam's problem on searching with a lie, *J. Combin. Theory* (Ser. A) **44**, 129–140.

1060. A. Pelc [1988], Prefix search with a lie, *J. Combin. Theory* (Ser. A) **48**, 165–173.

1061. A. Pelc [1989], Detecting errors in searching games, *J. Combin. Theory* (Ser. A) **51**, 43–54.

1062. A. Pelc [2002], Searching games with errors — fifty years of coping with liars, *Theoret. Comput. Sci.* (Math Games) **270**, 71–109, MR1871067 (2002h:68040).

1063. D. H. Pelletier [1987], Merlin's magic square, *Amer. Math. Monthly* **94**, 143–150.

1064. H. Peng and C. H. Yan [1998], Balancing game with a buffer, *Adv. in Appl. Math.* **21**, 193–204.

1065. Y. Peres, O. Schramm, S. Sheffield and D. B. Wilson [2007], Random-turn Hex and other selection games, *Amer. Math. Monthly* **114**, 373–387.

1066. G. L. Peterson [1979], Press-Ups is Pspace-complete, Dept. of Computer Science, The University of Rochester, Rochester New York, 14627, unpublished manuscript.

1067. G. L. Peterson and J. H. Reif [1979], Multiple-person alternation, *Proc. 20th Ann. Symp. Foundations Computer Science* (San Juan, Puerto Rico, Oct. 1979), IEEE Computer Soc., Long Beach, CA, pp. 348–363.

1068. C. Pickover [2002], The fractal society, in: *Puzzler's Tribute: a Feast for the Mind*, pp. 377–381, honoring Martin Gardner (D. Wolfe and T. Rodgers, eds.), A K Peters, Natick, MA.

1069. O. Pikhurko [2003], Breaking symmetry on complete bipartite graphs of odd size, *Integers, Electr. J. of Combinat. Number Theory* **3**, #G4, 9 pp., Comb. Games Sect., MR2036488 (2004k:05117). http://www.integers-ejcnt.org/vol3.html

1070. N. Pippenger [1980], Pebbling, *Proc. 5th IBM Symp. Math. Foundations of Computer Science*, IBM, Japan, 19 pp.

1071. N. Pippenger [1982], Advances in pebbling, *Proc. 9th Internat. Colloq. Automata, Languages and Programming*, Lecture Notes in Computer Science

(M. Nielson and E. M. Schmidt, eds.), Vol. 140, Springer Verlag, New York, NY, pp. 407–417.

1072. T. E. Plambeck [1992], Daisies, Kayles, and the Sibert–Conway decomposition in misère octal games, *Theoret. Comput. Sci.* (Math Games) **96**, 361–388, MR1160551 (93c:90108).

1073. T. E. Plambeck [2005], Taming the wild in impartial combinatorial games, *Integers, Electr. J. of Combinat. Number Theory* **5**, #G5(1), 36 pp., Comb. Games Sect., MR2192253 (2006g:91044). http://www.integers-ejcnt.org/vol5.html

1074. T. Plambeck [2009], Advances in losing, in: *Games of No Chance III*, Proc. BIRS Workshop on Combinatorial Games, July, 2005, Banff, Alberta, Canada, MSRI Publ. (M. H. Albert and R. J. Nowakowski, eds.), Vol. 56, Cambridge University Press, Cambridge, pp. 57–89.

1075. V. Pless [1991], Games and codes, in: *Combinatorial Games,* Proc. Symp. Appl. Math. (R. K. Guy, ed.), Vol. 43, Amer. Math. Soc., Providence, RI, pp. 101–110.

1076. A. Pluhár [2002], The accelerated k-in-a-row game, *Theoret. Comput. Sci.* (Math Games) **270**, 865–875, MR1871100 (2002i:05017).

1077. R. Polizzi and F. Schaefer [1991], *Spin Again*, Chronicle Books.

1078. G. Pólya [1921], *Über die "doppelt-periodischen" Lösungen des n-Damen-Problems, in: W. Ahrens, Mathematische Unterhaltungen und Spiele, Vol.* 1, B.G. Teubner, Leipzig, 1918, pp. 364–374, also appeared in vol. IV of Pólya's Collected Works, MIT Press, Cambridge, London, 1984 (G.-C Rota ed.), pp. 237-247.

1079. D. Poole [1992], The bottleneck towers of Hanoi problem, *J. Recr. Math.* **24**, 203–207.

1080. D. G. Poole [1994], The towers and triangles of Professor Claus (or, Pascal knows Hanoi), *Math. Mag.* **67**, 323–344.

1081. O. Pretzel [2002], Characterization of simple edge-firing games, *Inform. Process. Lett.* **84**, 235–236, MR1931725 (2003i:05128).

1082. D. Pritchard [1994], *The Family Book of Games*, Sceptre Books, Time-Life Books, Brockhampton Press.

1083. J. Propp [1994], A new take-away game, in: *The Lighter Side of Mathematics,* Proc. E. Strens Memorial Conf. on Recr. Math. and its History, Calgary, 1986, Spectrum Series (R. K. Guy and R. E. Woodrow, eds.), Math. Assoc. of America, Washington, DC, pp. 212–221.

1084. J. Propp [1996], About David Richman (Prologue to the paper by J. D. Lazarus et al.), in: *Games of No Chance,* Proc. MSRI Workshop on Combinatorial Games, July, 1994, Berkeley, CA, MSRI Publ. (R. J. Nowakowski, ed.), Vol. 29, Cambridge University Press, Cambridge, p. 437.

1085. J. Propp [2000], Three-player impartial games, *Theoret. Comput. Sci.* (Math Games) **233**, 263–278, MR1732189 (2001i:91040).

1086. J. Propp and D. Ullman [1992], On the cookie game, *Internat. J. Game Theory* **20**, 313–324, MR1163932 (93b:90131).

1087. P. Pudlák and J. Sgall [1997], An upper bound for a communication game related to time-space tradeoffs, in: *The Mathematics of Paul Erdős* (R. L. Graham and J. Nešetřil, eds.), Vol. I, Springer, Berlin, pp. 393–399.

1088. A. Pultr and F. L. Morris [1984], Prohibiting repetitions makes playing games substantially harder, *Internat. J. Game Theory* **13**, 27–40.

1089. A. Pultr and J. Úlehla [1985], Remarks on strategies in combinatorial games, *Discrete Appl. Math.* **12**, 165–173.

1090. A. Quilliot [1983], Discrete pursuit games, *Congr. Numer.* **38**, 227–241, Proc. 13th Southeastern Conf. on Combinatorics, Graph Theory and Computing, Boca Raton, FL, 1982.

1091. A. Quilliot [1985], A short note about pursuit games played on a graph with a given genus, *J. Combin. Theory* (Ser. B) **38**, 89–92.

1092. M. O. Rabin [1957], Effective computability of winning strategies, *Ann. of Math. Stud.* (Contributions to the Theory of Games), Princeton **3**(39), 147–157.

1093. M. O. Rabin [1976], Probabilistic algorithms, *Proc. Symp. on New Directions and Recent Results in Algorithms and Complexity* (J. F. Traub, ed.), Carnegie-Mellon, Academic Press, New York, NY, pp. 21–39.

1094. D. Ratner and M. Warmuth [1990], The $(n^2 - 1)$-puzzle and related relocation problems, *J. Symbolic Comput.* **10**, 111–137.

1095. B. Ravikumar [2004], Peg-solitaire, string rewriting systems and finite automata, *Theoret. Comput. Sci.* **321**, 383–394, Appeared first under same title in Proc. 8th Internat. Symp. Algorithms and Computation, Singapore, 233-242, 1997, MR 2005e:68119, MR2076153 (2005e:68119).

1096. B. Ravikumar and K. B. Lakshmanan [1982], Coping with known patterns of lies in a search game, in: *Foundations of software technology and theoretical computer science (Bangalore, 1982)*, Nat. Centre Software Develop. Comput. Tech., Bombay, pp. 363–377, MR678245 (83m:68114).

1097. B. Ravikumar and K. B. Lakshmanan [1984], Coping with known patterns of lies in a search game, *Theoret. Comput. Sci.* **33**, 85–94.

1098. N. Reading [1999], Nim-regularity of graphs, *Electr. J. Combinatorics* **6**, #R11, 8 pp.

1099. M. Reichling [1987], A simplified solution of the N queens' problem, *Inform. Process. Lett.* **25**, 253–255.

1100. J. H. Reif [1984], The complexity of two-player games of incomplete information, *J. Comput. System Sci.* **29**, 274–301, earlier draft entitled Universal games of incomplete information, appeared in Proc. 11th Ann. ACM Symp. Theory of Computing (Atlanta, GA, 1979), Assoc. Comput. Mach., New York, NY, pp. 288–308.

1101. S. Reisch [1980], Gobang ist PSPACE-vollständig, *Acta Informatica* **13**, 59–66.

1102. S. Reisch [1981], Hex ist PSPACE-vollständig, *Acta Informatica* **15**, 167–191.

1103. M. Reiss [1857], Beiträge zur Theorie des Solitär-Spiels, *Crelles Journal* **54**, 344–379.

1104. P. Rendell [2001], Turing universality of the game of life, in: *Collision-Based Computing*, Springer Verlag, London, UK, pp. 513 – 539.

1105. C. S. ReVelle and K. E. Rosing [2000], Defendens imperium romanum [Defending the Roman Empire]: a classical problem in military strategy, *Amer. Math. Monthly* **107**, 585–594.

1106. M. Richardson [1953], Extension theorems for solutions of irreflexive relations, *Proc. Nat. Acad. Sci. USA* **39**, 649.

1107. M. Richardson [1953], Solutions of irreflexive relations, *Ann. of Math.* **58**, 573–590.

1108. R. D. Ringeisen [1974], Isolation, a game on a graph, *Math. Mag.* **47**, 132–138.

1109. R. L. Rivest, A. R. Meyer, D. J. Kleitman, K. Winklman and J. Spencer [1980], Coping with errors in binary search procedures, *J. Comput. System Sci.* **20**, 396–404.

1110. I. Rivin, I. Vardi and P. Zimmermann [1994], The *n*-queens problem, *Amer. Math. Monthly* **101**, 629–639.

1111. I. Rivin and R. Zabih [1992], A dynamic programming solution to the *N*-queens problem, *Inform. Process. Lett.* **41**, 253–256.

1112. E. Robertson and I. Munro [1978], NP-completeness, puzzles and games, *Utilitas Math.* **13**, 99–116.

1113. A. G. Robinson and A. J. Goldman [1989], The set coincidence game: complexity, attainability, and symmetric strategies, *J. Comput. System Sci.* **39**, 376–387.

1114. A. G. Robinson and A. J. Goldman [1990], On Ringeisen's isolation game, *Discrete Math.* **80**, 297–312.

1115. A. G. Robinson and A. J. Goldman [1990], On the set coincidence game, *Discrete Math.* **84**, 261–283.

1116. A. G. Robinson and A. J. Goldman [1991], On Ringeisen's isolation game, II, *Discrete Math.* **90**, 153–167.

1117. A. G. Robinson and A. J. Goldman [1993], The isolation game for regular graphs, *Discrete Math.* **112**, 173–184.

1118. J. M. Robson [1983], The complexity of Go, *Proc. Information Processing* 83 (R. E. A. Mason, ed.), Elsevier, Amsterdam, pp. 413–417.

1119. J. M. Robson [1984], Combinatorial games with exponential space complete decision problems, *Proc. 11th Symp. Math. Foundations of Computer Science,* Praha, Czechoslovakia, Lecture Notes in Computer Science (M. P. Chytil and V. Koubek, eds.), Vol. 176, Springer, Berlin, pp. 498–506.

1120. J. M. Robson [1984], *N* by *N* checkers is Exptime complete, *SIAM J. Comput.* **13**, 252–267.

1121. J. M. Robson [1985], Alternation with restrictions on looping, *Inform. and Control* **67**, 2–11.

1122. E. Y. Rodin [1989], A pursuit-evasion bibliography – version 2, *Comput. Math. Appl.* **18**, 245–320.

1123. J. S. Rohl [1983], A faster lexicographical *n*-queens algorithm, *Inform. Process. Lett.* **17**, 231–233.

1124. J. S. Rohl and T. D. Gedeon [1986], The Reve's puzzle, *Comput. J.* **29**, 187–188, Corrigendum, *Ibid.* **31** (1988), 190.

1125. I. Roizen and J. Pearl [1983], A minimax algorithm better than alpha-beta? Yes and no, *Artificial Intelligence* **21**, 199–220.

1126. I. Rosenholtz [1993], Solving some variations on a variant of Tic-Tac-Toe using invariant subsets, *J. Recr. Math.* **25**, 128–135.

1127. A. S. C. Ross [1953], The name of the game of Nim, Note 2334, *Math. Gaz.* **37**, 119–120.

1128. A. E. Roth [1978], A note concerning asymmetric games on graphs, *Naval Res. Logistics* **25**, 365–367.

1129. A. E. Roth [1978], Two-person games on graphs, *J. Combin. Theory* (Ser. B) **24**, 238–241.

1130. T. Roth [1974], The tower of Brahma revisited, *J. Recr. Math.* **7**, 116–119.

1131. E. M. Rounds and S. S. Yau [1974], A winning strategy for SIM, *J. Recr. Math.* **7**, 193–202.

1132. W. L. Ruzzo [1980], Tree-size bounded alternation, *J. Comput. Systems Sci.* **21**, 218–235.

1133. S. Sackson [1969], *A Gamut of Games*, Random House.

1134. I. Safro and L. Segel [2003], Collective stochastic versions of playable games as metaphors for complex biosystems: team Connect Four, *Complexity* **8** issue 5, 46–55, MR2018947.

1135. M. Saks and A. Wigderson [1986], Probabilistic Boolean decision trees and the complexity of evaluating game trees, *Proc. 27th Ann. Symp. Foundations of Computer Science* (Toronto, Ont., Canada), IEEE Computer Soc., Washington, DC, pp. 29–38.

1136. D. Samet, I. Samet and D. Schmeidler [2004], One observation behind two-envelope puzzles, *Amer. Math. Monthly* **111**, 347–351, MR2057189 (2005e:00005).

1137. A. Sapir [2004], The tower of Hanoi with forbidden moves, *Comput. J.* **47**, 20–24.

1138. U. K. Sarkar [2000], On the design of a constructive algorithm to solve the multi-peg towers of Hanoi problem, *Theoret. Comput. Sci.* (Math Games) **237**, 407–421, MR1756218 (2001k:05025).

1139. M. Sato [1972], Grundy functions and linear games, *Publ. Res. Inst. Math. Sciences*, Kyoto Univ. **7**, 645–658.

1140. F. Scarioni and H. G. Speranza [1984], A probabilistic analysis of an error-correcting algorithm for the Towers of Hanoi puzzle, *Inform. Process. Lett.* **18**, 99–103.

1141. W. L. Schaaf [1955, 1970, 1973, 1978], *A Bibliography of Recreational Mathematics*, Vol. I – IV, Nat'l. Council of Teachers of Mathematics, Reston, VA.

1142. J. Schaeffer, N. Burch, Y. Björnsson, A. Kishimoto, M. Müller, R. Lake, P. Lu and S. Sutphen [2007], Checkers is solved, *Science* **317**(5844), 1518–1522, MR2348441.

1143. J. Schaeffer and R. Lake [1996], Solving the game of checkers, in: *Games of No Chance*, Proc. MSRI Workshop on Combinatorial Games, July, 1994, Berkeley, CA, MSRI Publ. (R. J. Nowakowski, ed.), Vol. 29, Cambridge University Press,

Cambridge, pp. 119–133. http://www.msri.org/publications/books/Book29/files/schaeffer.ps.gz

1144. T. J. Schaefer [1976], Complexity of decision problems based on finite two-person perfect information games, *Proc. 8th Ann. ACM Symp. Theory of Computing* (Hershey, PA, 1976), Assoc. Comput. Mach., New York, NY, pp. 41–49.

1145. T. J. Schaefer [1978], On the complexity of some two-person perfect-information games, *J. Comput. System Sci.* **16**, 185–225.

1146. M. Scheepers [1994], Variations on a game of Gale (II): Markov strategies, *Theoret. Comput. Sci.* (Math Games) **129**, 385–396.

1147. D. Schleicher and M. Stoll [2006], An introduction to Conway's games and numbers, *Mosc. Math. J.* **6**(2), 359–388, 407, MR2270619 (2007m:91033).

1148. G. Schmidt and T. Ströhlein [1985], On kernels of graphs and solutions of games: a synopsis based on relations and fixpoints, *SIAM J. Alg. Disc. Math.* **6**, 54–65.

1149. R. W. Schmittberger [1992], *New Rules for Classic Games*, Wiley, New York.

1150. G. Schrage [1985], A two-dimensional generalization of Grundy's game, *Fibonacci Quart.* **23**, 325–329.

1151. H. Schubert [1953], *Mathematische Mussestunden*, De Gruyter, Berlin, neubearbeitet von F. Fitting, Elfte Auflage.

1152. F. Schuh [1952], Spel van delers (The game of divisors), *Nieuw Tijdschrift voor Wiskunde* **39**, 299–304.

1153. F. Schuh [1968], *The Master Book of Mathematical Recreations*, Dover, New York, NY, translated by F. Göbel, edited by T. H. O'Beirne.

1154. B. L. Schwartz [1971], Some extensions of Nim, *Math. Mag.* **44**, 252–257.

1155. B. L. Schwartz, ed. [1979], *Mathematical solitaires and games*, Baywood Publishing Company, Farmingdale, NY, pp. 37–81.

1156. A. J. Schwenk [1970], Take-away games, *Fibonacci Quart.* **8**, 225–234.

1157. A. J. Schwenk [1991], Which rectangular chessboards have a knight's tour?, *Math. Mag.* **64**, 325–332, MR1141559 (93c:05081).

1158. A. J. Schwenk [2000], What is the correct way to seed a knockout tournament?, *Amer. Math. Monthly* **107**, 140–150.

1159. R. S. Scorer, P. M. Grundy and C. A. B. Smith [1944], Some binary games, *Math. Gaz.* **280**, 96–103.

1160. M. Sereno [1998], Binary search with errors and variable cost queries, *Inform. Process. Lett.* **68**, 261–270, MR1664741 (99i:68026).

1161. Á. Seress [1992], On Hajnal's triangle-free game, *Graphs Combin.* **8**, 75–79.

1162. Á. Seress and T. Szabó [1999], On Erdös' Eulerian trail game, *Graphs Combin.* **15**, 233–237.

1163. J. Sgall [2001], Solution of David Gale's lion and man problem, *Theoret. Comput. Sci.* (Math Games) **259**, 663–670, MR1832815 (2002b:91025).

1164. L. E. Shader [1978], Another strategy for SIM, *Math. Mag.* **51**, 60–64.

1165. L. E. Shader and M. L. Cook [1980], A winning strategy for the second player in the game Tri-tip, *Proc. Tenth S.E. Conference on Computing, Combinatorics and Graph Theory*, Utilitas, Winnipeg.

1166. A. S. Shaki [1979], Algebraic solutions of partizan games with cycles, *Math. Proc. Camb. Phil. Soc.* **85**, 227–246.

1167. A. Shamir, R. L. Rivest and L. M. Adleman [1981], Mental Poker, in: *The Mathematical Gardner* (D. A. Klarner, ed.), Wadsworth Internat., Belmont, CA, pp. 37–43.

1168. A. Shankar and M. Sridharan [2005], New temparatures in Domineering, *Integers, Electr. J. of Combinat. Number Theory* **5**, #G04, 13 pp., Comb. Games Sect., MR2139167 (2006c:91036). http://www.integers-ejcnt.org/vol5.html

1169. C. E. Shannon [1950], Programming a computer for playing chess, *Philos. Mag.* (*Ser. 7*) **41**, 256–275, MR0034095 (11,543f).

1170. K. Shelton [2007], *The game of take turn*, de Gruyter, pp. 413–430, Proc. Integers Conference, Carrollton, Georgia, October 27-30,2005, in celebration of the 70th birthday of Ronald Graham. B. Landman, M. Nathanson, J. Nešetřil, R. Nowakowski, C. Pomerance eds., appeared also in *Integers, Electr. J. of Combinat. Number Theory* **7(20)**, special volume in honor of Ron Graham, #A31, 18 pp., MR2337065. http://www.integers-ejcnt.org/vol7(2).html

1171. A. Shen [2000], Lights out, *Math. Intelligencer* **22** issue 3, 20–21.

1172. R. Sheppard and J. Wilkinson [1995], *Strategy Games: A Collection Of 50 Games & Puzzles To Stimulate Mathematical Thinking*, Parkwest Publications.

1173. G. J. Sherman [1978], A child's game with permutations, *Math. Mag.* **51**, 67–68.

1174. Z. Shi, W. Goddard, S. T. Hedetniemi, K Kennedy, R. Laskar and A. McRae [2005], An algorithm for partial Grundy numbers on trees, *Discrete Math.* **304**, 108–116.

1175. W. L. Sibert and J. H. Conway [1992], Mathematical Kayles, *Internat. J. Game Theory* **20**, 237–246, MR1146325.

1176. G. Sicherman [2002], Theory and practice of Sylver coinage, *Integers, Electr. J. of Combinat. Number Theory* **2**, #G2, 8 pp., Comb. Games Sect., MR1934618 (2003h:05026). http://www.integers-ejcnt.org/vol2.html

1177. N. Sieben [2004], Snaky is a 41-dimensional winner, *Integers, Electr. J. of Combinat. Number Theory* **4**, #G5, 4 pp., Comb. Games Sect., MR2116011. http://www.integers-ejcnt.org/vol4.html

1178. N. Sieben and E. Deabay [2005], Polyomino weak achievement games on 3-dimensional rectangular boards, *Discrete Math.* **290**, 61–78, MR2117357 (2005h:05044).

1179. A. N. Siegel [2000], Combinatorial Game Suite, a software tool for investigating games. http://cgsuite.sourceforge.net/

1180. A. N. Siegel [2006], Reduced canonical forms of stoppers, *Electron. J. Combin.* **13**(1), #R57, 14 pp., MR2240763 (2007b:91030).

1181. A. N. Siegel [2009], Backsliding Toads and Frogs, in: *Games of No Chance III*, Proc. BIRS Workshop on Combinatorial Games, July, 2005, Banff, Alberta, Canada, MSRI Publ. (M. H. Albert and R. J. Nowakowski, eds.), Vol. 56, Cambridge University Press, Cambridge, pp. 197–214.

1182. A. N. Siegel [2009], Coping with cycles, in: *Games of No Chance III*, Proc. BIRS Workshop on Combinatorial Games, July, 2005, Banff, Alberta, Canada, MSRI

Publ. (M. H. Albert and R. J. Nowakowski, eds.), Vol. 56, Cambridge University Press, Cambridge, pp. 91–123.

1183. A. N. Siegel [2009], New results in loopy games, in: *Games of No Chance III*, Proc. BIRS Workshop on Combinatorial Games, July, 2005, Banff, Alberta, Canada, MSRI Publ. (M. H. Albert and R. J. Nowakowski, eds.), Vol. 56, Cambridge University Press, Cambridge, pp. 215–232.

1184. R. Silber [1976], A Fibonacci property of Wythoff pairs, *Fibonacci Quart.* **14**, 380–384.

1185. R. Silber [1977], Wythoff's Nim and Fibonacci representations, *Fibonacci Quart.* **15**, 85–88.

1186. J.-N. O. Silva [1993], Some game bounds depending on birthdays, *Portugaliae Math.* **50**, 353–358.

1187. R. Silver [1967], The group of automorphisms of the game of 3-dimensional ticktacktoe, *Amer. Math. Monthly* **74**, 247–254.

1188. D. L. Silverman [1971], *Your Move*, McGraw-Hill.

1189. G. J. Simmons [1969], The game of SIM, *J. Recr. Math.* **2**, 66.

1190. D. Singmaster [1981], Almost all games are first person games, *Eureka* **41**, 33–37.

1191. D. Singmaster [1982], Almost all partizan games are first person and almost all impartial games are maximal, *J. Combin. Inform. System Sci.* **7**, 270–274.

1192. D. Singmaster [1999], Some diophantine recreations, in: *The Mathemagician and Pied Puzzler*, honoring Martin Gardner; E. Berlekamp and T. Rodgers, eds., A K Peters, Natick, MA, pp. 219-235.

1193. W. Slany [2001], The complexity of graph Ramsey games, *Proc. Intern. Conference on Computers and Games CG'2000* (T. Marsland and I. Frank, eds.), Vol. 2063, Hamamatsu, Japan, Oct. 2000, Lecture Notes in Computer Science, Springer, pp. 186–203, MR1909610.

1194. W. Slany [2002], Endgame problems of Sim-like graph Ramsey avoidance games are PSPACE-complete, *Theoret. Comput. Sci.* (Math Games) **289**, 829–843, MR1933803 (2003h:05177).

1195. C. A. B. Smith [1966], Graphs and composite games, *J. Combin. Theory* **1**, 51–81, reprinted in slightly modified form in: *A Seminar on Graph Theory* (F. Harary, ed.), Holt, Rinehart and Winston, New York, NY, 1967, pp. 86-111.

1196. C. A. B. Smith [1968], Compound games with counters, *J. Recr. Math.* **1**, 67–77.

1197. C. A. B. Smith [1971], Simple game theory and its applications, *Bull. Inst. Math. Appl.* **7**, 352–357.

1198. D. E. Smith and C. C. Eaton [1911], Rithmomachia, the great medieval number game, *Amer. Math. Montly* **18**, 73–80.

1199. F. Smith and P. Stănică [2002], Comply/constrain games or games with a Muller twist, *Integers, Electr. J. of Combinat. Number Theory* **2**, #G3, 10 pp., Comb. Games Sect., MR1945951 (2003i:91029). http://www.integers-ejcnt.org/vol2.html

1200. R. Smullyan [2005], Gödelian puzzles, in: *Tribute to a Mathemagician*, honoring Martin Gardner (B. Cipra, E. D. Demaine, M. L. Demaine and T. Rodgers, eds.), A K Peters, Wellesley, MA, pp. 49-54.

1201. R. G. Snatzke [2002], Exhaustive search in Amazons, in: *More Games of No Chance*, Proc. MSRI Workshop on Combinatorial Games, July, 2000, Berkeley, CA, MSRI Publ. (R. J. Nowakowski, ed.), Vol. 42, Cambridge University Press, Cambridge, pp. 261–278, MR1973017.

1202. R. Sosic and J. Gu [1990], A polynomial time algorithm for the n-queens problem, *SIGART* 1 issue 3, 7–11.

1203. J. Spencer [1977], Balancing games, *J. Combin. Theory* (Ser. B) 23, 68–74.

1204. J. Spencer [1984], Guess a number with lying, *Math. Mag.* 57, 105–108.

1205. J. Spencer [1986], Balancing vectors in the max norm, *Combinatorica* 6, 55–65.

1206. J. Spencer [1991], Threshold spectra via the Ehrenfeucht game, *Discrete Appl. Math.* 30, 235–252.

1207. J. Spencer [1992], Ulam's searching game with a fixed number of lies, *Theoret. Comput. Sci.* (Math Games) 95, 307–321.

1208. J. Spencer [1994], Randomization, derandomization and untirandomization: three games, *Theoret. Comput. Sci.* (Math Games) 131, 415–429.

1209. W. L. Spight [2001], Extended thermography for multile kos in go, *Theoret. Comput. Sci.* 252, 23–43, special "Computers and Games" issue; first version appeared in Proc. 1st Intern. Conf. on Computer Games CG'98, Tsukuba, Japan, Nov. 1998, *Lecture Notes in Computer Science*, Vol. 1558, Springer, pp. 232-251, 1999., MR1806224 (2001k:91041).

1210. W. L. Spight [2002], Go thermography: the 4/2/98 Jiang-Rui endgame, in: *More Games of No Chance*, Proc. MSRI Workshop on Combinatorial Games, July, 2000, Berkeley, CA, MSRI Publ. (R. J. Nowakowski, ed.), Vol. 42, Cambridge University Press, Cambridge, pp. 89–105, MR1973007 (2004d:91050).

1211. E. L. Spitznagel, Jr. [1967], A new look at the fifteen puzzle, *Math. Mag.* 40, 171–174.

1212. E. L. Spitznagel, Jr. [1973], Properties of a game based on Euclid's algorithm, *Math. Mag.* 46, 87–92.

1213. R. Sprague [1935–36], Über mathematische Kampfspiele, *Tôhoku Math. J.* 41, 438–444.

1214. R. Sprague [1937], Über zwei Abarten von Nim, *Tôhoku Math. J.* 43, 351–354.

1215. R. Sprague [1947], Bemerkungen über eine spezielle Abelsche Gruppe, *Math. Z.* 51, 82–84.

1216. R. Sprague [1961], *Unterhaltsame Mathematik*, Vieweg and Sohn, Braunschweig, Paperback reprint, translation by T. H. O'Beirne: *Recreation in Mathematics*, Blackie, 1963.

1217. R. G. Stanton [2004], Some design theory games, *Bull. Inst. Combin. Appl.* 41, 61–63.

1218. H. Steinhaus [1960], Definitions for a theory of games and pursuit, *Naval Res. Logistics* 7, 105–108.

1219. V. N. Stepanenko [1975], Grundy games under conditions of semidefiniteness, *Cybernetics* **11**, 167–172 (trans. of *Kibernetika* **11** (1975) 145–149).

1220. B. M. Stewart [1939], Problem 3918 (k-peg tower of Hanoi), *Amer. Math. Monthly* **46**, 363, solution by J. S. Frame, *ibid.* **48** (1941) 216–217; by the proposer, *ibid.* 217–219.

1221. F. Stewart [2007], The sequential join of combinatorial games, *Integers, Electr. J. of Combinat. Number Theory* **7**, #G03, 10 pp., MR2299809. http://www.integers-ejcnt.org/vol7.html

1222. I. Stewart [1991], Concentration: a winning strategy, *Sci. Amer.* **265**, 167–172, October issue.

1223. I. Stewart [2000], Hex marks the spot, *Sci. Amer.* **283**, 100–103, September issue.

1224. L. Stiller [1988], Massively parallel retrograde analysis, Tech. Rep. BU-CS TR88–014, Comp. Sci. Dept., Boston University.

1225. L. Stiller [1989], Parallel analysis of certain endgames, *Internat. Comp. Chess Assoc. J.* **12**, 55–64.

1226. L. Stiller [1991], Group graphs and computational symmetry on massively parallel architecture, *J. Supercomputing* **5**, 99–117.

1227. L. Stiller [1996], Multilinear algebra and chess endgames, in: *Games of No Chance,* Proc. MSRI Workshop on Combinatorial Games, July, 1994, Berkeley, CA, MSRI Publ. (R. J. Nowakowski, ed.), Vol. 29, Cambridge University Press, Cambridge, pp. 151–192.

1228. D. L. Stock [1989], Merlin's magic square revisited, *Amer. Math. Monthly* **96**, 608–610.

1229. L. J. Stockmeyer and A. K. Chandra [1979], Provably difficult combinatorial games, *SIAM J. Comput.* **8**, 151–174.

1230. P. K. Stockmeyer [1994], Variations on the four-post Tower of Hanoi puzzle, *Congr. Numer.* **102**, 3–12, Proc. 25th Southeastern Internat. Conf. on Combinatorics, Graph Theory and Computing (Boca Raton, FL, 1994).

1231. P. K. Stockmeyer [1999], The average distance between nodes in the cyclic Tower of Hanoi digraph, *Combinatorics, Graph Theory, and Algorithms, Vol. I, II (Kalamazoo, MI, 1996)*, New Issues Press, Kalamazoo, MI, pp. 799–808.

1232. P. K. Stockmeyer [2005], The Tower of Hanoi: a bibliography, *Internet publication*, 38 pp. http://www.cs.wm.edu/~pkstoc/hpapers.html

1233. P. K. Stockmeyer, C. D. Bateman, J. W. Clark, C. R. Eyster, M. T. Harrison, N. A. Loehr, P. J. Rodriguez and J. R. Simons III [1995], Exchanging Disks in the Tower of Hanoi, *Intern. J. Computer Math.* **59**, 37–47.

1234. K. B. Stolarsky [1991], From Wythoff's Nim to Chebyshev's inequality, *Amer. Math. Monthly* **98**, 889–900.

1235. J. A. Storer [1983], On the complexity of chess, *J. Comput. System Sci.* **27**, 77–100.

1236. W. E. Story [1879], Notes on the "15" puzzle, *Amer. J. Math.* **2**, 399–404.

1237. P. D. Straffin, Jr. [1985], Three-person winner-take-all games with McCarthy's revenge rule, *College J. Math.* **16**, 386–394.

1238. P. D. Straffin [1993], *Game Theory and Strategy*, New Mathematical Library, MAA, Washington, DC.

1239. P. D. Straffin, Jr. [1995], Position graphs for Pong Hau K'i and Mu Torere, *Math. Mag.* **68**, 382–386.

1240. T. Ströhlein and L. Zagler [1977], Analyzing games by Boolean matrix iteration, *Discrete Math.* **19**, 183–193.

1241. W. Stromquist [2007], Winning paths in N-by-infinity hex, *Integers, Electr. J. of Combinat. Number Theory* **7**, #G05, 3 pp., MR2299811. http://www.integers-ejcnt.org/vol7.html

1242. W. Stromquist and D. Ullman [1993], Sequential compounds of combinatorial games, *Theoret. Comput. Sci.* (Math Games) **119**, 311–321.

1243. K. Sugihara and I. Suzuki [1989], Optimal algorithms for a pursuit-evasion problem in grids, *SIAM J. Disc. Math.* **2**, 126–143.

1244. X. Sun [2002], Improvements on Chomp, *Integers, Electr. J. of Combinat. Number Theory* **2**, #G1, 8 pp., Comb. Games Sect., MR1917957 (2003e:05015). http://www.integers-ejcnt.org/vol2.html

1245. X. Sun [2005], Wythoff's sequence and N-heap Wythoff's conjectures, *Discrete Math.* **300**, 180–195, MR2170125.

1246. X. Sun and D. Zeilberger [2004], On Fraenkel's N-heap Wythoff's conjectures, *Ann. Comb.* **8**, 225–238, MR2079933 (2005e:91039).

1247. K. Sutner [1988], On σ-automata, *Complex Systems* **2**, 1–28.

1248. K. Sutner [1989], Linear cellular automata and the Garden-of-Eden, *Math. Intelligencer* **11**, 49–53.

1249. K. Sutner [1990], The σ-game and cellular automata, *Amer. Math. Monthly* **97**, 24–34.

1250. K. Sutner [1995], On the computational complexity of finite cellular automata, *J. Comput. System Sci.* **50**, 87–97.

1251. K. J. Swanepoel [2000], Balancing unit vectors, *J. Combin. Theory* (Ser. A) **89**, 105–112.

1252. M. Szegedy [1999], In how many steps the k peg version of the towers of Hanoi game can be solved?, in: *STACS 99, Trier, Lecture Notes in Computer Science*, Vol. 1563, Springer, Berlin, pp. 356–361.

1253. G. J. Székely and M. L. Rizzo [2007], The uncertainty principle of game theory, *Amer. Math. Monthly* **114**(8), 688–702, MR2354439.

1254. L. A. Székely [1984], On two concepts of discrepancy in a class of combinatorial games in: *Finite and infinite sets, Vol. I, II, Colloq. Math. Soc. János Bolyai, 37*, North-Holland, Amsterdam, pp. 679–683.

1255. J. L. Szwarcfiter and G. Chaty [1994], Enumerating the kernels of a directed graph with no odd circuits, *Inform. Process. Lett.* **51**, 149–153.

1256. A. Takahashi, S. Ueno and Y. Kajitani [1995], Mixed searching and proper-path-width, *Theoret. Comput. Sci.* (Math Games) **137**, 253–268.

1257. T. Takizawa [2002], An application of mathematical game theory to Go endgames: some width-two-entrance rooms with and without kos, in: *More Games of No Chance*, Proc. MSRI Workshop on Combinatorial Games, July,

2000, Berkeley, CA, MSRI Publ. (R. J. Nowakowski, ed.), Vol. 42, Cambridge University Press, Cambridge, pp. 108–124, MR1973008.

1258. G. Tardos [1988], Polynomial bound for a chip firing game on graphs, *SIAM J. Disc. Math.* **1**, 397–398.

1259. M. Tarsi [1983], Optimal search on some game trees, *J. Assoc. Comput. Mach.* **30**, 389–396.

1260. R. Telgársky [1987], Topological games: on the 50th anniversary of the Banach-Mazur game, *Rocky Mountain J. Math.* **17**, 227–276.

1261. W. F. D. Theron and G. Geldenhuys [1998], Domination by queens on a square beehive, *Discrete Math.* **178**, 213–220.

1262. K. Thompson [1986], Retrograde analysis of certain endgames, *Internat. Comp. Chess Assoc. J.* **9**, 131–139.

1263. H. Tohyama and A. Adachi [2000], Complexity of path discovery problems, *Theoret. Comput. Sci.* (Math Games) **237**, 381–406.

1264. G. P. Tollisen and T. Lengyel [2000], Color switching games, *Ars Combin.* **56**, 223–234, MR1768618 (2001b:05094).

1265. I. Tomescu [1990], Almost all digraphs have a kernel, *Discrete Math.* **84**, 181–192.

1266. R. Tošić and S. Šćekić [1983], An analysis of some partizan graph games, *Proc. 4th Yugoslav Seminar on Graph Theory*, Novi Sad, pp. 311–319.

1267. S. P. Tung [1987], Computational complexities of Diophantine equations with parameters, *J. Algorithms* **8**, 324–336, MR905990 (88m:03063).

1268. A. M. Turing, M. A. Bates, B. V. Bowden and C. Strachey [1953], Digital computers applied to games, in: *Faster Than Thought* (B. V. Bowden, ed.), Pitman, London, pp. 286–310.

1269. R. Uehara and S. Iwata [1990], Generalized Hi-Q is NP-complete, *Trans. IEICE* **E73**, 270–273.

1270. J. Úlehla [1980], A complete analysis of von Neumann's Hackendot, *Internat. J. Game Theory* **9**, 107–113.

1271. D. Ullman [1992], More on the four-numbers game, *Math. Mag.* **65**, 170–174.

1272. S. Vajda [1992], *Mathematical Games and How to Play Them*, Ellis Horwood Series in Mathematics and its Applications, Chichester, England.

1273. H. J. van den Herik and I. S. Herschberg [1985], The construction of an omniscient endgame database, *Internat. Comp. Chess Assoc. J.* **8**, 66–87.

1274. J. van den Heuvel [2001], Algorithmic aspects of a chip-firing game, *Combin. Probab. Comput.* **10**, 505–529, MR1869843 (2002h:05154).

1275. B. L. van der Waerden [1974], The game of rectangles, *Ann. Mat. Pura Appl.* **98**, 63–75, MR0351854 (50 #4342).

1276. J. van Leeuwen [1976], Having a Grundy-numbering is NP-complete, Report No. 207, Computer Science Dept., Pennsylvania State University, University Park, PA.

1277. A. J. van Zanten [1990], The complexity of an optimal algorithm for the generalized tower of Hanoi problem, *Intern. J. Comput. Math.* **36**, 1–8.

1278. A. J. van Zanten [1991], An iterative optimal algorithm for the generalized tower of Hanoi problem, *Intern. J. Comput. Math.* **39**, 163–168.

1279. I. Vardi [1991], *Computational Recreations in Mathematica*, Addison Wesley.

1280. L. P. Varvak [1968], A generalization of the concept of p-sum of graphs, *Dopovīdī Akad. Nauk Ukraïn. RSR Ser. A* **1968**, 965–968, MR0245468 (39 #6776).

1281. L. P. Varvak [1968], Games on the sum of graphs, *Cybernetics* **4**, 49–51 (trans. of *Kibernetika (Kiev)* **4** (1968) 63–66), MR0299238 (45 #8287).

1282. L. P. Varvak [1970], A certain generalization of the Grundy function, *Kibernetika* (Kiev) (5), 112–116, MR0295968 (45 #5029).

1283. L. P. Varvak [1971], Game properties of generalized p-sums of graphs, *Dopovīdī Akad. Nauk Ukraïn. RSR Ser. A* pp. 1059–1061, 1148, MR0307731 (46 #6851).

1284. L. P. Varvak [1973], A certain generalization of the nucleus of a graph, *Ukrainian Math. J.* **25**, 78–81, MR0323649 (48 #2005).

1285. J. Veerasamy and I. Page [1994], On the towers of Hanoi problem with multiple spare pegs, *Intern. J. Comput. Math.* **52**, 17–22.

1286. H. Venkateswaran and M. Tompa [1989], A new pebble game that characterizes parallel complexity classes, *SIAM J. Comput.* **18**, 533–549.

1287. D. Viaud [1987], Une stratégie générale pour jouer au Master-Mind, *RAIRO Recherche opérationelle/Operations Research* **21**, 87–100.

1288. F. R. Villegas, L. Sadun and J. F. Voloch [2002], Blet: a mathematical puzzle, *Amer. Math. Monthly* **109**, 729–740.

1289. J. von Neumann [1928], Zur Theorie der Gesellschaftsspiele, *Math. Ann.* **100**, 295–320.

1290. J. von Neumann and O. Morgenstern [2004], *Theory of Games and Economic Behaviour*, Princeton University Press, Princeton, NJ, reprint of the 3rd 1980 edition; first edition appeared in 1944.

1291. R. G. Wahl [2005], The Butler University game, *Tribute to a Mathemagician*, A K Peters, Wellesley, pp. 37–40, honoring Martin Gardner.

1292. J. Waldmann [2002], Rewrite games, *Rewriting Techniques and Applications, Copenhagen, 22-24 July 2002* (S. Tison, ed.), Vol. LNCS 2378, Lecture Notes in Computer Science, Springer, pp. 144–158.

1293. C. T. C. Wall [1955], Nim-arithmetic, *Eureka* no. 18 pp. 3–7.

1294. J. L. Walsh [1953], The name of the game of Nim, Letter to the Editor, *Math. Gaz.* **37**, 290.

1295. M. Walsh [2003], A note on the Grundy number, *Bull. Inst. Combin. Appl.* **38**, 23–26.

1296. T. R. Walsh [1982], The towers of Hanoi revisited: moving the rings by counting the moves, *Inform. Process. Lett.* **15**, 64–67.

1297. T. R. Walsh [1983], Iteration strikes back at the cyclic towers of Hanoi, *Inform. Process. Lett.* **16**, 91–93.

1298. W.-F. Wang [2006], Edge-partitions of graphs of nonnegative characteristic and their game coloring numbers, *Discrete Math.* **306**, 262–270.

1299. I. M. Wanless [1997], On the periodicity of graph games, *Australas. J. Combin.* **16**, 113–123, MR1477524 (98i:05159).

1300. I. M. Wanless [2001], Path achievement games, *Australas. J. Combin.* **23**, 9–18, MR1814595 (2002k:91050).

1301. R. H. Warren [1996], Disks on a chessboard, *Amer. Math. Monthly* **103**, 305–307.

1302. A. Washburn [1990], Deterministic graphical games, *J. Math. Anal. Appl.* **153**, 84–96.

1303. J. J. Watkins [1997], Knight's tours on triangular honeycombs, *Congr. Numer.* **124**, 81–87, Proc. 28th Southeastern Internat. Conf. on Combinatorics, Graph Theory and Computing (Boca Raton, FL, 1997), MR1605097.

1304. J. J. Watkins [2000], Knight's tours on cylinders and other surfaces, *Congr. Numer.* **143**, 117–127, Proc. 31st Southeastern Internat. Conf. on Combinatorics, Graph Theory and Computing (Boca Raton, FL, 2000), MR1817914 (2001k:05136).

1305. J. J. Watkins [2004], *Across the Board: The Mathematics of Chessboard Problems*, Princeton University Press, Princeton and Oxford, MR2041306 (2004m:00002).

1306. J. J. Watkins and R. L. Hoenigman [1997], Knight's tours on a torus, *Math. Mag.* **70**, 175–184, MR1456114 (98i:00003).

1307. W. D. Weakley [1995], Domination in the queen's graph, in: *Graph theory, Combinatorics, and Algorithms,* Vol. **2** (Kalamazoo, MI, 1992) (Y. Alavi and A. J. Schwenk, eds.), Wiley, New York, pp. 1223–1232.

1308. W. D. Weakley [2002], A lower bound for domination numbers of the queen's graph, *J. Combin. Math. Combin. Comput.* **43**, 231–254.

1309. W. D. Weakley [2002], Upper bounds for domination numbers of the queen's graph, *Discrete Math.* **242**, 229–243.

1310. W. A. Webb [1982], The length of the four-number game, *Fibonacci Quart.* **20**, 33–35, MR0660757 (84e:10017).

1311. C. P. Welter [1952], The advancing operation in a special abelian group, *Nederl. Akad. Wetensch. Proc.* (Ser. A) **55** = *Indag. Math.* **14**, 304–314.

1312. C. P. Welter [1954], The theory of a class of games on a sequence of squares, in terms of the advancing operation in a special group, *Nederl. Akad. Wetensch. Proc.* (Ser. A) **57** = *Indag. Math.* **16**, 194-200.

1313. J. West [1996], Champion-level play of domineering, in: *Games of No Chance,* Proc. MSRI Workshop on Combinatorial Games, July, 1994, Berkeley, CA, MSRI Publ. (R. J. Nowakowski, ed.), Vol. 29, Cambridge University Press, Cambridge, pp. 85–91.

1314. J. West [1996], Champion-level play of dots-and-boxes, in: *Games of No Chance,* Proc. MSRI Workshop on Combinatorial Games, July, 1994, Berkeley, CA, MSRI Publ. (R. J. Nowakowski, ed.), Vol. 29, Cambridge University Press, Cambridge, pp. 79–84, MR1427958.

1315. J. West [1996], New values for Top Entails, in: *Games of No Chance,* Proc. MSRI Workshop on Combinatorial Games, July, 1994, Berkeley, CA, MSRI Publ. (R. J.

Nowakowski, ed.), Vol. 29, Cambridge University Press, Cambridge, pp. 345–350.

1316. M. J. Whinihan [1963], Fibonacci Nim, *Fibonacci Quart.* **1**(4), 9–13.

1317. R. Wilber [1988], White pebbles help, *J. Comput. System Sci.* **36**, 108–124.

1318. R. Wilfong [1991], Motion planning in the presence of movable obstacles. Algorithmic motion planning in robotics, *Ann. Math. Artificial Intelligence* **3**, 131–150.

1319. R. M. Wilson [1974], Graph puzzles, homotopy and the alternating group, *J. Combin. Theory* (Ser. B) **16**, 86–96.

1320. P. Winkler [2002], Games people don't play, in: *Puzzler's Tribute: a Feast for the Mind*, pp. 301–313, honoring Martin Gardner (D. Wolfe and T. Rodgers, eds.), A K Peters, Natick, MA, MR2034896 (2006c:00002).

1321. D. Wolfe [1993], Snakes in domineering games, *Theoret. Comput. Sci.* (Math Games) **119**, 323–329.

1322. D. Wolfe [1996], The gamesman's toolkit, in: *Games of No Chance,* Proc. MSRI Workshop on Combinatorial Games, July, 1994, Berkeley, CA, MSRI Publ. (R. J. Nowakowski, ed.), Vol. 29, Cambridge University Press, Cambridge, pp. 93–98. http://www.gustavus.edu/~wolfe/games/

1323. D. Wolfe [2002], Go endgames are PSPACE-hard, in: *More Games of No Chance,* Proc. MSRI Workshop on Combinatorial Games, July, 2000, Berkeley, CA, MSRI Publ. (R. J. Nowakowski, ed.), Vol. 42, Cambridge University Press, Cambridge, pp. 125–136, MR1973009 (2004b:91052).

1324. D. Wolfe [2009], On day *n*, in: *Games of No Chance III,* Proc. BIRS Workshop on Combinatorial Games, July, 2005, Banff, Alberta, Canada, MSRI Publ. (M. H. Albert and R. J. Nowakowski, eds.), Vol. 56, Cambridge University Press, Cambridge, pp. 125–131.

1325. D. Wolfe and W. Fraser [2004], Counting the number of games, *Theoret. Comp. Sci.* **313**, 527–532, special issue of Dagstuhl Seminar "Algorithmic Combinatorial Game Theory", Feb. 2002, MR2056944.

1326. D. Wood [1981], The towers of Brahma and Hanoi revisited, *J. Recr. Math.* **14**, 17–24.

1327. D. Wood [1983], Adjudicating a towers of Hanoi contest, *Intern. J. Comput. Math.* **14**, 199–207.

1328. J.-S. Wu and R.-J. Chen [1992], The towers of Hanoi problem with parallel moves, *Inform. Process. Lett.* **44**, 241–243.

1329. J.-S. Wu and R.-J. Chen [1993], The towers of Hanoi problem with cyclic parallel moves, *Inform. Process. Lett.* **46**, 1–6.

1330. J. Wu and X. Zhu [2006], Relaxed game chromatic number of outer planar graphs, *Ars Combin.* **81**, 359–367, MR2267825 (2007f:05067).

1331. W. A. Wythoff [1907], A modification of the game of Nim, *Nieuw Arch. Wisk.* **7**, 199–202.

1332. A. M. Yaglom and I. M. Yaglom [1967], *Challenging Mathematical Problems with Elementary Solutions*, Vol. II, Holden-Day, San Francisco, translated by J. McCawley, Jr., revised and edited by B. Gordon.

1333. Y. Yamasaki [1978], Theory of division games, *Publ. Res. Inst. Math. Sciences, Kyoto Univ.* **14**, 337–358.

1334. Y. Yamasaki [1980], On misère Nim-type games, *J. Math. Soc. Japan* **32**, 461–475.

1335. Y. Yamasaki [1981], The projectivity of *Y*-games, *Publ. Res. Inst. Math. Sciences, Kyoto Univ.* **17**, 245–248.

1336. Y. Yamasaki [1981], Theory of Connexes I, *Publ. Res. Inst. Math. Sciences, Kyoto Univ.* **17**, 777–812.

1337. Y. Yamasaki [1985], Theory of connexes II, *Publ. Res. Inst. Math. Sciences, Kyoto Univ.* **21**, 403–409.

1338. Y. Yamasaki [1989], *Combinatorial Games: Back and Front*, Springer Verlag, Tokyo (in Japanese).

1339. Y. Yamasaki [1989], Shannon switching games without terminals II, *Graphs Combin.* **5**, 275–282.

1340. Y. Yamasaki [1991], A difficulty in particular Shannon-like games, *Discrete Appl. Math.* **30**, 87–90.

1341. Y. Yamasaki [1992], Shannon switching games without terminals III, *Graphs Combin.* **8**, 291–297.

1342. Y. Yamasaki [1993], Shannon-like games are difficult, *Discrete Math.* **111**, 481–483.

1343. Y. Yamasaki [1997], The arithmetic of reversed positional games, *Theoret. Comput. Sci.* (Math Games) **174**, 247–249, also in *Discrete Math.* **165/166** (1997) 639–641.

1344. J. Yang, S. Liao and M. Pawlak [2001], On a decomposition method for finding winning strategy in Hex game, in: *Proceedings ADCOG: Internat. Conf. Application and Development of Computer Games* (A. L. W. Sing, W. H. Man and W. Wai, eds.), City University of Honkong, pp. 96–111.

1345. L. J. Yedwab [1985], On playing well in a sum of games, M.Sc. Thesis, MIT, MIT/LCS/TR-348.

1346. Y. N. Yeh [1995], A remarkable endofunction involving compositions, *Stud. Appl. Math* **95**, 419–432.

1347. Y. Yesha [1978], Theory of annihilation games, Ph.D. Thesis, Weizmann Institute of Science, Rehovot, Israel.

1348. Y. K. Yu and R. B. Banerji [1982], Periodicity of Sprague–Grundy function in graphs with decomposable nodes, *Cybernetics and Systems: An Internat. J.* **13**, 299–310.

1349. S. Zachos [1988], Probabilistic quantifiers and games, *J. Comput. System Sci.* **36**, 433–451.

1350. M. Zaker [2005], Grundy chromatic number of the complement of bipartite graphs, *Australas. J. Combin.* **31**, 325–329, MR2113717 (2005h:05087).

1351. M. Zaker [2006], Results on the Grundy chromatic number of graphs, *Discrete Math.* **306**(23), 3166–3173, MR2273147.

1352. D. Zeilberger [2001], Three-rowed Chomp, *Adv. in Appl. Math.* **26**, 168–179, MR1808446 (2001k:91009).

1353. D. Zeilberger [2004], Chomp, recurrences and chaos(?), *J. Difference Equ. Appl.* **10**, 1281–1293, MR2100728 (2005g:39014).

1354. E. Zermelo [1913], Über eine Anwendung der Mengenlehre auf die Theorie des Schachspiels, *Proc. 5th Int. Cong. Math. Cambridge 1912*, Vol. II, Cambridge University Press, pp. 501–504.

1355. X. Zhu [1999], The game coloring number of planar graphs, *J. Combin. Theory* (Ser. B) **75**, 245–258, MR1676891 (99m:05064).

1356. X. Zhu [2000], The game coloring number of pseudo partial k-trees, *Discrete Math.* **215**, 245–262, MR1746462 (2000m:05102).

1357. M. Zieve [1996], Take-away games, in: *Games of No Chance*, Proc. MSRI Workshop on Combinatorial Games, July, 1994, Berkeley, CA, MSRI Publ. (R. J. Nowakowski, ed.), Vol. 29, Cambridge University Press, Cambridge, pp. 351–361.

1358. U. Zwick and M. S. Paterson [1993], The memory game, *Theoret. Comput. Sci.* (Math Games) **110**, 169–196.

1359. U. Zwick and M. S. Paterson [1996], The complexity of mean payoff games on graphs, *Theoret. Comput. Sci.* (Math Games) **158**, 343–359.

1360. W. S. Zwicker [1987], Playing games with games: the hypergame paradox, *Amer. Math. Monthly* **94**, 507–514.

AVIEZRI S. FRAENKEL
DEPARTMENT OF APPLIED MATHEMATICS AND COMPUTER SCIENCE
WEIZMANN INSTITUTE OF SCIENCE
REHOVOT 76100
ISRAEL
fraenkel@wisdom.weizmann.ac.il

Printed in the United States
by Baker & Taylor Publisher Services

Printed in the United States
by Baker & Taylor Publisher Services